HISTORIES OF THE ELECTRON

Dibner Institute Studies in the History of Science and Technology
Jed Z. Buchwald, general editor, Evelyn Simha, governor

Jed Z. Buchwald and Andrew Warwick, editors, *Histories of the Electron: The Birth of Microphysics*

Jed Z. Buchwald and I. Bernard Cohen, editors, *Isaac Newton's Natural Philosophy*

Anthony Grafton and Nancy Siraisi, editors, *Natural Particulars: Nature and the Disciplines in Renaissance Europe*

Frederic L. Holmes and Trevor H. Levere, editors, *Instruments and Experimentation in the History of Chemistry*

Agatha C. Hughes and Thomas P. Hughes, editors, *Systems, Experts, and Computers: The Systems Approach in Management and Engineering, World War II and After*

N. M. Swerdlow, editor, *Ancient Astronomy and Celestial Divination*

HISTORIES OF THE ELECTRON

The Birth of Microphysics

———

edited by Jed Z. Buchwald and Andrew Warwick

The MIT Press
Cambridge, Massachusetts
London, England

© 2001 Massachusetts Institute of Technology
All rights reserved. No part of this book may be reproduced in any form or by any electronic or mechanical means (including photocopying recording, or information storage and retrieval) without permission in writing from the publisher.

This book was set in Bembo by Graphic Composition, Inc. in Athens, Georgia with QuarkXPress and printed and bound by in the United States of America

Library of Congress Cataloging-in-Publication Data
Histories of the electron : the birth of microphysics / edited by Jed Z. Buchwald and Andrew Warwick.
    p. cm. — (Dibner Institute studies in the history of science and technology)
    Includes bibliographical references and index.
    ISBN 0-262-02494-2 (hc. : alk. paper)
    1. Electrons—History.   2. Thomson, J.J. (Joseph John), Sir, 1856–1940.
 I. Buchwald, Jed Z.   II. Warwick, Andrew.   III. Series.

QC793.5.E62 H57 2001
539.7′2112′09—dc21

00-046580

CONTENTS

QC
793
.5
E62
H57
2001
PHYS

ACKNOWLEDGMENTS   vii

CONTRIBUTORS   ix

INTRODUCTION   1
Jed Z. Buchwald and Andrew Warwick

I **CORPUSCLES AND ELECTRONS   19**

1 J. J. THOMSON AND THE ELECTRON, 1897–1899   21
George E. Smith

2 CORPUSCLES TO ELECTRONS   77
Isobel Falconer

3 THE QUESTIONABLE MATTER OF ELECTRICITY: THE RECEPTION OF J. J. THOMSON'S "CORPUSCLE" AMONG ELECTRICAL THEORISTS AND TECHNOLOGISTS   101
Graeme Gooday

4 PAUL VILLARD, J. J. THOMSON, AND THE COMPOSITION OF CATHODE RAYS   135
Benoit Lelong

II **WHAT WAS THE NEWBORN ELECTRON GOOD FOR?   169**

5 THE ZEEMAN EFFECT AND THE DISCOVERY OF THE ELECTRON   171
Theodore Arabatzis

6 THE ELECTRON, THE PROTYLE, AND THE UNITY OF MATTER   195
Helge Kragh

7	O. W. Richardson and the Electron Theory of Matter, 1901–1916 227 Ole Knudsen
8	Electron Gas Theory of Metals: Free Electrons in Bulk Matter 255 Walter Kaiser
III	**Electrons Applied and Appropriated 305**
9	The Electron and the Nucleus 307 Laurie M. Brown
10	The Electron, the Hole, and the Transistor 327 Lillian Hoddeson and Michael Riordan
11	Remodeling a Classic: The Electron in Organic Chemistry, 1900–1940 339 Mary Jo Nye
12	The Physicists' Electron and Its Appropriation by the Chemists 363 Kostas Gavroglu
IV	**Philosophical Electrons 401**
13	Who Really Discovered the Electron? 403 Peter Achinstein
14	History and Metaphysics: On the Reality of Spin 425 Margaret Morrison
15	What Should Philosophers of Science Learn from the History of the Electron? 451 Jonathan Bain and John D. Norton
16	The Role of Theory in the Use of Instruments; or, How Much Do We Need to Know about Electrons to Do Science with an Electron Microscope? 467 Nicolas Rasmussen and Alan Chalmers

Index   503

Acknowledgments

The essays in this volume originated in discussions at two meetings held in 1997 to celebrate the centenary of the electron's discovery. The first, held jointly at the Royal Society and the Science Museum in London, was organized by Alan Morton and Andrew Warwick. The second, organized by the editors of this volume, was held at the Dibner Institute for the History of Science and Technology in Cambridge, Massachusetts. The editors thank the Royal Society and the Science Museum for the London meeting, and the Dibner Institute for the meeting in Cambridge. They also thank the chairpersons, respondents, and delegates who contributed to the lively discussions that took place at both meetings.

Contributors

Peter Achinstein is Professor of Philosophy at Johns Hopkins University. He is the author of *Concepts of Science* (1968), *Law and Explanation* (1971), *The Nature of Explanation* (1983), and *Particles and Waves* (1991), which received the 1993 Lakatos Award. He is completing a new work entitled *The Book of Evidence*.

Theodore Arabatzis is Lecturer in History and Philosophy of Science at the University of Athens. He is the author of several articles on the history of the concept of the electron and a forthcoming book, tentatively titled *Electrons and Reality: A Biographical Essay*.

Jonathan Bain is Assistant Professor of Philosophy at Polytechnic University in Brooklyn. He has published on topics in the philosophy of spacetime and the philosophy of quantum field theory.

Laurie M. Brown is Professor of Physics and Astronomy (emeritus) at Northwestern University. He is the author and editor of many books and articles on the history of modern physics, including *The Origin of the Concept of Nuclear Physics* (with Helmut Rechenberg) and *Twentieth Century Physics* (with Sir Brian Pippard and Abraham Pais).

Alan Chalmers is Research Fellow in the Philosophy Department, Flinders University, South Australia. He is the author of *What Is This Thing Called Science?* (3rd ed., 1999) and *Science and Its Fabrication* (1990).

Isobel Falconer is an independent scholar and the coauthor (with E. A. Davis) of *J. J. Thomson and the Discovery of the Electron* (1997).

Kostas Gavroglu is Professor of History of Science at the University of Athens. His books include *Fritz London: A Scientific Biography* (1995), and he is currently working on a book on the history of artificial cold. He is also the director of a project to digitize all the scientific and philosophical books and manuscripts written in Greek from the beginning of the seventeenth century to the beginning of the nineteenth (www.space.noa.gr/hellinomnimon).

Graeme Gooday is Lecturer in the History and Philosophy of Science at the University of Leeds. He has published on experimental science and electrical engineering in the nineteenth century, and his first book, *The Morals of Measurement: Accuracy, Irony and Trust in Late Victorian Electrical Practice,* will soon be published by Cambridge University Press.

Lillian Hoddeson is Professor of History at the University of Illinois at Urbana-Champaign and historian at Fermilab in Batavia, Illinois. Her most recent book, with Michael Riordan, is *Crystal Fire: The Birth of the Information Age* (1997). She is also a coauthor of *Critical Assembly: A History of Los Alamos During the Oppenheimer Years* (1993) and *Out of the Crystal Maze: Chapters from the History of Solid State Physics* (1992), as well as coeditor of *The Birth of Particle Physics* (1983), *Pions to Quarks* (1989), and *The Rise of the Standard Model* (1997).

Walter Kaiser holds the Chair for the History of Technology at the University of Technology of Aachen. He has written on the history of electrodynamics and solid state physics, and, more generally, on the history and philosophy of science and the history of technology. He is the author of *Theorien der Elektrodynamik im 19. Jahrhundert* (1981); has edited *Ludwig Boltzmann's Vorlesungen über Maxwells Theorie der Elektricität und des Lichtes* (1982); and has coauthored the *Propyläen Technikgeschichte,* vol. 5, 1914–1990 (1992 and 1997).

Ole Knudsen is Associate Professor of the History of Science at the University of Aarhus. He has written on nineteenth-century electromagnetic theory and is currently coeditor of a general history of ideas in western culture commissioned by the Danish publisher Gyldendal.

Helge Kragh is Professor of the History of Science at the University of Aarhus. His most recent book is *Quantum Generations: A History of Physics in the Twentieth Century* (1999).

Benoit Lelong is a researcher in the laboratory "Uses, Creativity, Ergonomics" of CNET/France Télécom R&D. He is currently working on innovation and the social uses of telecommunication, especially the integration of the internet into households.

Margaret Morrison is Professor of Philosophy at the University of Toronto. She is the author of *Unifying Scientific Theories: Physical Concepts and Mathematical Structures* (2000).

John D. Norton is Professor in the Department of History and Philosophy of Science, University of Pittsburgh. He studies the history and philosophy of physics (relativity, quantum theory, and statistical physics), with a special interest in general relativity, and has published extensively on the detailed steps of Einstein's discovery of general relativity and on its philosophical foundations. He also works in general philosophy of science, with emphasis on different approaches to confirmation theory, inconsistency in theories, and thought experiments.

Mary Jo Nye is Horning Professor of the Humanities and Professor of History at Oregon State University, where she teaches the history of science. Her most recent book is *Before Big Science: The Pursuit of Modern Chemistry and Physics, 1800–1940* (1996/1999).

Nicolas Rasmussen is Senior Lecturer in History of Science in the School of Science and Technology Studies, University of New South Wales (Sydney). His research deals with the relations between knowledge and technology in experimental life sciences, and he is author of *Picture Control: The Electron Microscope and the Transformation of Biology in America, 1940–1960* (1997).

Michael Riordan is Lecturer in Stanford University's Program on the History and Philosophy of Science and Adjunct Professor of Physics at the University of California, Santa Cruz. He is coauthor (with Lillian Hoddeson) of *Crystal Fire: The Birth of the Information Age* (1997), which was awarded the 1999 Sally Hacker Prize of the Society for the History of Technology.

George E. Smith is a member of the Philosophy Department of Tufts University. He has published papers on Newton's methodology in the *Principia*, as well as on Book 2, and he is coeditor of the *Cambridge Companion to Newton* (2000). He is also a practicing engineer, specializing in aerodynamically induced vibration and resulting metal fatigue failures in jet engines and other turbomachinery.

Histories of the Electron

Introduction

Jed Z. Buchwald and Andrew Warwick

Today the world of objects that are smaller than a wavelength of visible light—the microworld—belongs to everyday life. Consumer devices, ranging from integrated circuits to designer molecules, work directly with microobjects; engineers build nanometer-size motors; molecular biologists fabricate DNA strands; chemists synthesize molecular structures designed by computers, which themselves consist of microbased devices; high-energy physicists seek to smash microobjects into the smallest possible bits; and scientists at IBM inscribe the company logo in Xenon atoms. Determined skeptics will continue to debate atomic reality, but new universes of devices inevitably marginalize metaphysics, as they always have. Hobbes's seventeenth-century objections to the vacuum, for example, persuade only in a world where vacuum pumps are rare and expensive, work erratically, and require specialists to build and use them. When pumps can be bought in toy stores, persuasive rhetoric turns into eccentricity.

Until just past the end of the nineteenth century, material entities like electrons played almost no role in the practice of physicists. Indeed, such things had much less working life than did that catholic underpinning of all nature, the ether, even though the ether has today disappeared altogether from the life of science, whereas electrons and other particles are ubiquitous. Yet as late as 1909 the English physicist Oliver Lodge profitably offered a book entitled *The Ether of Space* in which he could write that the Ether (solemnly capitalized) is "not only uniformly present and all-pervading, but also massive and substantial beyond conception." More than that, Lodge's Ether turned out to be "by far the most substantial thing—perhaps the only substantial thing—in the material universe." This was printed four years after the publication of Einstein's special relativity theory, and well into the era of elaborate experimental work concerned with electron properties and radioactivity.

Compare Lodge's certainty about the ether with the following remark, which appears in a textbook on quantum mechanics printed in 1961:

"Consider as familiar an object as the hydrogen atom. The evidence that such an atom consists of a nucleus and an electron, bound to the former by forces of electrostatic attraction, is too well known to need recapitulation. The electron can be removed from the atom and identified by its charge, mass, and spin. . . . These are empirical facts" (Merzbacher 1961, 2). The "substantial" ether has evidently ceded pride of place to "familiar" things like hydrogen atoms and even electrons. How did this happen? How did comparatively marginal things (atoms) become familiar, while the very essence of substance (ether) disappeared altogether?

An easy answer is the realist one: atoms, and microphysical entities in general, occupy ground level in the contemporary physicist's practice just because the world is made up of such things. If ether filled the universe, the realist must argue, then we would still be using it today, and we would probably know lots more about it than people did in the late nineteenth century. There is a satisfying solidity to this way of thinking. Electrons and suchlike entities simply *are,* indeed they have always (or at least for eons) *been,* and they eventually made their presence known in the laboratory. It is nowadays fashionable to write of nonhuman beings as though they act rather like people. The contemporary realist would not like the hint of anthropomorphism in talk of that kind, but he or she might nevertheless find poetically congenial the evocative image of an electron as a tangible presence in the late-nineteenth-century laboratory, where, it might be said, they first learned to speak—or where, perhaps, they were no longer silenced.

Whether real objects or just useful devices, whether vocal or silent, electrons figure prominently in every contribution to this collection. They are "discovered" by J. J. Thomson at the Cavendish—or by someone else at some other place—or perhaps "electrons" were not discovered at all; they worked hard to make instruments operate—or they didn't do much work at all; electrons colonized the chemical world—or the chemical world just assimilated electrons. Every one of these not altogether compatible views is upheld by one or more of the authors included here, all of whom would nevertheless agree that something took place in laboratories of the 1890s and early 1900s that had an immense effect on the practices of physics and chemistry decades later, and on the technological world as well. The present collection casts light on the question of how the microworld became real by concentrating closely on the various ways in which "electrons" did, or did not, play central roles in the laboratory, on paper, and in the universe of devices, like the microscope named eponymously for them.

Corpuscles and Electrons

The essays in the first section concentrate on J. J. Thomson's role in the experimental production of the electron in the mid- and late-1890s. Since the early twentieth century, Thomson has generally been described, at least in the English-speaking world, as the electron's discoverer. Several recent studies, however, have noted that Thomson was not the only, nor even the first, experimenter to claim that cathode rays were composed of particles smaller than the hydrogen atom. Moreover, he always termed the particles he had produced "corpuscles," and he resisted (at least until the 1920s) identifying them with the massless "electrons" or "ions" posited in the respective theoretical writings of Joseph Larmor and H. A. Lorentz. What, then, was Thomson's contribution to the emergence of the electron as a recognizable entity in microphysics, and why has his name more than any other come to be associated with its discovery?

George Smith casts fresh light on the first of these questions by arguing that it is historically misleading to regard Thomson's work of the late 1890s as either directed toward or as constituting the discovery of the electron. According to Smith, Thomson's experimental researches of this period were primarily aimed at understanding the nature of electrical conduction, a problem that had dogged Maxwellian electromagnetic theory since the 1860s. Thomson's accomplishment in 1897 was to produce powerful experimental evidence in support of the hypothesis that cathode rays consisted of a stream of negatively charged particles—dubbed "corpuscles" by Thomson—that were subatomic constituents of all atoms. Smith also argues persuasively that Thomson's investigation was extremely influential for its introduction of a new and fundamental asymmetry into the theory of electricity. Prior to his work, it had been assumed that, whatever their ultimate nature, positive and negative forms of electricity were symmetrical in their physical origin and properties. According to Thomson's corpuscular theory, negative electricity was quantized and always came attached to discrete subatomic corpuscles of a fixed mass, whereas positive charge existed independently of matter and in the form of a continuous cloud. This asymmetrical approach to electrical charge was eventually made explicit by Thomson in the form of his so-called plum pudding model of the atom, in which the negatively charged corpuscles orbited in a finite cloud of continuous positive charge. On Smith's showing, Thomson's work as an experimenter, and not his own work in mathematical physics, helped to alter the agenda of electrical studies in Cambridge. Instead of developing theories that concentrated (before the 1890s) on finding appropriate energy functions for the ether in unusual circumstances, or in probing the connection between ether and matter (for a period after the early 1890s), mathematical physicists there became more concerned with the discrete nature of electrical charge and the structure of the atom.

Isobel Falconer uses a comparison between two histories of the electron, written respectively by Oliver Lodge and Walter Kaufmann, to explore different national perceptions of the electron's discovery in the early twentieth century. For Lodge, writing in Britain in 1902, the establishment of the electron as a physical entity represented the culmination of long traditions in British mathematical and experimental physics. The electromagnetic effects produced by a discrete moving charge had been explored by Thomson and by Oliver Heaviside in the 1880s and had been developed into a fully fledged "electron theory" of matter by George FitzGerald and, especially, Joseph Larmor in the 1890s. Lodge identified a parallel line of experimental evidence, beginning with Michael Faraday's work on electrolysis, that tended to confirm the existence of discrete units of both positive and negative electricity. According to Lodge, Thomson's experimental evidence in support of the particulate nature of cathode rays represented both a continuation of this line of experimentation and an empirical verification of an electronic theory of matter, such as Larmor's. German physicist Walter Kaufmann described the electron's origins in a very different way. He understood it to have emerged gradually from a much broader range of experimental and theoretical work, and he placed particular emphasis on the respective theoretical and experimental researches of the Dutch physicists H. A. Lorentz and Pieter Zeeman. According to this account, Thomson's main accomplishment at the Cavendish Laboratory in the mid-1890s was to succeed in deflecting a beam of cathode rays electrostatically; something that continental experimenters had failed to do. From a theoretical perspective, Kaufmann saw Thomson's results mainly as confirming theoretical speculations that were already being made at several sites in Europe. As Falconer points out, Thomson's work might well have become invisible had it not been preserved in the British context by the likes of Lodge. Conversely, as Falconer also suggests, it was likely the abbreviated accounts of Lodge's history, propagated by Thomson's students and many physics textbooks, that established Thomson as the electron's main discover in the English-speaking world.

The forbidding historical problem of disentangling what Thomson contributed to the discovery of the electron from what his contemporaries understood or claimed him to have contributed, is further explored in Graeme Gooday's essay. We noted above that Thomson himself always referred to his particles as "corpuscles" and was loathe to conflate them with Larmor's massless "electrons" and Lorentz's "ions." Gooday emphasizes that the meaning of Thomson's cathode-ray experiments continued to be contested throughout at least the first decade of the twentieth century and that electrical engineers had as much to say on this matter as did experimental and mathematical physicists. Contrary to many historical accounts of this period, Thomson's contemporaries were well aware of the competing corpuscular and electronic theories of cathode rays and keenly debated the merits of each. An important claim of Gooday's paper is that the establishment of the electron as a stable entity in physics around 1910 was not the outcome of any single set of experiments but occurred only gradually as such diverse areas as spectroscopy, electrical conduction, the thermal properties of matter, cathode rays, x-rays, and radioactivity were satisfactorily reworked in electron theory. As Gooday points out, those participating in these enterprises would have considered the singular importance subsequently attributed by some historians to Thomson's experiments of 1897 as little short of bizarre. Gooday also notes that although the establishment of the electron must be understood in its role in many areas of physics, its impact on the electro-technology of the period is often overstated (which, we shall see, resonates well with Rasmussen and Chalmers's claim in their essay that theories of the instrument are rarely necessary for its productive use). Like Falconer, Gooday points to Thomson's influential students as the likely source of the myths surrounding his role in the electron's discovery, but Gooday additionally speculates that this same group was responsible for overplaying the importance of "pure physics" in the invention of such devices as the cathode-ray oscilloscope and the thermionic valve.

Benoit Lelong's essay explores the importance of disciplinary traditions in establishing the nature of cathode rays. Shortly after Thomson published his corpuscular account of the rays in 1897, the French chemist Paul Villard claimed that they were composed of negatively charged hydrogen atoms. These claims were not contested in France, but, in the light of the emergent consensus concerning the subatomic nature of the rays, they were quietly dropped by the French scientific continuity in the early twentieth century and then ignored by historians of science. It has generally been assumed that Villard's experimental work must somehow have been flawed or that his interpretation of his results was badly mistaken. As Lelong points out, however, Villard was a recognized expert on hydrate chemistry and a competent manipulator of cathode rays who was elected to the Académie des Sciences in 1908. How could such a competent and experienced experimenter make such apparently grave errors in his scientific work, and why were they not picked up by other members of the French scientific community? Lelong tackles this problem not by seeking to establish the truth or falsity of Villard's claims, but by showing why he and his French colleagues considered his explanation of the rays' properties to be perfectly reasonable. Villard began his career as an analytical chemist who specialized in the chemical composition of gas hydrates. When in 1897 he became interested in discharge phenomena, he therefore tackled the problem of cathode rays as a fundamentally chemical one and focused on their reducing power. This line of research eventually led him to the conclusion that the rays were composed of charged hydrogen atoms originating in water deposited on the surface of the cathode ray tube. Lelong points out that Villard's claims were never explicitly disproved, but were simply ignored when a younger generation of French physicists began to adopt the apparatus and experimental methods used by Thomson and his young collaborators at the Cavendish Laboratory. Especially important in this respect was that the young Paul Langevin spent a year studying with Thomson at the Cavendish in 1902 before returning to Paris to build a research group in discharge phenomena. This group adopted Thomson's experimental practices and terminology, even to the extent of continuing to refer to cathode rays as composed of "corpuscles" after Thomson's own students had adopted the term "electrons."

What Was the Newborn Electron Good For?

Issues of priority and discovery have long provided fodder for the reminiscing scientist and the historian. Who discovered entity x, whether he or she preceded or followed investigator y, or whether a discovery was truly that, or not—these kinds of questions often preoccupy the original scientists, but should they bother historians? Theodore Arabatzis thinks they shouldn't. According to him, because issues of discovery appear to be indissolubly bound to issues of realism, the historian should remain a discovery-agnostic, just as he or she should remain a realism-agnostic. Instead, Arabatzis offers a consensus-based account of discovery, asserting that entity x can be said to have been discovered just when group y reaches consensus that it has been. On Arabatzis's account one might say, for example, that the ether was discovered in the early nineteenth century, only to have been undiscovered sometime around 1900. But Arabatzis does not intend absurd consequences; he simply wishes to concentrate on belief, not on reality. Accordingly, one might on his account reasonably say that the ether was believed to have been discovered at a certain time, with that belief having fractured a century later. This does raise many difficult questions concerning what aspects, if any, of the original discovery persist, questions that will bother discovery realists like Peter Achinstein, who holds a considerably different view from Arabatzis that will be discussed below.

Arabatzis provides an account of Zeeman's discovery of the electron, concentrating—in accordance with his consensus-based views—on how Zeeman's work was tightly bound by Lorentz, and then by Larmor, into existing theoretical systems, which rapidly seated the newly born electron within schemes that were already being extensively elaborated. Arabatzis's electron, one might say, was discovered just when Lorentz and Larmor connected it to persuasive theoretical systems that were quickly taken up by scientific communities. Different communities emphasized or worked with different aspects of the newfound entity, however, as Helge Kragh's contribution demonstrates.

Kragh identifies four different kinds of electrons before 1900: the electrochemical, the electrodynamic, the one associated with cathode-ray work, and the magneto-optical. In each of these regimes the electron served particular ends, and they were not brought substantially together until near the turn of the century. After that point, Kragh shows in detail, the electron in Britain was used in attempts to flesh out a complete picture for the structure of matter, one that would incorporate and connect both chemical and physical properties. The underlying notion of the electron as a fundamental building block of matter appealed particularly to J. J. Thomson and several others, who often thought of the electron as a sort of chemical protosubstance. The nature of the object itself remained obscure in this context, with some thinking of it as a locus of concentrated ether, the latter meaning—as it had during the last quarter of the nineteenth century—the ursubstance. Here, however, the very notion of materiality fractured, particularly as many scientists during the early 1900s conceived that matter itself gains inertia not from its essential substantiality but rather from the very fact of being constituted of charge-carrying objects. Inertia, that is, had for many evolved into an electromagnetic implication. Here one could also investigate questions concerning the electron's shape, the distribution of charge over its surface, the dependence of its (electromagnetic) mass on velocity, and so on. Some of this work tied strongly to laboratory procedures, but much of it did not, having its being in a world of highly abstract calculation. This contrasts quite markedly with, for example, Zeeman's tremendous experimental focus—described by Arabatzis—in ruling out all but magnetic-induced widening in his discovery of the Zeeman effect and the effect's linkage to Lorentz's scheme for electrodynamics. Certainly Abraham's and Lorentz's calculations of the velocity dependence of electron mass formed the basis of Kauffmann's experiments, but these experiments were done precisely to discriminate among complex theoretical systems in their respective, elaborate computations of the electron's properties per se.

Other uses for the electron concentrated neither upon its detailed structural properties nor upon its usefulness for reconstructing matter at the deepest possible level (including chemical properties). These latter uses, though occasionally connected to calculation, were nevertheless quite distanced from the laboratory. Ole Knudsen's and Walter Kaiser's papers open windows on a considerably different realm, one that depended neither on analyses of potential electron structures nor on speculations concerning the fundamental nature of matter. Knudsen's account of O. W. Richardson's work in thermionics illustrates just how rapidly it became possible for the electron to be deployed as a tool. Richardson, Knudsen shows, based his work on the assumption that conduction electrons evaporate through metal surfaces, and this led to a wealth of laboratory activity, together with the computation and mutual linking of experimentally determined parameters. Then, in 1914, Richardson published an influential text entitled *The Electron Theory of Matter*—a book for beginning graduate students designed to connect electron theory with experimental work. Though Richardson discussed most of the contemporary developments, he concentrated especially on using the electron to explain specific phenomena: that is, he used the electron as a tool for analyzing phenomena on a microphysical basis.

With Richardson we glimpse the new world of physics that had begun to emerge at the end of the nineteenth century, a world in which scientists would work with microphysical entities. Kaiser's extensive discussion of the electron theory of metals shows just how complex and difficult microphysical practice rapidly became. Issues surrounding metallic conduction had in many respects been central to the development of field theory in Britain, where the nature and behavior of electric sources constituted a problem to be ignored rather than a resource to be deployed in developing physical practice. By the early 1900s, Kaiser shows in detail, metallic conduction had become a central feature of a burgeoning microphysical practice, one that in this case sought to unify the electrodynamics of electric sources in metals with the model of colliding particles that underlay the kinetic theory of gases. The half-century-long series of investigations that Kaiser analyzes sought specifically to utilize conduction electrons for the purpose of calculating constants that are otherwise known only by measurement—calculations that brought together fundamental constants from the otherwise very different regimes of ideal gases and electrodynamics. Here—in the context of a specific, longstanding set of problems in metallic conduction—we find, fully developed, the kind of physics that has become a trademark of the twentieth century: a physics that generates knowledge by working with the electron, first citizen of the microworld.

ELECTRONS APPLIED AND APPROPRIATED

All of the theoretical and practical uses for the electron during the twentieth century had to be painstakingly developed, in some cases entailing the reformation or invention of an entire discipline around a new concept or practice. The essays in this section discuss the accommodation, or nonaccommodation, of the electron in nuclear physics, chemistry and electrical science. Each case provides a specific insight into the electron's gradual progress from cathode ray to ubiquitous subatomic particle and eponymous entity in one of the world's most important and commercially successful industries—electronics. Consider, for example, the role of the electron in the atomic nucleus. From the 1910s on most physicists assumed that electrons were present in the atomic nucleus, but the successful application of quantum mechanics to the nucleus required by the 1930s that electrons play almost no role in the nuclear world. Conversely, the use of the electron in the design of amplifiers and semiconductors not only produced the new discipline of electronics but eventually enabled the very absence of the electron in certain material structures to be reified as a new entity in its own right, the "hole." The appropriation of the electron by chemists highlights the difficulties that can arise when a new experimental entity is theorized in different ways in different disciplines. Where physicists saw quantum mechanics and the electron as a route for reducing chemistry to physics, chemists reasserted their autonomy by developing their own version of quantum chemistry for their own purposes.

In his essay, Laurie Brown explores the question of whether electrons exist inside the atomic nucleus. Shortly after Thomson proposed his plum-pudding model of the atom, other physicists suggested alternative models in which the negative electrons circled a tiny positively charged nucleus. These models were initially resisted as mechanically unstable by most physicists, but they were widely accepted after 1913 in the form of Niels Bohr's quantized nuclear atom. Bohr's model provided a means of stabilizing the electronic orbits and accurately predicting atomic spectra, but it raised the problem of what existed inside the small but massive atomic nucleus. In 1919, Ernest Rutherford showed experimentally that the nucleus contained "protons," particles that carried an equal and opposite charge to the electron but which had roughly the same mass as the hydrogen atom. To preserve electrical neutrality, it was necessary to assume that the atom contained at least as many protons as electrons, but this assumption alone accounted for only about half of the mass of most atoms. To account for the rest of the mass, it was widely assumed that the nucleus contained additional protons, or combinations of protons, whose charge was negated by the presence of an equal number of electrons—an idea that received powerful empirical support from the fact that the nucleus emitted electrons during nuclear decay. Brown argues that, although this assumption was widely accepted from 1920 until the discovery of the neutron, it became increasingly untenable to some theoreticians following the advent of quantum mechanics in 1926. According to quantum mechanics the respective spins and magnetic moments of nuclei and electrons were inconsistent, and electrons would in any case quickly escape from the nucleus by quantum tunneling. Brown describes how attempts by theoreticians to deal with these and other problems during the late 1920s and 1930s led to the conclusion that electrons could play only a very minor role in physics of the nucleus.

Mary Jo Nye compares the histories of physics and chemistry in the late nineteenth and early twentieth centuries, and she asks why the former but not the latter appeared to experience a second revolution after the turn of the century. The arrival of the electron transformed chemists' understanding of chemical structure and bonding, yet these major changes were not hailed as having revolutionized chemistry in the way that atomic theory, relativity theory and quantum mechanics did physics. This is, as Nye points out, the more surprising in that the electron's role in chemical bonding was often explicable only in quantum mechanics, the very theory that was seen as so revolutionary in the other discipline. Nye investigates this issue by tracing the electron's impact on chemistry through theories of chemical bonding between 1900 and 1940. It was during the 1910s that G. N. Lewis introduced the idea of a "shared pair" of electrons as constitutive of a stable chemical bond, the bond being, in Irving Langmuir's terms, "ionic" or "covalent" depending on the relative positions of the shared electrons and their parent atoms. These ideas were modified and hotly contested during the 1920s, but were gradually accepted during the 1930s as important in explaining the valence bond and reaction mechanisms in organic chemistry. The electronic theory of chemical bonding was also justified theoretically during this latter period in quantum mechanics and Bohr's theory of the atom. Nye concludes that the electron's impact on theoretical chemistry was not seen as revolutionary because it was understood to have buttressed and extended extant theories rather than to have spawned new ones. Even quantum-mechanical explanations of molecular stability were seen as justifications of older concepts such as resonance, rather than as totally new chemical theories. In the broader historical picture, the arrival of the electron and quantum physics in chemistry was seen as fulfilling the expectations of men like Lavoisier and Dalton who were understood to have been the driving forces of the first chemical revolution.

In the next essay in this section, Lillian Hoddeson and Michael Riordan display the intimate relationship that existed between the electron, the notion of a positively charged "hole" in electronics, and the invention of the transistor. They begin by tracing the electron's move during the 1910s from an object whose practical and conceptual importance was confined to the physics laboratory, to one that was of operational reality in the design of commercial electrical devices. We noted above that the electron was of little practical significance in electrotechnological design before about 1910, but this changed rapidly through the second decade of the twentieth century when the communications company AT&T sought to develop the technology necessary to build a coast-to-coast telephone system in the United States. To develop amplifiers that could preserve the strength and integrity of signals over long distances, electrical engineers began to conceptualize and design devices with reference to the physical properties of electrons. The commercial success of this enterprise through the 1920s led to the establishment of "electronics" as a well-defined area of technological endeavor. It was from the hybrid environment of university physics and the research facilities of such companies as AT&T, Bell Telephone Laboratories, and General Electric that a new concept, the "hole," was proposed and became an operational reality. The term was coined by Rudolf Peierls to describe an empty electron state near the top of an otherwise filled band of electrons. Peierls noted that this pseudoentity could be treated theoretically as if it were a particle with an equal and opposite charge to that carried by an electron. But, as Hoddeson and Riordan show, the hole only came to life in the laboratory when it was used to explain the successful operation of the transistor. Hoddeson and Riordan describe in careful detail the long process of interplay among reasoned experimentation, theoretical interpretation, and serendipity. Many signal modifications to the experimental apparatus were tried for no deep theoretical reasons, and they often produced strikingly different effects from those vaguely anticipated by the experimenters. Explaining these effects was generally a retrospective process, one that eventually enrolled the notion of the hole. According to Hoddeson and Riordan, it was the power of the concept of the hole to explain the operation of the transistor that gave it entity status and operational reality from the late 1940s.

Kostas Gavroglu concludes the section with an essay on the role of theory in chemistry, especially as compared to its role in physics. Contrary to some received histories, Gavroglu argues that the electron and the quantized Bohr atom were initially irenic entities that helped to unite physics and chemistry around a common understanding of atomic theory. The chemists were more interested in the electron's role in chemical bonding than were the physicists, but the physical properties of the electron were agreed upon by members of both disciplines. Following the advent of quantum mechanics in the mid-1920s, however, physicists came to regard the electron as a new entity whose role in physics and chemistry could only be understood properly in their new theory. From this point on the physicists felt that they, rather than the chemists, possessed the theoretical tools required to understand chemical bonding and that chemical theory would eventually be reduced to physical theory. Chemists naturally took a different view. Although they too would eventually find quantum mechanics a useful theoretical tool, they found the physicists' account of chemical bonding—in the quantum mechanics of the electron—of no practical use in their discipline.

Gavroglu illustrates these points through the work of Walter Heitler and Fritz London, two young physicists who used quantum mechanics to produce the first theory that successfully explained the stability of the hydrogen molecule. The theory proved satisfactory to the physicists, even though its exact analytical application was confined to molecules with one or two electrons. As Gavroglu points out, however, chemists had little use for theories that failed to explain the vast majority of interesting chemical reactions; and it was no consolation to them to be told that a solution was in principle (but not in practice) possible. This for Gavroglu illustrates the respectively different roles played by theory in physics and chemistry. Where physics sought a single theory that, in principle, was analytically exact in all cases, chemistry, a primarily laboratory-based science, sought one or more theories that were practically applicable to a wide range of empirical data. For Gavroglu, the attempts by chemists to reappropriate the electron in the 1930s, and to build an autonomous quantum chemistry, provide a rare opportunity for historians and philosophers of science to glimpse the particular role of theory in chemistry.

PHILOSOPHICAL ELECTRONS

In this final section we consider questions that bear on the electron as a philosophical object, or, better put, in respect to issues of instrumentalism, epistemology, and realism. Peter Achinstein begins with objections and a proposal: he accepts neither Arabatzis's consensus-based criterion for discovery, nor Falconer's Hacking-like emphasis on measurement and manipulation. He offers instead the assertion that "P discovered X if and only if P was the first person (in some group) to be in an epistemic state necessary for discovering X," where "epistemic state" has three criteria associated with it: that P knows X exists, that observations of X or its effects caused belief, and that P's reasons include observation of X or its direct effects. He additionally distinguishes a weak from a strong sense of discovery, but for our immediate purposes it is the strong sense, which requires satisfaction of all three criteria, that matters. Achinstein's arguments are quite compelling, for he is able nicely to illustrate how anyone who satisfies his criteria can reasonably be said to have discovered X. Of course, to accept Achinstein's scheme one must also accept, in his words (this being his second criterion) that "observations of X or its direct effects caused, or are among the things that caused that person to believe that X exists." This criterion would presumably be unacceptable to a nonrealist. If it is not reasonable to talk about the existence of X then it appears hardly appropriate to talk about X having caused belief. Beyond issues of realism stands historical complexity.

In the case of J. J. Thomson, his primary example, Achinstein acknowledges that "the historical facts about who knew what and when are complex"—perhaps too complex to permit a persuasive answer even under Achinstein's criteria. Still, his criteria serve nicely to distill elements of discovery that do appear critical to the process, and that can be used as clarifying factors in historical analysis—even if we are primarily concerned, like Arabatzis, with concentrating on the production of consensus, since it appears clear that significant aspects of Achinstein's criteria entered the process. Achinstein argues accordingly for "joint efforts of philosophers and historians of science." Desirable though this might be (if only for the likelihood of such a collaboration's bringing the sharp light of philosophical precision to bear on loose historical logic), discovery stories are not compelling for most contemporary historians of science because they are—when told by the discoverer and his or her contemporaries—rather the subject than the object of historical inquiry. Achinstein understands this, but he resists it, arguing for the place of admiration and honor in considering historical figures.

Margaret Morrison injects a different brand of philosophy into history: she is concerned neither with realism or discovery, nor even with experiments, but rather with the epistemology of theory, for it is here that, she argues—in considerable contrast to much contemporary historical literature—one can find many of the sources of commitment to scientific ontology. Theory, Morrison argues, binds metaphysical commitment, but not in the manner of Kuhnian holism or entanglement. Rather, Morrison sees commitment to realist belief arising specifically out of the increasing enlargement and elaboration of what one might call an entity's field of play. In her words, "the reality ascribed to entities is often the result of their evolution in a theoretical history." The history of belief intersects not only, Morrison argues, with Hacking-like laboratory manipulation, but also—perhaps even primarily—with the evolution of theoretical trajectories. To illustrate her claim, Morrison provides a detailed account of the manner in which electron spin became a signal part of quantum mechanics. That, Morrison shows, took place over time as the properties assigned to "spin" evolved in conjunction with the solution of specific problems and theoretical desiderata, such as consistency with relativistic demands. As "spin" took shape within a theoretical system, so did it become a property with a purchase on reality—a "fundamental feature of the electron." This despite the fact that "spin" always resisted interpretation in ways similar to those that, for example, assign velocity to the electron. According to Morrison, reality commitments are inevitably and inextricably bound to the functions that a theoretical entity serves within a given scheme, including the ways in which the entity satisfies demands for consistency between the scheme's several elements.

Jonathan Bain and John Norton's article intersects with Morrison's concern to show that reality commitments evolve along with theoretical development. But where Morrison is concerned to show how commitment to an electron property, namely spin, evolved, Bain and Norton want instead to concentrate on the correction and expansion of the set of electron properties, arguing that the list of its "historically stable" properties grew over time. The sequence of theories—of electron properties—corrects "errors of former members while preserving their successes and providing richer and improved representations of the electron." There is a familiar ring to Bain and Norton's claim, since most scientists believe something similar: namely, that over time the good drives out the bad, with what was true and correct in the past moving forward, perhaps in a new form, into the future. The particular novelty of their argument lies in its emphasis on corrigibility, which connects with Morrison's notion that metaphysical commitment grows along with fruitful and consistent theoretical elaboration. This licenses a move away from the notion that theories change altogether whenever important alterations in them occur, to concentrate instead on just what it is that does change, and what the changes signify for scientific practice in specific circumstances.

There may well be incommensurability—we think that in many instances there undoubtedly is—but one must analyze with great care and subtlety the particular characteristics of a given episode to identify just where and how fundamental change occurs. One might, for example, compare Bain and Norton's episode with the manner in which the optical ray changed radically in the early nineteenth century with the introduction of the wave theory of light. Many of its functional characteristics were recaptured by wave optics, which worked with different fundamental entities, and these accordingly required those properties of the ray that remained stable to be recaptured by means of various approximations and alterations in physical interpretation (for example, by reducing a ray's sole stable property to its marking the path along which energy flows, the latter to be deduced from wave principles).

Nicolas Rasmussen and Alan Chalmers are not concerned with the evolution of a theoretical entity's properties, stable or otherwise, but they are interested in asking whether the effective use of an instrument necessarily depends in any meaningful way on theories about the way in which the device functions. On the whole, they argue, it does not, but, where it does, there is no vicious circle of mutual support between theory and observation. Specifically, the electron microscope was fruitfully used in discovering the biological cell's endoplasmic reticulum without a theory of how it interacted with the object entering in any meaningful way at all—work went on by drawing comparisons between observations done in different ways and with different devices.

In the case of crystal dislocations, work proceeded in much the same way—by independent measurement with different processes and instruments—but, in addition, a theory of Bragg diffraction by lattice planes was deployed that did support claims concerning the crystal dislocation, and that (ipso facto) also sustained claims for the operation of the electron microscope. Rasmussen and Chalmers see no vicious circularity here—it is rather a case of effecting a solid web of mutual support between the instrument and the object under investigation by it. "If," they assert, "there is a match between the precise predictions of some speculative theory, which may be or include a theory of instrument/specimen interaction, and the interpretation of some reproducible but otherwise mysterious observations, then why should this match not be taken as confirming both the theory and the interpretation of the observations?" This occurs primarily in the case of physics, they argue, and even then only when the object is sufficiently simple in structure to make the construction of a supporting web worthwhile. They see an immense range of "interpretive methods" available to experimenters who are looking to understand the character of the object under investigation.

Conclusion

Taken as a group, the essays in this book cast light on what might be called the electron's "biography" during its first hundred years of "life." Though it is not easy to decide when, or by whom, the electron was originally discovered, nevertheless by the early years of the twentieth century a broad consensus prevailed among physicists on the new entity's existence and its fundamental properties. And, unlike the ether, which faded from the physicist's world just as the electron made its appearance, the new entity was fruitfully bound to numerous branches of theoretical, experimental, and applied science. Indeed, in identifying the corpuscle as the elementary carrier of negative charge and subconstituent of the atom, Thomson relocated microphysical properties that many British physicists—Thomson included—had previously hoped to find in the ether. The emergent electron gained workaday reality at the ether's expense.

Like almost all entities in science, the electron has not remained unchanged, for its properties have evolved. During roughly the first quarter of this century the electron was generally assumed to be a particle whose mass was entirely electromagnetic in origin (Kaye and Laby 1936). Both notions—that of its purely particle-like character and that of the electromagnetic character of its mass—were undermined from the mid-1920s following the conceptual upheavals attendant on widespread acceptance of the theories of special relativity and quantum mechanics. The special theory of relativity provided a new relationship between energy and mass that did not depend at all on electromagnetic theory proper. Quantum mechanics required that the electron possess wave-like as well as particle-like properties. This remarkable requirement challenged the notion that subatomic particles were simply microversions of macroscopic bodies and received convincing experimental confirmation in the form of electron diffraction in 1927. As Morrison shows in her essay, the theories of relativity and quantum mechanics were also implicated in debates during the mid-1920s that led to the electron being attributed a quantized spin of one-half, a quantity that has no classical analogue and for which it is difficult to develop a physical image. In fundamental particle physics, too, the electron has gradually altered its role over the second half of the twentieth century. Originally one of only two or three microparticles from which physicists tried to build consistent models of atoms, the electron has now become one among hundreds of particles whose very existence, stability, and dynamics await a unified theoretical explanation.

These changes aside, the electron has in many other respects remained remarkably stable during a century that has witnessed extraordinary developments in experimental and theoretical physics. Unlike the other two fundamental building blocks of atoms—the proton and the neutron, which are now believed to be composed of the more fundamental quarks—the electron has remained a truly elementary particle. It also remains by far the most accessible and manipulatable of subatomic particles, one that plays an enormous conceptual and experimental role in numerous branches of physics, chemistry, and electronics. The electron's fundamental characteristics—its charge and its mass—have also become better known over time. When Thomson first suggested that cathode rays were composed of corpuscles in May 1897, he offered only as a tentative "hypothesis" the claim that the charge they carried was the same as that on the hydrogen ion (Thomson 1897, p. 109). If this were accepted, he noted, the mass of the corpuscles must be around a thousand times less than that of the hydrogen ion. Today the electron's charge and mass are among the most fundamental quantities in physics and have been measured to well within one part in a million (Kaye and Laby 1995, p. 19). After a century of microphysics the electron's existence as a theoretical and experimental entity appears to be more secure than ever. The essays in this book contribute to our understanding of how the first subatomic particle achieved this remarkable status.

REFERENCES

Arabatzis, T. 1996. "Rethinking the 'Discovery' of the Electron." *Studies in the History and Philosophy of Modern Physics* 27: 405–435.

Buchwald, J. Z. 1985. *From Maxwell to Microphysics*. Chicago: University of Chicago Press.

Falconer, E. 1987. "Corpuscles, Electrons, and Cathode Rays: J. J. Thomson and the 'Discovery of the Electron.'" *British Journal for the History of Science* 20: 241–276.

Feffer, S. M. 1989. "Arthur Schuster, J. J. Thomson, and the Discovery of the Electron." *Historical Studies in the Physical and Biological Sciences* 20: 33–61.

Hunt, B. J. 1991. *The Maxwellians*. Ithaca: Cornell University Press.

Kaye, G. W. C., and T. H. Laby (1911–). *Tables of Physical and Chemical Constants,* 8th edition, 1936; 16th edition, 1995. Essex: Longman.

Merzbacher, E. 1961. *Quantum Mechanics*. New York: Wiley.

Thomson, J. J. 1897. "Cathode Rays." *The Electrician:* 104–109.

Thomson, G .P. 1965. *J. J. Thomson: Discoverer of the Electron*. London: Thomas Nelson (U.S. edition: New York, Anchor Books, 1966).

Warwick, A. C. 1991. "On the Role of the FitzGerald-Lorentz Contraction Hypothesis in the Development of Joseph Larmor's Electronic Theory of Matter." *Archive for History of Exact Sciences* 43: 29–91.

I

Corpuscles and Electrons

1

J. J. Thomson and the Electron, 1897–1899
George E. Smith

What, precisely, did J. J. Thomson contribute to the discovery of the electron? Because the electron was "discovered" in 1897, one naturally takes this to be a question about what Thomson claimed pertaining to the electron during 1897, and hence a question about his April 30 Friday Evening Discourse on cathode rays at the Royal Institution,[1] in which he first put the subatomic proposal forward, and his subsequent classic paper "Cathode Rays" in the October issue of *Philosophical Magazine*.[2] Restricting the question to 1897, however, gives one a seriously incomplete and consequently misleading answer to the question of what Thomson contributed. Further, it gives a picture of what he and his research students at the Cavendish Laboratory were up to at the time that they would have had trouble recognizing. Thomson's contribution to the discovery of the electron stretched over the next two years as well. His 1897 paper is the first in a sequence of three equally classic *Philosophical Magazine* papers presenting fundamental experimental results on the electron: the second, "On the Charge of Electricity carried by the Ions produced by Röntgen Rays," appeared in December 1898,[3] and the third, "On the Masses of the Ions in Gases at Low Pressures," in December 1899.[4] The last five pages of this 1899 paper put forward a new account of ionization and electrical conduction in gases. These five pages culminated Thomson's efforts on the electron. The purpose of the present chapter is to answer the question of what Thomson contributed by considering these three papers together, taking them as presenting consecutive results of a research effort on "the connexion between ordinary matter and the electrical charges on the atom"[5] that began taking shape in 1896.

The key experiments in the 1897 *Philosophical Magazine* paper proceeded from the working hypothesis that cathode rays consist of negatively charged particles to two complementary measures of the mass-to-charge ratio, m/e, of these particles. Thomson's data, however, were less than perfect, with more than a factor of 4 variation in the m/e values he obtained. Moreover, he was not alone in publishing m/e values for cathode rays in 1897. Emil

Wiechert had announced more or less the same value on 7 January 1897 in a talk in Königsberg,[6] weeks before Thomson, and Walter Kaufmann published a value that proved, in hindsight, to be more accurate than Thomson's.[7] The only way in which Thomson's experiments might be said to have accomplished more than Wiechert's and Kaufmann's lay in his offering two complementary measures of m/e for cathode rays and in his confirming so extensively that this quantity does not vary with the cathode material or the residual gas in the tube.

Thomson also differed from Wiechert and Kaufmann in the emphasis he put on the proposal that cathode rays consist of particles.[8] Indeed, Thomson's 1897 paper and his earlier talk both give the impression that his primary aim was to settle a dispute over whether cathode rays are particles, the view favored in Britain, or some sort of etherial process, the view favored on the Continent. The paper did achieve this aim, for within months opposition to the particle view died. In point of fact, however, the issue over cathode rays was not drawing much attention at the time, and Thomson himself had not done much with cathode rays before late 1896 and did little with them after 1897.[9] To single out Thomson over Wiechert and Kaufmann for championing the particle view of cathode rays is to attach more importance to this issue than it probably deserves.

The second announced aim of Thomson's 1897 paper was to answer the question, "What are these particles?" The increasing importance of this question to Thomson when writing the paper becomes clear from comparing it with the text of his April talk. George FitzGerald's commentary on this talk had focused almost exclusively on the proposal that these particles are subatomic.[10] Partly in response to this commentary, the paper advanced considerably more evidence than the talk in support of subatomic "corpuscles." Thomson was unique in drawing this conclusion from the 1897 m/e values for cathode rays. Nevertheless, in contrast to the rapid acceptance of the particle view of cathode rays, the subatomic claim, while attracting a great deal of attention, was not accepted until after his December 1899 paper. Perhaps Thomson receives more credit than he deserves for putting this proposal forward in 1897. We need to ask, what exactly did the October 1897 paper show about the particles forming cathode rays, and what remained to be shown to provide compelling grounds that they are subatomic?

Thomson's 1897 paper ends with conjectures on the structure of atoms and the relationship between his subatomic corpuscles and the periodic table. As is widely known, over the next decade Thomson attempted to develop a "plum-pudding" model of the atom in which the negatively charged corpuscles are at rest in a configuration of static equilibrium within a positively charged matrix. The resulting widely held picture of Thomson's 1897 achievement is that he discovered the electron and then went off on a garden path on the structure of the atom, leaving to Rutherford in 1911 and Bohr in 1913, not to mention Millikan, the task of completing the project he had begun.

Taking Thomson's 1897 paper together with those from 1898 and 1899 gives a very different picture of what he accomplished. As noted, his central concern at the time was with "the connexion between ordinary matter and the electrical charges on the atom."[11] Electrical phenomena in gases provided his experimental means for getting at this connection. His 1897 paper gave a rough m/e value for cathode rays that was independent of the residual gas in the tube and the material of the cathode; this result pointed to a single carrier of negative charge that might well be ubiquitous. His December 1898 paper gave a rough value for the charge on individual ions in gases ionized by x-rays, concluding that it may well be the same as the charge per hydrogen atom in electrolysis. His December 1899 paper reported that the m/e of both the electrical discharge in the photoelectric effect and the electrical discharge from incandescent filaments is the same as the m/e of cathode rays he had obtained in 1897, and the e in the photoelectric effect is the same as the charge per ion in gases ionized by x-rays he had obtained in 1898. From these results, joined with those his research students had obtained on the migration of ions in gases, Thomson concluded that there is no positively charged counterpart to his corpuscle entering into electrical phenomena in gases. The 1899 paper ends by putting forward a new "working hypothesis" for electrical phenomena in gases in which the negatively charged corpuscle is universal and fundamental, ionization results from the dissociation of a corpuscle from an atom, and electrical currents in gases at low pressures consist primarily of the migration of corpuscles.

The three Thomson papers thus form a unit. The sequence of novel experiments reported in them replaced conjecture about the microstructural mechanisms involved in the electrification of gases with a new, empirically driven picture of these mechanisms. At the heart of this picture is an asymmetry of charge in the mechanism of electrification. This asymmetry, which stood in direct opposition to almost all theoretical work preceding it, was Thomson's most important and unique contribution to the discovery of the electron. Commentators have often pointed out that he received the Nobel Prize in 1906 not for the discovery of the electron but for his research on electricity in gases. Drawing a contrast between these two in this way misses the point made in the first two sentences of the preface to the first (1903) edition of his *Conduction of Electricity through Gases:*

> I have endeavoured in this work to develop the view that the conduction of electricity through gases is due to the presence in the gas of small particles charged with electricity, called ions, which under the influence of electric forces move from one part of the gas to another. My object has been to show how the various phenomena exhibited when electricity passes through gases can be coordinated by this conception rather than to attempt to give a complete account of the very numerous investigations which have been made on the electrical properties of gases.[12]

The work for which Thomson received the Nobel Prize was a direct extension from and elaboration of the "working hypothesis" he put forward at the end of the December 1899 paper. The central element of this working hypothesis, established experimentally through the efforts from 1897 to 1899, is the subatomic electron and its asymmetric activity.

There are four reasons why this picture of Thomson's efforts on the electron is of more than passing importance for both historians and philosophers of science. First, this episode is a striking example of research in which experiment took the lead and theory at best lagged behind and at worst acted as an impediment. The key experiments reported in Thomson's three papers and the many supporting experiments of his research students were not done for the philosophically standard purpose of testing theoretical claims. The aim of virtually every one of these experiments was to measure some quantity or other, generally a microphysical quantity. The goal of the experiments taken together was to develop enough data about what was happening microphysically to allow sense to be made of the large array of experimental phenomena involving electricity in gases that had been accumulating for over half a century. Theory offered no way of getting at many of the discoveries that came out of these experiments. In particular, the asymmetry in the action of charge at the microphysical level could not have been discovered except through experiment. The two pertinent theories at the time—Lorentz's theory of the electrodynamics of point charges and Larmour's theory of the aetherial electron—both assumed fully symmetrical activity of positive and negative charges. Episodes like this in which experiment is forced to take the lead have not received the attention they deserve, especially among philosophers.

Second, even setting aside the dominant role of experiment, this episode is an example of a kind of science that has not received enough attention, namely research in which an evolving working hypothesis substitutes for established theory. The fundamental problem in doing science is turning data and observations into evidence. High quality evidence is difficult to extract from data in the absence of established theory, for data rarely carry their evidential import on their surface, and the intervening steps in reasoning from them to evidential conclusions threaten to be too tenuous when not mediated by independently supported theory. This poses an obvious challenge for research in the early stages of theory construction in any domain. A common way of trying to surmount this challenge is to ask a working hypothesis to serve in place of theory in mediating steps in evidential reasoning, hoping to extend and develop the initial working hypothesis step by step in a bootstrap fashion into a reasonably rich fragment of a theory.

While Thomson drew heavily from both classic electromagnetic theory and the kinetic theory of gases in this research, the then available conjectural theories of the microphysics of electricity were failing to open the way for effective experimental investigations. The series of experiments that he and his research students carried out from 1896 to 1899 allowed him to develop his initially limited working hypothesis that cathode rays consist of negatively charged particles into the working hypothesis presented in the final pages of the December 1899 paper. One thing that makes this episode an especially instructive example of research predicated on an evolving working hypothesis is that so much was accomplished before the theory that was ideally needed began to emerge some fourteen years later, with the Bohr model.

Looking on Thomson's efforts on the electron during these years as science built off a working hypothesis carries with it a corollary on his research style in these efforts. The experiments he and his research students carried out had a "rough draft" character. The measured values obtained from them were at best approximate, usually indicating only the order of magnitude of the quantity under investigation. The key experiments were remarkably complex, requiring several separate measurements—each with their own problems—to be combined to obtain the targeted quantity. Admittedly, these experiments were groundbreaking not just in their gaining experimental access to the microphysics of electricity for the first time, but also in their adding a good deal of new experimental technology to laboratory practice. Even so, the variances in his results are large enough to prompt questions about whether Thomson should not have done more to perfect the experiments before publishing and moving on. As we shall see below, such questions reflect a lack of appreciation for the kind of science Thomson was engaged in. The experimental style he adopted in these efforts is entirely appropriate when the goal is one of further elaborating a working hypothesis. Trying to perfect experiments prematurely will more often than not be a waste of time; everyone will be in a much better position to refine them once more of a theory is in place. Just the opposite of being open to criticism for not doing more to perfect his experiments and leaving too much for others to clean up, Thomson should be praised for the judgment he showed in developing the experiments only to the point where they gave him what he needed to carry on.

The third reason why this episode is important for historians and philosophers of science is that the contrast between it and Thomson's efforts on the plum-pudding model of the atom underscores a crucial requirement in this kind of science: the empirical world has to cooperate for the research to get anywhere. It is sometimes suggested that Thomson's efforts on his plum-pudding model show him in decline as a scientist. To the contrary, he was engaged in exactly the same kind of science in his efforts on atomic structure, groping for a working hypothesis that would provide the logical basis for extracting conclusions from experimental results that could extend and refine this hypothesis. None of the variants of his plum-pudding model enabled such a bootstrap process to get off the ground. But then too he had tried several dead-end working hypotheses on the electrification of gases before cathode rays gave him one that turned out to be amenable to systematic experimental development. Criticizing Thomson for being unable to intuit the planetary structure that the subsequent experiments by Rutherford, Marsden, and Geiger revealed makes sense only if one thinks that the difference between great and mediocre scientists is some sort of clairvoyance. Perhaps instead we should praise Thomson, as we praise Rutherford and Bohr, for insight in recognizing the faint possibility that the empirical world might cooperate with a certain line of thought and for his ingenuity and diligence in marshalling experimental results in then developing this line of thought.

Fourth, considering Thomson's paper on cathode rays as just the first in a sequence of three seminal papers clarifies the way in which this paper marks a watershed in the history of science. Surely, the 1897 paper was a watershed, for it was the first time experimental access was expressly gained to a subatomic particle. When viewed from the perspective of twentieth century atomic physics, however, Thomson's cathode ray paper appears at most a minor initial breakthrough, of no more importance than the breakthroughs of Becquerel, the Curies, and Rutherford during the next few years. Modern atomic physics appears to derive far more from Rutherford's 1911 and Bohr's 1913 papers than from Thomson's 1897 paper. This is true. What made Thomson's paper a watershed is not that it initiated modern atomic and elementary particle physics. It was a watershed because, together with the papers of the next two years, it freed the investigation of phenomena of electrical conduction, in metals and liquids as well as in gases, from aether theory and questions about the fundamental character of electricity. As such, it marked the end of the period Jed Buchwald describes in *From Maxwell to Microphysics*[13] and the start of a new era in electrical science.

Because Thomson himself was a central figure in the electrical science in which ether theory and questions about the fundamental character of electricity were at the forefront, he had to go through a personal version of the transition that his papers effected. For this reason, an examination of Thomson's three seminal papers needs to start a little before 1897.

SOME HISTORICAL BACKGROUND

One tends to forget how much clearer the fundamental importance of cathode rays is in retrospect than it was during the six decades of research on electrical discharges at reduced pressures prior to the last years of the nineteenth century. In contrast to the often spectacular displays elsewhere in evacuated tubes, cathode rays themselves are invisible. They were discovered by Julius Plücker only in 1859, after Heinrich Geissler's invention of the mercury vapor pump allowed a degree of rarifaction at which the fluorescence they produce stood out. This was a century and a half after Hauksbee had called attention to visible electrical phenomena in gases at reduced pressure and two decades after Faraday had carried out his experimental investigations of these phenomena. Cathode rays were in turn experimentally characterized in the late 1860s and the 1870s, first by J. W. Hittorf and then by Eugen Goldstein and William Crookes. None of their findings, however, linked the cathode rays with the visible discharge, which tends to disappear at rarifactions suitable for investigating the rays. It was thus easy in the early 1890s to regard cathode rays as a separate discharge phenomenon unto themselves, occurring in the special circumstance of extreme rarifaction.

The six decades of research on electrical discharges at reduced pressures had revealed a wide array of phenomena by 1890, but scarcely anything of value for theory construction—not even well-behaved regularities among measurable quantities of the sort that had been established for electricity in solids and liquids. Nevertheless, interest remained high. This was not merely because the microstructure of gases was better understood than that of liquids and solids. A further key reason was stated by J. J. Thomson in his *Notes on Recent Researches in Electricity and Magnetism* of 1893:

> The phenomena attending the electric discharge through gases are so beautiful and varied that they have attracted the attention of numerous observers. The attention given to these phenomena is not, however, due so much to the beauty of the experiments, as to the widespread conviction that there is perhaps no other branch of physics which affords us so promising an opportunity of penetrating the secret of electricity; for while the passage of this agent through a metal or an electrolyte is invisible, that

through a gas is accompanied by the most brilliantly luminous effects, which in many cases are so much influenced by changes in the conditions of the discharge as to give us many opportunities of testing any view we may take of the nature of electricity, of the electric discharge, and of the relation between electricity and matter.[14]

As will be pointed out, Thomson was not speaking of cathode rays in this passage.

In his President's Address to the Royal Society at the end of 1893, Lord Kelvin attached comparable importance to research on electrical discharges in gases, though for a reason that puts a little more emphasis on cathode rays.[15] Kelvin turned to the subject of electricity in gases by raising the question of the difference between positive and negative electricity:

> Fifty years ago it became strongly impressed on my mind that the difference of quality between vitreous and resinous electricity, conventionally called positive and negative, essentially ignored as it is in the mathematical theories of electricity and magnetism with which I was then much occupied (and in the whole science of magnetic waves as we have it now), must be studied if we are to learn anything of the nature of electricity and its place among the properties of matter.[16]

Calling attention to the great difference in the behavior of the positive and negative electrodes in gaseous discharges led him into a brief history of cathode ray research, with primary emphasis on Crookes's electrical and other experiments at extremely high rarifaction. Whether in Crookes's experiments or those of Arthur Schuster and J. J. Thomson on the passage of electricity through gases, he went on to say, molecules are essential, while "ether seems to have nothing to do except the humble function of showing to our eyes something of what the atoms and molecules are doing." He then concluded:

> It seems certainly true that without the molecules there would be no current, and that without the molecules electricity has no meaning. But in obedience to logic I must withdraw one expression I have used. We must not imagine that "presence of molecules is *the* essential." It is certainly *an* essential. Ether also is certainly *an* essential, and certainly has more to do than merely telegraph to our eyes to tell us of what the molecules and atoms are about. If a first step towards understanding the relations between ether and ponderable matter is to be made, it seems to me that the most hopeful foundation for it is knowledge derived from experiment on electricity at high vacuum.[17]

On the question of whether cathode rays consist of negatively charged molecules, as Crookes had proposed, or some sort of wave-like disturbance of the ether, Kelvin in his presidential address considered the issue settled: "This explanation has been repeatedly and strenuously attacked by many other able investigators, but Crookes has defended it, and thoroughly established it by what I believe is irrefragable evidence."[18] Crookes had published his proposal in 1879,[19] and Goldstein had attacked it in 1880, raising a series of objections, including mean-free-path worries.[20] The case against the particle view was reinforced by Heinrich Hertz in 1883.[21] In one set of experiments designed for the purpose, Hertz was unable to detect any sign of the cathode discharge being discontinuous. When he moved the anode out of the direct stream of the cathode rays in a second set of experiments, he found that the current departed from the rays, leading him to conclude that the rays do not carry an electric charge. In a third set of experiments he was unable to deflect cathode rays electrostatically, from which he concluded that the only way there could be streams of charged particles was for their velocity "to exceed eleven earth-quadrants per second—a speed which will scarcely be regarded as probable."[22]

The Continental objections did not deter Schuster from putting forward a different version of the charged particle hypothesis in his Bakerian Lecture of 1884.[23] Schuster's experiments had persuaded him that intact molecules cannot receive a charge from contact with the cathode. He proposed instead that the emanations from the cathode consist of negatively charged atoms generated at it when molecules are torn apart by the fields produced by the interaction between it and positive ions migrating to it. He proceeded to formulate an algebraic relationship between the m/e and the velocity of these atoms implied by their curved trajectory in a magnetic field, arguing that measurements of this trajectory would allow a determination of the magnitude of m/e sufficient to corroborate his claim. In 1890 he used this relationship and such measurements, supplemented by assumptions giving estimates of the velocity, to calculate upper and lower bounds for this m/e.[24]

Kelvin's outspokenness notwithstanding, the issue of whether cathode rays consist of negatively charged particles or are a disturbance of the ether had, of course, not really been settled by the end of 1893, for figures on both sides were still advancing new evidence against the other. Hertz had augmented his argument against the particle hypothesis in 1892 when he found that cathode rays appear to pass through thin films of gold leaf.[25] In a footnote added in press to *Recent Researches,* Thomson had dismissed this finding, arguing that the cathode rays striking the film had turned it into a cathode with new rays generated from it.[26] Hertz's protégé, Phillip Lenard, then carried out extensive investigations of the rays external to the tube—which the British came to call "Lenard rays"—publishing the results in 1894.[27] In addition to showing that these rays do not propagate perpendicularly from the thin film in the way cathode rays propagate from electrodes, he added to the mean-free-path objection by determining the depth to which the rays outside the tube penetrate various gases at different densities.

During these same years Thomson advanced a similarly confuting line of argument against the aetherial-disturbance hypothesis, contending that the propagation velocity of cathode rays is orders of magnitude less than that of electromagnetic waves. The first version of this argument appeared in *Recent Researches*. Deriving basically the same relationship between m/e, velocity, and the curved trajectory of cathode rays in a magnetic field as Schuster, and adopting for e the value for hydrogen from electrolysis, Thomson concluded from Hittorf's published values for the curvature that the corresponding velocity of the cathode rays is no greater than "six times the velocity of sound."[28] The trouble with this argument was that it rather begged the question by assuming atomic values for m. Thomson published a second, seemingly more forceful version of this line of argument in 1894, obtaining comparably low values of velocity more directly from experiments using rotating mirrors.[29] This is the one set of experiments that Thomson himself conducted on cathode rays before 1896. His concern at the time appears to have been not so much with the properties of cathode rays as with the complications to ether theory that would be entailed by the magnetic deflection of these rays if they were some sort of electromagnetic waves.[30]

Thomson had succeeded Lord Rayleigh as the third Cavendish Professor and head of the Cavendish Laboratory in 1884, at the age of twenty-eight. After training first in engineering and then in physics and mathematics at Owens College in Manchester, where Schuster was one of his teachers, he matriculated at Cambridge, graduating in 1880. Although he was not a student of Maxwell's, his research between 1880 and 1896 was in the tradition of Maxwell's work in electricity and magnetism. The title page of *Recent Researches* includes as subtitle, "Intended as a Sequel to Professor Clerk-Maxwell's Treatise on Electricity and Magnetism." In surveying progress made in the field in the twenty years after Maxwell's *Treatise,* Thomson's book was no less committed than Maxwell's to combining abstract mathematical theory and experiment with concrete physical models of mechanisms and processes underlying electric and magnetic phenomena.[31] The physical model dominating Thomson's book is not the ether as such, but the Faraday tube[32]—"tubes of electric force, or rather of electrostatic induction, . . . stretching from positive to negative electricity."[33] Thomson introduces unit tubes all of the same strength, saying "we shall see reasons for believing that this strength is such that when they terminate on a conductor there is at the end of the tube a charge of negative electricity equal to that which in the theory of electrolysis we associate with an atom of a monovalent element such as chlorine."[34]

Thomson's introductory chapter on Faraday tubes ends with a proposed approach to the conduction of electricity generally in which a view of electrolysis takes the lead. The troubling issue of the interaction between electricity and matter that Maxwell's equations had left open included questions about electrical conduction and the contrasting conductivities of different substances.[35]

Chapter II of *Recent Researches* presents 154 pages covering research on "the passage of electricity through gases," including his own investigations on electrodeless tubes. The chapter surveys the full range of experiments on electricity in gases: circumstances in which gases can and cannot be electrified at normal pressures, the spark discharge, electrical discharges at reduced pressures, first in electrodeless tubes, then in tubes with electrodes, and the arc discharge; it ends with a 19 page section entitled "Theory of the Electric Discharge." The chapter is thus ideally suited for comparison with the first (1903) edition of *Conduction of Electricity Through Gases* to see just what difference the three seminal papers of 1897 to 1899 made.

Thomson reviews too many experiments in the chapter to cover them all here. Let me merely highlight some main points. The chapter calls attention to numerous asymmetries between electrical phenomena in gases at negatively and positively charged surfaces. It concludes early on, in keeping with Schuster, that molecules do not become charged, so that electrification of gases involves chemical dissociation:

> When electricity passes through a gas otherwise than by convection [i.e. such as by electrified dust particles], free atoms, or something chemically equivalent to them, must be present. It should be noticed that on this view the molecules even of a hot gas do not get charged, it is the *atoms* and not the molecules which are instrumental in carrying the discharge.[36]

Thomson cites Schuster in concluding that cathode rays—or "negative rays" as he here called them—consist of negatively charged, dissociated atoms; he responds to mean-free-path worries by suggesting that the charged atoms form "something analogous to the 'electrical wind.'"[37] Although he reviews cathode ray results thoroughly, he dismisses them as of secondary importance: "Strikingly beautiful as the phenomena connected with these 'negative rays' are, it seems most probable that the rays are merely a local effect, and play but a small part in carrying the current through the gas."[38] He lists a number of reasons for holding this, the key being the low velocity inferred from their magnetic curvature. The primary phenomenon is instead the striated positive column, the luminosity of which he concludes travels from the anode toward the cathode at a velocity of the same order of magnitude as that of light, with the striae forming a sequence of separate discharges.

The section on theory, which opens with the remarks on why electricity in gases is important quoted above, offers not a detailed theory but a "working hypothesis by which they [the phenomena] can be coordinated . . . to a very considerable extent."[39] Not surprisingly, this working hypothesis focuses predominately on the visible "positive discharge." It proceeds from two basic tenets:

> That the passage of electricity through a gas as well as through an electrolyte, and as we hold through a metal as well, is accompanied and effected by chemical changes; also that 'chemical decomposition is not to be considered merely as an accidental attendant on the electrical discharge, but as an essential feature of the discharge without which it could not occur'. (*Phil. Mag.* [5], 15, p. 432, 1883)[40]

The electric field between the anode and cathode, Thomson goes on to argue, is not sufficient to break up molecules, nor can the convection of dissociated charged atoms produce the great velocity of the discharge from the anode. Instead, the electric field polarizes the molecules spatially in the manner shown on top in figure 1.1, allowing them to form chains of the sort Grotthus had proposed for electrolysis. So aligned, interaction with the Faraday tube extending from anode to cathode is sufficient to dissociate the end atom, allowing it to combine with a charged atom at the anode, in the process contracting the Faraday tube and reinitiating the sequence. "The shortening of a tube of electrostatic induction is equivalent to the passage of electricity through the conductor."[41] The individual striae are bundles of such chains in parallel, so that the scale of electrical action in gases is not the mean-free-path from kinetic theory but the length of these chains, as dictated by conditions in the gas.

As Thomson indicates, this is a working hypothesis in the broad sense, a coordinated way of conceptualizing electrical phenomena in gases. In contrast to working hypotheses in the more narrow sense emphasized later, it does not enter constitutively into either the design or the formulation of the results of any of the experiments discussed in this chapter. As such, it is more a strategic approach for constructing a detailed theory than it is an initial fragment of such a theory that further experiments can extend and enrich. Thomson appears perfectly aware of this. At several points he tries to develop specific relationships out of his working hypothesis of a kind that might be

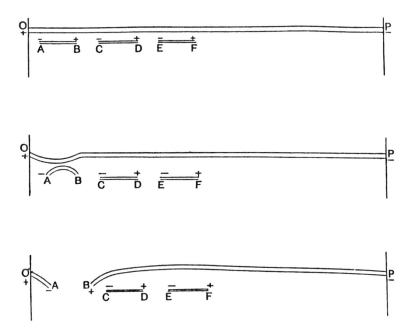

Figure 1.1
A schematic representation of Thomson's view of electrical conduction in gases (and liquids and solids) in *Recent Researches*. The electric field aligns the molecules AB, CD, and EF in a chain-like pattern. The interaction of the Faraday tubes OP and AB causes the molecule AB to dissociate, with the atom A combining with O, thereby shortening the Faraday tube OP to BP, reinitiating the sequence.

systematically tied to experimentally observable quantities, but he never sees a way of integrating any of these relationships into experiments. This is not to say that Thomson did not believe the working hypothesis he put forward. Rather, the question whether he believed it or not is largely beside the point so long as his goal was to formulate a comprehensive, detailed theory thoroughly tied to experiment and the hypothesis was unmistakably not yet enabling him to achieve progress toward that goal.

In presenting this working hypothesis—as well as earlier in this chapter and in the discussion of conduction at the end of the preceding chapter—Thomson puts special emphasis on "a remarkable investigation made more than thirty years by Adolphe Perrot, which does not seem to have attracted the attention it merits, and which would well repay repetition."[42] Perrot's experiments had shown that chemical equivalents of hydrogen and oxygen are released respectively at the cathode and anode when electricity passes through steam, just as in electrolysis. Thomson viewed these experiments as coming closer than any others to exhibiting a phenomenon whose interpretation is as "unequivocal as some in electrolysis."[43] He repeated and extended these experiments himself, publishing a paper on them in 1893[44] and including an appendix to *Recent Researches* devoted to them. A sign of how radically his view of electricity in gases changed with the three papers of 1897 to 1899 is that no mention whatever of Perrot or his experiments occurs in any of the editions of *Conduction of Electricity Through Gases*.

One shortcoming of Thomson's working hypothesis, which he noted near the end of the chapter, was that it offered nothing toward accounting for the various asymmetries of electricity in gases, in particular "the difference between the appearances presented by the discharge at the cathode and anode of a vacuum tube." Thomson's long theoretical paper of December 1895, "The Relation between the Atom and the Charge of Electricity carried by it," took a step in this direction.[45] Here too he emphasized the conjectural character of his proposals:

> The connexion between ordinary matter and the electrical charges on the atom is evidently a matter of fundamental importance, and one which must be closely related to a good many of the most important chemical as well as electrical phenomena. In fact, a complete explanation of this connexion would probably go a long way towards establishing a theory of the constitution of matter as well as of the mechanism of the electric field. It seems therefore to be of interest to look on this question from as many points of view as possible, and to consider the consequences which might be expected to follow from any method of explaining, or rather illustrating, the preference which some elements show for one kind of electricity rather than the other.[46]

In Thomson's view at the time, a molecule of hydrogen, for example, had to consist of a positively charged and a negatively charged atom of hydrogen, with a Faraday tube between them. Yet no hydrogen at all is released at the anode during electrolysis, implying that somehow all the atoms of hydrogen take on a positive charge in the process. The body of the December 1895 paper extends the working hypothesis of *Recent Researches,* taking Faraday tubes to consist of vortex filaments in the aether and trying gyroscope-like analogies, with their directional asymmetries, to account for the difference between electropositive atoms like hydrogen and electronegative atoms like chlorine and oxygen.

December 1895 was more notable for the publication of Wilhelm Röntgen's paper announcing the discovery of x-rays.[47] Since Röntgen's rays were generated by cathode ray tubes, his paper stimulated new interest in and experimentation with these tubes. Of more initial importance to Thomson was an effect of x-rays: "The facility with which a gas, by the application and removal of Röntgen rays, can be changed from a conductor to an insulator makes the use of these rays a valuable means of studying the conduction of electricity through gases."[48] 1895 was also the year in which Cambridge University first began admitting graduates of other universities as "research students."[49] Ernest Rutherford from New Zealand and John Townsend and J. A. McClelland from Ireland became research students at Cavendish at the end of 1895, joining C. T. R. Wilson, a Cambridge graduate, who had already begun his research on the condensation of moist air, having started at Cavendish early in the year. Thomson and McClelland carried out a series of investigations of x-rays and their effects in early 1896, immediately following Röntgen's announcement.[50] Thomson and Rutherford worked together on a series of experiments on gases electrified by x-rays during the first half of 1896 and Rutherford continued this effort into 1897.[51]

Sometime late in 1896 Thomson, without involving any of the research students, began experiments on cathode rays. Nothing indicates why he decided to do this, although several factors may have contributed. The efficacy of x-rays in ionizing gases implied energy levels for them, and hence for the cathode rays that generated them, that may have raised some doubts about the values for the velocity of the rays that he had published. Lenard's paper of 1894 had not changed Thomson's mind about the thin-film acting as a secondary cathode source, but the results it presented on the penetration and absorption of the rays external to the evacuated tube may have given him occasion to reconsider the mean-free-path worries. Recall that he had appealed to an "electric wind" to duck these worries in *Recent Researches*. Another factor that surely made a difference was a paper published by Jean Perrin in late 1895 reporting an experiment in which, contrary to Hertz, the negative electric charge does accompany the cathode rays.[52] Specifically, Perrin had measured an accumulation of negative charge as cathode rays strike a collector. Thomson was fully aware of the relationship between the m/e and the velocity implied by the curved trajectory of cathode rays under a magnetic field, for he had used it together with assumptions about the value of m/e in his 1893 estimates of the velocity and he knew of Schuster's similarly using it together with assumptions about the velocity in his 1890 estimates of m/e. The problem in both cases was that the magnetically curved trajectory provides a single experimental relationship between two unknowns. Perhaps what most of all got Thomson going on his cathode ray experiments in late 1896 was his seeing the possibility of Perrin's experiment yielding a second relationship between these two unknowns.

J. J. Thomson on Cathode Rays—1897

The first public indication that Thomson was doing experiments on cathode rays was in a February 8 talk he gave to the Cambridge Philosophical Society, reported a month later in *Nature*.[53] There, Thomson presented his results from experiments on the magnetic deflection of cathode rays and a refined version of Perrin's experiment from 1895. He appears to have made no mention of the subatomic. The occasion for his April 30 talk was a Friday Evening Discourse at the Royal Institution in London. Most of this lecture-with-demonstrations was again devoted to these experiments, but what made news was the subatomic hypothesis he placed before his distinguished audience at the end. The tenor of the reaction can be seen in an editorial remark in *The Electrician* three months later: "Prof. J. J. Thomson's explanation of certain cathode ray phenomena by the assumption of the divisibility of the chemical atom leads to so many transcendentally important and interesting conclusions that one cannot but wish to see the hypothesis verified at an early date by some crucial experiment."[54]

The text of the April 30 talk appeared in the May 21 issue of *The Electrician,* immediately following FitzGerald's commentary on it. After a brief review of the history of cathode rays, Thomson presented some experiments displaying the deflection of the rays in magnetic fields, in the process providing visible evidence that their trajectory in a uniform field is circular. He then demonstrated his version of Perrin's experiment and described some related experiments showing that cathode rays carry a charge. Along the way he pointed out that cathode rays turn the residual gas in the tube into a conductor, and he appealed to this to explain Hertz's failure to deflect the rays electrostatically. Finally, he demonstrated Lenard's result of rays outside the tube and reviewed Lenard's absorption data, agreeing that these data show that the distance the rays travel depends only on the density of the medium. This led him to the question of "the size of the carriers of the electric charge. . . . Are they or are they not of the dimensions of ordinary matter?" A mean-free-path argument gave him the answer: "they must be small compared with the dimensions of ordinary atoms or molecules."[55]

Thomson adopted a cautious tone in putting the "somewhat startling" subatomic hypothesis forward in the talk. It doubtlessly would have been passed over as nothing more than an interesting conjecture were it not for his having given an experimentally determined value of m/e for the cathode ray particles at the end of the talk. The single value he gave, 1.6×10^{-7} (in electromagnetic units), was inferred by combining the accumulation of charge and heat at the collector in a further variant of Perrin's experiment with the product ρH, where ρ is the radius of curvature of the rays deflected by a magnetic field of strength H. Not much could be made of the precise magnitude of this single value. (In fact, it falls entirely outside the range of values Thomson gives in his subsequent paper.) The point Thomson stressed was that this value is three orders of magnitude less than the m/e inferred for hydrogen from electrolysis and this favors "the hypothesis that the carriers of the charges are smaller than the atoms of hydrogen."[56] He closed his talk by noting that his m/e agrees in order of magnitude with the m/e Pieter Zeeman had inferred for charged particles within the atom in a recent paper on the magnetic splitting of lines in the absorption spectrum of sodium.[57]

As the title, "Dissociation of Atoms," suggests, FitzGerald's comments focused entirely on the subatomic proposal, ignoring the first three-quarters of Thomson's talk. It would be wrong to say that FitzGerald's response was dismissive. His concluding paragraph underscored the potential importance of Thomson's proposal:

> In conclusion, I may express a hope that Prof. J. J. Thomson is quite right in his by no means impossible hypothesis. It would be the beginning of great advances in science, and the results it would be likely to lead to in the near future might easily eclipse most of the other great discoveries of the nineteenth century, and be a magnificent scientific contribution to this Jubilee year.[58]

The stance FitzGerald adopted was that the potential importance of the proposal demanded that alternative interpretations of Thomson's experimental evidence be considered. The state of the field—FitzGerald expressly noted how little was known "about the inner nature of conduction and the transference of electricity from one atom of matter to another"—made other interpretations not hard to find. The alternative line of interpretation FitzGerald developed was that cathode rays consist of aetherial "free electrons" and the mass in Thomson's m/e measurement was entirely "effective" or quasi-mass from the electromagnetic inertia exhibited by a moving charge.[59]

Something needs to be said here about the word "electron." Thomson eschewed the term even as late as the second edition of his *Conduction of Electricity Through Gases* in 1906, when virtually everyone else was using it to refer to his corpuscles. Thomson chose "corpuscle" to refer to the material carrier of negative electric charge constituting cathode rays. G. Johnstone Stoney had introduced "electron" two decades earlier to refer to a putative physically fundamental unit of charge, positive and negative. He did this as part of a general argument that physically constituted units are preferable to arbitrary ones, proposing in the case of charge that the laws of electrolysis pointed to a fundamental unit, which at the time he calculated to be 10^{-20} electromagnetic units.[60] In the early 1890s Joseph Larmor, of Cambridge, had adopted the term at FitzGerald's instigation for the unit "twists" of ether comprising the atom in his theory of atomic structure.[61] (Larmor's proposal was that the quasi-mass of positive and negative electrons formed the mass of the atom; his original value for the electron quasi-mass corresponded to the mass of the hydrogen ion, but he reduced this in response to Zeeman's result.) Lorentz, who in 1892 had developed his version of Maxwell's equations, allowing for charged particles, did not adopt "electron" until 1899. Zeeman, who had turned to Lorentz, his former teacher, for the calculation of m/e, also did not use "electron." FitzGerald's "free electron" was adapted from Larmor. It refers to an aetherial unit charge, positive or negative, liberated from the atom, and was thus expressly intended to contrast with Thomson's "corpuscle." A compelling empirical basis for identifying Thomson's corpuscle with Stoney's unit charge emerged only with Thomson's December 1899 paper.

The influence of FitzGerald's commentary on Thomson is evident in the respects in which his October 1897 paper extends beyond his April 30 talk. In the results in the paper Thomson uses more than one material for the cathode, just as FitzGerald had suggested. The m/e experiment is repeated several times in different configurations, offering some response to FitzGerald's worries about the measurement of charge and heat accumulation. More importantly, a second way of determining m/e is added in which the charge and heat measurement is replaced by electrostatic deflection of the cathode rays. Thomson and his assistant encountered a good deal of difficulty in achieving stable electrostatic deflections of cathode rays.[62] Because the rays liberated gas from the walls of the tube, the rays had to be run in the tube and the tube then reevacuated several times to eliminate sufficiently the nullifying effects of ions in the residual gas.

Thomson submitted his paper on 7 August 1897, three months after his first going public with the subatomic hypothesis. The paper has three principal parts. After posing the particle versus ether-disturbance issue, the first part presents results of qualitative experiments supporting the particle hypothesis, including electrostatic deflection. The carefully phrased transition from the first to the second part is worth quoting:

> As the cathode rays carry a charge of negative electricity, are deflected by an electrostatic force as if they were negatively electrified, and are acted on by a magnetic force in just the way in which this force would act on a neg-

atively electrified body moving along the path of these rays, I can see no escape from the conclusion that they are charges of negative electricity carried by particles of matter. The question next arises, What are these particles? are they atoms, or molecules, or matter in a still finer state of subdivision? To throw some light on this point, I have made a series of measurements of the ratio of the mass of these particles to the charge carried by it.[63]

The second part presents the results of the two ways of determining m/e. The third part opens by laying out the subatomic hypothesis, stated finally as:

Thus on this view we have in the cathode rays matter in a new state in which the subdivision of matter is carried very much further than in the ordinary gaseous state: a state in which all matter—that is, matter derived from different sources such as hydrogen, oxygen, &c.—is of one and the same kind; this matter being the substance from which all the chemical elements are built up.[64]

The remainder of the third part offers conjectures about atomic structure and the periodic table. The paper ends with brief remarks on the difference in the announced cathode ray velocities between this paper and the paper of 1894 and on effects observed with different cathode materials.

Thomson's opening sentence announces that "the experiments discussed in this paper were undertaken in the hope of gaining some information as to the nature of the Cathode Rays." If the paper is read in isolation from its historical context, the rhetorical flourish with which the charged particle versus aetherial-disturbance issue is laid out in the remainder of the first paragraph gives the impression that the question Thomson was most seeking to answer was whether cathode rays are particles. Given the view of this question at the time among his primary British audience, however, a more historically plausible reading of this first sentence is that the information he most hoped to gain bore on the questions posed at the outset of the second part of the paper quoted above: "What are these particles?" and so forth. The qualitative experiments discussed in the first part have a more important role than merely providing evidence that cathode rays are particles, the presupposition of these questions. They clear the way for using charge accumulation, electrostatic deflection, and magnetic curvature to obtain experimental values of m/e and v. They do this by obviating worries about whether the accumulation of charge being measured is that of the cathode rays, whether the failure to obtain electrostatic deflection at anything but extraordinary levels of evacuation is truly because the rays ionize the residual gas, and whether the specifically observed curvature of the trajectory is that

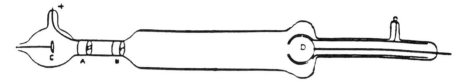

Figure 1.2
A schematic of one of the three kinds of tubes Thomson used in his first approach to measuring m/e for cathode rays (based on the description given in his October 1897 paper).

of the rays, in contrast to some secondary luminosity in the gas. (This fits the suggestion that what prompted Thomson to begin his experiments on cathode rays in late 1896 was the prospect of a fully experimental determination of m/e and v; for, to this end, he would have first needed to gain mastery of the basic experiments and safeguard against the possibility that measurements made in them are misleading artifacts.)

Only the experiments for m/e in the second part of the paper merit much comment here. Figure 1.2 shows a schematic of one of the three types of tubes Thomson used with the first method. A narrow cathode ray beam passes through slits in the anode A and the plug B, striking the collector D unless it is magnetically deflected as a consequence of current flowing through a coil magnet located along the middle of the tube. From the expressions given in the paper for the charge Q accumulated at the collector, the kinetic energy W of the particles striking it, and the radius of curvature ρ of the beam under a uniform magnetic field H, Thomson obtains the following expressions for the m/e and the velocity v of the particles:

$$\frac{m}{e} = \frac{H^2\rho^2 Q}{2W}, \qquad v = \frac{2W}{QH\rho}.$$

An electrometer was used to measure Q, W was inferred from the temperature rise at the collector (measured by a thermocouple), H was inferred by measuring the current in the coils, and ρ was inferred from the length of the magnetic field and the displaced location of the point of fluorescence on the glass tube. The design of the experiment is thus opening the way to obtaining values of microphysical quantities from macrophysical measurements.

In the second method, shown schematically in figure 2 of Thomson's paper,[65] electrostatic deflection of the beam replaces the accumulation of charge and heat at the collector. Thomson derives expressions for the angle θ to which the beam is deflected as it leaves the uniform electric field of strength F between plates of length l, and the angle θ to which it is deflected by the magnetic field H of the same length. In the version of the experiment reported in the paper, the magnetic field was superimposed on the electric field, and its strength H was varied until the electrostatically displaced spot was restored to its original location. In this case:

$$\frac{m}{e} = \frac{H^2 l}{F\theta}, \quad v = \frac{F}{H},$$

where θ was inferred from the displaced location of the fluorescent spot when only the electric field was present and F was inferred from the voltage drop applied to the plates. This method also thus involves only macrophysical measurements.

Thomson's presentation proceeds so smoothly, and the crossed-field approach with cathode rays has become so familiar, that readers can easily fail to notice the complexity of the logic lying behind these m/e experiments. The derivations of the two expressions giving m/e, along with the instruments used to obtain values of the parameters in them, presuppose a number of laws from physics; many of these had been discovered within the living memory of some of Thomson's colleagues and hence were less firmly entrenched in 1897 than they are now. The derivations also presuppose a variety of further assumptions. Some of these serve only to simplify the math. For example, in deriving the angular displacement of the beam in a magnetic field in the crossed-field experiment, Thomson implicitly assumes that the velocity of the beam is great enough that he can treat the magnetic force as unidirectional, just like the electrostatic force. He could easily have derived a more complicated expression, taking into account that the magnetic force is always normal to the direction of the beam. Similar to this are some assumptions in which he idealizes the experimental setup. He assumes, for example, that the collector is perfectly insulated thermally so that no heat leaks from it, and he assumes that the magnetic and electric fields extend only across the length l, ignoring the small field effects extending beyond the edges of the plates and the coils. He could easily have introduced corrections for these effects, complicating the math a little.[66]

Beyond these are such assumptions as the particles all have the same m/e and, in any one experiment, the same constant velocity both across the length of the magnetic and electric fields and downstream at the collector. These assumptions have a more wishful character. Because they concern the unknown quantities that are being measured, they are not so readily amenable to corrections. The main safeguard against being misled by them lies in the quality of the data. The falsity of any of them should show up in the form of poorly behaved data when the experiments are repeated with different field strengths, anode-to-cathode voltage drops, and tube configurations.

Some difficulties in executing the experiments complicated matters still further. Because the cathode rays ionized the residual gas in the tube, the leak of charge from the collector became increasingly significant as the total charge accumulated. As a consequence, the charge accumulation experiment had to be run over short time durations, entailing small temperature rises and hence greater sensitivity to small inaccuracies in measurement. Far worse was the so-called "magnetic spectrum." Birkeland had called attention to the fact that the fluorescent spot spreads out when displaced magnetically, generally forming a sequence of spots with darker regions between them. Thomson found the same thing with electrostatic deflection. The magnetic spectrum was prima facie evidence against all the particles having the same m/e. In the April 30 talk Thomson suggested that this effect might be from two or more corpuscles clumping together. In the October paper, however, he makes no mention of this possibility. Instead, the magnetic and electric displacements are identified with the brightest spot in the spectrum, if there is one, and with their middle, if there is not.

The magnetic and electrostatic "spectra" were in fact experimental artifacts, caused by different velocities among the particles resulting from Thomson's use of an induction coil to produce the anode-to-cathode voltage drops instead of a continuous source, such as a stack of batteries. This was established roughly a year later by Lord Rayleigh's son, R. J. Strutt, while still an undergraduate at Trinity College, Cambridge, and it was announced in a paper in the November 1899 issue of *Philosophical Magazine*.[67] No one at Cavendish appears to have repeated the cathode ray m/e measurements when this discovery was made.

The pivotal assumption underlying the m/e experiments is that cathode rays are streams of particles. One can think of this as a working hypothesis, with the results of the qualitative experiments presented in the first part of Thomson's paper providing the justification for predicating further research on it. A failure to come up with well-behaved results for m/e in the experiments would be evidence against it. Conversely, evidence would accrue to it from the experiments presupposing it to the extent that (1) the value of m/e obtained from each method remains stable as the field strengths, the anode-to-cathode voltage, and other things are varied and (2) the values obtained from the two methods are convergent with one another. This is typical of the way in which successful theory-mediated measurements of fundamental quantities have always provided supporting evidence for the theory presupposed in them.

How stable and convergent were Thomson's results? Here the logic becomes subtle. On the one hand, the data fall far short of yielding a precise value for m/e. His values for m/e from the first method range from a low of

0.31 × 10⁻⁷ to a high of 1.0 × 10⁻⁷, and from the second method, from a low of 1.1 × 10⁻⁷ to a high of 1.5 × 10⁻⁷.[68] Looking at his *m/e* numbers by themselves, therefore, one can legitimately question whether the results were all that stable or convergent. On the other hand, the *m/e* values are all three orders of magnitude less than the smallest theretofore known value, the *m/e* of the hydrogen ion. When viewed in this light, the results at the very least provided strong additional evidence for predicating further research on the hypothesis that cathode rays consist of negatively charged particles.

Because of the "rough draft" character of the *m/e* experiments, as well as the confounding factor of Birkeland's spectrum, Thomson's 1897 paper did not settle the question of whether all the particles forming cathode rays have the same *m/e*. The one feature of the data supporting a single, universal particle was the absence of systematic variation in *m/e* with the gas in the tube and the material of the cathode. This was enough for Thomson to proceed further under the extended working hypothesis that all cathode rays consist of corpuscles with a mass-to-charge ratio around 10^{-7} emu—presumably subatomic corpuscles of a single, universal type. He set the question whether there is a single value of *m/e* for cathode rays and, if so, what precisely it is to one side, turning instead to other questions raised by the paper.[69] The paper announces two questions: (1) is the very small *m/e* a consequence of a small *m*, a large *e*, or a combination of the two?; and (2) how many corpuscles are there in an atom, and how do they fit into it? Judging from his research over the next two years, however, the question most on his mind was, (3) how do the cathode ray corpuscles enter into other electrical phenomena?

Three final points need to be made about the 1897 paper. First, the experiments reported in it do not in themselves refute the view that cathode rays are wave-like. The velocities Thomson obtained varied with the cathode-to-anode voltage, ranging from a low of 2.2 × 10⁹ to a high of 1.3 × 10¹⁰ cm/sec—that is, from roughly 7 to 43 percent of the speed of light.[70] This difference from the speed of light was enough to accomplish Thomson's 1894 objective of refuting the proposal that cathode rays are a type of electromagnetic wave propagation, but not enough to show that they are not waves. The only way of proceeding from Thomson's results to the conclusion that cathode rays have no wave-like character is via the tacit premise that anything consisting of particles cannot have a wave-like character. However much this premise was an ingrained article of belief at the time, it was not presupposed by the experiments themselves. Consequently, nothing in the experiments of the 1897 paper, or subsequent refined versions of them, required any correction or adjustment when the wave-like character of electrons was established three decades later.[71]

Second, the premise that cathode rays consist of charged particles—or at least constituents that are sufficiently particle-like for laws governing charged particles to hold—is presupposed by the experiments. It is a constitutive element in the experiments and hence a working hypothesis in the narrow sense to which I alluded in the preceding section: a proposition of conjectural status that enters indispensably into a train of evidential reasoning leading from observations to the statement of the results of an experiment. Consider what the two m/e experiments would amount to without this premise. Ignoring the unlikelihood that someone would still have pursued the investigation, each would have shown only that a certain algebraic relationship among some macroscopic variables retains more or less the same numerical value when conditions involving cathode ray tubes are varied. Worse, without it the only reason to have taken the two algebraic relationships to perhaps be representing the same thing would have been the degree to which their roughly invariant values matched one another, which in fact was not all that great. The charged-particle working hypothesis, joined with the relevant laws from prior science and the various simplifying assumptions, put Thomson in a position where the empirical world could provide answers to such questions about the nature of cathode rays as, what is the mass-to-charge ratio of their constituents?, How, if at all, does this ratio vary with the gas in the tube, the electrode material, and the voltage drop from cathode to anode?, and how does it compare with other known values of m/e? Evidence—or at least grounds for predicating further research on it—accrued to the particle hypothesis from the extent to which these answers were well-behaved, allowing experiment to replace conjecture in extending it.

Third, as indicated earlier, Thomson was not the only one measuring m/e for cathode rays at the time. Both Emil Wiechert[72] and Walter Kaufmann[73] in Germany were independently obtaining more or less the same m/e values as Thomson by combining magnetic deflection with eV, the upper bound for the kinetic energy particles of charge e would acquire in falling through a potential difference V between the cathode and anode. Wiechert, in particular, had announced his results on 7 January 1897 in a talk in Königsberg, stating that the mass of the particle is between 2000 and 4000 times smaller than that of a hydrogen atom, having first assumed that the charge is one "electron"—that is, the charge per hydrogen atom in electrolysis, inferred from existing estimates of Avogadro's number. Thomson's 1897 work was nonetheless distinctive in three respects. First, he went beyond the others in the extent to which he determined that m/e is independent of the gas in the tube and the material of the cathode. Second, he was alone in devising two complementary measures, thereby adding a good deal of support for the underlying working hypothesis that the constituents of cathode rays are particle-like. Third, he alone immediately proposed that the charged particles in question are dissociated constituents of atoms.

J. J. Thomson on the Charge of Ions—1898

The results of several experiments supporting Thomson's m/e results for cathode rays, including more refined experiments by Kaufmann and by Lenard, were published in 1897 and 1898. In 1898 Lenard also announced that the m/e for the rays outside the cathode tube that were being named after him is the same as for cathode rays.[74] In 1886 Goldstein had noted faint rays passing through holes in the cathode into the space on the opposite side of it from the anode, seemingly symmetric counterparts of cathode rays. Wilhelm Wien used magnetic and electric deflection to determine that these rays, called "Canalstrahlen," were positively charged with a mass-to-charge ratio around three orders of magnitude greater than that of cathode rays; he announced the distinctive contrast between these and cathode rays in 1898.[75] By contrast, while others were pursuing refined measures of m/e for cathode and related rays, Thomson, though noting their results,[76] shifted the focus of his research away from these rays.

Thomson published two papers in *Philosophical Magazine* in 1898. The first, "A Theory of the Connexion between Cathode and Röntgen Rays," appeared in February.[77] In it Thomson derives theoretical expressions for the magnetic force and electric intensity that propagate when a moving electrified particle is stopped suddenly—more specifically, a particle moving at a velocity high enough that the square of the ratio of it to the speed of light can no longer be neglected. At the end of the paper he calls attention to the high velocity he had obtained for the negatively charged particles forming cathode rays, concluding that Röntgen rays are most likely impulses generated by the sudden stoppage of these particles, and not waves of very short wave-length.

The second paper, "On the Charge of Electricity carried by the Ions produced by Röntgen Rays," appeared in December 1898.[78] It reports the results of an elaborate experiment for determining the charge e of the negative ions produced when x-rays pass through a gas. The relationship between these negative ions and Thomson's corpuscle is left an entirely open question throughout this paper. The basic idea behind the experiment is to infer the charge per ion from the amount of electricity (per unit area per unit time) passing through the ionized gas under an electromotive force. Assuming all ions have the same magnitude of charge, e, this quantity of electricity is simply neu, where n is the number of ions per unit volume and u is the mean velocity of the positive and negative ions under the electromotive force. The charge per ion can be thus be inferred from a determination of n and u.

Three separate results published by Thomson's research students during 1897 opened the way to determining n and u. First, Rutherford's research on the conduction of electricity in gases ionized by x-rays had culminated in a paper published in *Philosophical Magazine* in November 1897, entitled "The Velocity and Rate of Recombination of the Ions of Gases exposed to Röntgen Radiation."[79] In an experiment that was fairly elaborate in its own right, Rutherford had determined ion velocities for a number of gases. In particular, the velocity of both the negative and the positive ions that he found in the case of atmospheric air was around 1.6 cm/sec per volt/cm potential gradient (i.e. 480 cm/sec per unit potential gradient in the esu units Thomson chose to use at the time); and the velocity he found in the case of hydrogen was around three times greater than this. Thomson assumed these values in his experiment.

Second, Wilson had established that, when x-rays pass through dust-free, saturated damp air and the air is then suddenly expanded, a cloud is produced by a degree of adiabatic expansion that produces no cloud when the air has not been subjected to x-rays.[80] The presumption was that the ions act as nuclei around which droplets of water form. Wilson had devised means for determining, through calculation, the total volume of water formed, so that the number of droplets—and hence the number of ions—per unit volume could be inferred if the radius of the presumably spherical droplets could be determined. The one tricky element, which Wilson had also found a way of handling, was to gain some assurance that a droplet forms on every available ion.

The remaining problem was to determine the radius of the droplets. For this Thomson ended up adopting an approach Townsend had devised in determining an approximate value for the charge on positive and negative ions of oxygen released in electrolysis.[81] Townsend too had relied on the formation of water droplets, in his case droplets that formed after the gases given off in electrolysis were bubbled through water. To determine the size of the droplets, he had measured their velocity in fall under their own weight and had then inferred their radius from Stokes's theoretical law for the purely viscous resistance force acting on small moving spheres.

As should be evident by this point, the logic underlying Thomson's method for measuring the charge of the ions is even more complicated than the logic underlying his methods for measuring m/e for cathode rays. Some of the assumptions entering into the method are not stated in his paper but are instead buried in the papers of his research students. On top of this, the experiment itself is complicated, involving three distinct parts: an irradiation part in which a quantity of gas is subjected to x-rays of an appropriate intensity; an electrical part in which the amount of electricity passing through the ionized gas under an electromotive force is determined; and a gaseous-expansion part in which the velocity of the water droplets is measured and the total amount of water is inferred from a measurement of temperature change.

Not surprisingly, the apparatus for the experiment (shown schematically in figure 1.3) has a distinctly Rube Goldberg character. The ionized gas is contained in the vessel A, which is covered by a grounded aluminum plate and

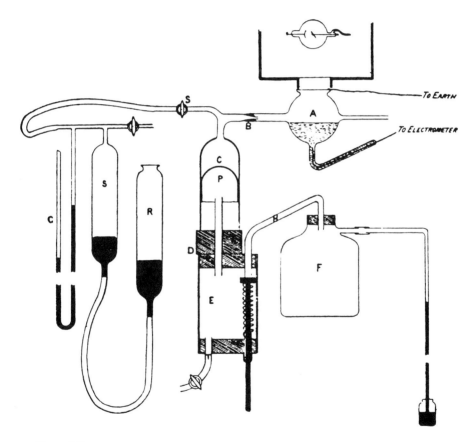

Figure 1.3
Thomson's schematic of his experiment to measure the charge per ion in gas ionized by x-rays. The gas to be ionized is contained in vessel A, below the cathode ray tube used to generate the x-rays. Most of the rest of the apparatus serves to effect the controlled expansion required for droplets to form on individual ions in a well-behaved fashion.

contains a pool of water electrically charged by a battery. The aluminum plate serves to limit the intensity of the x-rays reaching the gas. The expansion of the gas is effected by the piston P; all the paraphernalia attached to it, as well as the tubes R and S, serve to control the expansion. One pair of quadrants of an electrometer are connected to the tank and the aluminum plate, and the other pair are connected to the water. The tank, the aluminum plate, the water, the electrometer, and the wires connecting them form a system with an electric capacity that can be measured. Given this capacity, the amount of electricity passing through the ionized gas is determined by measuring the rate of charge leaking from the electrometer when the gas is irradiated.

Thomson's paper falls into six parts. The first presents the basic ideas underlying the experiment. The second describes precautions taken to assure that the level of radiation and the amount of expansion were appropriate. The third describes the apparatus and the method used for measuring the amount of electricity passing through the gas—that is, CV, where C is the measured electric capacity of the system and V the voltage change observed for it with the electrometer. The fourth part goes through the process of calculating, in sequence, the total amount of water q, the droplet radius a, the number of droplets n, and finally the charge-per-ion e from measured values for one trial of the experiment. The fifth part presents the results for e obtained from several trials for air and then for hydrogen. The last part offers concluding remarks, first in defense of an assumption and then on comparisons between the value obtained for e, the value of unit charge inferred from electrolysis, and the value Lorentz had recently inferred from the splitting of spectral lines.

The entire approach presupposes that there is some definite charge per ion when a gas is ionized by x-rays. Because so little was known about gaseous ions, the only way of defending this assumption was to appeal to regularities observed in electrolysis, the microphysical basis for which was still largely a matter of conjecture. This assumption accordingly fell mostly into the category of wishful thinking. It is akin to what is called "taking a position" in the card game contract bridge: if the only way to make a contract is for a particular card to be in a particular hand, then the best approach is to postulate that the card is in that hand and draw further inferences under this assumption, taken as a working hypothesis. If the only prospect for coming up with a telling experiment is to assume that nature is simple in some specific way, then the best approach may be to make this assumption and see what comes out of the experiment. This is especially true in the early stages of scientific research into a domain that cannot be observed comparatively directly. Thomson could have adopted a weaker assumption in this experiment: there is a consistent *average* charge per ion when a gas is ionized by x-rays. But if one is going to engage in wishful thinking, why adopt a less desirable line until the data give one reason to?

As with the *m/e* experiments, the most immediate safeguard against being misled by an experiment predicated on a tenuous assumption lies in the quality of the data obtained as the experiment is repeated in varying conditions. Thomson found it necessary to introduce two corrections to his raw data. The first correction, applied to the value of *e* obtained in each trial, served to compensate for the fact that some droplets form even in gas not radiated by x-rays.[82] (Cosmic rays, which were discovered in 1911, were causing some ionization, confounding the experiment.) The second correction, applied to the mean value of *e* obtained over the series of trials, compensated for electric conduction in the film of moisture coating the walls of the vessel. Neither of these corrections appears to have been introduced solely to make the data appear better behaved.

The values Thomson reports for *e* in air have a range about their mean of roughly ±16 percent. His corrected mean value for air is 6.5×10^{-10} electrostatic units, around 35 percent above the current value for the electron charge. The measurements with hydrogen involved greater uncertainty so that Thomson does not bother to carry through the corrections to the raw data. The range of the raw data is nevertheless about the same as in air. Thomson concludes that "the experiments seem to show that the charge on the ion in hydrogen is the same as in air. This result has very evident bearings on the theory of the ionization of gases produced by Röntgen rays."[83] The thrust of this last remark is that a single fundamental quantity of electricity per ion is involved when gases are ionized by x-rays, regardless of the chemical composition of the gas. (The comparison between the results for air and hydrogen might be more accurately summarized by saying that the experiments do not show that the charge on the ion in hydrogen is not the same as in air. The element of wishful thinking is carrying over into the extended working hypothesis that Thomson is extracting from the results of this experiment.)

The element of wishful thinking is also evident when he compares his 6.5×10^{-10} with the value of *e* inferred from the total quantity of electricity in electrolysis, using Avogadro's number—or, as Thomson preferred, the number of molecules per cubic centimeter at standard conditions. Thomson's value of charge, together with the total electricity per cubic centimeter of hydrogen released in electrolysis, gives a value of 20×10^{18} molecules per cubic centimeter. He compares this with the value of 21×10^{18} obtained from experiments on the viscosity of air. (Our modern value is 27×10^{18}.) The values at the time ranged far more widely than Thomson's comparison would suggest. A prominent 1899 textbook in kinetic theory, for example, gave 60×10^{18} as the value.[84] The conclusion Thomson draws from his comparison is suitably qualified: the agreement "is consistent with the value we have found for e being equal to, or at any rate of the same order as, the charge carried by the hydrogen ion in electrolysis."[85]

Thomson's experiments for determining e in ionization by x-rays were logically independent of his experiments for determining m/e for cathode rays. Even so, these 1898 experiments, more complicated though they were, evince the same research style as the 1897 experiments. The hypothesis that there is some characteristic value of ion-charge when a gas is ionized by x-rays is a constitutive element in the experiments, presupposed in inferring the value for e from the measured current neu. This working hypothesis, joined with experimental techniques and results from his research students, relevant laws from prior science, and some simplifying assumptions, allowed Thomson to design experiments in which the empirical world could give answers not only to the question of the magnitude of this e but also to whether it varies with the conditions under which a given gas is ionized by x-rays, whether it varies from one gas to another, and how it compares with the e of electrolysis.

Finally, just as with his m/e experiments, the achievement of Thomson's e experiment was not so much to establish a definite value for e as it was to license a working hypothesis for ongoing research: the same fundamental quantity of electricity is involved in both electrolysis and the ionization of gases by x-rays, and this quantity is of the order of magnitude of 6.5×10^{-10} esu. Thomson was struggling to find experiments involving macrophysical measurements that would yield some reasonably dependable conclusions about microphysical processes. In this early stage of research, working hypotheses had to stand in for established theory in the logical design of experiments. The results of his e experiment, in principle, could have provided good reasons for abandoning the wishful thought that nature is simple in the way the working hypothesis says it is. They did not. Instead, in spite of their roughness and uncertainty, his results showed the working hypothesis to have sufficient promise to warrant predicating further research on it. To see the role it ended up playing in this further research, we need to turn to his December 1899 paper.

THE ELECTRON AND IONIZATION—1899

Again in 1899 Thomson published two papers in *Philosophical Magazine:* "On the Theory of the Conduction of Electricity through Gases by Charged

Ions" in March,[86] and "On the Masses of Ions in Gases at Low Pressures" in December.[87] The first of these takes off from results obtained by Thomson's research students on the velocities of ions: by Rutherford and John Zeleny for gases exposed to x-rays; by Rutherford for gases exposed to uranium radiation and to the photoelectric discharge produced by ultraviolet light;[88] by McClelland and Harold Wilson for the ions in flames; and by McClelland for the ions in gases near incandescent metals and gases exposed to arc discharges.

> A remarkable result of the determination of the velocities acquired by the ions under the electric field is that the velocity acquired by the negative ion under a given potential gradient is greater than (except in a few exceptional cases when it is equal to) the velocity acquired by the positive ion. Greatly as the velocities of the ions produced in different ways differ from each other, yet they all show this peculiarity.[89]

Under the assumption that current in gases consists of migrating ions that have not yet recombined to form an electrically neutral molecule, Thomson derives a differential equation relating ion velocity to current. He is able to integrate this equation only under a simplifying assumption. He nevertheless proceeds in this way to develop an expression for the flow of electricity in gases of the form, $V = Ai^2 + Bi$, where V is the potential difference across a pair of plates, i is the current, and expressions for A and B are formulated in terms of properties of the ions, including their charge. The paper ends by considering various asymmetries between negative and positive electricity in the light of Thomson's mathematical theory and the observed asymmetry in ion velocities.

The paper immediately following Thomson's in the March issue of *Philosophical Magazine* is by William Sutherland, entitled "Cathode, Lenard, and Röntgen Rays."[90] This entire paper is in response to Thomson's subatomic proposal: "Before a theory of such momentous importance should be entertained, it is necessary to examine whether the facts to be explained by it are not better accounted for by the logical development of established or widely accepted principles of electrical science."[91] The principles Sutherland has in mind are those of ether theory and Larmor's etherial electron. He summarizes his alternative theory in two propositions: "The cathode and Lenard rays are streams, not of ions, but of free negative electrons. The Röntgen rays are caused by the internal vibrations of free electrons."[92] Negatively charged free electrons are generated when an immaterial "neutron" consisting of a positively and negatively charged pair becomes dissociated.

In a curt reply published in the following month's issue,[93] Thomson points to questions about whether an impacting quasi-mass is sufficient to produce x-rays and to questions about how aetherial electricity can be distributed within the atom, invoking the Zeeman effect to suggest that "the electron thus appears to act as a satellite to the atom." Thomson summarizes the situation from his point of view:

> As far as I can see the only advantage of the electron view is that it avoids the necessity of supposing the atoms to be split up: it has the disadvantage that to explain any property of the cathode rays such as Lenard's law of absorption, which follows directly from the other view, hypothesis after hypothesis has to be made: it supposes that a charge of electricity can exist apart from matter, of which there is as little evidence as of the divisibility of the atom: and it leads to the view that cathode rays can be produced without the interposition of matter at all by splitting up neutrons into electrons.[94]

Thomson's other 1899 *Philosophical Magazine* paper was originally presented at a meeting of the British Association a few months earlier. The published version, the next to last paper in the December issue, would have been a fitting final word of the nineteenth century from this journal. The paper consists of five parts. The first summarizes the findings of the paper, concluding, "we have clear proof that the ions have a very much smaller mass than ordinary atoms; so that in the convection of negative electricity at low pressures we have something smaller even than the atom, something which involves the splitting up of the atom, inasmuch as we have taken from it a part, though only a small one, of its mass."[95] The second part presents a novel method for measuring e/m of the electric discharge in the photoelectric effect, the results from which indicate that this discharge has the same m/e as Thomson's cathode ray corpuscles. The third part uses essentially the same method to determine the e/m of the electrical discharge from incandescent filaments, showing this too is the same. The fourth part uses the method of the December 1898 paper to obtain the charge e of the ions discharged in the photoelectric effect, concluding it agrees with the value obtained in that paper. The final part first draws conclusions about the fundamental character of this quantity of electricity and about the mass of the particle in cathode rays and these discharges (holding open the question whether it is quasi-mass); it then draws on the findings of this and related papers to elaborate a new "working hypothesis" about the microphysical mechanisms underlying not only electrical phenomena in gases but also electrolysis and ionic bonding.

Because the photoelectric and incandescent-filament discharges could not readily be collimated into beams that fluoresce glass, neither of the methods Thomson had used to determine m/e for cathode rays was applicable to

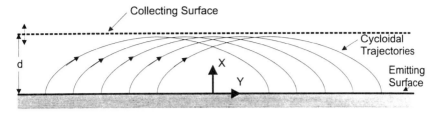

Figure 1.4
The cycloidal path of the photoelectric discharge under the action of an electric force parallel to the x-axis and a magnetic force parallel to the z-axis. For an appropriate combination of electric and magnetic force, the particles will cease reaching the collecting plate at a distance *d* from the emitting surface. (The same approach was used in measuring the *e/m* of the incandescent discharge.)

them. His new method employs crossed magnetic and electric fields to a different effect. Let the *x*-axis be normal to the surface producing the discharge, and let the electric force be parallel to the *x*-axis and the magnetic force be parallel to the *z*-axis (figure 1.4). Thomson shows that the trajectory of a negatively charged particle starting at rest on the emitting surface will then be a cycloid. Let a plate be located parallel to the emitting surface a short distance away from it. So long as the electric force is great enough, all the emitted charged particles will reach the plate. As the electric force is reduced, however, a value will be reached where the number of charged particles reaching the plate will abruptly diminish. If *V* is the voltage between the emitting surface and the plate at which the amount of charge reaching the plate drops, *H* is the magnetic field, and *d* is the distance between the emitting surface and the plate, then:

$$\frac{e}{m} = \frac{2V}{d^2 H^2}.$$

According to this theory, there should be a sharp cutoff point where the charges cease to reach the plate. In practice Thomson found this not to be the case. He consequently modified the approach a little. He still varied the voltage, but he now compared the amount of charge reaching the plate with and without the magnet on, searching for the voltage where this comparison would first show a difference. The formula for *e/m* remained the same.[96]

In the case of the photoelectric discharge, the paper gives the results of seven trials of the experiment with different distances *d*. With the exception of one slight outlier, the values obtained for *e/m* show relatively little variation. Inverted to ease comparison with the *m/e* values obtained for cathode rays, these values all lie between 1.17×10^{-7} and 1.43×10^{-7} except for one at 1.74×10^{-7}. Save for this exception, then, the range of these values falls within the range of the cathode ray m/e values Thomson had reported for the crossfield method. The same is true of the five e/m values obtained in the case of the incandescent filament discharge. Again inverted for ease of comparison, they all lie between 1.04×10^{-7} and 1.36×10^{-7} except for one at 0.88×10^{-7}.[97] Thomson concludes "that the particles which carry the negative electrification in this case are of the same nature as those which carry it in the cathode rays and in the electrification arising from the action of ultraviolet light."[98]

The experiments for measuring e/m of the incandescent filament discharge had initially been confounded by positively charged ions of gas released from the filament. These positively charged particles behaved quite differently from the negatively charged discharge, giving Thomson occasion to mention Wien's results for Canalstrahlen in reaching a further conclusion: "the carriers of positive electricity at low pressures seem to be ordinary molecules, while the carriers of negative electricity are very much smaller."[99]

Two results by Thomson's research students lay behind his determining the charge e of the photoelectric discharge. First, C. T. R. Wilson had shown that this discharge produces cloud formation once an electric field is applied to the discharge so that it moves away from the emitting surface.[100] Second, as noted earlier, Rutherford had measured the velocity of the discharge particles per unit electromotive force, thereby giving the value u needed in order to infer e from neu.[101] In developing the technique for cloud formation with the photoelectric discharge, Wilson had found that, just as with x-rays, the determination of the number of droplets n was best done with ultraviolet light of limited intensity. This, together with the relatively long times of ultraviolet irradiation required for measuring e, made the measurement sensitive to nonuniformities in the ultraviolet intensity. Thomson blames this for the larger variation in the values of e obtained here than in those in his 1898 paper.

Still, the variation in Thomson's results for the photoelectric e is not all that large, and more importantly their mean, 6.8×10^{-10}, is close to the 6.5×10^{-10} he had obtained for the ions produced by x-rays. A series of no less complex experiments on the diffusion of ions in gases that were being carried out at Cavendish by Townsend had in the meantime provided stronger evidence than Thomson had given at the end of the 1898 paper that the charge on the ions produced by x-rays is the same as the charge on an atom of hydrogen in electrolysis.[102] Thomson concludes from these results "that the charge on the ion produced by ultraviolet light is the same as that on the hydrogen ion in ordinary electrolysis."[103]

Thomson then joins the e/m and e results presented in this paper with the m/e results for cathode rays of the October 1897 paper to draw two major conclusions:

> In gases at low pressures negative electrification, though it may be produced by very different means, is made up of units each having a charge of electricity of definite size; the magnitude of this negative charge is about 6×10^{-10} electrostatic units, and is equal to the positive charge carried by the hydrogen atom in the electrolysis of solutions.
>
> In gases at low pressures these units of negative electric charge are always associated with carriers of a definite mass. This mass is exceedingly small, being only about 1.4×10^{-3} of that of the hydrogen ion, the smallest mass hitherto recognized as capable of a separate existence. The production of negative electrification thus involves the splitting up of an atom, as from a collection of atoms something is detached whose mass is less than that of a single atom.[104]

In a very real sense, then, the experimental results of this paper complete the line of argument that Thomson had first laid out tentatively in the 30 April 1897 talk before the Royal Institution.

A brief pause is required here to consider the logic of this line of argument—more especially, the way in which the conclusions Thomson reached in the October 1897 and December 1898 paper are entering into the reasoning. I have called these conclusions "extended working hypotheses" because each extended the basic working hypothesis underlying the key experiments presented in the paper by appending a value, admittedly rough, to it: the first, a value of m/e for the particles forming cathode rays, and the second, a value of e for the distinctive quantity of electricity involved in the ionization of gases by x-rays. My further point in calling them extended working hypotheses was that, while Thomson had not established their truth, he had provided strong grounds for predicating ongoing research on them. We can now see the way in which they entered his ongoing research. They did not play the role of assumptions in the experiments presented in the December 1899 paper. Rather, they functioned as premises in the evidential reasoning yielding the conclusions quoted above. Further research was predicated on them in the sense that they made a line of evidential reasoning possible that would have had the character of pure conjecture without them. In effect, Thomson is invoking a version of one of Newton's four rules for inductive reasoning in science, same effect, same cause. The version here is, same distinctive value for a characteristic property of two things, two things of a single kind—or, more precisely, same distinctive order of magnitude for the value of a characteristic property of two things, two things of a single kind.[105] Because the values Thomson is invoking are precise at best only to their order of magnitude, his evidential argument does not establish once and for all either of the conclusions quoted above. Nevertheless, it does provide compelling grounds for accepting them provisionally for purposes of continuing research.

The next sentence in the second of the paragraphs quoted above is, "We have not yet data for determining whether the mass of the negative atom is entirely due to its charge."[106] Thomson is backing off his earlier insistence that the mass is not quasi-mass, most likely because the magnitude of mass he has now obtained would entail, if taken to be quasi-mass, a radius of the corpuscle of the order of 10^{-13} cm, a not altogether implausible value. Typical of the style he has evidenced throughout the three papers included here, he is prepared to leave the question of mass versus quasi-mass for subsequent experimental investigation, suggesting one possible line of experiment himself.

The transition to the final segment of the paper, which considers the electrification of gases generally and not just at low pressure, is effected by Thomson's noting the three different kinds of carriers of charge in gases that experiments have revealed: a carrier of negative charge, with mass three orders of magnitude less than that of the hydrogen atom; carriers of positive charge with masses equal to or greater than that of the hydrogen atom; and carriers of negative charge with masses equal to or greater than that of the hydrogen atom. The first of these dominates electrical conduction in gases at low pressures, and the other two dominate it at higher pressures. Glaringly absent is a carrier of positive charge with small mass, a counterpart to Thomson's corpuscle. This gives his corpuscle a special status which, when joined with the fact that its charge is the characteristic charge of the more massive carriers of both kinds, leads him to the following proposal:

> These results, taken in conjunction with the measurements of the negative ion, suggest that the ionization of a gas consists in the detachment from the atom of a negative ion; this negative ion being the same for all gases, while the mass of the ion is only a small fraction of the mass of an atom of hydrogen.
>
> From what we have seen, this negative ion must be a quantity of fundamental importance in any theory of electrical action; indeed, it seems not improbable that it is the fundamental quantity in terms of which all electrical processes can be expressed. For, as we have seen, its mass and its charge are invariable, independent both of the processes by which the electrification is produced and of the gas from which the ions are set free. It thus possesses the characteristics of being a fundamental conception of

electricity; and it seems desirable to adopt some view of electrical action which brings this conception into prominence.[107]

Thomson is still resisting the term "electron," doubtlessly because of Larmor's use of the word to cover both positive and negative immaterial centers of charge. Nonetheless, the conclusion of this paper is that the negative ion Thomson is here referring to fulfills the requirements of Stoney's electron, so that the shift to this term had clearly become appropriate at this point.

The second of the paragraphs just quoted ends with the sentence, "These considerations have led me to take as a working hypothesis the following method of regarding the electrification of a gas, or indeed matter in any state." The three pages that follow are richer in detail than the listing of main points I offer here can indicate:

1. All atoms contain negatively charged corpuscles "equal to each other," with a mass around 3×10^{-27} grams, a very small fraction of the mass of any atom.[108] These corpuscles are somehow neutralized in the normal atom.

2. Electrification of a gas involves the detachment of a corpuscle from some of the atoms, turning these atoms into positive ions; negative ions result from a free corpuscle attaching to an atom.

3. In the release of anions and cations at the electrodes during electrolysis of solutions, "the ion with the positive charge is neutralized by a corpuscle moving from the electrode to the ion, while the ion with the negative charge is neutralized by a corpuscle passing from the ion to the electrode. The corpuscles are the vehicles by which electricity is carried from one atom to another."[109]

4. Assuming the hydrogen atom has the positive and the chlorine atom the negative charge in a molecule of HCl, the mass of the hydrogen atom in this molecule is less and the mass of the chlorine atom is greater than their nominal values. The extent to which the mass of an atom can vary from association and dissociation of corpuscles in known processes is proportional to the valence of the atom.

5. In the ionization of gases by x-rays and uranium rays, it appears that no more than one corpuscle can be detached. But the many lines of the spectrum in the Zeeman effect are evidence that atoms generally contain more than one corpuscle, raising the possibility that a process with sufficient energy can tear more than one corpuscle from an atom.

Needless to say, Thomson is calling this a "working hypothesis" not in the narrow sense that I have been using, but in the customary broad sense of a manner of conceptualizing the phenomena in question by which, in the phrasing of *Recent Researches,* "they can be coordinated." Even so, this working hypothesis differs radically in logical status as well as in substance from the one in *Recent Researches*. It is not just a conjecture that can be made qualitatively consistent with known experimental results. It is anchored to a core that has grown from the two premises on which the m/e and e measurements were predicated: (1) cathode rays and other negative discharges consist of charged particles with a distinct mass-to-charge ratio, and (2) the ionized atoms in an electrified gas have a characteristic magnitude of charge. The results of these m/e and e measurements, supplemented by the measurements Thomson and his research students carried out on velocities of ions in electrified gases, had yielded experimentally dictated extensions and refinements of these two initially narrow premises. Moreover, the experimental techniques and laboratory technology employed in these measurements were opening the way to further empirically driven extensions and refinements of this core. The extent to which the working hypotheses—in my narrow sense—forming its core had been fleshed out by experiments designed to answer specific questions was the most compelling reason to think that Thomson's new working hypothesis was on the right track.

Four other points about the new working hypothesis should be noted. First, even though the available evidence was indicating that all ionization involves liberation or attachment of a single corpuscle, the magnetic splitting of lines in the spectrum was indicating more than one corpuscle per atom. Thomson leaves the question of the number of corpuscles per atom open for subsequent investigation. Indeed, the new working hypothesis leaves all questions about atomic structure open.

Second, even though Thomson extends his working hypothesis beyond gases to the electrolysis of liquids and ionic bonding, and he says at the outset that it holds for electrification of matter generally, he does not here expressly extend it to conduction in metals. A few months later, at an international conference in Paris, he did propose a free-electron-based account of electrical conduction in metals along the lines that came to be called the Drude theory.[110] By the time he delivered the lectures at the Royal Institution in 1906 that became *The Corpuscular Theory of Matter,* however, he had backed off this view. The problem of the conduction of electricity in metals involved special phenomena, like the Hall effect, that the electron by itself did not shed much immediate light on.[111]

Third, a more conspicuous element missing from the new working hypothesis is any mention of the electrical phenomena in gases on which *Recent Researches* had placed primary emphasis, namely electrical breakdown and the spark discharge at normal pressures and the visible discharge, especially the striated positive column, at reduced pressures. Thomson rectified this by extending his working hypothesis in a paper read to the Cambridge Philosophical Society in February 1900 and published that September in *Philosophical Magazine* under the title, "On the Genesis of the Ions in the Discharge of Electricity through Gases."[112] The central idea of this paper is that corpuscles, when sufficiently accelerated by an electric field, produce further corpuscles either directly when they strike molecules or indirectly from the x-rays then generated. Electric breakdown and the spark discharge occur when corpuscles are liberated in a cascading fashion at high voltages—a proposal Thomson shows is consistent with observed trends, like the electrical force required for breakdown being roughly proportional to the density of the gas. In the case of evacuated tubes, experiments at Cavendish reported in Thomson's paper of March 1899 had led to "the conclusion that there is one centre of ionization close to the cathode, and another in the negative glow."[113] Corpuscles accelerated away from the cathode produce ionization in the negative glow, and corpuscles liberated in it produce the striated positive column. The luminous striae are regions where corpuscles have reached accelerations sufficient to produce ionization, which then reduces the electric force locally, slowing their acceleration; in the dark regions the energy reached by the accelerating corpuscles is below that required for ionization. The asymmetry between phenomena at the anode and cathode result from corpuscles being so much more effective than positive ions in producing ionization.[114]

Fourth, one should note the absence of the ether—more precisely, the ether continuum—in the working hypothesis elaborated in the three pages. The negatively charged electron, not some state or process in the ether, is doing the work. Needless to say, Thomson's experiments had not shown anything about the constitution of electricity in its own right. This is why Thomson speaks carefully of the "carriers of charge." Rather, what the working hypothesis was implying was that a theory covering a wide array of electrical phenomena could be developed without having to address the question of the ultimate constitution of electricity at all. The ether had ceased having a role to play in ongoing research in the areas Thomson was concerned with.

Earlier I remarked that his December 1899 paper would have been a fitting final word of the nineteenth century for *Philosophical Magazine*. The experiments reported in the three papers examined above are very much a product of nineteenth century science. The scientific laws underlying them and the instruments used in them, as well as the various phenomena they exploit and the laboratory practices followed in dealing with these phenomena, are almost entirely products of the nineteenth century where science had reached a position that allowed Thomson, with the help of two working hypotheses, to penetrate experimentally into the microphysics of electrical phenomena.

Aftermath—The Next Decade

The working hypothesis Thomson elaborates at the end of his December 1899 paper comprised only an initial fragment of a theory. A huge amount of experimental work remained to flesh this fragment out in detail, to pin points down, and to revise and refine it where needed. Thomson's order-of-magnitude numbers had generated promissory notes that would remain outstanding until precise values for m/e, e, and m had been determined. Only then would his insistence on their uniqueness be fully justified. Several advances were made in the immediately following years on m/e. In 1900 Henri Becquerel used crossed magnetic and electric fields to determine that the m/e of the uranium discharge is around 10^{-7}. The velocity he found in the experiments exceeded 60 percent of the speed of light. This led Kaufmann to develop much more precise measures of m/e of these particles in 1901–02, correcting for the theoretical change of mass with velocity implied by the Lorentz-FitzGerald equations. The value of e/m he zeroed in on was 1.77×10^7 or, inverted, an m/e of 0.565×10^{-7}. By the end of the decade values were being given to as many as four significant figures.[115]

Progress on e came more slowly. Thomson and his cadre at Cavendish recognized the uncertainties in their 1898 and 1899 results better than anyone, including uncertainties beyond those noted in the papers and above, such as the possible confounding effects of droplet evaporation. C. T. R. Wilson continued to refine techniques in using cloud formation, among other things determining an expansion ratio for which droplets would form almost exclusively on negatively charged ions. Thomson redid the 1898 measurement taking advantage of these advances and using uranium instead of x-rays as the radiation source to achieve a more uniform intensity of irradiation. These results, which he published in 1903 dropped his value of e from 6.5×10^{-10} to 3.4×10^{-10}. In the same year Harold Wilson added the further refinement of an electric field aimed vertically upward, counteracting the effects of gravity on the droplets. The values he published ranged from 2×10^{-10} to 4.4×10^{-10}, with a mean of 3.1×10^{-10}.

R. A. Millikan picked up from where Wilson left off, first with water drops, then a single water drop, and finally switching to oil drops to eliminate worries about evaporation. His single-water-drop experiments, published in 1909, gave comparatively stable values clustering around 4.6×10^{-10}. With the oil-drop experiments, which he initiated in 1909, he zeroed in on the tight value of 4.774×10^{-10}, published in 1913 and tightened further in 1917. Even though this value had to be refined two decades later to eliminate a systematic error arising from an inaccuracy in the viscosity for air, the tightness of Millikan's results rightly settled almost all questions about, in his words, "the atomicity of electricity."[116]

Some may want to accuse Thomson of having overreached the earlier data in saying that his corpuscles all have the same m/e and e. One thing that can be said in reply is that his taking m/e and e to be uniquely valued, rather than merely having characteristic orders of magnitude, involved little risk. Neither the results of his experiments nor the evidential reasoning on electricity in gases issuing from these results would have been undercut if electrons had later turned out to have several different values of m/e and e, all of the same order of magnitude.

Moreover, Thomson's stance can be defended as a sound approach to empirical research, reminiscent of Newton's first rule of inductive reasoning: No more causes of natural thing should be admitted than are both true and sufficient to explain their phenomena, restated for the case at hand as, No more complexity or degrees of freedom should be granted inferred entities than is dictated by the phenomena from which they are being inferred. What lies behind this dictum is more than just a blind faith in the simplicity of nature. The simpler a domain of nature is, the easier it is not only to develop a theory of it, but also to marshal high quality evidence bearing on the theory. Where nature is not simple, the best hope for developing a theory and marshalling evidence may be to proceed by successive approximations, starting with the most simple construal of the domain that shows promise of allowing experimental results to extend and refine it in a step by step fashion. Introducing more degrees of freedom in the early stages of theory construction than are absolutely needed runs the risk of having misleading ways of accommodating further experimental findings, heading the theory development process off on a garden path. It is safer to insist that further degrees of freedom and other complexities be added only when clearly forced by experimental results. Something of this general sort happened historically when electron spin proved necessary for the free-electron theory of conduction in metals.[117] No experimental results on conduction in gases and liquids had given reason to grant corpuscles spin, and the subsequent addition of spin in no way undercut any of the evidential reasoning that had issued from these results.

Thomson published the first edition of *Conduction of Electricity Through Gases* in 1903, well before Millikan's results. With the exception of a section on radioactivity, this book amounts to a rewrite of the long chapter on the subject in *Recent Researches* from ten years earlier, but now reflecting the new working hypothesis from December 1899 and the huge body of experimental research attendant to it. The second edition of the book appeared three years later. Even though it dropped the section on radioactivity, leaving that subject to Rutherford's *Radioactivity,* published a year earlier, more recent research expanded the new edition to 670 pages. Remarkably much of this second edition went over almost intact into the third edition two decades later, which Thomson authored jointly with his son. The Bohr model, quantum theory, and the wave character of the electron necessitated less revision of the account of electric conduction in gases than one might think, though they added immensely to it, expanding the work to two volumes and 1,100 pages. In the same year that the second edition was published, 1906, J. J. Thomson received the Nobel Prize for his research on electricity in gases.

That year also marked the first full year of his experimental research on Canalstrahlen or, as he renamed them, rays of positive electricity. He used strong crossed electric and magnetic fields to measure e/m, initially managing to get clean results only for hydrogen and helium, which he published in a *Philosophical Magazine* paper in 1907. He continued to develop the techniques involved in these experiments, joined in the effort by his new experimental assistant, F. W. Aston, in 1910. By 1913, the year in which Thomson's *Rays of Positive Electricity* appeared, they had established two distinct values of e/m for neon, corresponding to atomic weights of 20 and 22, though at that time the interpretation of these results was still very much up in the air. Aston continued this work after WWI, developing the mass spectrograph, which enabled him first to make a decisive case that these were two distinct isotopes of neon and then to distinguish isotopes of a great number of other nonradioactive elements.

Thomson had begun research on rays of positive electricity at the end of 1905 to obtain additional experimental basis for elaborating his "plum-pudding" model of the atom. Much of his effort in the first decade of the twentieth century went into this model. He published two books in which the subject of atomic structure is central during these years, both initially series of lectures, *Electricity and Matter* at Yale in 1903 and *The Corpuscular Theory of Matter* at the Royal Institution in 1906.[118] Both of these books hark back to the hope expressed in the passage from his 1895 paper "The Relation between the Atom and the Charge of Electricity carried by it" quoted earlier: an explanation of the connection between ordinary matter and the electrical charges on the atom should go a long way toward establishing a theory of the constitution of matter. Both books hark back to his earlier work in other ways, too, including the role played by Faraday tubes, especially prominent in the first. For Thomson the plum-pudding model was more than just a hypothesis about atomic structure; it was an attempt at a grand synthesis of his life's work.

When read today, both of these books on atomic structure have far more the flavor of unfettered conjecture than do the three seminal papers of 1897–99, even after adjustments are made for our awareness that the plum-pudding model led nowhere. This gives an impression that Thomson somehow became less a scientist in the years immediately following these papers. This is wrong. No less than before, Thomson was trying to open a pathway that would enable experimental research to develop a detailed theory:

> From the point of view of the physicist, a theory of matter is a policy rather than a creed; its object is to connect or coordinate apparently diverse phenomena, and above all to suggest, stimulate, and direct experiment. It ought to furnish a compass which, if followed, will lead the observer further and further into previously unexplored regions. Whether these regions will be barren or fertile experience alone will decide; but, at any rate, one who is guided in this way will travel onward in a definite direction, and will not wander aimlessly to and fro.[119]

The difference in the case of atomic structure lies in Thomson's failure to find even a fragment of a theory that lent itself to continuing elaboration and refinement through experimental research. This was accomplished by the Danish physicist Niels Bohr, who worked briefly with Thomson in Cambridge before going on to Manchester to work with Rutherford. Manchester provided an atmosphere conducive to Bohr's theoretical approach, and it was there in 1913 that he developed his model of the atom. The most telling piece of evidence Bohr offers for his model in his 1913 *Philosophical Magazine* paper is his purely theoretical calculation of the Rydberg constant:

$$\frac{2\pi^2 me^4}{h^3} = \frac{2\pi^2 e^5}{h^3}\left(\frac{m}{e}\right) = 3.1 \times 10^{15}.$$

Bohr used 4.7×10^{-10} (esu) for e and 1.77×10^7 (emu) for e/m in this calculation, obtaining a value within 6 percent of the observed value.[120]

Thomson contributed to the Bohr model in one other respect, albeit indirect. Starting while he was Thomson's research student at Cavendish, C. G. Barkla carried out extensive investigations of x-ray scattering during the decade, establishing a wide range of results, including that these rays are transverse electromagnetic waves. Thomson had published a theoretical formula for x-ray scattering in the first edition of *Conduction of Electricity through Gases*, adapting Larmor's old theory of radiation from an accelerated electron. In 1904 Barkla used this formula to infer from scattering results that the number of corpuscles per molecule of air is between 100 and 200. In 1906 Thomson published a paper, "On the Number of Corpuscles in an Atom," in which he concludes on the basis of a refined version of Barkla's result and two other methods that this number is the same as the atomic weight.[121] Looked at carefully, the most that can be said for Thomson's reasoning here is that the number implied by scattering, using then available values of the relevant quantities, was closer to the atomic weight than to any other salient number. While his conclusion misled Thomson in one respect in his work on the atom, it did not in another, for it showed that almost all the mass of the atom is due to something other than corpuscles. Barkla corrected the situation in 1911: "Using the more recently determined values of e/m, e, and n (the number of molecules per cubic centimetre of gas), the calculation gives the number of scattering electrons per atom as about half the atomic weight of the element."[122] Bohr cites Barkla on this in 1913.[123]

THOMSON'S CONTRIBUTION TO THE ELECTRON

The lesser part of Thomson's contribution to the discovery of the electron was his order-of-magnitude measurement of m/e for cathode rays and the proposal that the particle in these rays is subatomic. The major part of his contribution was his characterization of this particle as the asymmetrically acting, fundamental factor in ionization and electrical discharges. This part of the contribution, which dates from 1899 and culminates the effort on ionized gases begun by Thomson and his research students early in 1896, had the consequence of redirecting research on electrical conduction and related phenomena by indicating that a detailed theory of these phenomena could likely be developed without having to address questions about the fundamental character of electricity. From the point of view of the history of research into atomic structure, what Thomson's December 1899 paper contributed was primarily an experimentally determined order-of-magnitude for the electron mass, adding support for the subatomic thesis. This explains why most discussions of the discovery of the electron put comparatively little emphasis on this paper, for, viewed from that standpoint, it appears not much more than an addendum to the 1897 paper. From the point of view of the history of research on electrical conduction and the electrification of gases, however, the 1899 paper is most important. Only with it did it become clear that the electron is fundamental to ionization and a variety of electrical discharges and that no positively charged counterpart to it enters into any of these phenomena.

In a sense of the term that has not received the attention it deserves, the December 1899 paper established these claims about the electron. Of course, given the limited extent and quality of Thomson's data, this paper did not establish them once and for all. But it did provide decisive grounds for accepting them as an initial fragment of a theory and, pending evidence to the contrary, taking them for granted in further research aimed at extending this fragment. The success of the continuing further research—both before, but even more so after Bohr added his corresponding initial fragment of a theory of atomic structure—resulted en passant in the increasingly deep entrenchment of Thomson's claims. Nothing has been more central to twentieth century science than the electron. Thomson's 1899 paper has strong claim to being the point of departure for most strands of this effort.

Neither of the limited working hypotheses from which Thomson started—that cathode rays consist of charged particles and that ionization involves a characteristic magnitude of charge—originated with him. Nor did the idea that ions form when a unit charge becomes dissociated from atoms or molecules. What was original in Thomson's contribution was the design of a series of complex experiments predicated on these working hypotheses, enabling order-of-magnitude values of microphysical quantities to be inferred from macrophysical measurements. These values provided the basis for the claims made in the 1899 paper about the fundamental, asymmetric action of the electron. Save perhaps for the subatomic thesis, Thomson's work during this period is not marked by bold proposals. Even the extraordinary conclusion about the asymmetric role of the electron was less a bold proposal than it was a straightforward inference from experimental results. Thomson's contribution in these years thus lies not so much in the conceptual history of science as in the history of evidence. With the work at Cavendish from 1896 to 1899, effective empirical access was gained for the first time to the microphysics of electricity.

Society's predilections in judging the importance of advances in science incline one to underestimate Thomson's achievement with the electron. He put forth no mathematical theory, nor even any lasting laws. His discovery of the asymmetry of charge required no deep insight, and, anyway, this asymmetry is so second nature to us now that we have trouble appreciating how contrary to expectation it was. The experimental evidence Thomson and his research students produced has long since been supplanted by a vast array of higher quality, more definitive evidence, leaving no reason to appeal to it. Indeed, the only one of his experiments from the 1897 to 1899 period that still gets mentioned in physics textbooks is the cross-field experiment on cathode rays, usually with the misleading implication that the modern technology of cathode ray tubes dates from this experiment; in fact, Ferdinand Braun had published his paper describing the cathode-ray oscilloscope, from which CRT technology grew, on 15 February 1897, months before Thomson's experiment.[124]

What considerations like these overlook is how difficult and, even more so, how important to the history of science it is to get a sustained, experimentally driven process of theory elaboration off the ground. This is what Thomson accomplished. The crucial respect in which he went beyond Wiechert, Kaufmann, and others at the time was his successful pursuit of further experiments in 1898 and 1899 to answer questions about the role the electron plays in electrical phenomena.

ACKNOWLEDGEMENTS

Much of the research for this essay was done while the author was a visiting fellow in 1995–96 at the Dibner Institute for the History of Science and Technology. Thanks are due to Jed Buchwald for several helpful suggestions made at that time, and also to I. Bernard Cohen, Allan Franklin, and Eric Schliesser for comments on earlier drafts.

NOTES

1. The text of this lecture appeared under the title "Cathode Rays" three weeks later in the May 21, 1897 issue of *The Electrician* 39 (1897): 104–109.

2. J. J. Thomson, "Cathode Rays," *Philosophical Magazine* 44 (1897): 293–316.

3. J. J. Thomson, "On the Charge carried by the Ions produced by Röntgen Rays," *Philosophical Magazine* 46 (1898): 528–545.

4. J. J. Thomson, "On the Masses of the Ions in Gases at Low Pressures," *Philosophical Magazine* 48 (1899): 547–567. This paper had been presented a few months earlier at the Dover meeting of the British Association. These three papers, the text of the April talk, and FitzGerald's commentary on it have been reissued, with a historical introduction from which the present paper is adapted, in *The Chemical Educator* 2, n. 6 (1997), i.p. S1430–4171(97)06149–4, avail. url:http//journals.Springer-NY.com/chedr.

5. J. J. Thomson, "The Relation between the Atom and the Charge of Electricity carried by it," *Philosophical Magazine* 40 (December 1895): 512.

6. Emil Wiechert, "Ergebniss einer Messung der Geschwindigkeit der Kathodenstrahlen," *Schriften der physikalischökonomisch Gesellschaft zu Königsberg* 38 (1897): 3–16.

7. Walter Kaufmann, "Die magnetische Ablenkbarkeit der Kathodenstrahlen und ihre Abhängigkeit vom Entladungspotential," *Annalen der Physik und Chemie* 61 (1897): 544–552.

8. Wiechert thought he was measuring the m/e of immaterial electrons, and Kaufmann expressly concluded that "the hypothesis which assumes the cathode rays to be charged particles shot from the cathode is insufficient."

9. Isobel Falconer has made a convincing case that the issue over cathode rays was not drawing much attention at the time of Thomson's talk and that Thomson's interest in cath-

ode rays stemmed from other concerns. See her "Corpuscles, Electrons and Cathode Rays: J. J. Thomson and the 'Discovery of the Electron'," *British Journal for the History of Science* 20 (1987): 241–276. I have been helped by this paper in several places, as well as by John Heilbron's unpublished doctoral dissertation, *A History of the Problem of Atomic Structure from the Discovery of the Electron to the Beginning of Quantum Mechanics,* University of California, Berkeley, 1964 (available from University Microfilms, Inc. Ann Arbor, Michigan). I should also mention David L. Anderson's *The Discovery of the Electron* (Princeton: Van Nostrand, 1964), which first drew my attention to this episode in the history of science and is unfortunately now out of print, and Steven Weinberg's *The Discovery of Subatomic Particles* (New York: Freeman, 1983). The main difference between all of these and the present paper is that they treat Thomson's work primarily from the perspective of the problem of atomic structure, not from that of the problems of the electrification of gases and the conduction of electricity. Finally, I should acknowledge philosophic insights I gained from reading Peter Achinstein's chapters on cathode rays and the electron in his *Particles and Waves: Historical Essays in the Philosophy of Science* (New York: Oxford University Press, 1991), 279–333.

10. George F. FitzGerald, "Dissociation of Atoms," *The Electrician* 39 (21 May 1897): 103–104.

11. J. J. Thomson, "The Relation between the Atom and the Charge of Electricity carried by it,"cited in n. 5, 512.

12. J. J. Thomson, *Conduction of Electricity Through Gases* (Cambridge: Cambridge University Press, 1903), v. (Thomson uses the term "ion" in all three editions of this book to refer to electrons as well as to charged atoms and molecules.) This preface was retained in the second edition of 1906 but dropped from the third edition of 1928 and 1933. Its second paragraph nicely summarizes Thomson's conception of what had been achieved by 1903:

The study of the electrical properties of gases seems to offer the most promising field for investigating the Nature of Electricity and the Constitution of Matter, for thanks to the Kinetic Theory of Gases our conceptions of the processes other than electrical which occur in gases are much more vivid and definite than they are for liquids or solids; in consequence of this the subject has advanced very rapidly and I think it may now fairly be claimed that our knowledge of and insight into the processes going on when electricity passes through a gas is greater than it is in either of solids or liquids. The possession of a charge by the ions increases so much the ease with which they can be traced and their properties studied that, as the reader will see, we know far more about the ion than we do the uncharged molecule.

13. Jed Z. Buchwald, *From Maxwell to Microphysics: Aspects of Electromagnetic Theory in the Last Quarter of the Nineteenth Century* (Chicago: University of Chicago Press, 1985).

14. J. J. Thomson, *Notes on Recent Researches in Electricity and Magnetism* (Oxford: Oxford at the Clarendon Press, 1893), 189. Hereafter, *Recent Researches.*

15. Lord Kelvin, "Presidential Address," *Proceedings of the Royal Society,* 54 (1893): 376–394.

16. Ibid., 386.

17. Ibid., 389. The passage, ending the address, continues: "and if, as I believe is true, there is good reason for hoping to see this step made, we owe a debt of gratitude to the able and persevering workers of the last forty years who have given us the knowledge we have: and we may hope for more and more from some of themselves and from others encouraged by the fruitfulness of their labours to persevere in the work." For details on the problem of the interaction between electricity and matter and its impact on the history of physics at the end of the nineteenth century, see Buchwald, *From Maxwell to Microphysics*.

18. Ibid., 388. Kelvin goes on to list several of Crookes's findings: "the non-importance of the position of the positive electrode; the projection of the torrent *perpendicularly* from the surface of the negative electrode; its convergence to a focus and divergence thenceforward when the surface is slightly convex; the slight but perceptible repulsion between two parallel torrents due, according to Crookes, to negative electrifications of their constituent molecules; the change of direction of the molecular torrent by a neighboring magnet; the tremendous heating effect of the torrent from a concave electrode when glass, metal, or any ponderable substance is placed in the focus; the phosphorescence produced on a plate coated with sensitive paint by a molecular torrent skirting along it; the brilliant colours—turquoise-blue, emerald, orange, ruby-red—with which grey colourless objects and clear colourless crystals glow on their struck faces when lying separately or piled up in a heap in the course of a molecular torrent; 'electrical evaporation' of negatively electrified liquids and solids; the seemingly red hot glow, but with no heat conducted inwards from the surface, of cool solid silver kept negatively electrified in a vacuum of 1/1,000,000 of an atmosphere, and thereby caused to rapidly evaporate."

19. William Crookes, "On the Illumination of Lines of Molecular Pressure, and the Trajectory of Molecules," *Philosophical Transactions of the Royal Society,* A 170 (1879): 87–134.

20. Eugen Goldstein, "On the Electric Discharge in Rarefied Gases," *Philosophical Magazine,* 10 (1880): part I, 173–190, and part II, 234–247. This is the English translation of a paper that had initially appeared in German. The discussion of Crookes is in part II.

21. "Versuche über die Glimmentladung," *Annalen der Physik und Chemie* 19 (1883): 782–816; translated as "Experiments on the cathode discharge," in Heinrich Hertz, *Miscellaneous Papers* (London: Macmillan and Company, 1896), 224–254.

22. Ibid., 253.

23. Arthur Schuster, "Experiments on the Discharge of Electricity through Gases. Sketch of a Theory," The Bakerian Lecture, *Proceedings of the Royal Society* 37 (1884): 317–339.

24. Arthur Schuster, "The Discharge of Electricity through Gases," *Proceedings of the Royal Society* 47 (1890): 527–559. Schuster's bounds for m/e were 10^{-6} and 10^{-3}, in electromagnetic units.

25. Heinrich Hertz, "Uber den Durchgang der Kathodenstrahlen durch dünne Metallschichten," *Annalen der Physik und Chemie* 45 (1892): 28–32.

26. *Recent Researches,* 126.

27. Phillip Lenard, "Uber Kathodenstrahlen in Gasen von atmosphärischen Druck un in áussersten Vacuum," *Annalen der Physik und Chemie* 51 (1894): 225–267; and "Uber die magnetische Ablenkung der Kathodenstrahlen," 52 (1894): 23–33.

28. *Recent Researches,* 136ff. Thomson does not cite Schuster's Bakerian Lecture here, though he does so elsewhere in the book.

29. J. J. Thomson, "On the Velocity of the Cathode-Rays," *Philosophical Magazine* 38 (1894): 358–365.

30. Ibid., 359. Specifically, Thomson argues that, if the view that cathode rays are aetherial waves "is admitted, it follows that the aether must have a structure either in time or space." See also Heilbron, *A History of the Problem of Atomic Structure,* p. 67.

31. Reminiscent of the preface of Maxwell's book, Thomson remarks in the Preface of *Recent Researches* (vi), "The physical method has all the advantages in vividness which arise from the use of concrete quantities instead of abstract symbols to represent the state of the electric field; it is more easily wielded, and is thus more suitable for obtaining rapidly the main features of any problem; when, however, the problem has to be worked out in all its details, the analytical method is necessary."

32. At the end of the first chapter, entitled "Electric Displacment and Faraday Tubes of Force," Thomson notes: "The theory of Faraday tubes which we have been considering is, as far as we have taken it, geometrical rather than dynamical; we have not attempted any theory of the constitution of these tubes, though the analogies which exist between their properties and those of tubes of vortex motion irresistibly suggest that we should look to a rotatory motion in the ether for their explanation." (Ibid., 52.)

33. Ibid., 2.

34. Ibid., 3.

35. See Buchwald, *From Maxwell to Microphysics,* for details on the problems faced at the time in trying to offer an account of electric conduction.

36. *Recent Researches,* 56.

37. Ibid., 128.

38. Ibid.

39. Ibid., 189.

40. Ibid. In the second clause Thomson is quoting his "On a Theory of the Electric Discharge in Gases," *Philosophical Magazine* 15 (1883): 427–434, where he outlined an approach to ionization based on his vortex atom.

41. p. 190. For more details on Thomson's treatment of electrical conduction in *Recent Researches,* see Buchwald *From Maxwell to Microphysics,* 49–53, and Falconer, "Corpuscles, Electrons and Cathode Rays," 255f.

42. Ibid., 181.

43. Ibid., 190.

44. J. J. Thomson, "On the Effect of Electrification and Chemical Action on a Steam-Jet, and of Water-Vapour on the Discharge of Electricity through Gases," *Philosophical Magazine* 36 (October 1893): 313–327.

45. J. J. Thomson, "The Relation between the Atom and the Charge of Electricity carried by it," *Philosophical Magazine* 40 (December,1895): 511–544.

46. Ibid., 512.

47. Wilhelm Röntgen, *Sitzungsberichte der Würzburger Physikalischen-Medicinischen Gesellschaft*, December 28, 1895. An English translation appeared immediately in *Nature* 53 (January 1896): 274.

48. J. J. Thomson and E. Rutherford, "On the Passage of Electricity through Gases exposed to Röntgen Rays," *Philosophical Magazine* 42 (November 1896): 392. Thomson had announced this effect of Röntgen rays in "On the discharge of electricity produced by the Röntgen rays," *Proceedings of the Royal Society* 59 (February 1896): 391.

49. J. G. Crowther, *The Cavendish Laboratory: 1874–1974* (New York: Science History Publications, 1974): 121. Two years of residence and a thesis on their research work gave these students a Cambridge M.A.; later this became a PhD.

50. J. J. Thomson and J. A. McClelland, "On the leakage of electricity through dialectrics traversed by Röntgen rays," *Proceedings of the Cambridge Philosophical Society* 9 (1896): 126; also J. J. Thomson, "The Röntgen Rays," *Nature,* 53 (February 27, 1896): 391–392.

51. The paper cited in note 48 was first presented at a meeting of the British Academy in November 1896. Rutherford published two further papers in *Philosophical Magazine* the next year: "On the Electrification of Gases exposed to Röntgen Rays, and the Absorption of Röntgen Radiation by Gases and Vapours," 43 (April 1897): 241–255; and "The Velocity and Rate of Recombination of the Ions of Gases exposed to Röntgen Radiation," 44 (November 1897): 422–440. Thomson appended a short note to the former of these, proposing that if x-rays are a form of electromagnetic radiation, they can be regarded as groups of "Faraday tubes travelling outwards through space;" these cause molecules to dissociate, and an ion then forms when precisely one tube becomes detached from its group and its ends become anchored to dissociated parts of a molecule.

52. Jean Perrin, "Nouvelles propriétés des rayons cathodiques," *Comptes Rendus* 121 (1895): 1130–1134.

53. "On the Cathode Rays," *Proceedings of the Cambridge Philosophical Society* 9 (February 1897): 243–244; and *Nature* 55 (March 11, 1897): 453. Townsend immediately followed Thomson's talk with a presentation on electricity in gases and the formation of clouds in charged gases, indicating that "the gases, given off when certain chemical actions are going on, have sometimes a very large electrostatic charge" (ibid., 244–258). I will return to this paper below.

54. *The Electrician* 39 (July 2, 1897): 299.

55. "Cathode Rays," cited in n. 1, 108.

56. Ibid., 109.

57. "On the Influence of Magnetism on the Nature of the Light emitted by a Substance," *Philosophical Magazine* 43 (March 1897): 226–239. Zeeman had announced the effect that bears his name in November 1896 in Holland; an account appeared in the February 11, 1897 issue of *Nature*. Zeeman followed his *Philosophical Magazine* paper with a second that added support to his m/e calculation, "Doublets and Triplets in the Spectrum produced by External Magnetic Forces," ibid., 44 (September 1897): 55–60 and 255–259, and a third that began to complicate matters, "Measurements concerning Radiation-Phenomena in the Magnetic Field," ibid., 45 (February 1898): 197–201.

58. "Dissociation of Atoms," cited in n. 10, 104. 1897 was the sixtieth year of Queen Victoria's reign—hence FitzGerald's reference to "this Jubilee year."

59. Thomson was the first to call attention to the electromagnetic inertia of a moving charge in his "On the Electric and Magnetic Effects produced by the Motion of Electrified Bodies," *Philosophical Magazine,* 11 (1881): 229–249—the paper that first brought him to prominence. FitzGerald had made important additions to this finding.

60. G. Johnstone Stoney, "On the Physical Units of Nature," *Philosophical Magazine* 11 (1881): 381–390; see specifically page 387. This paper had been presented to the British Association in 1874. The term "electron" does not occur in the paper, but Stoney had apparently begun using it, and others had picked it up from him.

61. Joseph Larmor, "A Dynamical Theory of the Electric and Luminiferous Medium," *Philosophical Transactions of the Royal Society,* A 185 (1894): 719–822. See Buchwald, *From Maxwell to Microphysics,* for details of this theory, which was based largely on emission and absorption spectra.

62. Thomson's assistant at the time was Ebenezer Everett (incorrectly spelled 'Everitt' at the end of the 1897 paper). Thomson was legendarily inept in the laboratory, and Everett apparently always endeavored to keep him away from the apparatus. The experiments were therefore most likely carried out by Everett.

63. "Cathode Rays," cited in n. 2, 302.

64. Ibid., 312.

65. Ibid., 296.

66. Thomson introduced just such corrections in *Conduction of Electricity Through Gases;* see, for example, the second edition (1906), chapter V, 118–121, as well as elsewhere in this chapter.

67. R. J. Strutt, "The Dispersion of the Cathode Rays by Magnetic Force," *Philosophical Magazine* 48 (November 1899): 478–480. One must wonder whether this paper would have been so readily accepted had it not been communicated to the journal by Lord Rayleigh.

68. Our current value for m/e for the electron is $0.56856314 \times 10^{-7}$ emu, corresponding to a mass of $9.1093897 \times 10^{-28}$ grams and a charge of $1.60217731 \times 10^{-19}$ coulombs

(4.80653193 × 10^{-10} esu). See W. N. Cottingham, "The Isolated Electron," in *Electron: A Centenary Volume* (Cambridge: Cambridge University Press, 1997): 24–38, for a discussion of current methods of measurement.

69. Four decades later, Thomson said, "These experiments were of an exploratory nature; the apparatus was of a simple character and not designed to get the most accurate results. . . . These results were so surprising that it seemed more important to make a general survey of the subject than to endeavour to improve the determination of the exact value of the ratio of the mass of the particle to the mass of the hydrogen atom." J. J. Thomson, *Recollections and Reflections* (New York: Macmillan, 1937), 337f.

70. The fact that these values are well below Hertz's eleven earth-quadrants per second is further evidence that the electric fields in his attempted electrostatic-displacement experiments were lower than he thought.

71. An often remarked irony of this episode in the history of science is that Thomson's son George shared in the Nobel Prize given for establishing the wave-like character of electrons.

72. "Ergebniss einer Messung der Geschwindigkeit der Kathodenstrahlen," cited in n. 6. A year later Wiechert went a step further in determining m/e, using a pair of high frequency coils oscillating in phase to determine the velocity of the cathode rays from the timing required for the second coil to cancel the deflection of the first; see "Experimentelle Untersuchungen über die Geschwindigkeit und die magnetische Ablenkbarkeit der Kathodenstrahlen," *Annalen der Physik und Chemie* 69 (1899): 739–766.

73. "Die magnetische Ablenkbarkeit der Kathodenstrahlen und ihre Abhängigkeit vom Entladungspotential," cited in n. 7. Like Thomson, Kaufmann also determined that m/e does not vary with the gas in the tube or the material of the electrode. Unlike Thomson, he put a good deal of subsequent effort into determining more precise values for m/e. See Kaufmann and E. Aschkinass, "Uber die Deflexion der Kathodenstrahlen," *Annalen der Physik und Chemie* 62 (1897): 588–595; and Kaufmann, "Nachtrag zu der Abhandlung: 'Die magnetische Ablenkbarkeit der Kathodenstrahlen und ihre Abhängigkeit vom Entladungspotential'," *Annalen der Physik und Chemie* 62 (1897): 596–598; "Die magnetische Ablenkbarkeit electrostatisch beeinflusster Kathodenstrahlen," *Annalen der Physik und Chemie* 65 (1898): 431–439; and "Bemerkungen zu der Mittheilung von A. Schuster: 'Die magnetische Ablenkung der Kathodenstrahlen'," *Annalen der Physik und Chemie* 66 (1898): 649–651.

74. Phillip Lenard, "Uber die electrostatischen Eigenschaften der Kathodenstrahlen," *Annalen der Physik und Chemie* 64 (1898): 279–289.

75. Wilhelm Wien, "Untersuchungen über die elektrische Entladung in verdünnten Gasen," *Annalen der Physik und Chemie* 65 (1898): 440–452.

76. For a review of measurements of m/e in these early years, see J. J. Thomson and G. P. Thomson, *Conduction of Electricity Through Gases,* 3rd ed., vol. I (New York: Dover, 1969), chapter VI, 229–290. (Volume I of the original of this Dover republication appeared in 1928 and volume II in 1933.)

77. *Philosophical Magazine* 45 (February 1898): 172–183.

78. Cited in n. 3.

79. *Philosophical Magazine* 44 (November 1897): 422–440. This paper was submitted on July 19, one month before Thomson submitted his cathode ray paper.

80. C. T. R. Wilson, "Condensation of Water Vapour in the Presence of Dust-free Air and other Gases," *Philosophical Transactions of the Royal Society,* A 189 (1897): 265–307. For a detailed discussion of Wilson's development of the cloud chamber, both in the period of 1895–1900 and subsequently, see Peter Galison and Alexei Assmus, "Artificial Clouds, Real Particles," in *The Uses of Experiment: Studies in the Natural Sciences,* ed. David Gooding, Trevor Pinch, and Simon Schaffer (Cambridge: Cambridge University Press, 1989), 225–274. J. J. Thomson had published a key theoretical result underlying the use of droplet formation to detect ions two years before Wilson joined the Cavendish Laboratory in "On the Effect of Electrification and Chemical Action on a Steam-Jet, and of Water-Vapour on the Discharge of Electricity through Gases," *Philosophical Magazine* 36 (1893): 313–327.

81. John Townsend, "On Electricity in Gases and the formation of Clouds in Charged Gases," *Proceedings of the Cambridge Philosophical Society* 9 (February 1897): 244–258. Townsend obtained a value of 2.8×10^{-10} for the positively charged ion of oxygen and 3.1×10^{-10} for the negatively charged ion. As remarked in n. 53, Townsend's main point in this paper, presented some ten weeks before Thomson's first announcement of m/e for cathode rays, was that the gases released in electrolysis, contrary to what had been thought before, are electrified. Nothing in the paper indicates that Townsend was trying to measure a fundamental unit of charge; the paper does not compare the result he obtained with the charge per atom in electrolysis.

82. This correction is presented more clearly in Thomson's December 1899 paper, cited in n. 4, p. 562. For details of C. T. R. Wilson's wrestling with the electrification produced by cosmic rays, see Galison and Assmus, *The Uses of Experiment,* 254–257.

83. "On the Charge carried by the Ions produced by Röntgen Rays," cited in n. 3, 543.

84. R. A. Millikan's *The Electron* (Chicago: University of Chicago Press, 1924), 31, is the source for this claim. The text he refers to is O. E. Meyer's *Kinetische Theorie der Gase* (335, 1899).

85. "On the Charge carried by the Ions produced by Röntgen Rays," cited in n. 3, 544.

86. *Philosophical Magazine* 47 (March 1899): 253–268.

87. Cited in n. 4.

88. Rutherford's measurements on gases electrified by uranium radiation led him into the research on transmutation for which he won the Nobel Prize.

89. "On the Theory of Conduction of Electricity through Gases by Charged Ions," cited in n. 86, 254.

90. William Sutherland, "Cathode, Lenard, and Röntgen Rays," *Philosophical Magazine* 47 (March 1899): 268–284.

91. Ibid., 269. Sutherland goes on to acknowledge the experimental work of Thomson and Kaufmann: "Whatever proves to be the right theory of the nature of the cathode rays, the quantitative results which these experimenters [Thomson and Kaufmann] have obtained (as did also Lenard), in a region, where, amid a bewildering wealth of qualitative work, the quantitative appeared as if unattainable, must constitute a firm stretch of the roadway to the truth."

92. Ibid., 284.

93. "Note on Mr Sutherland's Paper on the Cathode Rays," *Philosophical Magazine* 47 (April 1899): 415–416.

94. Ibid., 416.

95. "On the Masses of Ions in Gases at Low Pressures," cited in n. 4, 548.

96. This was not the only complication in these measurements. The schematic shown in figure 1.4 is misleading in hiding from view the special efforts required in isolating and controlling the discharges sufficiently to allow meaningful measurements.

97. There appears to be a misprint in the table of results for the incandescent filament. The value of V in the last row should be 100×10^8, not 120×10^8. This is not the only misprint in this paper. A more egregious error occurs on line 4 of p. 563, where the exponent should be -10, not -8.

98. Ibid., 556.

99. Ibid., 557.

100. C. T. R. Wilson, "On the Condensation Nuclei produced in Gases by the Action of Röntgen Rays, Uranium Rays, Ultra-violet Light, and other Agents," *Philosophical Transactions of the Royal Society* A 192 (1899): 403–453.

101. Ernest Rutherford, "The Discharge of Electrification by Ultra-violet Light," *Proceedings of the Cambridge Philosophical Society* 9 (1898): 401–417.

102. John Townsend, "The Diffusion of Ions into Gases," *Philosophical Transactions of the Royal Society,* A 193 (1899): 129–158; and also "The Diffusion of Ions produced in Air by the Action of a Radio-active Substance, Ultra-violet Light and Point Discharges," ibid., 195 (1901): 259–278.

The first of these papers is remarkable in its own right, for like Thomson's seminal papers, it presents a difficult, theory-laden experiment and then combines the results of this experiment with other Cavendish results to draw several basic conclusions. The experiment, predicated on Maxwell's diffusion theory, determined values for the coefficient of diffusion for x-ray generated ions of different gases by having the ions pass down a long, narrow tube and measuring the rate at which they became neutralized by contact with the metal walls of the tube. From this value of the coefficient of diffusion, together with Rutherford's previously determined value of the velocity u of such ions under a potential gradient (see n. 79), Townsend inferred a magnitude for Ne, where N is the number of molecules per cubic centimeter under standard conditions. The uniformity of this magnitude for ions of different gases and its close correspondence to the value for NE from electrolysis (where E is the charge per hydrogen atom), then allowed Townsend to conclude, *independently of any specific value of e or N*, that the charge per ion, when generated by x-rays, is the same as the charge on the hydrogen atom in electrolysis. Adopting Thomson's 1898 value of 6.5×10^{-10} for e, Townsend goes on in the paper to infer, among other things, values of 20×10^{18} for N and 4.5×10^{-24} grams for the mass of the hydrogen molecule. The subsequent paper extends the results to ions produced in other ways.

103. "On the Masses of the Ions in Gases at Low Pressures," cited in n. 4, 563.

104. Ibid.

105. It goes without saying that conclusions reached by means of this rule, whether in this or Newton's original form, are not guaranteed to be true. Newton recognized this in his fourth, and last, rule of reasoning: "In experimental philosophy, propositions gathered from phenomena by induction should be considered either exactly or very nearly true notwithstanding any contrary hypotheses, until yet other phenomena make such propositions either more exact or liable to exceptions."

106. Ibid.

107. Ibid. 565.

108. The text of the paper gives as value of the mass "about 3×10^{-26} of a gramme." (ibid.). This is an obvious error insofar as the number is being obtained from an e of 6.5×10^{-10} esu (i.e. 2.17×10^{-20} emu) and an e/m of 7×10^6. Whether Thomson himself made the error in calculation is unclear. The value might well have been given as '.3 $\times 10^{-26}$' in the manuscript. The uncorrected typographical errors noted in n. 97 above are clear evidence that Thomson either did not proof read the galley proofs of this paper at all or did so extremely hurriedly.

109. Ibid., 566.

110. J. J. Thomson, "Indications relatives à la constitution de la matiere par les recherches récentes sur le passage de l'électricité à travers les gas," *Congres International de Physique* 3 (Gauthier-Villars, Paris): 138.

111. Paul Drude's initial theory of 1900 allowed both positively and negatively charged electrons ("Zur elektronen Theorie der Metalle," *Annalen der Physik und Chemie* 1 (1900): 566–613 and 3 (1900): 369–402. This further underscores the importance of Thomson's December 1899 paper, for it is where he first presents the evidence for the asymmetric role of the negatively charged corpuscle in electric conduction. For more details of early free-electron theories of electrical conduction in metals, see A. B. Pippard, "J. J. Thomson and the discovery of the electron," *Electron: A Centenary Volume*, ed. Michael Springford (Cambridge: Cambridge University Press, 1997), 1–23, especially 14–17. For a discussion of the Drude theory, its limitations, and the subsequent quantum free-electron model, see Brian K. Tanner, *Introduction to the Physics of Electrons in Solids* (Cambridge: Cambridge University Press, 1995).

112. J. J. Thomson, "On the Genesis of the Ions in the Discharge of Electricity through Gases," *Philosophical Magazine* 50 (September 1900): 278–283. Thomson opens the paper by reminding readers of his Grotthus chains in *Recent Researches* and then noting: "Since

that was written, many investigations have been made which have proved that where electrified particles move through a gas ions are produced under certain circumstances, at any rate if the particle is negatively electrified." (279)

113. Ibid., 282.

114. I have not assigned the same importance to this paper of 1900 as to the three seminal papers of 1897 to 1899 because, unlike them, it presents no experiments yielding new results, but only proposes an extension of the working hypothesis that culminates the 1899 paper.

115. The advances discussed here and below involved so many papers that I have generally chosen not to give citations. References can be found in the readily available Thomson and Thomson, *Conduction of Electricity*. Details on Thomson's work during the decade can be found in George Thomson's *J. J. Thomson: Discoverer of the Electron* (Garden City: Anchor, 1966).

116. For Millikan's version of all of this, see his book, cited in n. 84.

117. See Tanner, *Introduction to the Physics of the Electrons*.

118. *Electricity and Matter* (New Haven: Yale, 1904) and *The Copuscular Theory of Matter* (New York: Charles Scribner's Sons, 1907). The latter is especially concerned with the conduction of electricity in metals.

119. *The Corpuscular Theory of Matter*, 1f.

120. Our current value for e, $4.80653193 \times 10^{-10}$ esu, is 2.27 percent greater than the value Bohr used, versus the 0.64 percent difference between his and our current value, 1.7588×10^7 emu, for e/m. The discrepancy in e, taken to the fifth power, amounts to 11.86 percent, which was partly, but not completely compensated by his using 6.5×10^{-27} for h, 1.91 percent below our currrent value.

121. *Philosophical Magazine* 11 (1906): 769–781. See also *The Corpuscular Theory of Matter*, chapter VII.

122. Charles G. Barkla, "Note on the Energy of Scattered X-radiation," *Philosophical Magazine* 21 (1911): 648–652. Ironically, this paper was in response to J. G. Crowther, a research student working under Thomson, who had argued that the number of scattering electrons in aluminum is greater, not less, than its atomic weight. The specific values Barkla used in 1911 were: $e/m = 1.73 \times 10^7$ (from Bucherer), $e = 4.65 \times 10^{-10}$ (from Rutherford and Geiger), and $n = 28 \times 10^{18}$ (from Rutherford); Thomson had used $e/m = 1.7 \times 10^7$ and $e = 3.6 \times 10^{-10}$ in 1906.

123. "On the Constitution of Atoms and Molecules," part II, *Philosophical Magazine* 26 (September 1913): 29.

124. Friederich Kurylo and Charles Susskind, *Ferdinand Braun: A Life of the Nobel Prizewinner and Inventor of the Cathode-Ray Oscilliscope* (Cambridge: MIT Press, 1981), 90f. Braun used crossed magnetic fields. The modern cathode ray tube uses crossed electric fields, and hence Thomson's experiment did contribute something of value to this technology.

2

Corpuscles to Electrons
Isobel Falconer

On 30 April 1897, J. J. Thomson, the Cavendish Professor of Physics at Cambridge, announced the results of his pr evious four months experiments on cathode rays.[1] The rays, he suggested, were negatively charged subatomic particles that were a universal constituent of matter and whose arrangement determined the chemistry of the element. He called the particles "corpuscles," but they became known as "electrons," and Thomson has been hailed as their "discoverer."[2]

I have argued elsewhere that Thomson's work was not the outcome of a concern with the nature of cathode rays but of a much more general interest in the nature of gaseous conduction.[3] In this chapter, I discuss the acceptance of Thomson's corpuscle theory.

In recent years an attributional account of discovery has become widespread. While my discussion may lend credence to such a model of the "discovery" of the electron, it is distinct in at least two important ways.[4] First, it is agnostic as to whether there was actual a "discovery" and of what that discovery was constituted; it makes explicit that what we are considering is opinions. Second, it avoids some of the connotations of "discovery" that seek to locate discovery in a specific place, time, and actor or team; "acceptance" accommodates easily an episode that extends over several years, involves a variety of workers, and is a subject for debate.

Thomson later recalled that his corpuscle theory was not generally accepted until two years later when he spoke of it again at the British Association Meeting in 1899.[5] By 1900 also, the existence of electrons was becoming fairly widely accepted, and a whole new electromagnetic world view was being developed on this basis by H. A. Lorentz, J. Larmor, E. Wiechert, W. Kaufmann, and others.[6] But were these "electrons" the same as Thomson's "corpuscles," and how important were Thomson's experiments in establishing the existence of electrons?

In examining the acceptance of a theory we need to look at the evidence other scientists considered important in its favor. I have chosen two accounts of the development of the electron hypothesis: one British, one German. I look first at their accounts of the development of the electron idea, up to the point at which they declare that the electron exists, then at their accounts of the acceptance of the electron idea and of the role of the Cavendish experiments in this. This comparative approach highlights clearly, but crudely, the complexity of what was going on in the 1890s. In particular, it demonstrates how differing traditions led to different concepts of the electron, and how identical experiments meant different things within these traditions.[7]

The British account is Oliver Lodge's book *Electrons,* based on lectures given in 1902 but published in 1907.[8] Lodge was a leading British physicist, professor at Liverpool and later principle of Birmingham University. He was comparatively independent, owing no allegiance to Cambridge or the Cavendish, but it is worth noting that his book is dedicated to Thomson. The German account is Walter Kaufmann's "The Development of the Electron Idea," a lecture given to the seventy-third Naturforscher Versammlung at Hamburg in 1901.[9] Kaufmann was at the time assistant at the Physics Institute at Göttingen; he later became director of the Physics Institute at Königsberg. He was to make his name by his accurate experiments on the mass of the electron. Since both men were experimentalists, rather than theoreticians, one might naively expect that, allowing for nationalistic bias and personal credit seeking, their accounts would be broadly similar.

Neither account pinpoints a "discovery" or "discoverer" of the electron; both are reconstructions that attempt to trace how the idea grew and what evidence was important in its favor. Nevertheless, in both accounts, there comes a point at which the author considers that the evidence is sufficient, that the electron has a real existence, and in this sense that it has been discovered. Their accounts might thus help resolve what appears to be the weak point in Achinstein's model of discovery, that is, how one defines when an actor knows enough to have "discovered" an entity.[10]

THE EXISTENCE OF ELECTRONS: LODGE'S ACCOUNT

By 1902 the development of an electromagnetic view of nature was well under way and this is the main thrust of Lodge's book. The first ninety pages, however, cover electron theory up to 1900. Table 2.1 summarizes Lodge's account of the discovery of the electron. He starts with the properties of a charged particle in motion, reviewing rapidly Heaviside's work on the state of the surrounding ether, Poynting's on the transmission of energy, and Larmor's on the radiated energy of such particles. This leads up to a chapter on J. J. Thomson's formulation of the concept of electromagnetic mass in 1881, "one of the most remarkable physical memoirs of our time."[11] This was the

Table 2.1
Lodge's account of the development of the electron

Evidence	Cavendish	Worker elsewhere
Theory of motion of charged particles		Heaviside Poynting Larmor
Electromagnetic mass	Thomson	
Faraday's laws imply a unit of electricity, the "electron"		Stoney Loschmidt Kelvin
Cathode rays, attempts to explain		Crookes Goldstein Lenard Perrin
Mobility of carriers in gaseous conduction	Townsend	Schuster
1897. m/e for cathode rays, suggestion rays are "corpuscles"	Thomson	
	Electron/corpuscle exists	

idea that a moving charged particle has extra inertia associated with it that depends on its velocity. It later proved fundamental to the electromagnetic worldview.

Lodge next turns to tracing the idea of an indivisible unit of electric charge, starting with Faraday's laws of electrolysis. He credits Johnstone Stoney with naming this unit "the electron," and he derives the ratio of mass to charge for the hydrogen ion, citing experiments by Stoney, Loschmidt, and Kelvin.[12] Here Lodge slips in, implicitly, the idea that the electron might be a particle rather than simply a set amount of charge.

Lodge then moves on to the problems of understanding the nature of cathode rays. The general belief was that they were negatively charged particles. But particles of atomic dimensions would be too big to pass through thin metal foil, as cathode rays did, or to have the observed long mean free path in air. Moreover, Arthur Schuster, and later J. S. Townsend, had observed that the carriers of negative electricity in a discharge tube were highly mobile, implying a very small size.[13] Lodge suggests that they might be isolated charges or "electrons."

Lodge summarizes, "The magnitudes which need experimental determination in connection with cathode rays, in order to settle the question and determine their real nature, are the speed, the electric charge, and if possible the mass, of the flying particles."[14] It is worth noting this evidence for Lodge's unquestioned adherence to the mechanical philosophy—the belief that all phenomena could be reduced to matter in motion and described by their mass and velocity. In this he was typical of most British physicists.

The scene was thus set for J. J. Thomson's experiments of April 1897 in which he measured the velocity and ratio of mass to charge for cathode rays by his first method. This involved combining the magnetic deflection of the rays with their heating effect on a thermocouple.[15] He found velocities of up to one-tenth that of light, and mass to charge ratios only one-thousandth that for the hydrogen ion. Furthermore, the mass to charge ratio proved independent of the nature of matter present (i.e., of the gas in the discharge tube or the nature of the electrodes). It appeared likely, according to Lodge, that the mass associated with the cathode ray particle must be 1,000 times smaller than the hydrogen atom, and the particles might be the "detached and hitherto hypothetical individual electrons."[16]

For Lodge, then, by the end of April 1897 the existence of the electron had been established through experiments on cathode rays. Note that this was before Thomson had found the charge to mass ratio by his classic method using electric and magnetic deflections.[17] Lodge's account, in increasingly abbreviated form, is the one that has been included in British textbooks ever since.[18]

The Existence of Electrons: Kaufmann's Account

Kaufmann's account is summarized in Table 2.2. We might be forgiven for thinking we were talking about a different entity. Kaufmann starts with Weber's electromagnetic theory of the 1860s and 1870s, of electric atoms acting at a distance. It had, Kaufmann said, described the electrodynamical phenomena known at the time. Weber, however, had made no attempt to calculate the size of the electrical atom. Then Faraday and Maxwell had suggested that a finite rate of propagation should replace Weber's action at a distance. Hertz's confirmation of Maxwell's theory in 1887 appeared to spell the end for Weber's views. Maxwell's formulae were wholly void of any atomistic conceptions, could explain fundamental phenomena as well as Weber's, and were the only way of representing Hertz's waves.[19]

Kaufmann thought, however, that physicists were in danger of throwing out the baby with the bath water. The success of Maxwell's theory in explaining Hertz waves blinded them to its inability to explain some optical

Table 2.2
Kaufmann's account of the development of the electron

Evidence	Cavendish	Worker elsewhere
Electric atom theory of electromagnetism		Weber
Optical dispersion by mechanical oscillators		Helmholtz
Optical dispersion by electric oscillators		Lorentz
Faraday's laws imply unit of electricity, the "electron"		Helmholtz Stoney
Maxwell's continuum electromagnetic theory		Maxwell Hertz
Estimates of size of "electron"		Richarz Ebert Stoney
Reconciliation of Maxwell's and atomic theories of electromagnetism		Helholtz Lorentz
1986. magnetic splitting spectral lines		Zeeman Lorentz
	Electron exists	

phenomena such as deviations in predicted refractive indices and dependence of refractive index on color.

Helmholtz had tried to explain these by a mechanical theory of dispersion, founded on the vibrations of material molecules. In 1880 H. A. Lorentz laid the foundations of an analogous electromagnetic theory of dispersion that regarded every molecule as containing material points charged with electricity, the origin of electric vibrations of a definite period.[20]

Like Lodge, Kaufmann stresses that Faraday's laws of electrolysis provided evidence for the existence of electric atoms. These, Kaufmann claims, must be the electric particles Lorentz postulated. Hertz's demonstration of electromagnetic waves in 1887 stimulated physicists to try to reconcile the two opposing theories of electromagnetism. Between 1890 and 1893 works by F. Richarz, H. Ebert, and Johnstone Stoney attempted to determine the magnitude of the elementary electrical quantity, which Stoney named "electrons." Most of these dealt with the emission mechanism of luminous vapors, and calculations were based on the kinetic theory of gases. Ebert showed that the size of the electron might be very small compared with the molecular diameter. The charge on an electron was determined by electrolysis.[21]

Kaufmann continues: "The edifice of the electromagnetic theory of light" was completed in 1892 by Lorentz, who showed "how the assumption of vibrating charged particles in transparent bodies eliminates all the difficulties in the way of an adequate explanation of the propagation of light in moving bodies."[22]

Then, "In view of the facility with which Lorentz's theory explains the dispersion and observation phenomena, a direct proof of its truth was hardly required."[23] But in 1896 Zeeman's discovery of the splitting of spectral lines in a magnetic field provided this proof. The effect was predicted by Lorentz's theory and allowed, for the first time, a determination of the size of the vibrating charges. The negative charges proved to have a mass to charge ratio about 2,000 times smaller than the hydrogen ion, forcing the conclusion, Kaufmann said, that the vibration is that of the electron itself. Thus, for Kaufmann, the electron was formulated theoretically, and its existence was then established in the Zeeman effect in 1896.

Concepts of the Electron

These accounts by Lodge and Kaufmann are so entirely dissimilar that we are left searching for explanations. We might expect that Kaufmann, as a German, might value German contributions more highly than Lodge did. We might also expect that, as a rival of Thomson's for credit for measuring m/e for cathode rays, he might downplay Thomson's contribution, as indeed he does, relegating him to the role of a mere experimenter. He notes that "an unobjectionable explanation of the numerical results [for gaseous conduction], especially as obtained by J. J. Thomson and his followers, is only possible on the assumption of wandering particles within the gas,"[24] with no mention of Thomson as the author of this theory.

What we would not expect, judging by traditional accounts of the discovery of the electron, is an entirely different conceptual buildup to the electron. Like Lodge, Kaufmann was an experimentalist, yet the development he concentrates on was theoretical and formulated to answer an entirely different set of questions from those posed by Lodge. The question arises, was the outcome of these two developments the same? Was the "electron" whose existence Lorentz and Zeeman established in 1896, the same as Thomson demonstrated in 1897? The situation is further obscured by the fact that Lodge's and Kaufmann's accounts both talk of "electrons," whereas in 1896 Lorentz termed his particles "ions," while Thomson called his "corpuscles." Lorentz switched to "electrons" in 1899, while Thomson clung to "corpuscles" until 1911 or 1912.

Were either ions or corpuscles the same as the "electron" we now deem to have been discovered in 1897? If different, what was the origin of the differences and how did the views become unified? We must begin by considering the differing nature of German and British science, and by looking at the work of Joseph Larmor, a British physicist neglected in both accounts.[25]

The essential difference between British and German world views, according to McCormach and Buchwald, was that the Germans held to a particulate world view.[26] They were concerned with material particles embedded in a stationary ether, and, as Kautmann points out, they had a tradition of atomistic theories of electricity. The problem of trying to reconcile these views, and the phenomena they explained, with the apparent success of Maxwell's continuum theory loomed large. Lorentz succeeded in doing this in 1892 with his electric particles, which were material, charged, and embedded in a stationary ether.[27] These "ions" were elastically bound within the molecules and mediated the interaction between ether and matter, but the coupling mechanism was not specified and neither was the structure of the ether. Nor did Lorentz's theory give any indication of the size of the ions or a method of finding this. His terminology suggests that he thought them comparable to electrolytic ions. Following Zeeman's calculation of e/m for the ions, Lorentz briefly named his particles "lightions," thus distinguishing them from the ions of electrolysis,[28] before switching to "electrons" in 1899.[29]

In Britain similar problems with the inability of Maxwell's theory to explain some optical phenomena were occupying physicists. But they came from the opposite direction, that of continuum mechanics. At first reading, their work often appears more atomistic than the German work, and they seem preoccupied with reducing the world to matter in motion. But a second reading shows that, for them, matter is merely a structure of the ether, often a vortex ring or center of strain.[30] By 1894 Joseph Larmor had independently arrived at a theory of electric particles that addressed the same problems as Lorentz's theory.[31] Following FitzGerald's suggestion, Larmor named his particles "electrons," defining them as centers of radial strain in a rotationally elastic ether.[32] Larmor was the first to suggest that matter might be purely electromagnetic in origin, writing in the spring of 1895 that "material systems are built up solely out of singular points in the aether which we have called electrons and that atoms are simply very stable collocations of revolving electrons,"[33] although he constantly hedged his bets on this subject.[34] He had previously shown that if the mass was purely electromagnetic, then electrons must be capable of moving near the speed of light, and he had noted their possible connection with cathode rays.[35] Until the discovery of the Zeeman effect, Larmor assumed that his electron was associated with a mass at least as massive as the hydrogen atom. In 1897 he revised this assumption and identified his electron with the small oscillating charges postulated by Zeeman and Lorentz.[36]

Thomson worked within the same theoretical framework as Larmor and was familiar with Larmor's work, which he refereed. Like Larmor and Lorentz, he was deeply concerned about the interaction between the ether and matter, but his theory was formulated to answer a completely different set of questions from theirs. He was unique in seeing chemical effects as important and in seeking atomic models that would explain chemical, rather than optical or thermodynamic, phenomena.[37] For the previous fifteen years, he had seen gaseous discharge (but not cathode rays in particular) as the experimental key to untangling the matter-ether relationship. Throughout, he relied on an analogy between gaseous discharge and electrolysis, which thus placed the problems and concerns of electrochemistry in a central position in his program. By 1890, based on his discharge work, he had worked out qualitatively a view of discrete units of electricity, and by 1895 he had a tentative explanation of how these interacted with matter. It is worth examining Thomson's views of 1890–95 more closely, for they explain why he did not accept Larmor's theory, why he was in a unique position in 1897, and why his "corpuscle" differed from contemporary "electrons."

Like Larmor, Thomson was trained in Maxwell's electrodynamics, and his early beliefs belong to this tradition. Maxwell relegated electric charge and electric current to the status of secondary phenomena—they were the by-product of processes in the field. The Maxwellian view of electricity was of a strain state of the ether. The ether was continuous and pervaded all matter. The strain state was also continuous throughout any medium, but there was a discontinuity at the boundary between media, with different ratios of conductivity to dielectric permeability. Electric charge was a manifestation of this discontinuity. It was smeared uniformly over the boundary and could not exist anywhere except at the boundary.[38]

Around 1890 Thomson felt forced by the evidence from electrolysis, which he believed analogous to discharge, to recognize that charge must be discrete rather than continuous. The Faraday tube theory that he devised reconciled the experimentally found discrete charges with Maxwell's theory.[39] Based mainly on Poynting's work on the energy of the electromagnetic field, Thomson suggested that electromagnetic effects were propagated by the motion of "Faraday tubes," which carried electrostatic force. The tubes either formed closed loops or terminated on atoms. They were all of the same strength, corresponding to the charge of the electrolytic hydrogen ion. Thomson pictured these tubes as vortex filaments in the ether.

Faraday tubes were essentially discrete, and the electrification produced at the end of them was discrete also. Continuing the Maxwellian tradition, Thomson believed that a charge could exist only at the boundary of the dielectric and a conductor; that is, Faraday tubes could end only on matter. Blake and Sohncke's experiments had shown that molecules could not be charged, hence Thomson concluded that Faraday tubes could end only on atoms.[40] By 1895 he had developed this conclusion into a theory to account for the differing attractions that different chemical atoms had for electricity.[41] He suggested that the atom behaved as though it contained a large number of outward-pointing "gyrostats." An incident ethereal vortex Faraday tube would modify the motion of the gyrostats depending on whether the tube and gyrostats were rotating the same or opposite ways. In one case the energy of the atom would be lowered, in the other raised. Different atoms might have differently rotating gyrostats and thus have a preference for one particular type of vortex tube or charge.

For our purposes, the essential feature of this theory is that charge remained a boundary effect between matter and ether. Both chemical atom and vortex tube had to be present before a charge could exist. This may account for Thomson's remark that he did not find Larmor's (purely electromagnetic) theory very useful[42] and certainly explains his emphatic statement in 1896 that "the idea of charge need not arise, in fact does not arise, as long as we deal with the ether alone."[43] Furthermore, the particular structure and chemistry of atoms was implicated in the nature of electric charges.

This belief placed Thomson in a unique position among physicists. When he identified cathode rays as small, negatively charged "corpuscles," he made their structural implications clear, citing Prout's and Lockyer's chemical ideas of divisible atoms as precedents, rather than Lorentz's or Larmor's electromagnetic theories (though he did point out that his results were in broad agreement with Zeeman's).[44] Two months later Thomson proposed an atomic structure based on the stable grouping of corpuscles in a uniform sphere of positive electrification.[45] Although he was not explicit about the nature of a corpuscle, he continued to treat it on occasion as the locus of interaction between the end of a vortex tube and some material part of the atom, which might have no more extension than a mathematical point. The whole entity, matter plus boundary plus vortex, however, was an essential part of the atom.

Thus Lorentz's ion was different from Thomson's corpuscle and was different again from Larmor's electron. Table 2.3 summarizes the characteristics of all three. The later idea of an electron took elements from all three theories.

Table 2.3
Summary of features of Lorentz's, Larmor's, and Thomson's theories of 1897

Lorentz	Larmor	Thomson
Stationary ether	stationary, rotationally elastic ether	state of ether not mentioned
Material, particle	ethereal, strain center electron	boundary effect between either vortex and atom
Electron embedded in matter but separate from it	electron provides ethereal origin of matter	corpuscle a building block of chemical atoms

ACCEPTANCE OF THE ELECTRON

Given these differences, how did Thomson's corpuscle theory become accepted and transmuted into the later electron? If we return to our two accounts, there is more general agreement about the acceptance of electron theory than about its origin, but there are still some significant differences. Tables 2.4 and 2.5 summarize the accounts. We have two aspects of electron theory to consider: first, the electric particles of Lorentz and Larmor, whether ethereal or not, which explained optical phenomena, and then Thomson's corpuscle, which also explained atomic structure.

Lorentz and Larmor both had theories with far-reaching implications but a dearth of definite experimental evidence to back them up. They had both seized on Zeeman's results as support for their theory and were seeking further support.[46] Thomson's measurement of the mass to charge ratio for cathode rays provided this. George FitzGerald realized the implications for Larmor's theory immediately. Writing in the same issue of *The Electrician* in which Thomson's results were published, he suggested that Thomson's measurements be reinterpreted as showing that cathode rays were "free electrons."[47]

Thus FitzGerald rejected the importance of corpuscles for atomic structure and shifted the context of Thomson's results to Larmor's electron theory. He ensured that the term "electron" was associated with Thomson's experimental work several years before there was full assent to Thomson's theory. That "electrons" were originally proposed as an alternative interpretation of the cathode ray results to "corpuscles" was forgotten.

The continental situation was similar, except that here Thomson was seen as just one of many who determined the mass to charge ratio for cathode rays and not necessarily the most reliable. Kaufmann's measurements were generally deemed the most accurate.[48] Kaufmann credits Emil Wiechert with

Table 2.4
Lodge's account of the acceptance of the electron

Evidence	Cavendish	Worker elsewhere
m/e for cathode rays		Lenard
		Kaufmann
m/e for Lenard rays		Lenard
Velocity of cathode rays		Wiechert
m/e for photoelectric carriers	Thomson	Lenard
Ionization by incandescent metals	Thomson	Branly
	McClelland	Preece
	H. A. Wilson	Fleming
	Richardson	
	Owen	
Ions in flames	H. A. Wilson	
	Gold	
Number of ions in a conducting gas	Thomson	Lenard
	Rutherford	Righi
	Zeleny	Beattie
	McClelland	De Smolan
	McLennan	
	Richardson	
	H. A. Wilson	
	Owen	
Mobilities of ions	Townsend	
	Zeleny	
Measurement of e	Thomson	
	H. A. Wilson	

first suggesting that the cathode ray particles and Lorentz's ions were the same.[49] For Lorentz, the existence of a direct means of experimenting on ions was immensely insignificant, and he recast his whole theory in terms of individual particles, now called "electrons," rather than averages over many ions.[50]

What both accounts show is that the ultimate success of Lorentz's and Larmor's electron theories depended on their potential for unification. A wide variety of hitherto unrelated experimental phenomena could be encompassed. And the suggestion that all matter might be electromagnetic in origin, first made by Larmor, promised fundamental advances in physics. Kaufmannn stated, "Although much may appear hypothetical, it is clear . . .

Table 2.5
Kaufmann's account of the acceptance of the electron

Evidence	Cavendish	Worker elsewhere
m/e for cathod rays	Thomson	Wiechert Aschkinass Kaufmann Lenard Des Coudres
Suggestion cathode rays are electrons		Wiechert
Metallic conduction		Riecke Drude
m/e for photoelectric arriers		Lenard
Gaseous conduction	Thomson et al.	
Measurement of e	Thomson	
m/e for rays		Becquerel Dorn Kaufmann
Electromagnetic view of nature	Thomson	Lorentz Wien

that these electrons are one of the most important foundations of our whole world structure," while Lodge, ever more florid in style, agrees that "[w]e are now beginning to have some hope of obtaining unexpected answers to riddles—such as those concerning the fundamental properties of matter—which have proposed themselves for solution throughout the history of civilization."[51]

Both accounts suggest that Thomson played a major role in achieving this unification. Throughout the diverse branches of physics that were brought within the orbit of electron theory, Thomson's name crops up as having made significant contributions. Philip Lenard is the only other physicist whose name occurs so universally, and it is noteworthy that Lenard received his Nobel Prize in 1905 for his work on cathode rays, the year before Thomson received his for his work on "conductivity of gases." Neither citation mentioned electrons.

The major difference between the two accounts is the importance they assign to other work on gaseous conductivity, largely done at the Cavendish. For Lodge, the idea of an electron had arisen from investigations of gaseous conduction. Electron theory and Thomson's conductivity theory were mutually self-supporting; the success of one depended critically on the success of the other. For Kaufmann, gaseous conduction was merely another corroboration of a theory derived from and supported by advances in electrodynamics.

This difference shows most clearly in their attitude to Thomson's experiment of 1898 that measured the charge on a gaseous ion, and later a photoelectric particle, directly.[52] For the British, two lingering doubts had remained: for Thomson, that the small value of the mass to charge ratio might be due as much to a large charge as to a small mass[53]; for FitzGerald, Larmor, and probably Lodge, that the corpuscle might not be the same as the electron.[54] When Thomson established for the first time the actual value of the charge, all doubts as to the smallness of the mass and the equality of charge on corpuscle and electron were removed. His results were later refined by his student H. A. Wilson.[55] This experiment was, for the British, so fundamental that Lodge wrote, "it seems to me one of the most brilliant things that has recently been done in experimental physics. Indeed I should not need much urging to cancel the 'recently' from this sentence."[56]

Kaufmann, conversely, dismisses the experiment with a one-liner, "J. J. Thomson has even succeeded by observation of conducting gases in measuring the absolute magnitude of the charge of a single ion, and found good agreement with the elementary quantity previously obtained."[57] He added that Planck had also derived the charge from black-body radiation. Kaufmann evidently felt the value of the electronic charge sufficiently well established from electrolysis. The experiment appears to have had significance only for the British. Ramsay was still stressing it in 1912 as was O. W. Richardson in 1916. For the continentals, however, it appeared unimportant. In his *Theory of Electrons* of 1909, Lorentz did not discuss it at all.[58]

Thus, the first aspect of Thomson's corpuscle—that it was a very small electrified particle—appears to have been accepted readily, explicitly because it supported Lorentz's and Larmor's theories. Disagreement continued over whether the particle was material or ethereal and how it was structured. This difference was brought into focus when Kaufmann attempted to discover whether the electron had purely electromagnetic inertia.[59] He measured the masses of beta rays traveling at various velocities approaching that of light and compared them with theoretical values for electromagnetic inertia developed by Thomson and O. Heaviside. He initially used Searle's model of the electron as a spherical shell over which charge is uniformly spread, and he obtained the result that only one-fourth to one-third of the mass was electromagnetic. Dissatisfied with this result, Max Abraham revised Searle's analysis on the assumption that the electron was a conducting sphere.

Thomson also took up Kaufmann's results, but he applied his own ideas, treating the particle as a mathematical point (the center of the tubes of force). Both Abraham and Thomson found the entire mass to be electromagnetic. This result was physically preferable because, to quote Lodge, "it enables us to progress and is definite,"[60] and Kaufmann revised his analysis. Interestingly, Thomson's own ideas vacillated on this point and by 1907, while agreeing that the corpuscle had purely electromagnetic mass, he emphatically refused to speculate about its ethereal structure or about the distinction between matter and nonmatter.[61]

What of the second aspect of Thomson's corpuscle—that it was a building block of a divisible atom? This was much harder for physicists to entertain. It is not clear from Lodge's account at what point he, and the British, accepted it. It is evident, however, that initially they all rejected it. A divisible atom smacked of alchemy. If corpuscles were a building block of a divisible atom, then their production involved disrupting or dissociating the atom, and it appeared that this should change the chemical nature of the atom and also allow the reaggregation of corpuscles into new atoms. FitzGerald was clear that this was his objection to the corpuscle theory, writing that the free electron hypothesis "is somewhat like Prof. J. J. Thomson's hypothesis, except that it does not assume the electron to be a constituent part of an atom, nor that we are dissociating atoms, nor consequently that we are on the track of the alchemists."[62]

Thomson's experiments were sufficient to support electron theory, with which they intersected neatly, but not to establish corpuscle theory. An editorial in *The Electrician* on 2 July 1897 bears this out. It acknowledges the implications of corpuscle theory but would "wish to see the hypothesis verified at an early date by some crucial experiment." Such an experiment was not forthcoming, at least from Thomson.

Although the increasing power of electron theory added prestige to Thomson's experiments, physicists remained uncertain about the constituent role of corpuscles in atoms. Indeed Lodge in 1906 appears totally confused, writing, "While the units of negative charge appear in some cases with a separate existence,—perhaps carrying with them part of the atom, in which case they might be called corpuscles, having a material nucleus; perhaps pure disembodied electricity, whatever that may be—an electrical charge detached from matter—a mere complexity in the ether, in which case they would correspond with those hypothetical entities familiar in theoretical and mathematical treatment as 'electrons.'"[63]

There are three things to note about this quotation. First, Lodge deems electrons "familiar" while corpuscles were not. Second, and most significant here, he still has not understood the distinction between Larmor's "electrons" and Thomson's "corpuscles," nor the constituent role of corpuscles. Despite his advocacy of the electronic theory of matter, he here divorces electrons from the matter of which Larmor claimed they were the origin. He speaks of negative charges "carrying with them" some part of the atom rather than actually of being an integral part of the atom, as Thomson would have it. Third, Lodge was unable to make his attempted distinction stick, and he failed to adhere to it through the rest of the book, betraying further confusion.

It appears that even in 1906 and in Britain, the corpuscle's constituent role was far from firmly established, and Thomson's theory might have disappeared into oblivion were it not for the discovery of radioactivity. Becquerel showed that beta rays could be deflected magnetically, and Dorn demonstrated their electric deflection. Becquerel, and then Kaufmann himself (not Thomson), showed that their mass to charge ratio was the same as for cathode rays, thus identifying them with electrons or corpuscles.[64]

Kaufmann's account suggests that this was a turning point.[65] Here was the crucial evidence that atoms might emit corpuscles without any external influence. Corpuscles were not an artifact of the interaction of atoms and the electric field, but must have been contained within the atom. Equally important, in 1903 Rutherford and Soddy argued that in radioactive decay atoms did change their chemical nature.[66] Physicists were on the track of the alchemists. Thus the corpuscle's constituent role was finally accepted, although by now it was almost universally known as an "electron" and this terminology stuck. Indeed, Kaufmann's beta ray experiments gave additional momentum to electron theory, enabling his experiments on electromagnetic mass to which I referred earlier. These ensured the success of the electromagnetic view for several years to come.

CONCLUSION

The story I have been telling traces two parallel and apparently quite similar theoretical developments by Lorentz and Larmor, although Larmor's is now largely submerged. Yet they were based on fundamentally different concepts of nature. Intertwined was a series of experiments that were ultimately successful largely because they got hijacked by both theoretical camps. The existence of the phenomena demonstrated by Thomson was sufficient evidence for Lorentz and especially Larmor, but the quality of the experiments was not sufficient to establish Thomson's own corpuscular theory in opposition to the electron theories. The potential unifying power of the electromagnetic view of nature concentrated attention on the electron's charge and mass, and these became its defining characteristics .[67] The one respect in which Thomson does seem to have been before others is in deflecting cathode rays electrostatically. His cross-field e/m method, said to involve fewer assumptions than Wiechert's or Kaufmann's original measurements, came to exemplify the new physics.

In this process, a significant historical contingency is that Lodge's account, which set the tone for many later histories, was delivered to the Institution of Electrical Engineers. As Gooday points out, electrical engineers were a far larger community than academic physicists and were also intimately familiar with the history and potential of vacuum tube technology.[68] Lodge's decision to present the electron development through a familiar technology rather than a more abstruse theoretical path was well received and was perpetuated by a wide audience. Thus, even "acceptance" begins to look more complex than it at first appeared for, as well as the background concepts of the author, we have to take into account the potential influence of the intended audience.

Both accounts agree that cathode rays were particularly compelling evidence for the existence of electrons. Even Kaufmann, who placed the reality of electrons prior to 1897, considered that "[w]e have in the cathode rays the electrons—which in optical phenomena lead a somewhat obscure existence—bodily before us so to speak."[69] In Britain, Thomson was the first to produce this evidence, while in Germany, Wiechert performed a similar role. That Wiechert is now largely forgotten while Thomson is remembered as "*the* discover of the electron" is due to more than the contingency that Thomson had a large and increasingly powerful group of former research students who extolled his work. It is due in part to the nature of Thomson's corpuscle suggestion. In speculating about the role of the corpuscle in the structure of the chemical atom, Thomson initiated a research program in subatomic physics among these students that was to dominate British physics in the first half of the twentieth century. By the 1920s the ethereal concepts in which Thomson's work was founded were outmoded, yet his ideas underpinned subatomic physics and his successors needed to justify their belief in them. His students, unable to accept his concepts, transformed his experiments into a paradigm of pure physics research. They thus used his cathode ray work to make their own enterprise acceptable (and fundable).[70]

Ultimately Thomson's corpuscle added an important property to electron theory, expanding its evidential context to the chemical atom.[71] But the accurate, precise, and sometimes crucial experiments were done by many different workers. The weight attached to these experiments depended on the differing metaphysical orientation of the physicists concerned and highlights the interplay of the differing traditions.

ACKNOWLEDGMENT

An earlier version of this paper appeared in D. Hoffmann, F. Bevilacqua, and R. Stuewer (eds.), *The Emergence of Modern Physics,* Proceedings of a Conference, Berlin 22–24 March 1995, (1996) Pavia, 217–232.

NOTES

1. Thomson, "Cathode-rays," *Electrician* 39 (1897): 104–109.

2. This account has been much criticized in recent years, in particular in: T. Arabatzis, "Rethinking the 'Discovery' of the Electron," *Studies in the History of Modern Physics* 27 (1996): 405–435; N. Robotti and F. Pastorino, "Zeeman's discovery and the mass of the electron," *Annals of Science* 55 (1998): 161–183; and Graeme Gooday (this volume).

3. I. Falconer, "Corpuscles, Electrons and Cathode Rays: J. J. Thomson and the 'Discovery of the Electron,'" *British Journal for the History of Science* 20 (1987): 241–276.

4. Note that, while I avoid the issue in this paper, I am not advocating abandoning the attempt to define "discovery." The classic formulation of an attributional model of discovery is A. Brannigan, *The Social Basis of Scientific Discoveries* (Cambridge, 1981); the idea was developed in S. Schaffer, "Scientific Discoveries and the End of Natural Philosophy," *Social Studies of Science* 16 (1986): 387–420, who situates discovery accounts firmly in the "local practices of contemporary research communities." This model is implicitly adopted by Gooday (this volume), who does, however, document one of its problems—the possibility that "local practices" may further subdivide into individual practices, and the whole account become too messy to be useful—a danger indicated in S. Shapin, "Discipline and Bounding: The History and Sociology of Science as Seen through the Externalism-Internalism Debate," *History of Science* 30 (1992), 333–369, esp. 353–354. Arabatzis has tried to avoid this problem by concentrating on the realism of the entity discovered rather than the actor making the discovery. In his account discovery is still socially negotiated, but "an entity has been discovered only when consensus has been reached with respect to its reality" ("Rethinking the 'Discovery' of the Electron," 406). While sidestepping some of the problem by providing a terminus ad quem for a discovery rather than attempting to pinpoint a specific locus at which it was made, his approach appears to be fraught with the difficulty of individualism when different scientists attach different concepts to the same word, as is the case with "electron." Conversely, if the consensus of all is required, then significant differences in local practice may be lost.

5. J. J. Thomson, "On the Masses of the Ions in Gases at Low Pressures," *Philosophical Magazine* 48 (1899): 547–567; *Recollections and Reflections* (London: Bell, 1936); for a fuller account of the proceedings at the British Association Meeting, see "Physics at the British Association," *Nature* 60 (1899): 585–587; and Gooday (this volume).

6. J. Buchwald, *From Maxwell to Microphysics* (Chicago: University of Chicago Press, 1985); R. McCormach, "H. A. Lorentz and the Electromagnetic View of Nature," *Isis* 61 (1970): 459–497; A. Warwick, "On the Role of the FitzGerald-Lorentz Contraction Hypothesis in the Development of Joseph Larmor's Electronic Theory of Matter," *Archive for History of Exact Sciences* 43 (1991): 29–91. Warwick distinguishes sharply between Lorentz's electromagnetic view of nature and Larmor's electronic theory of matter, a distinction based on the foundational role Larmor assigned to the ether. Robotti implies that it was Thomson's measurement of the electronic charge independently of the mass that swayed the argument by 1900: N. Robotti, "J. J. Thomson at the Cavendish Laboratory: The History of an Electric Charge Measurement," *Annals of Science* 52 (1995): 265–284.

7. The institutional basis of the German electrodynamic tradition is extensively described in C. Jungnickel and R. McCormmach, *Intellectual Mastery of Nature: Theoretical Physics from Ohm to Einstein*, 2 volumes (Chicago: University of Chicago Press, 1986). The Maxwellian tradition in both Britain and Germany is described in Buchwald, *From Maxwell to Microphysics*, and within Britain in B. Hunt, *The Maxwellians* (Ithaca: Cornell University Press, 1991). Buchwald and Hon have both given detailed studies of episodes in the history of cathode rays or the electron that show how experiments may mean different things in different traditions: J. Buchwald, *The Creation of Scientific Effects* (Chicago: University of Chicago Press, 1994), especially chapter 10; "Why Hertz was Right about Cathode Rays," in J. Buchwald (ed.), *Scientific Practice* (Chicago: University of Chicago Press, 1995), 151–169; G. Hon. "Is the Identification of Experimental Error Contextually Dependent?: The Case of Kaufmann's Experiment and Its Varied Reception," ibid., 170–223; "H. Hertz: 'The electrostatic and electromagnetic properties of the cathode are either nil or very feeble' (1883): A case study of an experimental error," *Historical Studies in the Physical Sciences* 18 (1987): 367–382.

8. O. Lodge, *Electrons* (London: Bell, 1907).

9. W. Kaufmann, "The Development of the Electron Idea," *Electrician*, 8 Nov. 1901, 95–97.

10. P. Achinstein (this volume) suggests three components for discovery: the existence of what is discovered; the knowledge that it exists; and priority. The problem is in defining who "knows" that the entity exists.

11. Lodge, *Electrons*, 17.

12. Discussed in J. O'Hara, "George Johnstone Stoney and the conceptual discovery of the electron," in *Stoney and the Electron* (Royal Dublin Society, 1993), 5–28.

13. This passage shows the reconstructive nature of Lodge's work. Although Lenard had stressed that cathode rays did pass through metal foils, this was not clear to many other physicists, J. J. Thomson included. A large part of Thomson's April 1897 lecture "Cathode-rays" was devoted to showing that Lenard rays could be a secondary effect, and as late as 1900 he was surprised that beta particles could pass through matter: P. Lenard, "Uber Kathodenstrahlen," *Verhandlunger der Gesellschaft Deutsche Natforscher und Aerzte* 2 (1893), 36; "Uber Kathodenstrahlen in Gasen von atmospharischern Druck und im aussersten Vacuum," *Annalen der Physik und Chemie* 51 (1894), 225; "Uber die magnetische Ablenkung der Kathodenstrahlen," *Annalen der Physik und Chemie* 52 (1894), 23; Rayleigh,

The Life of Sir J. J. Thomson (Cambridge, 1942), 133. Moreover, Townsend's experiments on the velocity of ions postdated Thomson's on the size of cathode rays: J. S. Townsend, "The Diffusion of Ions into Gases," *Philosophical Transactions of the Royal Society* 193 (1899), 259.

14. Lodge, *Electrons,* 39.

15. Thomson, "Cathode-rays."

16. Lodge, *Electrons,* p. 47

17. J. J. Thomson, "Cathode-rays," *Philosophical Magazine* 44 (1897): 269–316.

18. My account of Lodge's book differs from Graeme Gooday's (this volume). Gooday states that "Lodge attached no special place to the contribution of J. J. Thomson . . . except in his prefatory dedication." While I agree that Lodge does not identify Thomson as the electron's discoverer, I argue that he does give Thomson very considerable pride of place. While Schuster is described as "skilfull," Kaufmann as having "great skill," Zeeman as having merely "skill," and Lenard as "indefatigable" Thomson's work is variously described as "brilliant," "epoch-making" and "remarkable." Even more important in forming later accounts is the fact that Lodge describes the route to the electron via a cathode ray research program *first*. Even Zeeman's work is relegated to later in the book and, in a remarkable piece of reconstruction, to later in time. Lodge places Zeeman's 1896 calculation of e/m for the charged particle causing the Zeeman effect several months after Thomson's April 1897 determination of e/m for cathode rays (112). This is noteworthy because in 1897 Lodge himself was the chief publicist of Zeeman's work in Britain, arranging for publication of Zeeman's papers in English, and had himself verified both the effect and the e/m calculation, all before April 1897: *Nature* 55 (1896): 192; P. Zeeman, "On the Influence of Magnetism on the Nature of the Light Emitted by a Substance," *Philosophical Magazine* 43 (1897): 226–239; O. Lodge, "The Latest Discovery in Physics," *Electrician,* 38 (1897): 568–570; "The Influence of a Magnetic Field on Radiation Frequency," *Proceedings of the Royal Society,* 60 (189?): 513–514; "A few notes on Zeeman's discovery," *Electrician,* 38 (1897): 643. For a discussion of this work see Arabatzis, "Rethinking the 'Discovery' of the Electron," esp. 423; and Robotti and Pastorino, "Zeeman's discovery and the mass of the electron," esp.172–175.

19. For a modern assessment of this period in German electrodynamics, see O. Darrigol, "The Electrodynamic Revolution in Germany as Documented by Early German Expositions of 'Maxwell's Theory,'" *Archive for History of Exact Sciences,* 45 (1993): 189–280; Buchwald, *From Maxwell to Microphysics.*

20. Discussed in B. Carazza and N. Robotti, "The First Molecular Models for an Electromagnetic Theory of Dispersion and Some Aspects of Physics at the End of the Nineteenth Century," *Annals of Science* 53 (1996): 587–607; O. Darrigol, "The Electron Theories of Larmor and Lorentz: A Comparative Study," *Historical Studies in the Physical and Biological Sciences* 24 (1994): 265–336; Buchwald, *From Maxwell to Microphysics.*

21. See Carazza and Robotti, "The First Molecular Models for an Electromagnetic Theory of Dispersion"; Robotti and Pastorino, "Zeeman's Discovery and the Mass of the Electron."

22. Kaufmann, "The Development of the Electron Idea," 96.

23. Ibid. In this context it is worth noting Carazza and Robotti's observation that "none of the authors [of German dispersion theories] examined seems worried about a direct comparison of the dispersion formula obtained with specific experimental data. . . . their primary interest is that of demonstrating the feasibility of a dispersion theory founded on electromagnetism, capable of acknowledging at first glance the qualitative aspects of the phenomenon" ("The First Molecular Models for an Electromagnetic Theory of Dispersion," 607).

24. Kaufmann, "The Development of the Electron Idea," 97.

25. Although Lodge does mention Larmor, his role is theoretical elucidation of experimental phenomena; his ideas are not presented as a coherent whole. The work of Lorentz and Larmor, and the influence of the traditions within which they were working, is compared very directly in Darrigol, "The Electrodynamic Revolution in Germany."

26. Buchwald, *From Maxwell to Microphysics;* Darrigol, "The Electrodynamic Revolution in Germany"; Jungnickel and McCormmach, *Intellectual Mastery of Nature;* McCormmach, "H. A. Lorentz and the Electromagnetic View of Nature."A comparison of styles of materialism in Britain, France, and Germany around the turn of the century may be found in H. Kragh, "The New Rays and the Failed Anti-materialistic Revolution," in D. Hoffmann, F. Bevilacqua, and R. Stuewer (eds.), *The Emergence of Modern Physics,* Proceedings of a conference, Berlin 22–24 March 1995 (Pavia, 1996), 61–77.

27. H. A. Lorentz. "La Théorie Electromagnetique de Maxwell et son Application au Corps Mouvants," *Archives Néerlandaises* 25 (1892), 363. Although, as Darrigol is at pains to point out, Lorentz was not German and took elements from many different traditions, he did base his electrodynamics on Helmholtz's exposition of Maxwell, and this places him firmly in the German tradition in this respect: Darrigol, "The Electrodynamic Revolution in Germany." Helmholtz's interpretation of Maxwell and the influence this had on German electrodynamics are discussed in Buchwald, *From Maxwell to Microphysics,* esp. chapters 21, 22.

28. H. A. Lorentz, "Optical Phenomena Connected with the Charge and Mass of the Ions" (1898), in P. Zeeman and A. D. Fokker (eds.), *H. A. Lorentz, Collected Papers,* 9 volumes (The Hague, 1935–1939), vol. 3, 12–39.

29. McCormach, "H. A. Lorentz and the Electromagnetic View of Nature."

30. Buchwald, *From Maxwell to Microphysics.*

31. J. Larmor, "A Dynamical Theory of the Electric and Luminiferous Medium," *Philosophical Transactions of the Royal Society,* part I: 185 (1894), 719–822; part II: 186 (1895), 695–743; part III: 190 (1897), 205–300; these papers are also contained in J. Larmor, *Mathematical and Physical Papers,* 2 volumes (Cambridge, 1929), vol. 1, 414–535, 543–597; vol. 2: 11–132 (further page references will be to this version). For detailed accounts of Larmor's work see: Buchwald, *From Maxwell to Microphysics;* Darrigol, "The Electrodynamic Revolution in Germany"; Hunt, *The Maxwellians;* Robotti and Pastorino, "Zeeman's Discovery and the Mass of the Electron"; Warwick, "On the Role of the FitzGerald-Lorentz Contraction Hypothesis in the Development of Joseph Larmor's

Electronic Theory of Matter." These accounts all point out the close relationship between Larmor and FitzGerald and, through him, with Stoney.

32. Buchwald, *From Maxwell to Microphysics,* chapters 19, 20.

33. Larmor, "Dynamical Theory," part II, 566.

34. For example: "the consideration of groups of electrons or permanent strain centres in the anther, which form a part of, or possibly the whole of, the constitution of the atoms of matter" ("Dynamical Theory," part II, 543 [May 1895]); "In this medium [the ether] unitary electric charges, or electrons, exist as point singularities, or centres of intrinsic strain, which can move about under their mutual actions; while atoms of matter are in whole or in part aggregations of electrons in stable orbital motion" ("Dynamical Theory," part III, 12 [April 1897]); "The question is fundamental how far we can proceed in physical theory on the basis that the material molecule is made up of revolving electrons and of nothing else" (J. Larmor, "The Influence of a Magnetic Field on Radiation Frequency," *Proceedings of the Royal Society* 60 [1897], 514 [February 1897]; in *Mathematical and Physical Papers,* vol. 2, 138–139.) For Larmor's progress towards a completely electronic theory of matter see Warwick, "On the Role of the FitzGerald-Lorentz Contraction Hypothesis."

35. Larmor, "Dynamical Theory," part I, 523–524.

36. J. Larmor, "On the Theory of the Magnetic Influence on Spectra and on the Radiation from Moving Ions," *Philosophical Magazine* 44 (1897): 503–512; also *Mathematical and Physical Papers,* vol. 2, 140–149. For a detailed account of this episode, see Robotti and Pastorino, "Zeeman's Discovery and the Mass of the Electron."Here Larmor reserves "electron" for charged particles with purely electromagnetic mass, using "ion" for those that might be attached to inertial matter. This continues a distinction he had made in 1894 (Larmor, "Dynamical Theory," part I, 523).

37. Thomson's interest in chemistry was first evident in his *Treatise on the Motion of Vortex Rings* (Adams prize essay 1882, published London 1883). It continued in a series of papers on discharge (see Falconer, "Corpuscles, Electrons and Cathode Rays" and references cited therein). A clear example of the different interests of Thomson and Larmor came in 1894. In elaborating his theory of vortex "monads," the precursors of "electrons," Larmor invoked Prout's hypothesis of primordial atoms. He noted as a difficulty, "why the molecule say of hydrochloric acid is always H+Cl–, and not sometimes H–Cl+," but dismissed it as not worthy of consideration: "This difficulty would however seem to equally beset any dynamical theory whatever of chemical combination which makes the difference between a positive and a negative atomic charge representable wholly by a difference of algebraic sign" (Larmor, "Dynamical Theory," part I, 475). At the same time, Thomson was elaborating a theory of atoms and electric charge that had precisely this aim: to explain the different electric affinities of hydrogen and chlorine (J. J. Thomson, "The Relation between the Atom and the Charge of Electricity Carried by It," *Philosophical Magazine* 40 (1895): 511–544.

38. Buchwald, *From Maxwell to Microphysics.*

39. J. J. Thomson, "On the Illustration of the Properties of the Electric Field by Means of Tubes of Electrostatic Induction," *Philosophical Magazine* 31 (1891): 150–171. The

theoretical ramifications of this, and Thomson's associated theory of conduction, are discussed in Buchwald, *From Maxwell to Microphysics*.

40. L. Blake, "Uber Electricitatsentwikelung bei der Verdampfung . . . ," *Annalen der Physik und Chemie* 19 (1883), 518, translation, *Philosophical Magazine* 16: 211–224; L. Sohncke, "Beitrage zur Theorie der Luftelectricitat," *Annalen der Physik und Chemie* 34 (1888): 925.

41. Thomson, "The Relation between the Atom and the Charge of Electricity Carried by It."

42. J. J. Thomson, "Referee's Report on Larmor's Paper," Royal Society MSS (6 February 1894), RR 12.160; (9 June 1897), RR 13.207.

43. J. J. Thomson, "Notes for Princeton Lectures" (1896), Cambridge University Library MS ADD 7654 NB40.

44. Thomson, "Cathode-rays," *Electrician* 39 (1897): 104–109.

45. Thomson, "Cathode-rays," *Philosophical Magazine* 44 (1897): 269–316.

46. Larmor, "Influence of a Magnetic Field on Radiation Frequency"; "Magnetic Influence on Spectra"; H. A. Lorentz, "Uber den Einfluss magnetischer Krafte auf die Emission des Lichtes," *Annalen der Physik* 63 (1897): 278–284; Zeeman, "On the Influence of Magnetism on the Nature of the Light Emitted by a Substance."

47. G. F. FitzGerald, "Dissociation of Atoms," *Electrician* 39 (1897): 103–104.

48. W. Kaufmann, "Die magnetische Ablenkbarkeit der Kathodenstrahlen und ihre Abhangigkeit vom Entladungspotential," *Annalen der Physik und Chemie* 61 (1897), 544; 62 (1897), 596. Both the Curies and Nicholas Oumoff (professor of physics at Moscow) recommended Kaufmann and Thomson jointly for the Nobel Prize, extolling the quality of Kaufmann's experiments: "Bien que les travaux de J. J. Thomson soient plus nombreux que ne le sont ceux de W. Kaufmann et embrassent un plus grand nombre de phenomenes, . . . nos connaissances dans ce demaine de physique n'auraient pas atteint en ce moment leur niveau actuel sans les recherches de W. Kaufmann" (Oumoff to the Nobel Foundation, 31 January 1904, Nobel Foundation archives); "Ces conceptions theorique on reçu diverses confirmations parmi lesquelles nous citerons les recherches recentes de Mr Kaufmann" (P. and M. Curie to the Nobel Foundation, 26 December 1904, Nobel Foundation archive).

49. E. Wiechert, "Ergebniss einer Messung der Geschwindigkeit der Kathodenstrahlen," *Schriften der physikalischökonomisch Gesellschaft zu Königsberg* 38 (1897), 3. Wiechert was, at the time, privatdozent at Köningsberg, but moved in 1897 to the observatory at Göttingen, subsequently establishing a world-famous school of geophysics there.

50. Lorentz, "Optical Phenomena Connected with the Charge and Mass of the Ions"; McCormmach, "H. A. Lorentz and the Electromagnetic View of Nature."

51. Kaufmann, "The Development of the Electron Idea," 97; Lodge, *Electrons*, xv.

52. J. J. Thomson, "On the Charge of Electricity Carried by the Ions Produced by Rontgen-rays," *Philosophical Magazine* 46 (1898): 528–545; "On the Masses of the Ions in Gases at Low Pressures," *Philosophical Magazine* 48 (1899): 547–567.

53. Thomson made this suggestion in his classic cathode ray paper of 1897, basing it on evidence from the specific inductive capacity of gases: Thomson, "Cathode-rays," *Philosophical Magazine* 44 (1897): 269–316.

54. Robotti, "J. J. Thomson at the Cavendish Laboratory."

55. H. A. Wilson, "A Determination of the Charge on the Ions Produced in Air by Rontgen Rays," *Philosophical Magazine* 5 (1903), 429–441.

56. Lodge, *Electrons,* 79.

57. Kaufmann, "The Development of the Electron Idea," 97.

58. H. A. Lorentz, *Theory of Electrons* (Leipzig: Teubner, 1909); W. Ramsay, *Elements and Electrons* (London: Harper, 1912); O. W. Richardson, *The Electron Theory of Matter,* 2nd ed. (Cambridge: Cambridge University Press, 1916).

59. W. Kaufmann, "Die magnetische und electrische Ablenkarbeit der Becquerelstrahlen und die scheinbare Masse der Elektronen," *Göttingen Nachrichten* (1901) 143–155. This whole episode, and the evidence it supplies for the interaction between theoretical and experimental interpretation, is well discussed in Hon, "Is the Identification of Experimental Error Contextually Dependent?"

60. Lodge, *Electrons,* 96; M. Abraham, "Prinzipein der Dynamik des Elektrons," *Annalen der Physik* 10 (1903), 105–179; A. H. Bucherer, *Mathematische Einfuhrung in die Elektronentheorie* (Leipzig: Teubner, 1904).

61. J. J. Thomson, *The Corpuscular Theory of Matter* (London: Constable, 1907), 1–2, 28.

62. FitzGerald, "Dissociation of Atoms," 104.

63. Lodge, *Electrons,* 69.

64. H. Becquerel, "Influence d'un Champ Magnetique sur le Raynonnements des Corps Radio-actifs," *Comptes Rendus* 129 (1899): 996–1001; "Contribution à L'étude du Rayonnement du Radium," *Comptes Rendus* 130 (1900): 206–211; E. Dorn, *Abhandlungen der Naturforschenden Gesellschaft in Halle* 22 (1900): 47; *Physikalische Zeitschrift* 1 (1900), 337; Kaufmann, "Die Magnetische und Electrische Ablenkarbeit der Becquerelstrahlen und die Scheinbare Masse der Elektronen."

65. Kaufmann, "The Development of the Electron Idea," 97.

66. E. Rutherford and F. Soddy, "Radioactive Change," *Philosophical Magazine* 5 (1903): 576–591.

67. That these were not, of necessity, the cathode rays' defining characteristics can be seen by considering the amount of time Thomson devoted in his Royal Institution lecture to the chemical and thermoluminescent changes caused by cathode rays, and by Paul Villard's attempts to define them by their chemical reducing properties: B. Lelong, in this volume.

68. Gooday, in this volume.

69. Kaufmann, "The Development of the Electron Idea," 97.

70. Alan Morton has shown the manner in which Thomson's e/m tube was used to exemplify pure physics in the British Empire Exhibition of 1924, where it formed the centerpiece of the physics section. This section was put together by Blackett and Chadwick, two of Rutherford's closest students: A. Morton, talk given at the Dibner Workshop, "The Birth of Microphysics," May 1997. Rutherford was Thomson's most powerful student, held to a particularly empirical philosophy, and appears influential in the transformation of accounts of Thomson's work.

71. Mary Jo Nye (in this volume) suggests that Thomson's lectures at Yale in 1903, in which he not only outlined his atomic model but also discussed the role of Faraday tubes in chemical bonding, was influential in stimulating the adoption of the electron into chemistry.

3

The Questionable Matter of Electricity: The Reception of J. J. Thomson's "Corpuscle" among Electrical Theorists and Technologists

Graeme Gooday

[W]hen I brought these results before the meeting of the British Association at Dover in 1899, . . . I think I made a good many converts.

—J. J. Thomson, autobiographical reminiscence, 1936[1]

No scientific discovery of prime importance has been announced during the recent meeting.

—*Electrician* editorial on the BAAS meeting at Dover, September 1899[2]

What is an electron, and what are its properties? This, we conceive is the most pressing question at the moment for the physicist.

—"The Theory of Electrons," *Electrician* editorial on the BAAS meeting at Bradford, September 1900[3]

I do not know what electricity is, and I do not know what matter is.

—J. J. Thomson at Institution of Electrical Engineers, 1907[4]

In this chapter I look at the ways in which J. J. Thomson's "corpuscle" was—or indeed was not—taken up by practitioners in the complex overlapping domain between physics and electrical engineering in the decade following 1897. It will be shown that J. J. Thomson's allegedly crucial measurements of corpuscle mass to charge ratios, published in 1897 and revised in 1899, were neither sufficiently convincing nor even strictly necessary for his contemporaries to incorporate some of his results into their working practices. Instead, I will argue that the reception of J. J. Thomson's claims among physicists and electrical engineers involved diverse agendas and contexts of application and exploration with comparatively little weight being attached to purely quantitative evidence. I show that the reception—or perhaps better the "appropriation"—of Thomson's researches on corpuscles was prolonged and complex. It was a process that intersected with debates over spectroscopy, cathode rays, x-rays, wireless telegraphy, radioactivity and metallic conduction, and was coextensive with deliberations over the long vexing question of what *constituted* "electricity." I contend that J. J. Thomson's work was only accorded a great significance from around 1900 onwards when a number of *other* British and continental practitioners deployed the somewhat heterogeneous notion of the "electron" to a wide range of theoretical contexts into which Thomson's results were then widely assimilated. Yet I shall emphasize that notably few, if any, of Thomson's contemporaries interpreted his work in precisely his own "corpuscular" terms, and indeed some were perplexed by Thomson's persistent ambivalence about the relationship between the corpuscle and the "electron."

In broader terms I shall suggest that it was not Thomson, nor at first even his Cavendish students[5], but contemporaries in the world of electrical engineering who were chiefly responsible for the assimilation of the "electron" into the laboratory, workshop, and the theoretical treatise. Accordingly I will focus on some of the major characters who worked at the interface of physics and electrical engineering, dubbed by Sungook Hong as the "scientist-engineers."[6] These included John Ambrose Fleming (University College London), Oliver Lodge (University College Liverpool, later Principal of the University of Birmingham), Silvanus Phillips Thompson (Finsbury Technical College), and Elihu Thomson (General Electric in the United States); passing reference will also be made to important figures in Germany, including Wiechert, Kauffman, Drude and Braun. I will cover in detail the role of the British journal most conspicuously devoted to the interlinked issues of electrical science and technology, *The Electrician,* and its columnist Fournier D'Albe. Much attention will also be given to other interested and active national organizations: Section A of the British Association, the Royal Institution, the Institution of Electrical Engineers, and the Society of Arts. By way of an ironic technological counterpoint, I will also suggest that Thomson's corpuscular researches were not particularly important for early developments in twentieth century electronics such as the thermionic valve and cathode ray oscilloscope, however much later electron-centered stories of their genesis might have anachronistically suggested otherwise.

It need hardly be said that I shall not offer any support to the historiographically problematic claim that J. J. Thomson "discovered" the electron in 1897. Indeed, in my conclusion I suggest how the "discovery" story so familiar from Thomson's own autobiography[7] and perpetuated by his protégés was promoted much later in the 1920s and 1930s in creating a local folk history of the Cavendish laboratory in Cambridge. Popular accounts unhelpfully continue today to privilege 1897 as a chronological watershed, to fetishize Thomson as a unique individual discoverer, to describe his work as referring unambiguously to the "electron," and to claim that the electronics of valves and oscilloscopes were direct "applications" of his putative discovery.[8] I would suggest rather that informed observers of electrical research in the two decades after 1897 would have considered it bizarre to attach any such singular significance to the rather coarse "measurements" on the mass to charge ratio of cathode ray particles that Thomson undertook in 1897. They might well have been more concerned, along with George FitzGerald, Elihu Thomson, and H. E. Armstrong, with the implications of Thomson's somewhat heretical—even alchemical—claim that atoms were divisible by the loss of corpuscles.

From what I present here it should be obvious that commentators at the turn of the century would have explained the early years of the "electron" with reference to the prior spectroscopic identification of the "electron's" mass to charge ratio in late 1896 by the Dutch physicist Pieter Zeeman, the earlier cathode rays researches of William Crookes and Arthur Schuster, and the contemporaneous measurements of mass to charge ratios undertaken by Kauffman and Wiechert. Certainly I will show that this is how historiographies of the "electron" presented the story in the decade up to 1907, notwithstanding the tendency of later scientists and historians to downgrade the significance of such work by casting it as mere "supporting" evidence for Thomson's heroic "discovery." Historians in this book, most notably Isobel Falconer and Theodore Arabatzis[9], have of course been careful to refine long-entrenched popular mythologies on this point, recognizing that they can only be upheld by oversimplified narratives and distorted chronologies. Yet the ramifications of challenging the premises of this older tradition of Thomson-centered and discovery-oriented historiography of the "electron" have not yet been fully examined. Having challenged these assumptions explicitly, my strategy in this chapter will be to explore the reception of Thomson's researches among physicists and electrical engineers on a somewhat different basis.

THE ELECTRICIAN AND THE QUESTION CONCERNING ELECTRICITY

> UNSOLVED QUESTIONS—Our knowledge of what electricity *will* do is still daily advancing, and we are justified in predicting that this will be of immense service in the future; but when we ask what electricity *is,* we have to confess that very little is known about it.
>
> —Review of G. R. Wormell, *Magnetism and Electricity: An Elementary Textbook for Students,* in *The Electrician,* July 1882[10]

> A schoolmaster once said to one of his boys, "Can you tell me what Electricity is?" The boy replied "Please, Sir, I have heard, but I have forgotten." "Alas," said the Schoolmaster, "what a misfortune! The only person who

ever knew what electricity was, has forgotten it!" Thirty years ago we were all in the same state of ignorance, but now, thanks to the researches of eminent men, we do know something at least about electricity and its nature.

—J. Ambrose Fleming, "Electricity and its Manifestations," 1931[11]

At the end of the nineteenth century there was a positive Babel of voices proposing theories on the nature and action of electricity. Texts written by physicists, electricians, and electrical engineers offered, with varying degrees of anachronism or neologism, multifarious accounts of electricity depicted as a kind of field, or a force, or as an atomic particle, a mode of motion, an imponderable fluid or pair of fluids, a form of energy or as a strain in the electromagnetic ether. This question remained an open one for James Clerk Maxwell throughout his *Treatise on Electricity and Magnetism* of 1873, although he did reserve particular skepticism for claims that electricity could be equated with "energy" and for the "two fluid" theory of positive and negative electricities.[12] Speculation on such matters was tellingly eschewed by the young J. J. Thomson in his lengthy analysis of the five distinct mathematical accounts of electrical action that he identified in a detailed report for the British Association for the Advancement of Science in 1885. While he concluded that theories which took account of the dielectric such as did Maxwell's and Hemholtz's[13], were better supported by empirical evidence than their rivals, no reader of either of their research publications could easily have found an explicit account of what electricity *was*. Even as the new "electron" theory was being taken up by many physicists and electrical engineers in 1902, its chief promoters were cautious about whether it would finally provide an answer. John Ambrose Fleming, for example, was most reluctant to specify that the notion of "electron" offered any more than a "hypothesis" to the great perennial question "What is electricity?"[14]

The persistent disagreements about the basic nature of electricity had not inhibited electrical engineers and physicists from effectively harnessing it to develop the electromagnetic technologies of global telegraphy, power generation, lighting, traction, and wireless transmission. Not even the novel "electronic" devices, notably the thermionic valve developed by Fleming and Marconi, and the cathode ray oscilloscope that emerged from the work of Braun and others in the early 1900s, were obviously applications of new corpuscular or electronic theories of matter. The communities involved in electrical technology were not merely passive secondary recipients of physicists' newfangled electrons and corpuscles but rather provided important arenas and tribunals for debating the import of these parvenu entities. Thus one of the most important British periodicals devoted to "electrical engineering industry and science" in the late nineteenth century, *The Electrician,* published articles ranging from the mundane specifications of domestic electrical meters to Oliver Heaviside's loftier mathematical excursions into Maxwell's theory of electromagnetism. It was also one of the principal arenas for debating the recurrent question "What is electricity?"

While not a few teachers of electrical science and engineering were pragmatically agnostic on the subject, others—including some writers for *The Electrician,* shared the view proposed by Maxwell in his *Treatise on Electricity and Magnetism* of 1873 that a closer examination of the electrical discharge of rarefied gases would "probably throw great light" on the "nature" of electricity.[15] At the BAAS meeting in 1879 William Crookes's had followed up Maxwell's suggestion by showing how the results of magnetically manipulating cathode rays could ground speculations that such rays were made up of a new fourth, "radiant" particulate state of matter.[16] Throughout the following two decades, *The Electrician* gave broad if intermittent coverage to a wide range of researches on the use of cathode ray tubes, reporting for example Crookes's presidential address[17] on this subject to the Institution of Electrical Engineers in 1891. Although highly deferential in tone, *The Electrician*'s comments were not entirely uncritical: following Maxwell's precedent eighteen years before, it challenged the terminology of "positive" and "negative" electricity that Crookes had chosen to employ. Particularly objectionable was the implication that these were two different things rather than "two converse manifestations of one and the same entity."[18] Further challenges to Crookes's claims came from Germans such as Goldman, Hertz and his student Lenard that cathode rays were not particulate but were etherial vibrations. These were also reported without chauvinism in *The Electrician*'s weekly column, "Contemporary Electrical Science" by polyglot journalist Edmund Edward Fournier D'Albe.[19]

Although of passing interest to many, the character of cathode rays only became a pressing topic of *general* interest to physicists and engineers, however, after Röntgen announced that he had used them to produce a new kind of rays, x-rays, in late 1895. The ensuing international flurry of interest is reflected in the mild explosion of articles published on related topics in *The Electrician* during the next two years. On 10 July 1896, for example, the journal reported at some length the electrostatic deviation of cathode rays by G. Jaumann working under Lecher in Bohemia, thereby challenging the etherialist claims of Hertz that such deviation had proved impossible.[20] On 1 January 1897, it reported the finding of John Ambrose Fleming, Professor of Electrical Engineering at University College London, that an electromagnet could induce some interesting spiral effects on Crookes's famous Maltese cross experiment.[21] Adjacent to Fleming's article was a piece by Elihu Thomson, chief research engineer at General Electric in the United States on the extraordinary fluorescent effects of directing Röntgen rays on Crookes's tubes.[22]

Among all of this excitement, *The Electrician* also gave attention to spectroscopic developments and theoretical speculations concerning the "electron." This was a term invented by Irishman George Johnstone Stoney in 1891 to account for the double lines in gas spectra, postulating a tiny charged particle that emitted electromagnetic radiation as it rotated around an atom to which it was inseparably bound.[23] Soon afterward the "electron" was appropriated and recontextualized by theorists of the electromagnetic ether: the "Maxwellian" Joseph Larmor in Cambridge used it to label the little vortex "knots" of ether by which he sought to explain interactions between ether and matter.[24] The Dutch physicist Hendrik Lorentz used the more familiar name "ion" to refer to a rather different conception of the tiny charged particles in matter;[25] but it was Lorentz's student Pieter Zeeman who stimulated a reconciliation of these two research traditions by his magnetospectral experiments.[26] In October 1896 Zeeman announced that he had succeeded (where many had failed before) in splitting the spectrum of vaporized sodium by the application of a strong magnetic field. Following Lorentz's theory he explained this splitting as the effect of the field in modifying the vibrational frequency of (some) sodium "ions," and from the angular size of the spectrum splitting inferred that these ions had a remarkably large charge to mass ratio.[27] This result was quickly communicated by Larmor's co-Maxwellian Oliver Lodge to the *Philosophical Magazine,*[28] and as "The Latest Discovery in Physics" to *The Electrician* of 26 February 1897.[29]

In his first account of this "discovery" for the readership of *The Electrician,* Lodge claimed that Zeeman's result had a rather fundamental importance for contemporary understanding of the electromagnetic ether. Transmuting Zeeman's "ion" into the Larmorian electron, and drawing upon the researches of arch-Maxwellian Oliver Heaviside and the theoretical work of the young Professor of Experimental Physics at Cambridge, Lodge promoted the view that Zeeman's research clarified the mechanism of the atomic generation of radiation:

> The importance of the discovery lies, of course, in its theoretical bearing, in the evidence it can furnish as to the nature of the motions which enable matter at high temperature to disturb the ether, [and] in the conclusion that can be drawn from it as to the physical nature of a radiating or absorbing body . . . It has for some time now appeared likely that radiation could only

be excited by the motion of electrified particles . . . [but] some philosophers have doubted about the existence or necessity for any material nucleus beyond the electric charge itself; such a charge, when in motion, would behave as if it had inertia, in accordance with well known electrical laws, as worked out by Mr Heaviside, Prof. J. J. Thomson and others; and accordingly the idea of radiation excited by the motion of electrons pure and simple has been steadily gaining ground.[30]

Whatever else might be debatable about this phenomenon, Lodge was convinced that the electron-theorist's mechanism for electromagnetic radiation would come to be regarded as "substantiated" and "established" by Zeeman's research. Thus for Lodge it was clearly *not* the case that the electron needed to be discovered: Zeeman's putative discovery of the empirical effect was just ammunition to support what he—and in his view also J. J. Thomson—knew on (Larmorian) theoretical grounds must already exist anyway. What was radically new for Lodge, however, was the extraordinarily high charge to mass ratio of 10^7 (electromagnetic units) that Zeeman had inferred for the vibrating subatomic "electron." As Lodge reported in a follow up piece on March 12, this value was roughly a thousand times greater than the "customary" value for the ion found in electrolytic data, a conclusion which evidently left him and fellow Maxwellian[31] George FitzGerald in Dublin struggling somewhat to account for this in terms of electromagnetic theory.[32]

Notwithstanding Lodge's allusion to J. J. Thomson as a fellow electromagnetic theorist[33], Thomson appears at this stage to have had no comparable interest in the Zeeman effect. As Isobel Falconer has pointed out, however, by this time Thomson had, like many other physicists, taken a keen interest in another, earlier unexpected result from experiments on radiation: the phenomenon of x-rays. And it was the publicity surrounding the extraordinary properties of x-rays that emerged in 1895–96 which turned Thomson's attention for the first time to the constitution of cathode rays.[34] As Thomson happened to be president of BAAS Section A that year, a large part of his presidential address was devoted to cathode rays and x-rays. Thomson promoted the Crookesian view of cathode rays as particulate against the apparent counterevidence presented by Phillip Lenard that such particles were not stopped by an encounter with metal; indeed under Thomson's chairmanship, the debate that following Lenard's own paper at the Section A meeting was accordingly rather one-sided.[35] By house custom Thomson's address was printed in *The Electrician,* on September 18th 1896 along with a portrait (figure 3.1) and biography by his old friend John Henry Poynting,[36] in which Thomson was dubbed "electrician" for his work on electrical discharge of gases.

Figure 3.1
Engraved portrait of Professor J. J. Thomson in *The Electrician* for 18 September 1896, accompanying a reproduction of his Address to section A of the British Association for the Advancement of Science in Liverpool. Source: The Electrician, 37 (1896), opposite 672.

Thomson's expertise evidently made some impression on the editor of *The Electrician,* W. G. Bond. Three weeks later Bond wrote to Thomson inviting him to write a book—or at least a series of articles—on "Electric Discharge in Rarefied Gases."[37] His suggestion was that the book cover the research undertaken by himself, Crookes, Arthur Schuster, Röntgen, Lenard and, other German researchers, and that it would complement other well-established series of texts for electrical technologists published by *The Electrician,* notably Ambrose Fleming on a.c. transformers, Oliver Heaviside on electromagnetic theory of induction, and a treatise on the magnetic induction of iron published by J.Alfred Ewing, Thomson's colleague as Cambridge Professor of Engineering.[38] Even though Thomson did not take up this offer, it is striking that after he gave his lecture on "cathode rays" at the Royal Institution on 30 April 1897, *The Electrician* moved with great alacrity to publish this lecture a mere three weeks later on May 21.[39] Unlike Lodge's account of the Zeeman effect, however, the journal did not announce Thomson's claims for the existence of corpuscles as the "latest discovery in physics." And indeed *The Electrician* was the *only* national journal that took the trouble to publish Thomson's lecture or comment at length upon it.[40] The next section will examine why this was the case.

Dissociation and Indifference: Varied Responses to Thomson's 1897 Researches

Much in the vein of his comments on cathode rays in his presidential address to Section A of the BAAS in the previous year, Thomson's Royal Institution lecture was devoted to attacking the arguments that German etherealists had wielded against the British particle theorists. Thus, as Falconer,[41] Arabatzis,[42] and Feffer[43] have pointed out, much of Thomson's April lecture held few surprises for anyone familiar with the previous few years of cathode ray research, or to Stoney's and Zeeman's researches on elemental spectra, and one might add to the electrostatic deflection experiments of Jaumann. The major novelty of Thomson's paper was to invert the force of Lenard's evidential claims about the ability of cathode rays to pass through thin metal sheets: the considerable mean distance that they could travel outside the cathode tube was entirely compatible with his thesis that cathode rays were made up of particles much smaller than air molecules. The only other apparent novelty was the somewhat peremptory and unexpected coda in which he used a magnetic deflection method to establish a mass-charge ratio for the cathode-ray particle of 1.6×10^{-7} (much as Arthur Schuster had tried to do in 1890). Interestingly, he marshaled a hand-waving order of magnitude agreement with Pieter Zeeman's spectroscopically inferred value of the electron's mass to charge ratio—albeit with the m/e value confused with e/m—to confirm the existence of what he called "corpuscles," not Zeeman's electrons.[44]

Given the slender resources marshaled by Thomson for his claims, it is not surprising to read in his autobiography of 1936 that one distinguished physicist present at the lecture had thought Thomson had been "pulling" his audience's legs.[45] Since Thomson had not paid much attention to contemporary protocols of precision research in physics or engineering, such a reaction is not hard to explain. First Thomson—or rather his assistants—had apparently only bothered to make *one* measurement of the mass/charge ratio. Second, he used engineering instruments—ammeters and voltmeters—in rather cavalier ways of which many physicists would not have approved.[46] Furthermore, the figures that he gave for the mass to charge were only to two significant digits, with no estimates of his errors—a point that, as Kathy

Olesko would emphasize, cannot have greatly impressed his German opponents.[47] He drew support for his figure merely from an order of magnitude agreement with Zeeman's work and, given that contemporary measurements research on for example the mechanical equivalent of heat, the gravitational constant or the BA's electrical measurement standards had been replicated by many experimenters on hundreds of experiments undertaken to four or sometimes five significant figures, Thomson's results cannot have looked compelling.[48]

It is important to bear in mind here that the student Thomson, unlike so many of his contemporary physicists, had not been trained in the rigorous practices of precision laboratory measurement. Indeed at the BAAS meeting the previous summer, Thomson had poured scorn on the tradition of British physics pedagogy that mercilessly drilled its trainees in the minutiae of exact measurement[49] and produced practitioners that were as dulled and unimaginative as the white knight in *Alice's Adventures in Wonderland*. Such physicists commenced their career "knowing how to measure or weigh every physical quantity under the sun, but with little desire or enthusiasm to have anything to do with any of them."[50] Given that in 1897 the Cavendish Laboratory was by no means the world center of research that it became in the following decade, it was not surprising that the scientific press paid little attention to this minor result from Free School Lane, especially given the ongoing diversions provided by the fascinating novelties of Becquerel rays, x-rays, and wireless telegraphy. For example, while it did not acknowledge Thomson's paper, *Chemical News* showed great interest in the results of Campbell Swinton's work on cathode rays, reporting in full his paper given at the Royal Society on 17 March 1897,[51] also printing the abstract of Silvanus P. Thompson's paper on the same subject given to the Royal Society on 17 June 1897.[52] The generalist journal *Nature* made no comment whatever, despite showing an interest in J. J. Thomson's work on several other occasions during that year.[53]

The editorial staff members of *The Electrician,* however, were clearly sufficiently interested in the implications of Thomson's lecture; they appear to have solicited a commentary on it from George FitzGerald which was published immediately *preceding* the transcript of Thomson's lecture. While FitzGerald did indeed suggest that Thomson's corpuscles could—in the jargon of his "Maxwellian" friend Larmor—be identified as "free electrons,"[54] it would be a gross distortion to see this as FitzGerald's only relevant reading of Thomson's work. The *title* of his piece was after all, "The Dissociation of Atoms," and indeed for the large bulk of his rather convoluted and ambivalent argument, FitzGerald addressed Thomson's main claim as being rather that "atoms are divisible into much smaller parts and are so divided in cathode rays"—quite a long way from the discourse of Larmorian ether theory.[55] Indeed, alluding to contemporary debates on the mutability of chemical elements, he suggested that Thomson "ought to be able to transmute any substance into any other he desired by passing it through the furnace of the cathode rays." This alchemical theme is one to which I shall return shortly.

Usually a journal of very forthright editorials, *The Electrician* made no direct comment on J. J. Thomson's work until 2 July 1897. Its comment then was effectively a response to a qualitative speculation on the "cause of Röntgen rays" published in the same issue of the journal (and in its main rival, *The Electrical Review*) by Elihu Thomson of the U.S. company General Electric. The American Thomson hinted rather heavily that he too might wish to claim priority for the discovery of the "breaking down of the elements"—an interpretation of Thomson's work previously pursued by FitzGerald—that notably did not focus primarily on the character of the alleged new particle.[56] In announcing his suggestion that the high frequency vibrations of J. J. Thomson's corpuscles might account for the generation of x-rays, Elihu Thomson wrote

> Since so eminent a physicist as Prof. J. J. Thomson has in a recent Paper put forward the hypothesis of the breaking down of what we have been accustomed to call the "elements," and has shown a reasonable basis for such a hypothesis, the writer deems it not improper to state that a similar view had quite independently arisen in his own mind, the origin and progress of which may, it is thought, be interesting to others working and thinking in the same field.

In a small note, an *Electrician* editorial commented that Thomson's explanation of certain cathode-ray phenomena by the "assumption of the divisibility of the chemical atom" led to so many "transcendentally important" conclusions that they could "not but wish to see the hypothesis verified" at an early date by some crucial experiment.[57] Evidently the single mass to charge measurement that Thomson had undertaken in the spring of 1897 was *not* thereby deemed to be "crucially" persuasive at all. Moreover, alluding to Elihu Thomson's and FitzGerald's distinctly *chemical* reading of J. J. Thomson's experiments it added:

> An hypothesis that threatens to lead us to the alchemist's bourne, the transmutation of metals, must needs give us pause, but seeing that it is put forward so cogently and yet so modestly by natural philosophers of no small insight, it is entitled to respectful consideration at the hands of even the most sceptical.[58]

What is interesting in this discussion of J. J. Thomson's work is that although his putative measurement of the corpuscle's mass-to-charge ratio is not taken to be conclusive, its very inconclusiveness is made all the more important to resolve in view of the apparent fertility of the "corpuscular hypothesis" as a resource to apply to contexts of research beyond that of just cathode rays. Thus in a concise summary of Elihu Thomson's work on x-rays, the editorial note added that Elihu Thomson's "suggestive" contribution showed how, given a divisible atom, he could account for the origin and idiosyncrasies of Röntgen rays, the increased vacuum observed in x-ray tubes, the dark rifts in the photosphere in the sun, their electrical effect upon mundane affairs, and the "puzzle of the sun's well-sustained temperature."[59] Such fertile results, *The Electrician*'s columnist inferred, by no means decreased the intensity of the general "desire" for a crucial experiment on J. J. Thomson's contentions.

It is notable then that, although both J. J. Thomson and two German researchers—Wiechert and Kauffman—attempted to improve upon Thomson's measurement of the mass/charge ratios, none of them appear to have been sufficiently "crucial"—qua measurement experiments—to win over a large consensus in favor of Thomson's views. For example, when J. J. Thomson published more detailed results in the *Philosophical Magazine* for October 1897,[60] the results for his magnetic method ranged wildly from a smallest value of the m/e ratio of 0.32×10^{-7} to a largest value that was about five times greater, viz. 1.5×10^{-7}—not perhaps the most convincing evidence for a universal constant. And the mean of the values he accomplished by electromagnetic and electrostatic deviation differed by as much as 20 percent. In his biography of J. J. Thomson, the younger Lord Rayleigh pointed out that this level of "uncertainty" would hardly have been acceptable for a transaction in a contemporary grocers' shop—let alone in a physics laboratory.[61]

Converts and Appropriators at the BAAS Meeting of 1899

For the next two years and with the assistance of J. S. E. Townsend and C. T. R. Wilson, Thomson continued to try to get more convincing measurements, using cycloidal cathode-ray paths and oil-drop methods to measure the charge on the corpuscle. In his autobiography he claimed that bringing his results before the meeting of the British Association at Dover in 1899 brought "a good many converts" to his views.[62] It must be said, however, that Section A of the BAAS meeting 1899 was not that well attended by British physicists and electrical engineers since such luminaries as Lord Kelvin, William Ayrton, and Silvanus Thompson had traveled to Italy for the centenary celebrations of Volta. Indeed a reporter for *Nature* suggested that the attendance of physicists at the sectional meeting was "rather smaller than usual."[63] There was, nevertheless, some compensation for the Voltaic diversion in the form of a visiting contingent from the Association Française pour l'Avancement des Sciences which was holding its own meeting just over the channel at Boulogne, the British and French associations mustering more entente cordiale than their respective governments could manage at this time. According to a reporter for *The Electrician*, Thomson's lucid presentation of his paper "On the existence of masses smaller than atoms" made a striking impression on the French visitors.[64]

Thomson's paper summarized his research on cathode rays to date, emphasizing that the mass-to-charge of particles magnetically deflected in cathode ray tubes was about 1/1000 of the ratio attained from electrolytic data. According to the *Nature* reporter, Thomson's experiments with cycloidal paths to determine the charge on cathode ray particles offered the "simplest crucial experiment" to prove that this ratio arose from their comparatively small mass, not any large charge on these particles. Emphasizing that this ratio appeared to be the same in all gases used for the experiments, he then invoked Prout's claims on the nonuniformity of atomic masses between elements, as well as supporting spectroscopic evidence from Lockyer and others, to contend that electrification of all matter was a process that consisted universally in atoms losing a negatively charged corpuscle much small than the atom itself.[65] Yet from the only extant reportage in *Nature* of the discussion which followed Thomson's paper, it is not at all obvious that auditors offered a passive acquiescence to Thomson's corpuscular claims. Rather, they actively harnessed this opportunity to promote their own agendas, appropriating Thomson's results for their own ends.

M. Broca of the French contingent held forth on his own related spectroscopic observations of sparks in Crookes tubes, emphasizing a significant difference in the spectra obtained near his platinum electrodes and in the space between. Arthur Rücker of the Royal College of Science drew attention to Arthur Schuster's work on spectra, suggesting that this showed matter to be made up of "a complicated collection of units themselves similar."[66] Oliver Lodge, waxing somewhat Larmorian, averred that Thomson's investigations "might turn out to be the discovery of an electric inertia" and thus might lead to a "theory of mass"—an issue to which he had alluded in his March 1897 discussion of Zeeman's researches.[67] Sir Norman Lockyer, the former editor of *Nature*, rather predictably argued that his own prior spectrographic evidence on calcium, iron, magnesium, and copper had already shown that atoms of (at least) these elements had a somewhat complex internal constitution of subatomic particles.[68] The South Kensington chemist Henry Armstrong was somewhat critical of Thomson's claims to have demonstrated the separability of corpuscles from their respective atoms. Defending the integrity of the atom, as was Armstrong's disciplinary wont, he repeated his recurrent accusation that Thomson and his Cavendish school were "insufficiently instructed in chemistry" to judge such matters.[69] Although the physicists and electrical engineers present were evidently more enthusiastic about Thomson's claims, it is not clear that any of them subscribed to his own particular interpretation of these results, and certainly none attributed to him any priority in making a major "discovery" of a hitherto unknown particle.

The chief attraction of the B.A.A.S. meeting that year was not, in fact, Thomson's ingenious Cavendish experiments with cathode rays, but the first public demonstration of cross-channel wireless transmission accomplished by the new partnership of J. A. Fleming and Guilelmo Marconi. These transmissions between the BAAS meeting at Dover and the Association Française in Boulogne were a central part of a spectacular lecture on wireless telegraphy given by Fleming. Reporting on the comparative impact of the Thomson and Fleming papers an editorial in *The Electrician* on 22 September 1899 opined that:

> No scientific discovery of prime importance has been announced during the recent meeting, neither have the proceedings been marked by any event of unusual scientific interest, with the single exception of the remarkable extension of the Marconi system of wireless telegraphy, which has been carried out under the direction of Prof. J. A. Fleming. But although the meeting is thus unmarked by any event of great importance to the scientific world, two items in the week's proceedings stand above the level of general mediocrity and command special attention, These are respectively the admirable address given by Prof. J. J. THOMSON before Section A, on Saturday, on the subject of electrons[sic] and the ultimate constitution of matter, and the masterly discourse with which on Monday evening, Prof. J. A. FLEMING, charmed the entire Association . . .[70]

On J. J. Thomson's paper it said that if his views on the constitution of matter and the nature of electricity proved to be correct, they would "necessitate a complete revision" of the commonly accepted theories. For the editorial staff at *The Electrician,* it appeared that Thomson's cycloidal experiments were not so definitively crucial after all. Judging from the content of Fleming's lecture it appeared that he did not see great immediate significance in Thomson's results either. This lecture was entitled "The Centenary of the

Electric Current," and Fleming's narrative characteristically rendered his own experiments on Marconi's wireless system as if they were the climax of 100 years of research.[71] Nowhere in his account of the electric current are Thomson's corpuscles mentioned: Fleming still viewed current in distinctively Maxwellian terms as a secondary phenomenon due to "certain events in the space-filling aether," the only aspect being determined by "what we call the conductor" was the localization of these events.[72] Although this direct exposure to Thomson's researches at Dover did not immediately change Fleming's understanding of the electrical current, other developments in contemporary research soon inspired a radical transformation of the electrical engineer's views.

NEW TERRITORIES AND NEW AUDIENCES FOR THE ELECTRON AND CORPUSCLE

> It is somewhat strange that after the victory of the wave theory of light, a corpuscular theory of electricity should oust all its rivals. Yet such an occurrence appears about to take place.
> —Editorial note in *The Electrician,* 24 August 1900[73]

The year after the BAAS meeting at Dover, several individuals began actively to appropriate Thomson's corpuscles into other preexistent debates on the "electron," variously conceived in the terms of Larmor or his continental counterparts. Throughout much of 1900 there was vigorous activity at various locations across Europe of individuals drawing together these different theories of the electron with Thomson's researches in ways that drew rather more interest and conviction from wider audiences than had putative measurements of particulate mass-to-charge ratios. The German physicist and wireless expert Paul Drude published his first analysis of electronic conduction in metals,[74] Larmor's monograph *Aether and Matter* appeared in that year,[75] and there were several papers arguing for the common identity of Becquerel rays with cathode and Lenard rays.[76] In response to this, an editorial in *The Electrician* of 24 August 1900 commented "The conception of the electron has been worked out by THOMSON[sic], POYNTING[77], DRUDE, and others," presumably referring to Larmor and Lorentz, and "may now be said to be fairly concrete." It contended that "electrons" appeared to be able to account for "almost" every known electric phenomenon.[78]

And it was not just the editorial staff of this journal who were now customarily consolidating Thomson's results on corpuscles with specifically "electronic" theories of matter. A few weeks later at the meeting of British

Association Section A in Bradford, with Larmor in the presidential chair, the term "electron" was much bandied about in a lengthy discussion of George FitzGerald's paper on ionization in gases and liquids. Here FitzGerald interpreted explicitly Thomson's "corpuscle" as an electron on the grounds that it had the same apparent mass as the particle observed by Zeeman.[79] FitzGerald's paper stimulated an editorial discussion on electron theory in *The Electrician* which ranged across the many contexts to which various formulations of electron theory were now being applied from radioactivity to the production of x-rays. It concluded that further experimental evidence might yet establish that the electron theory was the key to solving that "great problem" of physics and chemistry—the "ultimate" constitution of matter. Interestingly, however, this editorial did not mention Thomson's 1897 and 1899 researches, indicating that they were not necessarily seen as central to the ongoing debate on electrons.[80]

By November 1900, Ambrose Fleming had started to attribute some significance to Thomson's gas discharge researches, albeit only in a fleeting mention in a lecture on "Electric waves and electrical oscillations" at the Society of Arts.[81] Like all others involved in the ongoing debate—except Thomson and his students—Fleming too did not differentiate strictly between the electron and Thomson's corpuscle. While borrowing occasionally from electronic theories himself, Thomson persisted in using the term "corpuscle" or referring to them collectively as "bodies smaller than atoms"; this indeed was the title of the article he wrote for the American journal *Popular Science Monthly* of August 1901.[82] In introducing his researches to a wider popular audience, Thomson passed rapidly over the tradition of researches on cathode ray particles that preceded his own and focused on the extraordinarily small mass to charge ratio that (implicitly) he and his Cavendish coworkers had found for them. The comparable values attained by Wiechert, Kauffman, and Lenard by different methods were invoked to support his evidence, although he subtly evaded potentially controversial issues of priority by omitting all chronological detail.[83] Indeed rather than defer to contemporary theories of the "electron," Thomson referred his readers to Franklin's one-fluid theory of the late eighteenth century as one to which his theory of negative "corpuscles" approximated "very closely."[84] Declaring next that his researches pointed "unmistakably" to a definite conception of the "nature of electricity," Thomson went on to show the importance of this in that phenomena as diverse as Röntgen rays, electrical conduction, radioactivity, and atmospheric auroras could all be correlated to the behavior of corpuscles.

Responding directly to Thomson's article, and evidently courting the same general lay market for such "popularizations," Fleming published a piece called "The Electronic Theory of Electricity" for the same journal in May 1902.[85] Fleming began by giving credit to the work of Thomson and others for showing that "so-called corpuscles" were projected from the "kathode" of platinum wires when they carried an electrical current in an evacuated glass vessel. Yet Fleming's genealogy of the "corpuscle" started not with Thomson's experiments, but rather a quarter of a century earlier. For Fleming the starting point of the "electronic theory" was William Crookes's illustration in the 1870s of how electrical discharges in high vacua evinced a particulate and inertia-bearing fourth state of "radiant matter." According to Fleming, the import of Thomson's 1897 work was largely that it furnished a "proof" of Crookes's contention by quantifying the comparative mass of particulate "corpuscles" as being a thousand times smaller than hydrogen atoms. Thereafter Fleming's account gave relatively little emphasis to Thomson's investigations, focusing instead on the "electron" as a tool for correlating the work of Faraday, Weber, Stoney, and Helmholtz, then moving successively through the atomic speculations of Maxwell, their relation to matters of chemical valency, on to the researches of Drude and Riecke on the conductivity of metals, and thence to the ether theories of Larmor, Lorentz, and Zeeman. Fleming now explicitly suggested that the electronic theory did offer, at last, a promising hypothesis that might yet answer the question "What is electricity?" So impressed was Fleming by this theory that he even specifically conceded that an electric current was "the regular free movements of electrons" in a conductor—somewhat in contrast to his Maxwellian claims at Dover in 1899.[86]

Later that month, when his patron Marconi was forced to postpone his lecture on wireless telegraphy at the Royal Institution, Fleming stood in at short notice and used the opportunity to air an extended account of his "electronic" theory of electricity. A reviewer for *The Electrician* among the "large audience" present was most impressed by Fleming's "surpassingly lucid" presentation and reported in detail the extraordinary fertility of the "electron" as a means of accounting for diverse phenomena. For example, Fleming was able to harness his theory, along with Crookes's well-known demonstrations on the luminosity of cathode ray tubes, to explain the hitherto mysterious mechanism of the connection between sunspots and auroras. He argued that gusts of electrons emitted by sun spots tended to wind spirally round the lines of magnetic force—as had been "observed in cathode ray experiments"—and so descended toward the earth's magnetic poles, producing luminous circular effects known as aurorae when they reach the layers where the air was at the density of a Crookesian low vacuum tube. Moreover, Fleming's interpretation of electron theory even provided a radically new answer for electrical engineers particularly vexed by the question of where the seat of the E.M.F. was located in the voltaic cell,[87] his radical suggestion being that it was located "equally in all parts of the circuit."[88]

The Autonomy of Hardware—or Not Harnessing the "Electron" in New Technologies

By 1903 Fleming was bringing the electronic theory of matter to bear on the wireless transmissions that he had undertaken with Marconi during the preceding three years.[89] Having just been (temporarily) disengaged as Marconi's technical consultant, as Hong has pointed out,[90] Fleming had plenty of time on his hands for theoretical analysis. Yet we find that in this phase of his work, his deployment of the somewhat protean electronic theory veered back to his original commitment to a Larmorian interpretation of etherial electrons. At a lecture on "Hertzian Wave Telegraphy" presented to the Society of Arts in July 1903, he applied a very Larmorian ether strain version of the electron theory to account for the means by which vibrating electrons generated Hertzian waves. Ironically, although etherial cathode rays had now effectively disappeared from the electron-corpuscle debate, the etherial electron was still much in evidence in wireless technology. Indeed Fleming described the future of electrical technology as that of "ether engineering"—significantly not as that of harnessing and manipulating material electrons.[91]

What then was the role of the "electron," if any, in the material practices of Fleming's electrical engineering researches? The thermionic rectifying valve that Fleming produced between 1904 and 1905 is certainly his most famous invention. As Sungook Hong has recently shown, however, the emergence of Fleming's valve owed little to Larmorian ether theory and was hardly an application of a Thomsonian "corpuscle" theory either. Rather it was a synthetic development of several phases of Fleming's researches in the preceding three decades. First it incorporated the work he shared with his old physics teacher Frederick Guthrie at South Kensington in the mid-1870s: the electrically asymmetrical discharge of incandescent bodies[92]. Also highly important, as many historians of electrical engineering have pointed out, was Fleming's examination of lightbulbs which revealed the curious shadows in the carbon discharges from filaments known after his employer as the "Edison effect."[93]

Less often emphasized by historians but nevertheless crucial was Fleming's harnessing of the techniques and agendas of a.c. power transmission and wireless telegraphy for which Fleming had acquired a high-ranking expertise.[94] Furthermore, Hong has pointed out, Fleming was determined to win back Marconi's interest in his work and also the lucrative consultancy fees that he had hitherto enjoyed. The arrival of the first material artifact of twentieth-century electronics soon after the development of theories of electrons of corpuscles is thus little more than a historical coincidence. Far from being an "application" of the electron as conventional histories of electrical engineering have seen them, Fleming's rectifying valve supplemented the cathode ray and x-ray tube as technologies for exploring and manipulating electrical discharge, whether conceived as electron flow or otherwise. Thus the only mention of the electron in Fleming's 1905 account of his new rectifying device to the Royal Society concerns a small matter of explication.[95] To see how the "electron" became a central feature of how Fleming accounted for the characteristics of his "valve," one has to look at his much later and rather notorious patent disputes over the "triode" with de Forest; that has not prevented historians of electrical engineering, however, from reading the story backwards to see the valve as the "application" of the electron to technological ends.[96]

The cathode ray oscilloscope is another oft-cited technological candidate for an early electronic device; indeed Thomson's son later claimed that an early version of this device was "in essence" the original apparatus that Thomson and his assistants had used in the 1897 laboratory determination of e/m.[97] On closer inspection, however, one finds that the origins of this device lay elsewhere in the broader Anglo-German tradition of cathode ray investigation of which Thomson's was only one parvenu strand. In 1897, Karl Ferdinand Braun was developing a new form of cathode ray tube that could be calibrated to convert beam deflections to give proportional readings of the strength of the electrostatic or magnetic field causing the deflections.[98] Collaborating with a Dr. Zenneck in 1902, Braun showed how the screen of a cathode ray tube attached to a timing device could be deployed to display the forms of alternate current waves.[99] This technique of using virtually inertia-less cathode rays to register field-strength was immediately taken up by W. Mansergh Varley at Heriot Watt College in Edinburgh.[100] He showed that this was far better adapted to the display of high frequency a.c. waves than the relatively insensitive mechanical oscillograph previously made by W. Dubois Duddell and W. E. Ayrton at the Central Technical College in South Kensington in 1898–1900.[101]

Three years later, Varley and his assistant W. H. F. Murdoch presented to the readers of *The Electrician* the first full account of the wide-range of uses to which the "Braun Cathode-Ray Tube" could be deployed in displaying phase differences, hysteresis, and oscillatory discharge.[102] It is telling that in these explorations of the nascent cathode ray oscilloscope no reference is made to any contemporary work by J. J. Thomson on corpuscles or electrons—the crucial issues appear to have been the ability to manipulate cathode rays and harness them to preexistent skills in constructing measuring apparatus. Thus the historian must conclude that the corpuscle and the cathode ray oscilloscope are related only as common products of a shared tradition of contemporary research in cathode rays. Later appropriations of the cathode ray oscilloscope as a pedagogical exemplification of Thomson's alleged "discovery" of 1897 are, however anachronistic, important vehicles for the perpetuation of this discovery story.[103]

THOMSON VS. THOMPSON: THE CONTESTED RELATION BETWEEN CORPUSCLES AND ELECTRONS

Having shown that technologists and physicists developed early cathode ray technologies without specific reliance on theories of electrons or corpuscles, one should not infer that such practitioners as electrical engineers were not interested in such theories. Far from it, they were in some respects the most critical and the most numerous constituency for research on the electron or corpuscle. In sheer demographic terms they constituted an audience for such work much larger than that of contemporary physicists—thousands of electrical engineers as compared to mere hundreds of physicists in the country as a whole were kept well abreast of recent publications in the field by Fournier D'Albe's columns on contemporary electrical science in *The Electrician*,[104] and later by his successive editions of *The Electron Theory* from 1906.[105] In this work, Fournier D'Albe, well acquainted with both the continental and the British literature,[106] presented an eloquent historiography of the electron, ignoring all Thomsonian talk of corpuscles, and telling a highly inclusive tale for collective research that was marked by no single crucial experiment or discovery. Looking back at the development over the preceding decade, he remarked tellingly on the "almost ominous silence" with which the new "electron theory" had made its appearance:

> It has not been heralded by a flourish of trumpets, nor has it been received with violent opposition from the older schools. No one man can claim the authorship of it. The electron dropped, so to speak, into the supersaturated solution of electrical facts and speculations, and furnished the condensation nucleus required for crystallization. One after another the molecules—the facts of electricity—fell into line, and one department of electrical science after another, crystal on crystal, clicked into its place, dispersion first, then electrolysis, then gas discharges, then radium rays, then metallic conduction, and lastly, magnetism.[107]

This putative crystallization of wisdom on the electron was not yet, according to D'Albe, "fully shaped." And it is notable that leading academics in the cross-disciplinary field of electrotechnology had been and continued to be spokespersons on the articulation and synthesis of the newly emerging "electronic" theory.

The origins of Oliver Lodge's 1906 book *Electrons* lay in a lecture that the Institution of Electrical Engineers had invited him to give five years previously (and indirectly in his account of the "Zeeman effect" in *The Electrician* four years before that).[108] According to *The Electrician,* his two hour marathon presentation in November 1902 went down so well that the electrical engineers present would "gladly" have listened for another hour to Lodge's "masterly" exposition. Those of his audience who had followed the development of the electron theory could "not fail to be impressed" by Lodge's "lucid" review of relevant research, and those members who had come with "no definite idea" as to the nature of the electron theory were certainly sent away "with a clear grasp of its true import."[109] Like D'Albe's book, Lodge's expanded text of 1906 captured the breadth and complexity of recent research on "electrons," albeit much more specifically located in the electromagnetic theory of the ether.[110] Equally much, except in a prefatory dedication, Lodge attached no special place to the contribution of J. J. Thomson[111] despite the latter's recent award of a Nobel prize for his researches on the electrical conductivity of gases.[112]

In the next year, newly adorned with his Nobel Prize, J. J. Thomson was invited for the first time to speak to the Institution of Electrical Engineers, lecturing on 21 February 1907 on the "The Modern Theory of Electrical Conductivity of Metals."[113] With another former Cavendish celebrity R. T. Glazebrook taking the presidential chair for the meeting, Thomson presented a semiqualitative version of Drude's theory, augmented by some investigations at the Cavendish Laboratory, translated back into his own language of corpuscles. In the discussion afterward, this complete avoidance of "electron" language brought quite a strong response from his near namesake, Silvanus Phillips Thompson, the principal and Professor of Electrical Engineering at Finsbury Technical College. Ever the forthright Quaker, the Finsbury Thompson said:

> I have found it one of the difficulties of the study of this branch of the subject that the different authorities who write and talk about it do not always speak in the same language. We have heard a good deal tonight about corpuscles; I do not think we have heard one thing about electrons. I want to know whether in the lecturer's usage those two words mean the same

thing. . . . do we understand, by that, which we have been hearing of tonight under the name of "corpuscle," a minute portion of matter much smaller than an atom and electrified? Or is there no matter at all in it? Is it simply a little bit of electricity? Is it a disembodied bit of electricity which acts as a corpuscle, or is it an electrified piece of matter? We desire something definite about the terms which are used, and precisely what they connote.[114]

Thompson continued by earnestly haranguing Thomson about his claim that electrical conduction could be explained by corpuscles being made to "jump" from one atom to another. This appeared to him to be a rather "new point" in physics that required further explanation to differentiate the basis of this new theory from that of older—implicitly Maxwellian—accounts of conduction.

Before Thomson could reply, the famous post office "practician" Sir William Preece[115] avowed an uncharacteristic deference to university learning in proposing his vote of thanks to the recent Nobel prizewinner. Yet while he also dryly hinted that Thomson had not attained the last word on the subject—at least not until he had answered the long-debated question of whether electricity was a form of energy, a form of matter, or "something sui generis." Although Thomson's reply to Thompson was confidently idiosyncratic on the terminological question, it is significant that the Cambridge professor was somewhat cagier about the ontological questions. Thomson defended his somewhat individualistic terminology, declaring:

> I prefer the corpuscle for two reasons: first of all it is my own child, and I have some kind of parental affection for it; and secondly I think it has one merit which the electron has not. We talk about positive and negative electrons . . . [but] from my point of view the difference between the negative and positive is essential, and much greater than I think would be suggested by the term positive electron and negative electron. Therefore I prefer to use a special term for the negative unit and call it a corpuscle. A corpuscle is just a negative electron.[116]

After this unusually explicit clarification from the patriarch of the corpuscle, the Finsbury Thompson asked "What do you call a positive electron," to which Thomson helpfully replied "I should call it a positive electron." The Cambridge professor liked the term electron "well enough" so long as usage of it was not liable to "run the positive and negative" into an "equality."[117] He then appeared to retreat further, however, in response to Thompson's yet more searching metaphysical enquiries:

> Professor Thompson went into some questions which, if I could answer, I should be very near solving the problem of the universe—the relation between electricity and matter, and whether a corpuscle was a bit of electricity or a bit of matter with a charge on it. I do not know what electricity is, and I do not know what matter is. . . . I think I should like to ask those people who talk about electricity and matter to try to think for themselves what they mean by matter and what they mean by electricity. If they do so they will not find it so easy to define the terms they mean.

After further evasions on the question of corpuscles' movements between atoms, Thomson ended his reply and returned to Cambridge to await the publication of his *Corpuscular Theory of Matter*.[118] This rather more confident title indicates, perhaps, that his anxieties about the nature of electricity and "matter" were rather contextually specific to the interrogative context of the Institution of Electrical Engineers. Certainly when commissioned to write some articles for the eleventh edition of the *Encyclopaedia Britannica* (1910–11), Thomson did not allow his own self-disqualification to prevent him from writing the article on "matter" based on his own "corpuscular" theory. He did make a significant concession to Thompson, however, in suggesting for the first time that the particles with the "smallest mass known to science" were called either "corpuscles" or "electrons."[119]

When Ambrose Fleming wrote the article on "electricity" for the same edition of the *Encyclopaedia Britannica* he referred to the "electronic theory of matter" despite drawing heavily upon Thomson's own corpuscle researches. Quite how readers of the *Encyclopaedia* dealt with this discrepancy in terminology is not at all clear.[120] Interestingly, however, Fleming concluded his account with a summary of views on the nature of electricity and matter at the beginning of the twentieth century by drawing together the somewhat heterogeneous contexts in which the "electron" had been deployed:

> the term electricity had come to be regarded, in part at least, as a collective name for electrons, which in turn must be considered as constituents of the chemical atom, furthermore as centres of certain lines of self-locked and permanent strain existing in the universal aether or electromagnetic medium.[121]

Having documented the many ways in which electrons now entered into explanations of electric current, electromotive force, and electric charge, he did not, however, privilege either Thomson or 1897 in his account of where matters stood. It was he said in "the hands" of Lorentz,

Drude, Thomson, Larmor, and "many others" that the "electronic hypothesis" of matter and electricity had been developed in great detail and represented the "outcome" of recent research researches on electrical phenomena. Strikingly, neither Fleming's nor Thomson's articles for the *Encyclopaedia Britannia* refer to anything like the "discovery" of the electron—such stories were only to come in the 1920s and 1930s with a collective postwar amnesia about the linguistic, theoretical, and experimental complexities of the electron theory's genesis.

EPILOGUE: THE INVENTION OF DISCOVERY STORIES FOR THE "ELECTRON"

By 1923 Thomson, now retired from the Cavendish Laboratory, was using different language: lecturing on "The electron in chemistry" at the Franklin Institute in Philadelphia, he spoke unequivocally of 1897 as the crucial year in which the "discovery of the electron" led to the first understanding of the structure of matter.[122] Notably, however, Thomson did not specify *who* exactly had made the discovery. In the following year, Robert Millikan wrote with much reverence on Thomson's work in the second edition of his *The Electron*, but not even he credited Thomson with the specific discovery of any new subatomic entity. According to Millikan's account, Thomson was responsible for creating a method of determining the mass-to-charge ratio of cathode rays more reliable than Schuster had used in 1890 and with the finding that e/m was the same for all residual gases in the discharge tube.[123] Yet Millikan otherwise appears to have no truck with simplistic discovery narratives focusing on merely one individual. And in this regard it is significant that Millikan was never directly a student of Thomson's nor had great deference to the Cambridge sphere of influence. Falconer and Davies have suggested that it was J. J. Thomson's own Cavendish students who were chiefly responsible for constructing and promulgating the narrowly specific claim that Thomson had been the sole discoverer of the electron in 1897.[124] While this is undoubtedly the case, little research has been done to prove exactly which students and through which media the story that J. J. Thomson achieved a heroic solo "discovery" in 1897 came to be canonical in textbooks of both science and the history of science.

The creation of this story appears to have been closely linked to the commonplace simplifications of pedagogy as well as interwar campaigns to create autonomous funding for "pure" research. Thus, for example, the first article encountered by any aspiring technologist reading Molloy's manual on *Practical Electrical Engineering* in 1931 would have been a piece entitled "How Research has Helped Electrical Engineering." Written by the eminent Cavendish graduate Sir Richard Tetley Glazebrook, this piece narrated the development of electrical technology with a palpable attempt to contrive a priority for "pure" research, as undertaken by "disinterested seeker[s] after the truth." After delving back into the distant past to show "those fundamental researches" on which the engineer's work "must be based," Glazebrook rounded off with an account of the "discovery of the electron" by J. J. Thomson, in 1897. Glazebrook concluded in a somewhat monocausal vein that the alleged "consequences of this discovery," such as the wireless valve, would need a whole volume by themselves to be documented properly. Yet rather than elaborate evidentially on this point, he contended that the relationship between the electron and the wireless valve was yet "another instance" of the way in which the selfless "desire" of physics researchers to "advance" natural knowledge had proved of "inestimable" value to the engineer.[125]

Yet such partisan and ideologically loaded accounts were by no means consensual. Certainly any account of the development of the Thomson "discovery" story would have to explain the long persistence of tales even by his own former students which do not single out his 1897 work as that of a solo "discovery." Such is the case with the *History of Science in its Relations with Philosophy and Religion* published in 1929 by one of Thomson's former Cavendish students, William Cecil Dampier Dampier-Whetham.[126] Despite the fact that, as fellow and senior tutor at Trinity College, Cambridge, Dampier-Whetham was biographically and geographically close to Thomson, he chose not to center his story of "the new era in physics" upon the cathode ray researches undertaken by the master of his college thirty years previously. Rather, his account of the "great discovery" of "ultra-atomic corpuscles" in 1897 began with Hittorf's study of cathode rays in 1869 and culminated in the combined work of Thomson, Wiechert, and Kauffman in equal measure, with Kauffman achieving the first significantly "accurate" value of e/m in December 1897.

Whereas Dampier-Whetham's study was a broad-ranging and inclusive history of science that took pains to incorporate the work of Thomson's contemporaries, the biographical genre of later years did not. The contrast between Dampier-Whetham's version and Thomson's own self-laudatory autobiographical reminiscence of 1936 (backed by that of his son G. P. Thomson in 1962) could not be more striking. Certainly from the analysis here, it should be obvious that the context for the creation of such heroic historiography must have been quite remote from that which generated comparable stories in the first three decades after the alleged "annus electronicus" of

1897. Following the suggestion of Davies and Falconer, one might infer that the instantiation of a "discovery" story centering exclusively upon J. J. Thomson's work in 1897 was probably the retrospective creation of his erstwhile Cavendish students in seeking to establish an institutional genealogy for their research on subatomic particles. Certainly an important insight into the nationalistic and pedagogical origin of such "discovery" stories can be gained by a brief international comparison. Nersessian and Arabatzis note that physicists at the University of Leiden claim (equally problematically) that it was their Dutch mentor and hero Hendrik Lorentz who should be recognized as the discoverer of the electron.[127]

If one is to pursue the arguments concerning the genesis of "discovery" stories that have been put forward by Brannigan and Schaffer,[128] it is likely that the only way to understand the creation of the Thomson "discovery" story is to look at the particular politics of the Cavendish laboratory and its diaspora in the 1920s and 1930s to see what important institutional and disciplinary purposes would have been served by the cultivation of such a myth.[129]

Conclusion: A New Uncertainty Principle

In this chapter I have argued that the reception of J. J. Thomson's researches into the corpuscle was a long-drawn out and complex affair that fits no discretized chronology of "discovery." While Thomson's own students were most likely responsible for creating later myths of such a singular discovery in 1897, it was the communities and journals concerned with electrical technology that were the more important vehicles for reporting upon and molding contemporary responses to Thomson's cathode ray researches. Over at least a ten-year period they worked hard to assimilate the many different strands of research from fields as diverse as spectroscopy, cathode ray research, x-rays, chemistry, and radioactivity into a coherent body of "electron theory"—a theory to which Thomson himself by no means unequivocally subscribed until after World War I. By contrast, the work of electrical technologists on new forms of hardware, such as the cathode ray oscilloscope and the thermionic valve, was undertaken relatively independently, despite the fact that those involved, such as J. A. Fleming, were fully aware of the contemporary electron and corpuscle debate. Thus the more closely one studies the diverse ways in which the "electron" or "corpuscle" was deployed in physics and electrical engineering between 1894 and 1911, the harder it becomes to agree with J. J. Thomson's rather overprecise retrospective claims that he made decisive experimental contributions in either 1897 or 1899.

NOTES

1. J. J. Thomson, *Recollections and Reflections* (London: G. Bell & Sons, 1936), 341.

2. [Editorial], "The British Association at Dover," *The Electrician* 43 (1899): 772–773.

3. [Editorial], "The Theory of Electrons," *The Electrician* 45 (1900): 818–819, quote on 818.

4. J. J. Thomson, in discussion of his "The Modern Theory of Electrical Conductivity of Metals," *Journal of the Institution of Electrical Engineers* 38 (1906–07): 455–468, quote on 467.

5. See E. A. Davis and I. Falconer, *J. J. Thomson and the Discovery of the Electron* (London: Taylor & Francis, 1997), 134.

6. S. Hong, "Forging Scientific Electrical Engineering: John Ambrose Fleming and the Ferranti effect," *Isis* 86 (1995): 30–51.

7. Thomson, *Recollections and Reflections,* 332–341; see also his son's biography: G. P. Thomson, *J. J. Thomson and the Cavendish Laboratory in His Day* (London: Nelson: 1964), 45–56.

8. Examples of popular hagiographic literature include N. Eldredge, R. C. Gallo, R. Hoffman, et al., *Scientific American: Triumph of Discovery* (London: Helicon Publishing, 1995), 75. Similar claims arise in populist histories of television, for example Ian Sinclair, *The Birth of the Box* (Wilmslow, Cheshire: Sigma, 1995), 35.

9. I. Falconer, "Corpuscles, Electrons and Cathode Rays: J. J. Thomson and the 'Discovery of the Electron.'" *British Journal for the History of Science* 20 (1987): 241–276; T. Arabtzis, "Rethinking the 'Discovery' of the Electron," *Studies in the History and Philosophy of Modern Physics* 27 (1997): 405–435.

10. [Anon], "Literature: Magnetism and Electricity," *The Electrician* 9 (1882): 184–185, quote on 185.

11. J. Ambrose Fleming, "Electricity and Its Manifestations," in E. Molloy (ed.), *Practical Electrical Engineering,* revised ed. (1931–32), vol. 1, 107–111, quote on 107.

12. J. C. Maxwell, *Treatise on Electricity and Magnetism,* vol. 1 (1873), third edition (posthumous), reprinted (New York: Dover, 1954), especially 38–43.

13. J. J. Thomson, "Report on Electrical Theories," *Report of the BAAS,* 1885, part 1: 97–155.

14. J. A. Fleming, "The Electronic Theory of Electricity," *Popular Science Monthly* 61 (1902): 5–23, esp. 8–9.

15. Maxwell, *A Treatise on Electricity and Magnetism,* p. 60; O. Lodge, *Electrons, Or the Nature and Properties of Negative Electricity* (London: G. Bell & Sons, 1906), xiii.

16. See *Nature* 20 (1879): 419–423, 436–440, and *Chemical News* 40 (1879): 91–93, 104–107, 127–131.

17. W. Crookes, "Electricity in Transitu: From Plenum to Vacuum," *Journal of the Institution of Electrical Engineers* 20 (1892): 4–49; reprinted in *The Electrician* 26 (1891): 323–327, 354–360, 389–392.

18. "In the Van of Progress," *The Electrician* 26 (1891): 328–330, quote on 329.

19. See other chapters in this volume.

20. G. Jaumann, "Electrostatic Deflection of Cathode Rays," *The Electrician* 37 (1896): 343–345. For Hertz see J. Z. Buchwald, *The Creation of Scientific Effects: Heinrich Hertz and Electric Waves* (Chicago: University of Chicago Press, 1994), and J. Z. Buchwald, "Why Hertz Was Right about Cathode Rays," in J. Z. Buchwald, *Scientific Practice: Theories and Stories of Doing Physics* (Chicago: University of Chicago Press, 1995), 151–169.

21. J. A. Fleming, "An Experiment Showing the Deflection of Cathode Rays by a Magnetic Field," *The Electrician* 38 (1896–97): 302.

22. E. Thomson, "Some Notes on Röntgen Rays," *The Electrician* 38 (1896–97): 302–303; for more on Elihu Thomson see W. B. Carlson, *Innovation as a Social Process: Elihu Thomson and the Rise of General Electric, 1870–1900* (Cambridge: Cambridge University Press), 1991, 311–331 on x-ray equipment.

23. J. G. O'Hara, "George Johnstone Stoney, F.R.S., and the Concept of the Electron," *Notes and Records of the Royal Society of London* 29 (1975): 265–276, G. J. Stoney, "On the Cause of Double Lines and of Equidistant Satellites in the Spectra of Gases," *Scientific Transactions of the Royal Dublin Society* 4 (1891): 563–607.

24. See Bruce Hunt, *The Maxwellians* (Ithaca: Cornell University Press, 1991), 217–222.

25. Ibid., 222, and C. Jungnickel and R. McCormmach (eds.), *Intellectual Mastery of Nature: Theoretical Physics from Einstein to Ohm* (Chicago: University of Chicago Press, 1986), vol. 2, 231–236.

26. For Zeeman's account of things see P. Zeeman, *Researches in Magneto-Optics with Special Reference to the Magnetic Resolution of Specturm Lines* (London: Macmillan, 1913).

27. See summary report of Zeeman's paper (delivered by Lorentz) at the Royal Academy of Sciences Amsterdam on October 31, 1896, and published on December 24 in *Nature* 55 (1896): 192.

28. P. Zeeman, "On the Influence of Magnetism on the Nature of the Light emitted by a Substance," *Philosophical Magazine* 5th series, 43 (1897): 226–239. Lodge was then Professor of Physics and expert on wireless telegraphy at University College, Liverpool. For further discussion on Zeeman see Theodore Arabatzis, "The Discovery of the Zeeman Effect: A Case-Study of the Interplay between Theory and Experiment," *Studies in the History and Philosophy of Science* 23 (1992): 365–388. Arabatzis notes somewhat ironically that Lodge's 1906 book *Electrons* later claimed priority for Thomson's 1897 identification of a mass/charge ratio, a claim to which Zeeman objected vigorously; see the letter from Zeeman to Lodge of 3 August 1907, reproduced by Arabtzis, 423.

29. O. Lodge, "The Latest Discovery in Physics," *The Electrician* 38 (1897): 568–570.

30. Ibid., 569.

31. For the use of this term, see J. Z. Buchwald, *From Maxwell to Microphysics* (Chicago: University of Chicago Press, 1985), and Hunt, *The Maxwellians*.

32. O. Lodge, "A Few Notes on Zeeman's Discovery," *The Electrician* 38 (1897): 643.

33. For Thomson's research on the electromagnetic nature of mass and the theoretical properties of vortices see Davis and Falconer, *J. J. Thomson and the Discovery of the Electron:* 11–48.

34. Falconer, "Corpuscles, Electrons and Cathode Rays," 255–259; Davis and Falconer, *J. J. Thomson and the Discovery of the Electron,* 111–192.

35. For accusations of one-sidedness see "Notes," *The Electrician* 37 (1896): 685; an interesting commentary on Thomson's own address can be found on 653.

36. J. H. P[oynting], "Prof. J. J. Thomson," *The Electrician* 37 (1896): 657–658, engraved portrait of Thomson opposite 672. Poynting almost certainly knew Thomson from student days at Owens College, Manchester, in the early 1870s.

37. Isobel Falconer has kindly shown me a copy of the letter from the editor of *The Electrician,* W. G. Bond, to J. J. Thomson dated October 10, 1896; the original is in the possession of the Thomson family.

38. J. Ambrose Fleming, *The Alternating Current Transformer,* 2 vol., London: Electrician Company, 1889–1892; J. A. Ewing, *Magnetic Induction in Iron and Other Metals,* London: Electrician Company, 1892; O. Heaviside, *Electromagnetic Theory,* 3 vol., London: Electrician Company, 1893–1912.

39. J. J. Thomson, "Cathode Rays," *The Electrician* 39 (1897): 103–109.

40. It was, of course, published in the *Proceedings of the Royal Institution* for 1897, 1–14, reproduced in Davis and Falconer on 139–152. Thomson's lecture was advertised in both *The Electrical Review,* "Royal Institution," 40 (1897): 501, and *Chemical News,* "Royal Institution," 74 (1897): 180, but neither journal reported on the lecture afterward. See below for further discussion of this point.

41. Falconer, "Corpuscles, Electrons and Cathode Rays."

42. T. Arabatzis, "Rethinking the 'Discovery' of the Electron," *Studies in the History and Philosophy of Modern Physics* 20 (1987): 241–276.

43. S. M. Feffer, "Arthur Schuster, J. J. Thomson and the Discovery of the Electron," *Historical Studies in the Physical and Biological Sciences,* 20 (1989): 33–61.

44. Thomson, "Cathode Rays."

45. Thomson, *Recollections and Reflections,* 341.

46. See G. Gooday, "The Morals of Energy Metering: Constructing and Deconstructing the Precision of the Electrical Engineers' Ammeter," in M. N. Wise (ed.), *The Values of Precision* (Princeton: Princeton University Press, 1995), 239–282.

47. K. Olesko, "Precision, Tolerance and Consensus: Local Cultures in German and British Resistance Standards," in J. Z. Buchwald (ed.), *Archimedes: Scientific Credibility and Technical Standards* (Dordrecht: Kluwer, 1996), 117–156.

48. See Gooday, "The Morals of Energy Metering."

49. See G. Gooday, "The Genesis of Physics Teaching Laboratories in Victorian Britain," *British Journal for the History of Science* 23 (1990): 25–51. For further comments on Thomson's indifference to precisionist research, see Isobel Falconer, "Theory and Experiment in J. J. Thomson's Work on Gaseous Discharge," unpublished PhD thesis, University of Bath, 1985. , especially 73–102.

50. J. J. Thomson, "Address to Section A," *Report of the British Association for the Advancement of Science,* 1896, part 2, 699–706; reprinted in *The Electrician* 37 (1896): 672–675, quote from 673.

51. A. C. C. Swinton, "Some Experiments with Cathode Rays," *Chemical News* 75 (1897): 218.

52. S. P. Thompson, "Cathode Rays and Some Analogous Rays," *Chemical News* 76 (1897): 4–5.

53. See, for example, "Cambridge: Philosophical Society," *Nature* 55 (1897): 453; "Cambridge: Philosophical Society," *Nature* 57 (1897): 119.

54. G. F. FitzGerald, "The Dissociation of Atoms," *The Electrician* 39 (1897): 103–104. Note that the notion of a "free electron" was entirely at odds with Stoney's notion of an atom-bound electron.

55. Ibid., 103; for details of Larmor's electron theory see Hunt, *The Maxwellians,* 210–239.

56. E. Thomson, "A Speculation Regarding the Cause of Röntgen Rays," *The Electrician* 39 (1897): 317–318, also published in *The Electrical Review* 41 (1897): 64–65.

57. "Notes," *The Electrician* 39 (1897): 299.

58. Ibid. Contrast the claim made by William Crookes in the conclusion to his 1891 presidential address to the IEE on cathode ray research that "Science has emerged from its childish days. It has shed many delusions and impostures. It has discarded magic, alchemy and astrology": "Electricity in Transitu: From Plenum to Vacuum," *The Electrician* 26 (1891): 392.

59. "Notes," *The Electrician* 39 (1897): 299.

60. Thomson, "Cathode Rays"; see Falconer, "Theory and Experiment in J. J. Thomson's Work on Gaseous Discharge."

61. Rayleigh, 4th Lord, *The Life of Sir J. J. Thomson* (Cambridge: Cambridge University Press, 1942), 91.

62. Thomson, *Recollections and Reflections,* 341.

63. "Physics at the British Association," *Nature* 60 (1899): 585–587, quote on 585.

64. Ibid., 586.

65. Ibid.

66. See Feffer, "Arthur Schuster, J. J. Thomson and the Discovery of the Electron," for more discussion of Schuster's work.

67. Lodge, "A Few Notes on Zeeman's Discovery," 643.

68. "Physics at the British Association."

69. Reported in Rayleigh, *The Life of Sir J. J. Thomson,* 113–114; eighteen months later in a letter from Montreal, Ernest Rutherford intimated to Thomson that chemists at McGill University were also opposed to Thomson's "corpuscular theory"; see the letter of March 26, 1901, reproduced in A. S. Eve, *Rutherford* (Cambridge: Cambridge University Press, 1939), 77.

70. [Editorial], "The British Association at Dover," *The Electrician* 43 (1899): 772–773.

71. J. A. Fleming, "The Centenary of the Electric Current, 1799–1899," *The Electrician* 43 (1899): 764–768.

72. Ibid., 768. For further discussion of Fleming's lecture see S. Hong, "Styles and Credit in Early Radio Engineering: Fleming and Marconi on the First Transatlantic Wireless Telegraphy," *Annals of Science* 53 (1996): 431–465.

73. "Notes," *The Electrician* 45 (1900): 655–656.

74. P. Drude, "Zur Elektronen Theorie der Metalle," *Annalen der Phys. Chem.* 1 (1900): 566–613 and 3 (1900–01), 369–402. For further discussion see Buchwald, *From Maxwell to Microphysics,* 215–232.

75. J. Larmor, *Aether and Matter* (Cambridge: Cambridge University Press, 1900).

76. See Fournier D'Albe, *The Electron Theory,* 1906.

77. In 1895, J. H. Poynting had published in *The Electrician* a series of articles entitled "Molecular Electricity," 35 (1895): 644–647, 668–671, 708–712, 741–743; reproduced in J. H. Poynting, *Collected Scientific Papers* (Cambridge: Cambridge University Press, 1920), 269–298. Poynting was the only British physicist I have been able to locate who adopted Thomson's language of corpuscles; see his 1902 piece for *The Inquirer,* "Molecules, Atoms and Corpuscles," reproduced in *Collected Scientific Papers,* 664–672.

78. "Notes" *The Electrician* 45 (1900): 656.

79. See "The British Association Meeting in Bradford," *The Electrician* 45 (1900): 782–784.

80. [Editorial,] "The Theory of Electrons," *The Electrician* 45 (1900): 818–819, quote on 818.

81. J. A. Fleming, "Electric Waves and Electrical Oscillations," reproduced in *The Electrician* 46 (1901): 514–516, 551–553, 588–591.

82. J. J. Thomson, "Bodies Smaller than Atoms," *Popular Science Monthly* 59 (1901): 323–335.

83. Ibid., 324–326.

84. For Franklin and the debate over his theories see J. Heilbron, *Elements of Early Modern Physics* (Berkeley: University of California Press, 1982), 187–240.

85. J. A. Fleming, "The Electronic Theory of Electricity," *Popular Science Monthly* 61 (1902): 5–23.

86. Ibid., 22; compare his 1899 BAAS speech at Dover.

87. See S. Hong, "Controversy over Voltaic Contact Phenomena, 1862–1900," *Archive for History of Exact Sciences* 47 (1994): 233–289.

88. "Notes," *The Electrician* 49 (1902): 250–251. Fleming had been associated with this topic ever since he addressed it in giving the first paper to the Physical Society of London in 1874; see Hong "Controversy over Voltaic Contact Phenomena."

89. Hong, "Styles and Credit in Early Radio Engineering."

90. S. Hong, "From Effect to Artifact (II): The Case of the Thermionic Vvalve," forthcoming.

91. J. A. Fleming, "Hertzian Wave Telegraphy, *Journal of the Society of Arts* 60 (1902–03): 713ff.

92. On Guthrie's work on thermoelectric emission see G. Gooday, "Teaching Telegraphy and Electrotechnics in the Physics Laboratory," *History of Technology* 13 (1991): 73–114, and J. T. McGregor-Morris, *The Inventor of the Valve: A Biography of Sir Ambrose Fleming* (London: The Television Society, 1954), 65–68.

93. See J. A. Fleming, *Memories of a Scientific Life,* London, 1934, 134–138.

94. See Hong, "From Effect to Artifact (II)."

95. J. A. Fleming, "n the Conversion of Electric Oscillations into Continuous Currents by Means of a Vacuum Valve, *Proceedings of the Royal Society of London* 74 (1905): 476–487.

96. P. Dunsheath, *A History of Electrical Engineering* (London: Faber & Faber, 1962), 266.

97. G. P. Thomson contended that the "much-used cathode ray oscillograph" was "in essence" his father's 1897 e/m apparatus: G. P. Thomson, *J. J. Thomson and the Cavendish Laboratory in His Day,* 166.

98. K. F. Braun, *Wiedemann's Annalen* 60 (1897): 552; also Dunsheath, *A History of Electrical Engineering,* 283–284.

99. K. F. Braun and Zenneck, *Annalen der Physik* 9 (1902): 497.

100. W. M. Varley, *Philosophical Magazine* 6th series, 3 (1902): 497.

101. See Dunsheath, *A History of Electrical Engineering,* 283–285, and Gooday, "The Premisses of Premises," in C. W. Smith and J. Agar (eds.), *Making Space for Science* (Basingstoke: Macmillan, 1998).

102. W. M. Varley and W. H. F. Murdoch, "Some Applications of the Braun Cathode-Ray Tube," *The Electrician* (1905): 335–336; see also the letter by them published on June 30, 430.

103. M. Nelkon and P. Parker, *Advanced Level Physics,* 6th edition (London: Heinemann Educational, 1987).

104. For more on D'Albe see A. Warwick, "The Electrodynamics of Moving Bodies and the Principle of Relativity in British Physics, 1894–1919," PhD dissertation, University of Cambridge, 1989.

105. Fournier D'Albe, *The Electron Theory,* first edition 1906, second edition 1907.

106. S. Simon's research on the "Charge and Mass of Cathode Particles" as summarized in *The Electrician* (December 1899): 189.

107. D'Albe, *The Electron Theory,* 4.

108. O. Lodge, "On Electrons," *Journal of the Institution of Electrical Engineers* 32 (1902–03): 45–116. In his concluding remarks the president commented "Electrons are hypothetical entities which help us think straight," 115.

109. "Notes," *The Electrician* 50 (1902): 253–254.

110. See Isobel Falconer, "From Corpuscles to Electrons," in this volume, and Hunt, *The Maxwellians,* 231.

111. Lodge, *Electrons.*

112. Thomson's Nobel Prize was awarded in 1906 for his work on discharge in gases rather than for any single specific "discovery."

113. J. J. Thomson, "The Modern Theory of Electrical Conductivity of Metals," *Journal of the Institution of Electrical Engineers* 38 (1906–07): 455–468, discussion on 466–468.

114. See S. P. Thompson's contribution to the discussion: ibid., 466–467.

115. For Preece see Hunt, *The Maxwellians.*

116. J. J. Thomson, "The Modern Theory," discussion at the IEE, 467–468.

117. The use of the word "electron" for both the negative particle and the as yet unobserved positive particle long proved contentious. In 1924 Robert Millikan contended that "most authoritative writers"—namely Thomson, Rutherford, Campbell, Richardson—had retained the "original significance of the word 'electron' instead of using it to denote *solely* the free negative electron." That is, he and they considered George Johnstone Stoney's coinage of the term "electron" in 1891 not only to have no reference to

physical inertia, but specifically to refer to both negative and positive varieties. Millikan's support for this view was that it followed the manner in which the word "man" served "admirably to designate the genus 'homo'" as well as to "denote the male representative of that genus," the female form being on this account "differentiated by the use of a prefix." Robert Millikan, *The Electron,* 2nd ed. (Chicago: Chicago University Press, 1924), 26–27. For Stoney's original position see J. G. O'Hara, "George Johnstone Stoney, F. R. S., and the Concept of the Electron," *Notes and Records of the Royal Society of London* 29 (1975): 265–276.

118. J. J. Thomson, *The Corpuscular Theory of Matter* (London: Archibald Constable & Co., 1907).

119. J. J. Thomson, "Matter," *Encylopaedia Britannica* (1910–11), 11th ed., 891–895, quote from 892.

120. J. A. Fleming, "Electricity," *Encylopaedia Britannica* (1910–11), 11th ed., 191–193.

121. Ibid., 192.

122. J. J. Thomson, *The Electron in Chemistry* (London: Chapman & Hall, 1923), 2.

123. Millikan, *The Electron,* 42–43.

124. Davis and Falconer, *J. J. Thomson and the Discovery of the Electron,* 134.

125. R. T. Glazebrook, "How Research Has Helped Electrical Engineering," in E. Molloy (ed.), *Practical Electrical Engineering,* vol. 1, revised ed., 1931, 3–7, quotes from 7.

126. William Cecil Dampier Dampier-Whetham, *History of Science in Its Relations with Philosophy and Religion* (Cambridge: Cambridge University Press, 1929).

127. Arabatzis, 424, citing N. Nersessian, "Why Wasn't Lorentz Einstein?" *Centaurus* 29 (1986): 205–242, note 64, 209.

128. Augustine Brannigan, *The Social Basis of Scientific Discoveries* (Cambridge: Cambridge University Press, 1982); S. Schaffer, "Scientific Discoveries and the End of Natural Philosophy, *Social Studies of Science* 16 (1986): 387–420.

129. I thank David Cahan for pointing out that by the late 1920s most of Thomson's theoretical researches had been discredited by his successors and for suggesting that the now-retired Professor Thomson needed some such accolade as the "discoverer" of the electron to preserve his prestige in the "Rutherford" era.

4

Paul Villard, J. J. Thomson, and the Composition of Cathode Rays

Benoit Lelong

The experiments conducted by J. J. Thomson in 1897 on cathode rays are known now as the "discovery of the electron." Until recently, these experiments were considered to have definitively established the composition of the rays.[1] Falconer, Feffer, and Robotti have contested this interpretation, however, arguing that there were in fact in 1897 *several* concepts of elementary particles of electricity. They have shown that, at the time, the precise identification of the particle supposedly present in cathode rays was highly problematic. Thomson himself asserted that cathode rays were made of "corpuscles" in motion and he remained resolutely opposed to the identification of his "corpuscles" with the "electrons" introduced by the theoreticians of electromagnetism. To describe Thomson's experiments as a moment of "discovery" now appears mistaken, for it attributes a decisive character to these events that they certainly did not possess in 1897 nor in the immediately following years. The studies by Falconer, Feffer, and Robotti clearly indicate that, far from being established by a definitive and individual stroke of genius, the composition of cathode rays only became the object of a consensus amongst physicists after a long series of controversies.[2]

This calls for the study of *other* researches on the composition of cathode rays contemporary to Thomson's that produced different results. This chapter deals with the experiments of the physicist Paul Villard, who claimed in 1898 that the rays were made of charged particles of hydrogen. This was later considered to be an "error," and Villard's work disappeared from the history of science. The fact that Villard's work failed to raise any interest with historians implies that Thomson was right, Villard wrong, and thus that the latter's work does not deserve any scholarly attention. Yet Villard's position was hardly contested in France in his own time and he was a recognized authority on the subject for a decade. He was even elected to the Académie des Sciences in 1908.[3]

This chapter is divided into three parts. The first examines Villard's work from his first researches in 1888 on gas hydrates to his work on the chemical effects of cathode rays. Next, the approaches adopted by Villard and

Thomson in their work on cathode rays are contrasted and the differences in their experimental aims and methods are analyzed. Villard's and Thomson's intellectual, material, and professional worlds are then reconstructed; these are shown to be both deeply incompatible with the new electronic physics that later prevailed. The final part deals with the period following 1898, after the publication of Villard's and Thomson's researches. The composition of cathode rays gave rise to much discussion among physicists in different places. Both scientists' results were carefully weighed, commented, and discussed, with Villard and Thomson themselves taking an active part in these debates. It will be seen that there was no open confrontation between their respective views; in the end they were both replaced by a third interpretation, according to which "cathode rays are electrons." It was only after the closure of this process that their results were retrospectively reinterpreted: Thomson's as "the discovery of the electron" and Villard's as an "experimental error." This rereading took place in a practical and conceptual context utterly alien to both Thomson and Villard, and the way in which they had actually worked was forgotten.

This chapter seeks firstly to reestablish a *symmetry* between Thomson's and Villard's work. Each scientist's work was known and valued for a certain time by a particular community of physicists, before being transformed or forgotten. From a historical perspective, these are sufficient reasons for giving equal amounts of attention to both scientists. This temporary symmetry led to a radical *asymmetry*, however, one of the outcomes becoming a "discovery" and thus acceding to permanent notoriety, while the other became an "error" and fell into oblivion. The second objective of this chapter is to account for the progressive constitution of this asymmetry.

Paul Villard and the Chemical Analysis of Substances

Paul Villard was born in Lyon on 28 September 1860. He was admitted to the Ecole Normale Supérieure in 1881 and obtained his physics *agrégation* in 1884. He then became a professor in several provincial *lycées,* before finally settling down in Montpellier. He applied for and obtained, thanks to Berthelot's support, a lectureship in physics at the faculty of sciences in this city. Villard soon stopped teaching, however, his wealth being sufficiently high to dispense him from seeking another post in public education.[4] He was later admitted as a *travailleur libre,* (i.e., financially independent) to the Ecole Normale Supérieure chemistry laboratory directed by Henri Jules Debray. Debray's group was, with François Raoult's in Grenoble, one of the rare research teams in France to specialize in physical chemistry.[5] Villard undertook with Forcrand researches on gas hydrates.[6] He continued working on the topic on his own from June 1888, publishing nine articles on the physico-chemical properties of gas hydrates between 1888 and 1896.

It was already known in 1888 that certain gases could combine with water to form gas hydrates. It had only been possible until then to form them from easily liquefied and water-soluble gases, such as chlorine or carbonic acid. On 4 June 1888 Villard announced that he had succeeded in combining water with methane, ethane, ethylene, acetylene, and nitrous oxide, but that he had failed with nitrogen, oxygen, and a dozen other gases. At the end of his paper, he proposed to "determine precisely the main properties of these hydrates and, if possible, to establish their composition."[7] On 19 March 1894 Villard published the composition of the nitrous oxide hydrate.[8] He had decomposed the crystals completely by heating them, and he measured the quantities of water and nitrous oxide liberated. He then calculated the proportion in which they had previously combined. He concluded: "It is thus likely that the exact formula would be N^2O, $6 H^2O$." Villard published on 6 August 1894 the composition of a second hydrate, carbonic hydrate.[9] His experiments showed that "the exact composition seems to be represented by the following formula: CO^2 $6 H^2O$, which is similar to that of nitrous oxide hydrate N^2O, $6 H^2O$."[10]

At the end of the nineteenth century the atomic notation had by no means been unanimously accepted by French chemists.[11] Was Villard an atomist? The majority of the French chemists who promoted the atomic notation claimed that it was a conventional representation, no more than a useful fiction.[12] Our sources give no indication as to Villard's beliefs on the existence of atoms. But they do indicate how he *used* atoms practically in his scientific work and in his publications. The terms "atom," "molecule," and "ion" are completely *absent* from Villard's publications between 1888 and 1896. This contrasts strongly with the work done in Wurtz's laboratory by Boutlerov, Couper, and Wurtz. In the 1880s and 1890s they were busy building three-dimensional representations of organic molecules; they were seeking to understand the geometrical and spatial organization of the atoms constituting these molecules and the nature of the bonds linking them (length, solidity). They studied how these atoms were exchanged or replaced each other during a chemical reaction.[13] Villard's work differed widely from this. For him, producing knowledge about a gas hydrate did not mean trying to determine the atomic structure of its molecule. It meant simply measuring its dissociating tension, the heat necessary for its formation, its densities in the liquid and gaseous states, or studying the effect of its crystals on polarized light. Villard had taken the habit of writing the formula of water

H^2O, nitrous oxide N^2O, and its hydrate N^2O, 6 H^2O, which signified that water and nitrous oxide combined in a proportion of 6 to 1. For Villard, determining the composition of a gas hydrate was thus no more than producing such a formula.

Villard cast aside physical chemistry at the end of 1897 and turned to the study of rays. A note on the effect of x-rays on photographic plates appeared in the *Comptes Rendus* on 26 July 1897.[14] A year later he started publishing regularly on the topic of cathode rays. The first paper appeared on 9 May 1898. Villard asked himself where the matter that constituted the rays originated from. He stated that the base of the beam (the part closest to the cathode) was repelled by a positive charge but found this surprising as the rays were made of negatively charged particles. In addition, he found that the shape of the Crookes tubes used had an influence on the shape of the beam: the positively charged inside surfaces of the tubes tended to repel the rays. Villard concluded: "These results can be easily explained if one admits that the cathodic emission is sustained by an influx of positively charged matter coming from various parts of the tube." Villard meant by "cathodic emission" everything that was emitted by the cathode, which included not only the cathode rays, but also the Goldstein rays (or Canalstrahlen).[15] He explained in a second article published a few weeks later, on 31 May 1898, that he had attempted to identify the matter constituting this "cathodic influx" and the matter making up the cathodic emission:

> I have shown in a previous note that cathode rays as well as Goldstein rays are produced from an influx of positively charged matter which arrives to the cathode at a considerable velocity. It is possible to determine the nature of the matter in motion by placing on the path of these various currents carefully chosen obstacles.[16]

These obstacles were in fact screens that Villard had coated with chemical reagents (lead sulfate, oxidized copper). These chemicals always reacted in the same way to cathode rays: they were *reduced*. This remained true even under different conditions. Villard wrote:

> Cathode rays, Goldstein rays and the cathodic influx thus seem to arise from a matter possessing the consistent ability to reduce certain metallic oxides, independently of its electrical state (which is neutral in the case of Goldstein rays). This remains true whether a void is produced in the oxygen, or whether one operates without any particular precautions, i.e., in very variable conditions. It thus appears that cathode rays cannot arise from any gas. Besides, J. J. Thomson's recent experiments show clearly that the properties of cathode rays are independent of the gas in which a void is

made. Now the only reducing gas which can be found in a Crookes tube without electrodes, washed and brought to a very high temperature, is of course hydrogen. It is hardly possible to rid the tubes completely of water by using dessicating agents, and glass can produce water nearly indefinitely. I must add that when I used a tube fitted with mercury electrodes in which a void was made on boiling mercury, only the Geissler phenomenon appeared and no cathode rays were formed. The chemical and physical properties of hydrogen already separate it from the other substances in the series of elements; it is thus not difficult to admit that it, and it alone, has the ability to produce cathode rays.[17]

According to Villard, cathode rays were therefore made of hydrogen, which itself came from the tube's surfaces. The particles of hydrogen bombarded the cathode, received there a negative charge, and were subsequently violently repelled by the cathode, forming cathode rays. Villard repeated this interpretation in an article published on 25 July 1898 in which he studied the diffusion of cathode rays:

> It is likely that when the electrified particles which constitute the cathode rays hit an obstacle, they partially diffuse in every direction, retaining to some extent their charge and kinetic energy. New rays result from this diffusion, whose direction is quasi-rectilinear, because the field is very weak in the region where they are formed. These are identical to the direct rays in all respects save their specific mode of emission. Like direct rays, the secondary rays represent the trajectories of particles of electrified hydrogen.[18]

Villard continued to publish regularly after July 1898 on cathode rays. He published a synthesis of his researches on the topic in 1899.[19] He also worked in 1899 and 1900 on x-rays, on radium rays, and on Crookes tubes. He found, among others, new rays emitted by radium, soon named "Villard rays" or "gamma rays."[20]

J. J. Thomson and Paul Villard: Two Different Ways of Doing Physics

J. J. Thomson had expressed an interest in cathode rays before 1897. He published measurements of the speed of propagation of the rays in 1894.[21] He started new experiments at the end of 1896. On 8 February 1897 his experiments on the detection of the electric charge and the measurement of magnetic deviation of the rays were presented to the Cambridge Philosophical Society. He showed that this magnetic deviation did not depend upon the gas contained in the tube. A second article appeared on 30 April in which J. J.

Thomson suggested for the first time that the rays were composed of particles approximately a thousand times smaller than the hydrogen atom. He measured the ratio e/m of these particles. Thomson wrote a third article on the topic in August 1897 which was published in October of the same year. In it, he summarized the different British and German proposals about the nature of cathode rays. For some, he wrote, cathode rays were waves, for others they were particles. Thomson went on to give a series of theoretical and experimental arguments in favor of the particulate view and against the undulatory interpretation. Thomson was thus working in a specific research context, characterized by a conflict between the "undulatory theory" and the "emission theory."[22]

Villard's researches were patently conducted in a different environment, in *another* research context. Villard's articles show that he was convinced of the particulate nature of cathode rays, but that this was for him a marginal, secondary aspect.[23] He was concerned instead with the determination of the *substance*(s) that composed these particles. For him these substances were to be found among those used by chemists, that is, in the periodic table of elements. Villard followed a procedure in these experiments similar to the one used in his previous researches on gas hydrates.

Villard presented his results in 1898 as a *continuation* of the work of those, like Thomson or Perrin, who wrote that particles constituted cathode rays. He believed he was following in the footsteps of these scientists when he showed that these particles were composed of hydrogen. In most of his publications between 1898 and 1905, Villard explicitly sided with the particulate point of view and against the undulatory interpretation.

But Thomson actively *fought* the idea that cathode particles might be made of hydrogen. He did so as early as 1897, before Villard even started his own researches on cathode rays. In his 1897 article, Thomson asserted that cathode particles were "corpuscles." He believed that the atom was not impossible to break up and that it was made of elementary particles, or "primordial atoms." He explicitly rejected the idea that the substance constituting these subatomic particles might be hydrogen:

> The explanation which seems to me to account in the most simple and straightforward manner for the facts is founded on a view of the constitution of the chemical elements which has been favourably entertained by many chemists : this view is that the atoms of the different chemical elements are different aggregations of atoms of the same kind. In the form in which this hypothesis was enunciated by Prout, the atoms of the different elements were hydrogen atoms ; in this precise form the hypothesis is not

tenable, *but if we substitute for hydrogen some unknown primordial substance X, there is nothing known which is inconsistent with this hypothesis* . . .[24]

Thomson introduced the term "corpuscle" after this clarification. He wrote that, in tubes containing rarefied gases, "the molecules of the gas are dissociated and are split up, not into the ordinary chemical atoms, but into these primordial atoms, which we shall for brevity call corpuscles."[25]

He added:

> Thus on this view we have in the cathode rays matter in a new state, a state in which the subdivision of matter is carried very much further than in the ordinary gaseous state: a state in which all matter—that is, matter derived from different sources such as hydrogen, oxygen, &c—is of one and the same kind; this matter being the substance from which all the chemical elements are built up.[26]

Ultimately, Thomson rejected the idea that his corpuscles might be made of hydrogen because this would have meant, for him, that the corpuscles were made of *atoms* of hydrogen. And Thomson believed that corpuscles were not atoms but elementary constituents of the atom. For Thomson corpuscles existed at a level of subdivision of the atom where hydrogen simply did not exist. The question of the nature of the substance composing the corpuscles could thus not subsist : according to Thomson, there were not *several* substances (among which that which composed corpuscles could be found) but *one* substance only. The difference between oxygen and nitrogen was less a difference in the substance than in the number and arrangement of the corpuscles constituting their atoms. "Substance" thus represented different things for Thomson and for Villard: for the former it was not part of the intellectual resources which could be used to describe matter at its most fundamental level.

This is revealing of another difference between Thomson and Villard on the status of chemistry compared to physics. For Thomson, chemistry was de jure if not de facto a simple consequence of the properties of matter that physics had the task of bringing to light. The prime interest in using corpuscles lay for him in their ability to explain the periodic table of elements. Thomson tried to explain in his 1897 paper how corpuscles associated to form atoms. The mathematical treatment of these arrangements was complex, but it was possible to draw an analogy with magnets floating at the surface of a liquid. These floating magnets spontaneously formed concentric circles; the more magnets were added, the more circles appeared. Thomson noticed that the increase in the number of circles was periodic, and that this period was also the period of the table of chemical elements. Falconer has written about Thomson's constant desire to use physical theories to account for chemical phenomena in general and the periodic table of elements in particular. This tendency can already be found in Thomson's first publication in 1882, where he proposed an atomic theory based on hydrodynamic analogies such as the whirlpool. This particular approach distinguished Thomson from the other theoreticians of elementary particles of electricity, such as Lorentz, FitzGerald, or Larmor. The latter attempted to explain electromagnetic rather than chemical phenomena.[27]

Thomson practiced what would now be termed "reductionist" physics, which sought to explain phenomena by reference to the microconstituents of matter. This had important consequences for the values that shape the work of the physicist. Thomson rated the value of a theory in terms of its explanatory power: the greater the scope the better the theory. Thomson himself always sought to build universal theories.[28] He also believed that the most interesting researches were those dealing with the ultimate constituents of matter. Thomson wrote with a patronizing tone about C. T. R. Wilson's research on the formation of clouds and rain. This was for him too far from what he called "transcendental physics."[29]

It would be vain to look for these implicit hierarchies in Villard's work. His publications on cathode rays were rather like an enumeration of the rays' properties: they were deviated by the tube's surfaces, they had no effect on each other, they were emitted at right angles to the cathode, they were reflected by a metallic plate, they were propagated in a straight line, they were refracted by a thin aluminum plate, and so forth. The fact that the rays were made of hydrogen was simply one of many properties; it was not meant to explain these either. This proposition thus did not possess a status more fundamental than the others. This can be seen in Villard's way of justifying his results and organizing his argument. The fact that cathode rays were made of hydrogen was presented by him as a *consequence* of experimental data: he never sought to support his results with entities that a theory based on them would explain in turn. Villard reiterated the typical argument of the French physicists against theories based on the atomic hypothesis. Such an explanation might appear "extremely attractive," he wrote, but there would always remain "certain experimental details which this explanation would be incapable of clarifying."[30] These theories, according to Villard, could not account for the complexity of experimental facts. He thus definitely preferred the production and description of experimental facts to such theorizing.

Another essential difference between the two men resided in their use of different experimental technologies. Thomson commonly employed electromagnets, condensors, electrometers, galvanometers, and cells—all of which only functioned in electric circuits. He was familiar with these instruments, and he had been using them on a daily basis for the previous decade. Villard, in contrast, was a newcomer to the field of electrical discharges. He did not possess Thomson's expertise: the only electrical instruments he used were those needed to work his Crookes tubes.

Thomson and Villard both experimented directly on cathode rays, studying the effect of different influences on them, but both scientists did not use the same kinds of devices. Thomson chose to manipulate and to deviate cathode rays using electrical and magnetic fields while Villard privileged mechanical obstacles. Villard made cathode rays bounce off screens, pass through metallic plates and diffuse against the edge of a plate. He also studied the role played by the shape of the surfaces and the electrodes on the form of the beam, the trajectories and on the point of origin of the rays.

Thomson's work centered on electrical charges. He constantly used quadrant electrometers to detect and measure electrical charges and produce numerical data. The experimental arrangements that he imagined and built were always designed to produce displacements or accumulations of electrical charges. The concepts of particle that he used and transformed were always derivations of the concept of *ion,* and were always electrically charged. Thomson's experimental work only rarely dealt with *electrically neutral* particles, whether atoms or molecules. Significantly, Galison and Assmus describe his work as "ion physics."[31] The mathematical models that Thomson used and formulated enabled him to calculate the movements of punctual charges in electromagnetic fields. The physical constants that he sought to measure experimentally were the electrical charges of elementary particles or the ratio of their charge to their mass. All these elements were used in combination: the corpuscle's e/m ratio, for instance, was derived from mathematical models that supplied relationships between the quantitative properties of the ion on the one hand, and the empirical data read on electrometer and galvanometer scales on the other hand.

In his article, Villard appeared convinced that cathode rays were electrified, but it was not a particularly interesting property to him. Cathode rays as well as Goldstein rays and more generally everything produced by the cathode (what he called "cathodic emission") was for Villard made of hydrogen. This hydrogen was negatively charged in the case of cathode rays, but it was in other electrical states (positive or neutral) in different parts of the tube. In addition, for him, the hydrogen composing cathode rays originated from the surfaces of the tube, which meant that it was initially positively charged.[32] Villard, in contrast to Thomson, thus believed that the electrical charge carried by cathode rays was not at all an intrinsic and unalterable property of the matter constituting them.

Villard's experimental technology focused less on electric charges than on chemical effects. His method of determining the chemical composition of cathode rays consisted in exposing them to substances of known composition to obtain characteristic reactions. He used for this purpose metallic plates coated with reagents such as lead sulfate. Sometimes the material of the plate itself was used as reagent (platinum, crystal, or oxidized copper). The darkening of platinum plates by the rays was interpreted by Villard as the result of a chemical reduction. When he presented his results to the Société de Physique, Villard circulated these darkened plates among the audience; this is revealing of the status and of the conclusive power that he attributed to these plates.[33] Two years later, Villard again used these typically chemical techniques of experimental investigation to study x-rays. X-rays were then known to have an effect on photographic plates. Villard studied the chemical changes that x-rays produced on platino-baryum cyanide salts and gelatino-silver bromide salts which coated photographic plates. Here again, the chemical effects of radiations were studied and used in turn to grasp the very nature of these rays.[34]

Thomson's and Villard's scientific backgrounds account in part for all the differences noted above. Thomson was trained in Cambridge and learned there its typical combination of mathematical physics, Maxwellian theories, and the use of electromagnetic instruments. Villard, in contrast, was a chemist and an expert at the manipulation of reagents, the determination of substances and composition formulae, and the production of chemical reactions. When both scientists turned to the composition of cathode rays, each used the materials, tools, and the intellectual and material skills they were accustomed to, and for which they were acknowledged experts.

Other factors also contributed to these differences, in particular, each scientists' *working environments*. Thomson directed the Cavendish Laboratory, Cambridge, while Villard was an independent worker at the Ecole Normale Supérieure chemistry laboratory. Each worker was steeped in a specific local culture, with its characteristic norms and values. Each of them had access to a particular range of instruments and materials, and more generally to distinct cultural, intellectual, material, human and financial resources.[35]

Second, both physicists addressed very different *publics*. A reader of the *Philosophical Magazine,* or a member of the Royal Society would have been familiar with the debate on cathode rays. One could not present them with numerical data on cathode particles without including results that explicitly invalidated the undulatory hypothesis. On the other hand, this public was sufficiently in favor of atomic hypotheses to consider the possibility of splittable atoms or atoms made of corpuscles. A reader of the *Comptes Rendus de l'Académie des Sciences* or a member of the Société Française de Physique would have very likely remained rather more sceptical about this. This is suggested by Langevin's precautionary justifications for his use of atoms and ions when he presented the ionisation of gases in 1900 as well as his own researches on gaseous ions in 1902.[36] Each audience tended to be more sympathetic to certain results, certain experimental methods, and certain ways of presenting evidence and arguments. This also plays a significant role in the work of the physicist busy experimenting, elaborating hypotheses, writing articles, or preparing lectures.

AFTER 1898: THE CIRCULATION AND TRANSFORMATION OF SCIENTIFIC FACTS

The conclusions that Thomson published in 1897 received a mixed welcome. Campbell Swinton and others ignored the subatomic character of the corpuscle but retained the fact that the particulate nature of the rays had been conclusively proven. FitzGerald reinterpreted Thomson's experiments, claiming that he had detected "free electrons" rather than corpuscles. FitzGerald understood "electrons" to be the elementary entities that Larmor (on FitzGerald's advice) had recently added to the electromagnetic ether. Thomson contested this reformulation of his work and he continued referring to corpuscles until 1913. These divergences took their origin in differing interpretations of Maxwellian methodology.[37] This confusion in terminology did not pass unnoticed in France. Gustave Le Bon wrote for instance in 1904: "Some physicists, like Lorentz, use indifferently the terms ion and electron which, for others, have very distinct meanings. J. J. Thomson calls corpuscles the electrical atoms which Larmor and others recognize as electrons, etc."[38] To clarify the situation, Alex de Hemptinne proposed in 1905 to cease using the term "ions" to describe electrified gas particles. This term would henceforth be reserved for ions in aqueous solutions. In gases, electrified particles of subatomic dimensions were to be called *electrons* and the ones of atomic dimensions *electrions*.[39]

In Cambridge, Thomson and his students at the Cavendish attempted to measure e/m for the particles emitted by incandescent metals (later the "thermionic effect") and by metals exposed to ultraviolet rays (what was then the "Hertz phenomenon," later the "photoelectric effect"). They published values close to the e/m value for cathode rays and wrote that metals must thus also emit "corpuscles" under very different experimental conditions. Note here that it was the experimental value obtained for e/m that enabled them to claim that the particle emitted by the metal was the same as the cathode ray corpuscle; e/m had thus to some extent become for the Cantabrigians the corpuscle's "signature." This work established the existence of Thomson's "corpuscle" outside cathode rays. It also was intended to confront the critics who refused to accept that the cathodic particle might be an atomic constituent. In this context, the photoelectric effect took a considerable importance since Thomson succeeded in measuring with it e as well as e/m. This enabled him to calculate the corpuscle's mass m, which turned out to be a thousand times smaller than the atomic mass for hydrogen. This was meant to show that the corpuscle was considerably smaller than the atom and that it was likely even to be a fragment of it.[40]

In their scientific publications, J. J. Thomson's old students avoided the use of the term "electron," preferring Thomson's "corpuscle." Langevin wrote for instance in his doctoral thesis, submitted in 1902: "We will retain the name *corpuscle* for these cathodic particles, which was first proposed by Professor J. J. Thomson. The term *electron* is also used, albeit often in a slightly different sense."[41] Langevin pointed out that the term "electron" described the electrical charge itself, while the term "corpuscle" referred to the particle that carried this charge.

The physicists taught by Thomson gradually adopted the practices of the theoreticians of electromagnetism. One such instance of this development was the preparation by Langevin and Rutherford of their communication to the Saint-Louis International Congress in 1904. Rutherford wrote to Langevin in 1904 from Montreal. They had been entrusted to write a common communication by the committee responsible for the organization of the congress. Langevin replied to Rutherford, and on 11 July 1904 Rutherford suggested to present not one, but *two* separate communications. He proposed to write himself a report on radioactivity and invited Langevin to prepare a presentation on "elektrons."[42] He also sent him a possible outline for this paper.[43]

Langevin described in it the new physics of the electron as "a new America, where one can breathe freely, which invites all sorts of activities, and which can teach much to the Old World." According to him, "the concept of electron [. . .] has developed dramatically in the past few years, shattering the structure of the old physics and overthrowing the established order of concepts and laws. This will lead to a reorganisation which one predicts will be simple, harmonious and fecund." He added that the electron was a nonmaterial particle and that the new physics imagined that the atoms themselves were solely composed of "electrons of both charges." In this new approach, matter in its traditional meaning had disappeared; only the electrical charge remained, itself nothing else than a particular structure of the ether. One thus hoped to build a synthesis of physics that would do away with the concept of mechanical mass; a synthesis in which all the energy would have an electromagnetic origin and which would be solely based on "these two concepts of electron and ether."[44]

An old student of J. J. Thomson's at the Cavendish, J. S. Townsend, published in 1915 a synthesis and compilation on the ionization of gases. He presented it as an update of J. J. Thomson's treatise on the *Conduction of Electricity Through Gases,* last published in 1906 (second edition). The term "corpuscle" did not appear in Townsend's new book. The word "electron" described indifferently the negatively charged ions in rarefied gases, the particles emitted by metals in the photoelectric and thermionic effects, the particles composing cathode rays and β-rays, the theoretical particles of electromagnetic dynamics, as well as the negatively charged elementary subatomic particles. The measurement of e/m was presented as the best means of detecting electrons. In Townsend's book, the rays composed of electrons in motion were several times referred to as "corpuscular radiation."[45]

In the years 1896–1905 several French scientists researched cathode rays: Jean Perrin, André Broca, Henri Deslandres, Henri Pellat, and Villard himself.[46] All these physicists worked in Paris, and they met regularly and formally at the sessions of the Societé Française de Physique. Their researches followed individual paths, but they read and quoted each other's papers. Their articles show a common interest in the magnetic properties of cathode rays.[47]

With the exception of Broca, none of these physicists reacted in any significant way to Thomson's or to Villard's results. It must be said that the composition of cathode rays was not a topic directly relevant to their researches. These Frenchmen were concerned instead with identifying the empirically observable properties of the rays. They did not, save for Broca and Perrin, seek to interpret them in terms of microscopic particles. They were not interested either in the chemical composition of the rays. These issues were for these physicists neither legitimate nor even interesting. Deslandres published in 1897 an article where he mentioned that the "composition" of cathode rays was "complex": this meant for him that a cathode ray deviated by an obstacle separated into several rays deviated at unequal angles. For Deslandres, establishing that the composition of the rays was complex did not mean (as it did for Langevin or Thomson) showing that they were composed of different particles; it meant instead that the rays constituting cathode rays had different velocities, temperatures, angles of deviation, or emission origins.[48]

Only one reaction to Villard's results was found in the writings of French physicists working on cathode rays. In an article dated 1899 on anode rays and the sparks produced in Crookes tubes, Broca wrote that:

> In M. Villard's opinion, cathode rays are due to hydrogen. He bases this view on their reducing action, which he has undeniably established. In the tube that I have just described, there are at the same time cathode rays and anode rays. The anode rays are due to a metal: it is very surprising that hydrogen is distinguished from the metals by its electrical properties. This issue must be raised, but with the greatest caution, for there are in these tubes a great number of unknown phenomena.[49]

Broca was clearly sceptical. It is significant that it was precisely Broca, among the Frenchmen mentioned above, who was the only one to express an interest in ions and electrons in his publications. He systematically interpreted his results using corpuscular theories, comparing them to the work done by Thomson's team. He confronted all his results on the influence of magnetic fields to Lorentz's work on the Zeeman effect. It is likely that Broca was aware of the incompatibility between Thomson's conceptions and Villard's idea that cathode rays were made of hydrogen. This would explain why he reacted in such a doubtful manner.[50] Perrin was also favorable to Thomson's views, but he had abandoned the study of rays in June 1897 after submitting his doctoral thesis on cathode rays and x-rays.

In April 1900, Villard presented to the Societé Internationale des Electriciens the latest researches of physicists on cathode rays, x-rays, and uranium rays.[51] He also wrote before July 1900 a report on cathode rays for the International Congress of Physics which was to take place in Paris in August 1900.[52] These two texts were quite different in nature, but Villard's point of view was argued in both places in a similar fashion. Three aspects are particularly significant.

Villard considered first that Perrin's experiments, and not Thomson's, had resolved the controversy over cathode rays; this attitude contrasted sharply with the British scientists' attitude:

> The ballistic hypothesis explains very satisfactorily the properties [of cathode rays]: it lends itself to the interpretation of magnetic and electric effects much better than the undulatory hypothesis, but it was only clearly confirmed by the rigorous demonstration which M. J. Perrin gave of the electrification of the rays.[53]

In December 1895, Perrin had announced that cathode rays carried an electrical charge. He wrote in this article that this result was "difficult to reconcile" with the undulatory hypothesis, and it "agreed instead" with the particulate interpretation.[54] Thomson had read Perrin's article and reproduced his experiments in 1897 with an arrangement he thought would produce more decisive results. Thomson only rarely reproduced experiments that had been already published and Falconer explains this improvement-reproduction: on the one hand it was considered to be a crucial experiment by Thomson, and on the other, this was the first publication of a young Frenchman completely unknown in Britain.[55] Perrin first described this experiment in his doctoral thesis (submitted in June 1897) and it was published shortly after in the *Annales de Physique et de Chimie*.[56] It is to this article that Villard referred in a footnote. For Villard, Perrin's experiment was therefore the decisive one and not Thomson's.

The second important point is that Villard described J. J. Thomson's work in considerable detail, but all references to atoms (and all the more to corpuscles) were removed. For instance, Villard wrote of the particles in motion which constituted cathode rays that:

> M. J. J. Thomson was able to determine the velocity of these cathodic projectiles. The measurement of the deviation of the trajectories in a magnetic and in an electrostatic field yields indeed two relations in which the velocity and the ratio of a particle's mass and charge are represented. This ratio is approximately a thousand times greater than the ratio obtained for electrolytic phenomena. In other words a gram of hydrogen would carry *one hundred million coulombs*.[57]

One of the noteworthy characteristics of Villard's nonatomist reading of Thomson's work is that the term "corpuscle" was never used. Villard wrote only of "particles" and "projectiles." Moreover, as was pointed out above, Thomson put forward the enormous value of the e/m ratio to justify the idea that the corpuscle's mass was a thousand times smaller than hydrogen's.[58] Villard also expressed surprise at this value, but he obliterated in his papers all references to particles in this context. The value e/m did not represent for him the ratio between two properties of a particle, but a ratio enabling the calculation of the quantity of electricity carried by a given mass of hydrogen.

Third, Villard appropriated Thomson's results, like FitzGerald in 1897, and used them to support his own interpretation:

> I have just argued that the chemical effects produced by the cathodic projectiles give a first indication as to their nature [. . .] More recently, M. J. J. Thomson has established that the electric or magnetic deviation depends solely on the potential difference which sets the particles in motion. These

particles are thus always made from the same matter, most likely a simple body.[59]

Villard concluded on "the nature of radiating matter" that composed the rays:

> If one notes [. . .] that cathodic phenomena are independent of the nature of the gas contained in the bulbs, and that in particular the e/m ratio is invariable, one is led to accept the unity of radiating matter. Now, hydrogen is the only reducing gas known. It is precisely the gas whose spectrum is always, and often alone, visible in the luminescent layer which marks the arrival of the influx at the cathode. This element already possesses very special characteristics, such as its ability to pass through red-hot metals. Until another simple reducing gas is discovered, one must consider acceptable the hypothesis that hydrogen constitutes radiating matter.[60]

Thomson used the constancy of the e/m ratio to assert that the corpuscle was a universal particle. Villard also used this constancy but he concluded from it that the substance composing cathodic particles was always the same. He went on to determine this substance with characteristic chemical reactions.

Villard changed the way in which he read, understood, and presented Thomson's work after the Paris Physics Congress, which took place on 6–11 August 1900.[61] Rasmussen has emphasized the fact that this was the very first *international* physics congress. It is significant that it took place as late as 1900, because scientists in other disciplines had organized themselves earlier and had been meeting relatively regularly at international congresses for the previous half century. This congress in 1900 was thus the first occasion French physicists were given of meeting their foreign counterparts.[62] This meeting enabled them to become collectively aware of the importance that the work done at the Cavendish under Thomson had acquired abroad. The congress modified the classification initially planned by the organizers to include recent research on ions and corpuscles. This new classification was used from 1901 in the *Journal de Physique*.[63] The congress reports were the first texts available in the French language that described in detail the work of Thomson and the Cambridge physicists.[64] It is perhaps the reason why Villard presented Thomson's work differently in the book he subsequently wrote.[65] This book was published at the end of 1900 by Gauthier-Villars, in the Scientia collection. This collection was directed by Appell, d'Arsonval, Haller, Lippmann, Moissan, and Poincaré, all members of the Académie des Sciences. They wished to make of this collection "an account of and elaboration on the scientific questions of the day."

In contrast to what he had done in his two previous texts, Villard briefly presented in this book Thomson's ideas on the corpuscle and its substance:

> The ratio of the cathodic particles' charge to their mass is a constant; M. J. J. Thomson was thus led to the conclusion that cathode rays are made of one substance only, which exists at a state of division much greater than in gases. This substance is the same whether it comes from hydrogen or another body. It is the hypothesis of universal matter. Without going so far, it is worth asking if this single substance is not one of the simple bodies known.[66]

This distinguished itself from Villard's earlier texts, in that he now *explicitly* pointed to the differences between his own and Thomson's positions. Moreover, Villard no longer used Thomson's researches to legitimate his own interpretation. In this book, Villard wrote that "the flux of matter which circulates in the tube in a radiant state seems to be made of hydrogen." This time, though, he drew in his demonstration exclusively upon his own experiments in chemistry and no longer on Thomson's results.

Thomson's opposition in 1897 to the idea that cathode rays might be made of hydrogen has already been mentioned. He did not judge it necessary, however, at the time to reply to Villard. No mention of Villard's result was found in Thomson and his students' scientific publications between 1898 and 1902. Only one mention was found in J. J. Thomson's great treatise on the ionization of gases, published in 1903. The book was a vast compilation of current researches on conducting gases and the new rays. In his chapter on cathode rays, Thomson wrote:

> Villard found that cathode rays exert a reducing action ; thus if they fall upon an oxidised copper plate, the part exposed to the rays becomes bright. In considering the chemical effects produced by the rays we ought not to forget that the incidence of the rays is often accompanied by a great increase in temperature, and that some of the chemical changes may be secondary effects due to the heat produced by the rays.[67]

What Thomson failed to mention was that these chemical effects led Villard to the conclusion that the rays were made of hydrogen. Thomson questioned Villard's techniques of empirical investigation, and argued that the chemical effects were secondary, that is, they were not directly due to the nature of the rays and thus could not be applied to their understanding.

Villard was otherwise only mentioned once more in Thomson's treatise, with regard to his research on canal rays. Thomson pointed that Wien, Ewers, and Villard had all three attempted in 1899 "to detect the positive charge carried by the Canalstrahlen" but that "the aforesaid physicists differ in their interpretation of the results they obtain." They all agreed on the fact that the rays could, in certain cases, transfer a positive charge to a Faraday cylinder, "but while Wien and Ewers think that this charge is carried by the Canalstrahlen, Villard is of opinion that it is a secondary effect." Thomson wrote: "In spite of the indecisive results obtained by this experiment *(Villard's)*, the magnetic and electric deflections obtained by W. Wien seem conclusive evidence that the Canalstrahlen carry a positive charge." This account by Thomson of Villard's researches revealed again fundamental differences in their conception of electric charges. For Villard, the electric charge of the canal rays was not an inalterable property of the matter constituting them: it could be equally positive or neutral. In contrast, Thomson considered an experiment as having failed which did not detect a charge carried by canal rays. The same cognitive and cultural differences that characterized both men's attitudes to cathode rays were clearly at play here again.[68]

It must be remembered that Thomson's treatise rapidly became the work of reference for most physicists working in the field of conducting gases and radiation. It was in particular systematically quoted in the articles of the young physicists trained and supervised by Langevin at the Collège de France, who regularly published on the ionization of gases between 1902 and 1911. Two such scientists were Eugène Bloch and Maurice de Broglie, only to mention the two most important ones. The following edition of Thomson's treatise, which appeared in 1906, was heavily reworked and extended, but the two passages on Villard remained unchanged. This second edition was translated and published in French in 1912. In this work of reference, which was widely distributed, Villard was thus only mentioned twice and each time his results were severely criticized.[69]

It is thus likely that Villard's credibility was low in the international network of the physicists who recognized J. J. Thomson as their main scientific authority. This did not prevent him from being increasingly known among the elite of the French physicists. A long controversy opposed Villard to Pellat from February to December 1904. Pellat then occupied the chair of experimental physics at the Sorbonne, and he headed one of the three laboratories of the Paris Faculty of Sciences (Lippmann and Bouty heading the two others). Pellat reckoned he had discovered that cathode rays passing through a magnetic field were subject to anisotropic friction, a phenomenon he termed "magnetofriction."[70] Villard published an article contesting the existence of this friction, starting a controversy between the two scientists that lasted eleven months.[71] Villard and Pellat exchanged in the *Comptes Rendus* exceptionally long, argumented and illustrated notes.[72] The controversy ended with a brilliant victory for Villard. Pellat explicitly admitted in his later articles that there was no such thing as magnetofriction.[73] Villard and Pellat were then considered to be the two French "specialists" on cathode rays. They wrote far more articles on this topic than any other French physicist. They regularly spoke at the sessions of the Societé Française de Physique to present the results of their latest researches. Pellat's defeat in this controversy unquestionably made Villard the first scientific authority on cathode rays in France.

The members of the Académie des Sciences awarded Villard the Wilde prize on 19 December 1904. The jury included Levy, de Lapparent, Mascart, Berthelot, Darboux, Troost, and Loewy. As the tradition dictated, the report enumerated Villard's scientific researches. After an account of his work on gas hydrates, the report recalled that "another series of researches, no less important, relates to cathode rays, x-rays etc.," and that Villard "has shown that cathode rays always carry with them hydrogen."[74]

Langevin and Abraham's textbook, *Ions, Electrons, Corpuscules* helps one understand the situation of Villard and his results in 1905.[75] This book resembled Thomson's treatise in that it sought to become a work of reference on radiation, conducing gases, and elementary particles of electricity. Two-thirds of the articles reprinted in this compendium were by Thomson and his students, and dealt with the ionization of gases. On the subject of cathode rays, the authors presented the experimental researches of Crookes, Deslandres, Hittorf, Kaufmann, Lenard, Pellat, Perrin, Plücker, Simon, J. J. Thomson, Villard, and Wiechert. None of the articles written after 1898 and reprinted there mentioned Villard's results. In addition, all the British and German physicists who published on cathode rays in the two or three years preceding the publication of this book considered the rays to be made of "corpuscles" (for the British) or "electrons" (for the Germans).

Villard's article was specifically written for this book and presumably dates from 1905.[76] As he had done in 1900, Villard described again the chemical effects of cathode rays, concluding that "cathodic projectiles are nothing else than particles of electrified hydrogen."[77] The tone of this text was that of a textbook: the properties of cathode rays were described one after the other with no mention of the scientist who had found them. The lack of social elements is striking in this text, compared to Villard's previous papers. The contrast is particularly clear with Pellat's article in the same volume, which was also an original contribution. Pellat "converted" to ions and corpuscles in 1904. His article was a retrospective reformulation of the whole of his work on cathode rays since 1900; it was presented by the editors as "M. Pellat's researches on corpuscles." Contrarily to Villard, Pellat presented his researches in a chronological order, in a narrative mode, making references to the work of Thomson, Broca, and Villard himself.[78]

The reading of Langevin and Abraham's book suggests that Villard was held in high enough esteem to be unavoidable in this kind of compilation; and this despite the fact that his research was marginal to his contemporaries' work. Villard's ideas had become obsolete by 1905, but this probably did not displease the aged physicists of the moment (such as the academicians who had just awarded him the Wilde prize, thereby legitimating his claim that cathode rays were made of hydrogen). This state of affairs gives an indication of the gulf that was beginning to separate the older generation of academicians from the rising generation embodied by the young physicists working with and around Langevin at the Collège de France. The former probably approved of the absence of atoms from Villard's work, while the latter used corpuscles and ions constantly in their researches on ionized gases and radiation. It was precisely in 1905 that *Le Radium,* a journal initially devoted to the popularisation of science, was taken over by the Curies' and Langevin's groups. They changed the editorial policy, transforming it into a purely scientific journal. Between 1906 and 1914 the young physicists favorable to ions published predominantly and preferably in *Le Radium*. This editorial takeover was completed in 1918 when *Le Radium* merged with the *Journal de Physique* to form *Le Journal de Physique et Le Radium*.[79]

Villard's research practices changed fundamentally from 1906. For a start, the results he published were now dubbed "theories"; the first of which was a theory of aurora borealis which he developed between 1906 and 1908. It was presented in two parts. Villard focused initially on the shapes of the trajectories taken by corpuscles in a magnetic field. He considered several possibilities for the distribution of the field in space and he calculated the trajectory of a corpuscle in motion in these different cases. He obtained several helicoidal trajectories, the geometry of which varied in each case. Villard then supposed that the clouds in the polar atmosphere emitted corpuscles. He attempted to show that the Earth's magnetic disturbance of the corpuscles' motion could explain the appearance of aurorae, as well as their geographical configuration and their movements, or "dance of the rays."[80]

The production and publication of a *theory* based on *corpuscles* was a novelty in Villard's scientific work. To this was added the use of new literary devices to justify his results. In his papers, Villard now first presented his theoretical developments before moving on to discuss their experimental consequences. He then concluded that his experimental data conformed to the theory's predictions. The argumentation in these articles was based on prediction and confirmation instead of Villard's earlier empirical methods of justification.

Moreover, Villard also started in 1906 to use systematically the terms "ion" and "corpuscle" in his experimental articles. Whereas he had earlier described "rays," "influxes," "currents," "transfers of matter," and "particles," he started, occasionally at first, more and more often later, to write of "corpuscle emission," "collisions against molecules," and of "motion of positive ions." This change in terminology, this reading of his results using atomist words and concepts were radically new features of Villard's work.[81]

Villard nevertheless only ever spoke of "corpuscles," and never of "electrons" in 1906 and 1907. This changed from 1908 onwards: in the second edition of his book on cathode rays, published 1908, he wrote:

> In the first edition of this book, we stated that cathodic corpuscles were particles of electrified hydrogen. This hypothesis seemed to follow logically at a time when electricity was thought to be inseparable from matter. Indeed, it explained perfectly the reducing action of the rays and the constant presence of hydrogen close to the cathode.
>
> The remarkable researches of M. J. J. Thomson and M. Max Abraham have shown that the presence of a ponderable support is unnecessary and that the laws of electricity suffice to explain the main properties of cathode rays. Reversing to an old hypothesis, it is argued that there are electrical fluids. By analogy with ordinary matter, one supposes these fluids to be made of atoms of electricity, or *electrons,* whose absolute value, $1.13 \cdot 10^{-19}$ coulomb, is equal to the charge of the hydrogen atom in electrolysis. It follows that an electrical charge consists of an integer of these elementary indivisible units; this corresponds to an exact multiple of the unitary charge mentioned above. In this hypothesis, cathodic corpuscles are atoms of negative electricity, that is, negative electrons.[82]

Villard thus completely abandoned his position on the composition of cathode rays, along with the methodology that had guided his work on radiations since 1897. For him, cathode rays were now electrons and no longer particles of electrified hydrogen. In his book, Villard presented the new electronic theories in detail and the current attempts to produce an electromagnetic synthesis of the whole of physics. According to these theories, the electron's inertia would be of an entirely electromagnetic origin. Ordinary inertia could thus be accounted for "if one admits that matter is only made of electrons." Thanks to electrons, "the mechanical properties of cathode rays [. . .] can be explained perfectly simply." Villard pointed that there were slight problems with these new theories, but that: "Despite these anomalies, the electromagnetic theory of cathode phenomena constitutes a considerable progress, in fact the greatest advance to be made in this branch of physics since the work of Sir. W. Crookes."

On the chemical effects that had previously enabled him to argue that cathode rays were made of hydrogen, Villard only wrote: "As for reduction phenomena, the example of light shows us that these are possible without any input of hydrogen."[83]

On 16 November 1908, Becquerel was elected *secrétaire perpetuel* of the Académie des Sciences. His seat in the Physics Section of the Académie became vacant and elections were organized to fill it. The names proposed by the members of the Physics Section were ranked in decreasing order of preference. Bouty and Villard came first, followed by Berthelot, Branly, Broca, Cotton, Pellat, and Pérot. The election took place on 23 November 1908. Bouty won the seat with thirty-seven votes, while Villard received nine and Branly eight.[84] New elections took place again in the Physics Section on 21 December 1908 after Mascart's death. This time, Villard came out as the preferred candidate, followed by Berthelot, Branly, Broca, Cotton, Pellat, and Pérot. Villard won the seat, obtaining thirty-four votes against eighteen for Branly at the elections on 28 December 1908.[85]

This shows that Villard was respected within the Académie, and that he was systematically rated above all the other physicists named, with the exception of Bouty. Interestingly, he was also rated higher than the other cathode rays specialists, Broca and Pellat. Branly appeared to be the only physicist to present a serious challenge to him in the second election.[86] At the end of 1908 Villard was thus awarded one of the highest possible distinctions for a French physicist.[87]

In 1909, Villard wrote a manuscript note for the attention of the academicians presenting Langevin's latest researches.[88] He explained there at length the meaning of the term "ion" and their function in phenomena of gaseous conduction according to Langevin and Thomson. Villard had, at least on this occasion, become the spokesman of ions and electrons at the Académie des Sciences.

Elementary Particles, Intellectual Itineraries and Scientific Social Worlds

This chapter has attempted to restore a symmetry between Thomson and Villard. This has not only meant giving both equal amounts of attention, for an attempt was also made to describe *in the same terms* their different research practices. I thus hope to have reconstructed the coherence inherent to the scientific activity of each man, without trying to distinguish between right or wrong. I was especially careful to avoid considering Villard's work the way scientific "errors" are usually dealt with, that is, by looking for the misleading effect of "prejudice" or "social and cultural influences."

When Villard published his conclusions on the nature of cathode rays in 1898, he expressed a view which J. J. Thomson had described the preceding year as being incompatible with his own. Villard knew Thomson's work and mentioned him in his own articles; but he never mentioned this remark of Thomson's and never justified himself in reply to it. As for Thomson, he clearly did not seek to publish a response to Villard, if we except the few lines of his 1903 treatise, which are both very critical and allusive. Villard made explicit the opposition between his views and Thomson's in 1900, but he did not bother involving himself in a polemic on the matter. This supports the claim that both views coexisted without open conflict, despite having been explicitly recognized as antagonistic.

Yet Thomson's and Villard's writings were widely and carefully read, and sometimes led to violent controversies. But these debates took place in two separate scientific environments, and this is one essential aspect of this historical situation. Villard's researches on cathode rays were published in the *Comptes Rendus* and the *Journal de Physique*. They were described and debated at several sessions at the *Societé de Physique*. They were taken up or criticized by Broca, Deslandres, Gouy, and Pellat. In short, they were known and recognized in a particular intellectual and social "territory," that of the French physicists. Thomson's work was produced and considered in *another* context, in the English-speaking community of physicists. The relative disjunction, autonomy and indifference that separated both camps have been described; but several cases of transfer from one to the other have also been considered. An example of this was Thomson's reproduction-transformation in 1897 of Perrin's experiment on the charge of cathode rays. The extreme difference in the credibility of Thomson's and Perrin's experimental results in their respective national spaces was pointed to. What enabled the "pacific" coexistence of Villard's and Thomson's interpretations of the composition of cathode rays was the independence and distance between their two worlds.[89]

These two national spaces were far from being homogeneous. The reception of Villard's researches made possible a construction of a differential map of the French field. Three groups were thus isolated: the French physicists working on cathode rays, the academicians of the physics section, and the young physicists trained by Langevin at the Collège de France working on the ionization of gases. The members of these various groups took into account and used Villard's results according to their own professional interests and in different publications. The English-speaking context was socially and intellectually heterogeneous too—which led to different and conflicting interpretations of the electron and the nature of cathode rays.[90]

The arrangement of these various *milieux* was far from remaining static in the years 1895–1910. The spread of the international network of workers on ionization is perhaps the most rapid and dramatic one described in this study. In 1895 the experimental culture of the study of ions was localized in the Cavendish Laboratory. Five years later the network had grown to become international. Young physicists trained in Cambridge practiced a physics based on the production of conducting gases, electrometric measurements, and the detection and manipulation of corpuscles and ions. This was a reductionist and atomist physics, which hierarchized matter in successive levels of depth. These physicists enthusiastically supported Thomson's corpuscles for being the most elementary particles accessible to scientific investigation. These physicists found particularly interesting the attempts to unify physics based on ether and electrons, and they ended up abandoning Thomson's corpuscles for electrons, as was pointed above with regards to the St. Louis International Congress of 1904. These research practices typical of Cambridge thus transformed while they spread geographically.

These developments greatly contributed to the introduction of Cavendish Laboratory physics into the French milieu. Two episodes stand out as being particularly important in this process: the Paris International Congress of Physics of 1900, and Langevin's return to Paris after his one-year stay in Cambridge, and his building in 1902–5 of a research group working on ions and the corpuscles of conducting gases. The 1900 congress reports and Abraham and Langevin's textbook, *Ions, électrons, corpuscules,* published in 1905, were equally important: these publications which explained in French the latest research on ions were both widely distributed. More specifically, these books presented the ion physics of the Cavendish, not only atomist physics in general. Thus when Pellat, Villard, or Deslandres substituted Thomson's "corpuscles" for their old "particles," they did more than incorporate atoms and ions in their research practices. They adopted at the same time the experimental methods and terminology of Cambridge physics (even if these borrowings remained partial and localized). It is particularly significant in this respect that these Frenchmen started researching "corpuscles" at the very moment when the physicists trained at the Cavendish were moving on to "electrons." It is only after a similar intermediary phase of work on "corpuscles" that the French scientists later started in turn using "electrons."[91]

From 1900 onward, after the progressive "irruption" of the network of ionization on the international scene, the distance separating the French and British *milieux* was considerably reduced in several intellectual, institutional, and editorial spaces. The appearance of these spaces partly explains why the nonconflictual coexistence of Thomson's and Villard's results ended. The books mentioned above especially contributed to this process, by juxtaposing both interpretations in the same publication. They suddenly made the antagonism between the two scientists very visible, and brought this conflict to the attention of an audience wider than ever.

This chapter has concentrated on the activities of the different groups that "took hold" of the "electron," giving it a certain existence, a given shape in their own research practices. This was only done when such regrouping appeared valid, and we avoided dissolving scientists' individualities in their contexts when they obviously possessed original traits. In particular, Villard's and Thomson's roles in the process mentioned here were described. These two physicists never passively accepted the fate given to their results by other physicists. They took an active part in the determination of events post-1898, even if their range of action was necessarily limited.

It is worth noting that both took up and mentioned their own results, albeit in different ways. Thomson rapidly started using his own corpuscle in the rest of his own work. Corpuscles were everywhere in the synthesis of the phenomena of gaseous conduction produced by rays he wrote in 1899–1900. It has also been mentioned that Thomson succeeded in interesting most of the young physicists working in his laboratory in the identification and the manipulation of corpuscles. Villard reiterated his arguments to his own public, notably in 1900 and 1905. He even gave more detailed justifications then than the first time he published his views on the matter in 1898. But Villard only did so in a spirit of compilation and summary, never in his current research proper. Villard worked in the small room given to him in the chemistry laboratory of the Ecole Normale Supérieure, and all the articles he published between 1890 and 1914 were submitted by him alone. In the same period, Thomson published several researches done in close collaboration with his students or assistants (such as Mc Clelland, Rutherford, Aston). Villard's biographers have often described him as a solitary and withdrawn man, but this description appears to correspond better to his later years (1918–1937). Between 1897 and 1909, Villard was enthusiastically involved in the invention and improvement of instruments. He developed his inventions in close collaboration with the instrument builders Chabaud and Charpentier, such as a falling process for mercury tubes, an osmo-regulator for x-ray tubes, or a cathode valve. Villard's experimental practices belonged socially to the milieu of instrument-makers rather than of physicists.

Thomson actively resisted the reinterpretation of his work. He openly criticized the new electronic theories and he continued to call "corpuscles" the cathode particles even when his close collaborators switched to "electrons." He ended up giving the impression of an isolated and obsolete scientist. Villard, in contrast, never openly fought against electronic theories. He *ignored* them completely until 1906. And then he adopted their terminology in 1908, explicitly voicing his approval of the new theories on the nature of electricity. He asserted that it was no longer possible to consider cathode rays as particles of hydrogen. From this perspective, the situation in the years 1908–1913 appears quite paradoxical. We have on the one hand Thomson refusing to admit the electrons of which he was the acknowledged discoverer, yet rewarded for this work with a Nobel prize in 1906. On the other hand Villard not only replaced his "particles" with "electrons," using them in his own researches, but he also became its active propagandist in French physicists' circles.

ACKNOWLEDGMENTS

This paper is an based on the fourth chapter of my doctoral thesis, *Vapeurs, Foudres et Particules: les pratiques expérimentales de l'ionisation des gaz à Paris et à Cambridge, 1895–1914,* University of Paris VII, May 1995. I would like to thank Dominique Pestre, my Ph.D. supervisor, for his advice and Christine Blondel, Olivier Darrigol, Yves Gingras, Christian Licoppe, Simon Schaffer, and Andy Warwick for their comments on an early version of this paper, an earlier version of which was published as "Paul Villard, J. J. Thomson et la composition des rayons cathodiques" in the *Revue d'Histoire des Sciences* 50/1–2 (1997): 89–130.

NOTES

1. D. Anderson, *The Discovery of the Electron* (Princeton, N.J., 1964); G. Owen, "The Discovery of the Electron," *Annals of Science* 11 (1956): 172–182; E. Whittaker, *A History of the Theories of Aether and Electricity,* New York, 1952; B. Turpin, "The Discovery of the Electron: The Evolution of a Scientific Concept 1800–1899," unpublished doctoral thesis (University of Notre Dame, Indiana, 1980).

2. Falconer 1987, Feffer 1989, Robotti 1995. See the bibliography for the full references of the abbreviated works mentioned in the footnotes.

3. On Villard's work seen as an "error," see for instance E. Picard, *La Vie et l'Oeuvre de Paul Villard et de Georges Gouy, Lecture Faite en la Séance Publique Annuelle du 20 Décembre 1937, Institut de France,* p. 8. Villard's other biographers do not mention his work of 1898 (see the next footnote).

4. A. Blondel, Paul Villard, *Revue Générale de l'électricité* XXXV/13 (1934): 411–414; E. Scott Barr, "Anniversaries of Interest to Physicists in 1960," *American Journal of Physics* XXVIII/5 (1960): 462–475 (472–474); Sigalia Dostrovski, Villard, Paul, in C. C.

Gillispie (ed.), *Dictionary of Scientific Biography* (Charles Scribner's sons, New York, 1976), vol. XIV: 31–32.

5. On the chemistry laboratory of the Ecole Normale Supérieure, see Mary Jo Nye, *From Chemical Philosophy to Theoretical Chemistry. Dynamics of Matter and Dynamics of Discipline , 1800–1950* (University of California Press, Berkeley, 1993), 141, 145–148. On the beginnings of physical chemistry in France and in particular on François Raoult, see Mary Jo Nye, *Sciences in the Provinces, Scientific Communities and Provincial Leadership in France 1860–1930.* (University of California Press, 1986), 98–116.

6. Forcrand and Villard, "Sur l'Hydrate d'Hydrogène Sulfuré," *CRAS* CVI (1888) : 849–851; "Sur l'Hydrate de Chlorure de Méthyle," *Ibid.*, 1357–1459; "Sur la Composition des Hydrates d'Hydrogène Sulfuré," *Ibid.*, 1402–1405.

7. Paul Villard, "Sur Quelques Nouveaux Hydrates de Gaz," *Ibid.*, 1602–1603.

8. P. Villard, "Sur la Composition et la Chaleur de Formation de l'Hydrate de Protoxyde d'Azote," *CRAS* CXVIII (1894) : 646–649.

9. P. Villard, "Sur l'Hydrate Carbonique et la Composition des Hydrates de Gaz," *CRAS* CXIX (1894): 368–371.

10. *Ibid.*, 369.

11. The literature on this topic is abundant. See for instance Shinn 1980; Mary Jo Nye, "The Nineteenth-Century Atomic Debates," *Stud. Hist. Phil. Sci.*, VII (1976): 245–268; Bernadette Bensaude-Vincent, "Karlsruhe, Septembre 1860: L'atome en Congrès," *Relations Internationales* LXII (1990): 149–169; Jacques 1987, 71–83, 195–208.

12. For instance, Auguste Laurent in 1844 (Shinn 1980, 542) and Alfred Naquet in 1900 (Jacques 1987, 205).

13. Jacques 1987, 71.

14. P. Villard, "Sur le Voile Photographique en Radiographie," *CRAS* CXXV (1897): 232–234.

15. P. Villard, "Sur les Rayons Cathodiques," *CRAS* CXXVI (1898): 1339–1441. Quotation, 1340.

16. P. Villard, "Sur les Rayons Cathodiques," *CRAS* CXXVI (1898): 1564–1566, 1564.

17. *Ibid.*, 1566.

18. P. Villard, "Sur la Diffusion des Rayons Cathodiques," *CRAS* CXXVII (1898): 223–224.

19. Villard 1899.

20. P. Villard, "Redresseur Cathodique pour les Courants induits," *CRAS* CXXVIII (1899), 994; "Sur la Réflexion et la Réfraction des Rayons Cathodiques et des Rayons non Déviables du Radium," *CRAS* CXXX (1900): 1010; "Sur le Rayonnement du

Radium," *Ibid.*, 1178; "Sur les Rayons Cathodiques," *Ibid.*, 1614; "Sur la Discontinuité de l'Emission Cathodiques, *Ibid.*, 1750.

21. J. J. Thomson, "On the Velocity of Cathode-rays," *PM* V/38 (1894): 365.

22. J. J. Thomson 1897. The details of J. J. Thomson's experiments have not been given as they have been described at length elsewhere. See the references in the introduction for further information.

23. The wave/particle debate is not mentioned in Villard 1899.

24. J. J. Thomson 1897, 311. Emphasis mine.

25. *Ibid.*

26. *Ibid.*, 312.

27. Falconer 1987, 253; Falconer 1985; Buchwald 1985.

28. Falconer 1987, 253; Feffer 1989, 38.

29. Galison and Assmus 1989, 226.

30. These statements are taken from a reply of Villard to Perrin and Langevin at a session of the Société Française de Physique. Villard had just finished describing the formation of magnetocathode rays along cathode rays. Perrin and Langevin attempted to explain this phenomenon by the ionisation produced by the collision of cathodic corpuscles against molecules of the rarefied gas. *BSFP* (1905), 32.

31. Galison and Assmus 1989, 246–247.

32. P. Villard 1899, 13–16.

33. P. Villard, "Sur les Rayons Cathodiques," *BSFP* (1898): 17.

34. P. Villard, "Sur l'Action Chimique des Rayons X," *CRAS* CXXVIII (1899): 237–239; *CRAS* CXXIX (1899): 882–883.

35. On the Cavendish Laboratory, see Dong-Won Kim, "J. J. Thomson and the emergence of the Cavendish School, 1885–1900," *BJHS* 28 (1995): 191–226.

36. Langevin, "Sur l'Ionisation des Gaz," *BSFP* (1900): 39; "Recherches sur les Gaz Ionisés," *BSFP* (1902): 45–47; "Recherches sur les gaz ionisés," *Ibid.*, 49–51.

37. Falconer 1987. Buchwald 1985. Hunt 1991. Feffer 1989. See also J. Z. Buchwald, "The Abandonment of Maxwellian Electrodynamics: Joseph Larmor's Theory of the Electron," *Archives Internationales d'Histoire des Sciences* 31 (1981): 135–180, 373–438.

38. Gustave Le Bon, "La Dématérialisation de la Matière," *Revue scientifique* 5, 2 (1904): 609–617, 609.

39. Alex de Hemptinne, "Sur la terminologie relative aux phénomènes d'ionisation," *Le Radium* 2 (1905): 353–355.

40. J. J. Thomson, "On the masses of the ions in gases at low pressures," *PM,* 48 (1899), 547; C. T. R. Wilson, "On the condensation nuclei produced in gases by the action of Röntgen rays, uranium rays, ultraviolet light and other agents," *Philosophical Transactions of the Royal Society of London* 192 (1899): 403. Robotti 1995, 277, 281.

41. Langevin 1903, 350. Langevin's emphasis.

42. Rutherford wrote the word in quotation marks and used the German spelling.

43. Rutherford to Langevin, 15 June 1904, *FL,* L76/41. Rutherford to Langevin, 11 July 1904, *FL,* not referenced, file 76.

44. Paul Langevin, "La Physique des Électrons, Rapport au Congrès International des Sciences et des Arts, Saint-Louis, 22 February 1904," *Revue Générale des Sciences* (1905). This text was reprinted in *La Physique Depuis Vingt Ans* (Doin, Paris, 1923), 1–69. The quotations are 2, 69, 16. This text is a classic: It was analysed in detail by Paty and by Bensaude-Vincent. Michel Paty, "The scientific reception of relativity in France," in Thomas F. Glick (ed.), *The Comparative Reception of Relativity* (D. Reidel, 1987), 113–167. B. Bensaude-Vincent, *Langevin. Science et Vigilance* (Belin, Paris, 1987), 41–49.

45. John S. Townsend, *Electricity in Gases* (Clarendon Press, Oxford, 1915), 6, 284–285, 472. See also Lelong, "Townsend."

46. See for example: H. Deslandres, "Action Mutuelle des Électrodes et des Rayons Cathodiques Dans les Gaz Raréfiés," *CRAS* CXXIV (1897): 678–680, 945, 1297; *CRAS* CXXV (1897): 373; A. Broca, "Quelques Propriétés des Cathodes Placées dans un Champ Magnétique Puissant," *CRAS* CXXVI (1898): 736–738; "Quelques Propriétés des Décharges électriques Produites dans un Champ Magnétique," *Ibid.,* 823–826; H. Pellat, "Contribution à l'Étude des Stratifications," *CRAS* CXXX (1900): 323; "Contribution à l'Étude des Tubes de Geissler dans un Champ Magnétique," *CRAS* CXXXIII (1901): 1200.

47. The researches of these Frenchmen on the magnetic properties of cathode rays have been described in B. Carazza and H. Kragh, "Augusto Righi's Magnetic Rays: A Failed Research Program in Early 20th-century Physics," *HSPS* 21 (1990): 1–28.

48. H. Deslandres, "Propriété Nouvelle des Rayons Cathodiques qui Décèle leur Composition Complexe," *CRAS* CXIV (1897): 678.

49. A. Broca, "Décharges Disruptives dans le Vide. Formation de Rayons Anodiques," *CRAS* CXXVIII (1899): 356–358.

50. On Broca, see *Notice sur les travaux scientifiques de M. André Broca* (Gauthier-Villars, Paris, 1906).

51. Villard 1900a.

52. Villard 1900b.

53. Villard 1900b, 127. See also Villard 1900a, 176.

54. J. Perrin, "Nouvelle propriété des rayons cathodiques," *CRAS* CXXI (1895): 1130–1134, 1133.

55. Falconer 1985.

56. J. Perrin, "Rayons cathodiques et rayons de Roentgen," *ACP* XI (1897): 496–554.

57. Villard 1900a, 177. Villard's emphasis. cf. also Villard 1900b, 131.

58. J. J. Thomson 1897, 311.

59. Villard 1900a, 177.

60. Villard 1900b, 137

61. For what follows, see *Rapports Présentés au Congrès International de Physique*, (Gauthier-Villars, Paris, 1900), vol. I, 1–41.

62. Anne Rasmussen, "Jalons pour une Histoire des Congrès Internationaux au XIX° Siècle: Régulation Scientifique et Propagande Intellectuelle," *Relations Internationales* 62 (1990): 115–133.

63. Bernard Brunhes, "Avertissement pour l'Usage de la Table Analytique," in *Table Analytique et Table par Noms d'Auteurs des Trois Premières Séries du Journal de Physique 1872–1901* (Journal de Physique, Paris, 1901), v; *Journal de Physique*, III/10 (1901), 792.

64. cf. Bouty, *Cours de Physique de l'Ecole Polytechnique. Troisième Supplément. Radiations, Électricité, Ionisation* (Gauthier-Villars, Paris, 1906), vi.

65. P. Villard, *Les Rayons Cathodiques* (Gauthier-Villars, Paris, 1900).

66. *Ibid.*, p. 109.

67. J. J. Thomson, *Conduction of Electricity through Gases* (Cambridge University Press, Cambridge, 1903), 496.

68. *Ibid.*, 521–522.

69. J. J. Thomson, *Passage de l'Électricité à Travers les Gaz*, translation by R. Fric and A. Faure (Gauthier-Villars, Paris, 1912).

70. H. Pellat, "Etude de la Magnétofriction du Faisceau Anodique," *CRAS* CXXXV (1902): 1321; "Loi Générale de la Magnétofriction," *CRAS* CXXXVIII (1904): 618.

71. P. Villard, "Sur la Décharge Électrique dans les Gaz Raréfiés," *BSFP* (1904): 44–47.

72. H. Pellat, "Explication des Colorations Diverses que Présente un Tube à Gaz Raréfié," *CRAS* CXXXVIII (1904): 1206–1208; P. Villard, "Sur les Rayons Cathodiques," *BSFP* (1904): 51–52; P. Villard, "Sur les Rayons Cathodiques," *CRAS* CXXXVIII (1904): 1408–1411; H. Pellat, "Remarques au Sujet d'une Note de M. P. Villard sur les Rayons Magnéto-Cathodiques," *Ibid.*, 1593–1594; P. Villard, "Sur les Rayons Cathodiques. Réponse à la Note de M. Henri Pellat," *CRAS* CXXXIX (1904): 42–44; H. Pellat, "Sur les Rayons Cathodiques et la Magnétofriction. Réponse à la Note de M. Villard," *Ibid.*, 124–126; P. Villard, "Sur les Rayons Cathodiques et les Lois de l'Électromagnétisme," *Ibid.*, 1200–1202; H. Pellat, "Champ Magnétique Auquel est Soumis un Corps en Mouvement dans un Champ Électrique," *CRAS* CXL (1905): 229–231.

73. H. Pellat, *Complément de la Notice sur les Travaux Scientifiques de M. Henri Pellat* (Gauthier-Villars, Paris, 1906), 1.

74. *CRAS* CXXXIX (1904): 1123.

75. Abraham and Langevin 1905.

76. Paul Villard, "La Formation des Rayons Cathodiques," in Abraham and Langevin 1905, 1013–1028.

77. *Ibid.*, 1024.

78. H. Pellat, "Reproduction des Notes ou Mémoires de M. Pellat sur les Corpuscules," in Abraham and Langevin 1905, 547–557.

79. On this evolution of *Le Radium*, cf. John L. Davis, "Physics in France Circa 1850–1914. Its National Organisation, Characteristics and Content," unpublished doctoral thesis, (University of Kent, 1990), 269, 282.

80. P. Villard, "Sur l'Aurore Boréale," *CRAS* CLII (1906): 1330–1333; *CRAS* CLIII (1906): 143–145, 587–588; "Sur l'Induction Unipolaire et la Cause Probable des Aurores Polaires," *CRAS* CLVII (1908): 740–742.

81. See for instance: P. Villard, "Sur le Mécanisme de la Lumière Positive," *CRAS* CLII (1906): 706–709; "Sur Certains Rayons Cathodiques," *CRAS* CLIII (1906): 674–676.

82. P. Villard, *Les Rayons Cathodiques* (Gauthier-Villars, Paris, 1908), 103. Villard's emphasis.

83. *Ibid.*, 104.

84. *CRAS* CLVII (1908): 949, 960.

85. *Ibid.*, 1447, 1460.

86. On Branly, see the special issue: "Edouard Branly et la T.S.F: Tradition ou Innovation?," *Revue d'Histoire des Sciences*, 26 (1993).

87. For a vivid description of the election proceedings at the l'Académie des Sciences, cf. Robert Reid, *Marie Curie. Derrière la Légende* (Seuil, Paris, 1979), 186–188.

88. P. Villard, "Rapport lu Devant l'Académie des Sciences sur les Titres Scientifiques de M. Paul Langevin, Professeur au Collège de France, Comité secret du 8 février 1909," 4 pages manuscript, (*AS,* Paul Villard file).

89. For a comparison between both national contexts, see Lelong 1996.

90. Falconer 1997. A. Warwick, "Cambridge Mathematics and Cavendish Physics: Cunningham, Campbell and Einstein's relativity, 1905–1911," *Stud. Hist. Phil. Sci.* 23, 4 (1992): 625–656 and 24, 1 (1993): 1–25.

91. B. Lelong, "Constant Values, Real Particles: the Importation of Cambridge Metrology and Gaseous Ions to France, 1898–1906," unpublished paper, presented at the workshop

"Physics, Industry, Philosophy: 1900–1950," Department of History and Philosophy of Science, Cambridge University, 1 December 1995.

REFERENCES

Abbreviations used:
ACP *Annales de Chimie et de Physique*
AS   Archives of the Académie des Sciences, Paris
BJHS *British Journal for the History of Science*
BSFP *Bulletin des séances de la Societé Française de Physique*
CRAS *Comptes Rendus Hebdomadaires des séances de l'Académie des Sciences*
FL   Fonds Langevin, Ecole de Physique et de Chimie, Paris
HSPS *Historical Studies in the Physical Sciences*
JP   *Journal de Physique*
PM   *Philosophical Magazine*

Abraham, Henri, and Langevin, Paul. 1905. eds. *Les Quantités Élémentaires d'Électricité. Ions, Électrons, Corpuscules.* Paris: Gauthier-Villars.

Buchwald, Jed. 1985. *From Maxwell to Microphysics: Aspects of Electromagnetic Theory in the Last Quarter of the Nineteenth Century.* Chicago: University of Chicago Press.

Falconer, Isobel. 1985. "Theory and experiment in J. J. Thomson's Work on Gazeous Discharge." Doctoral Thesis. University of Bath.

Falconer, Isobel. 1987. "Corpuscles, Electrons, and Cathode rays: J. J. Thomson and the 'Discovery of the electron'." *BJHS* 20: 241–276.

Feffer, Stuart M. 1989. "Arthur Schuster, J. J. Thomson, and the Discovery of the Electron." *HSPS* 20: 33–61.

Galison, Peter, and Assmus, Alexi. 1989. "Artificial Clouds, Real Particles," in D. Gooding, T. Pinch, S. Schaffer eds. *The Uses of Experiment. Studies in the Natural Sciences.* Cambridge: Cambridge University Press, 225–274.

Hunt, Bruce. 1991. *The Maxwellians.* Ithaca & London: Cornell University Press.

Jacques, Jacques. 1987. *Berthelot. Autopsie d'un Mythe.* Paris: Berlin.

Langevin, Paul. 1903. "L'ionisation des Gaz." *ACP* VII/28: 289–389.

Lelong, Benoit. 1996. "X-rays and the Discharge of Electricity through Gases in Cambridge and Paris, 1896–1898: A Comparative Study of Experimental Practice and Laboratory Culture" in Dieter Hoffmann, Fabio Bevilacqua and Roger Stuewer eds. *The Emergence of Modern Physics.* Università degli Studi di Pavia, 267.

Lelong, Benoit, "Translating Ion Physics from Cambridge to Oxford: John Townsend and the Electrical Laboratory, 1900–1914," in Robert Fox and Graeme Gooday eds. *The Clarendon Laboratory and Oxford Physics to 1939.* Oxford (forthcoming).

Lelong, Benoit. *Langevin et les Ions. L'importation de la Physique de Cambridge en France, 1896–1914*. Paris: Editions des Archives Contemporaines. (forthcoming).

Robotti, Nadia. 1995. "J. J. Thomson at the Cavendish Laboratory: The History of an Electric Charge Measurement." *Annals of Science* LII, 3: 265–284.

Shinn, Terry. 1980. "Orthodoxy and Innovation in Science: The Atomist Controversy in French Chemistry." *Minerva* XVIII/4: 539–555.

Thomson, Joseph-John. 1897. "Cathode Rays." *PM* V/44: 293–316.

Villard, Paul. 1899. "Sur les Rayons Cathodiques." *JP* VIII: 5–16, 148–161.

Villard, Paul. 1900a. "Les Rayons Cathodiques et les Rayons Röentgen." *Bulletin de la Société Internationale des Electriciens* XVII/1 (1900): 169–184.

Villard, Paul. 1900b. "Les Rayons Cathodiques," in *Rapports présentés au Congrès International de Physique*. Paris: Gauthier-Villars, tome III: 115–137.

II

WHAT WAS THE NEWBORN ELECTRON GOOD FOR?

5

The Zeeman Effect and the Discovery of the Electron
Theodore Arabatzis

The discovery of the electron is usually attributed to J. J. Thomson and assigned a specific date and location. On this widely accepted view, the electron was discovered by Thomson in 1897, while he was experimenting on cathode rays at the Cavendish Laboratory.[1] This attribution is problematic, both from a philosophical and a historiographical point of view. On the philosophical side, it presupposes a realist perspective toward unobservable entities and requires a theory of scientific discovery that would support such a perspective. As far as I can tell, no such adequate theory has been developed. On the historiographical side, this attribution downplays several British and continental developments that were quite decisive for the gradual acceptance of the electron as a universal, subatomic constituent of matter. In this chapter I want to examine one of those developments, an experimental discovery (the magnetic splitting of spectral lines) by the Dutch physicist Pieter Zeeman, and its effect on the main electromagnetic theories of the time by H. A. Lorentz and Joseph Larmor. As I will show, Zeeman's discovery was crucial for the initial articulation of the concept of the electron within the theoretical framework provided by Lorentz and Larmor and played a very important role in convincing physicists of the reality of the electron. Furthermore, I will address the question of whether Zeeman should also be considered as a discoverer of the electron.

On Scientific Discovery

Before proceeding to the historical reconstruction, some methodological remarks about scientific discovery are in order. To talk about the discovery of an unobservable entity, like the electron, it is necessary to specify some criteria as to what constitutes a discovery of this kind. Antirealist philosophers would deny the possibility of finding such criteria, since from their point of view one has to be agnostic with respect to the existence of unobservable entities.[2] Realist philosophers, on the other hand, would have to suggest what constitutes an adequate demonstration for the existence of such entities. A realist would have to propose certain epistemological criteria whose satisfaction would provide adequate grounds for believing in the existence of an unobservable entity. Then he could reconstruct the discovery episode in question by showing how an individual or a group managed to meet the required criteria.

It is evident that the adequacy of the proposed way for deciding when something qualifies as a genuine discovery depends on the adequacy of the epistemological criteria for what constitutes unobservable reality. Any difficulties that might plague the latter would cast doubt on the adequacy of the former. Although this approach can be, in principle, realized, no adequate proposal of the kind outlined has been made so far. That is, no epistemological criteria have been formulated whose satisfaction would amount to an existence-proof of an unobservable entity.

Thus, the historical reconstruction of discovery episodes appears to require a resolution of one of the most intricate debates in philosophy of science. Rather than trying to resolve this debate, there is another way to approach discovery episodes that avoids philosophical pitfalls. One should simply try to adopt the perspective of the relevant historical actors, without worrying whether that perspective can be justified philosophically.[3] On this approach, the discovery of an entity amounts to the formation of consensus within the scientific community about its existence. Given the realist connotations of the term "discovery," one might even avoid using it when writing the history of a concept denoting an unobservable entity. In undertaking such a task, one would show how the given entity was introduced into the scientific literature and would reconstruct the experimental and theoretical arguments that were given in favor of its existence. The next step would be to trace the developmental process that followed the introduction of that entity and gradually transformed the concept associated with it. The evolution of any such concept resembles a process of gradual construction that takes place in several stages and, thus, can be periodized.[4] A realist might want to label the first stage of that process "the stage of discovery," but this would make no difference whatsoever with respect to the adequacy of the historical reconstruction.[5]

The main advantage of this approach is that it enables the reconstruction of past scientific episodes without presupposing the resolution of pressing philosophical issues. Since the debate on scientific realism goes on and has proved, so far, inconclusive, it is preferable to avoid historical narratives based, explicitly or implicitly, on realist premises. The intricacies of that debate suggest that an agnostic perspective is best suited for reconstructing the "discovery" of unobservable entities.

What I have said so far relies on the distinction between observable and unobservable entities, since my suggestion to avoid the category of discovery concerns unobservables. On the other hand, I do not wish to imply that the discovery of observable entities and phenomena should be treated in a similar agnostic fashion. In this case the category of discovery might be retained. It might be possible to specify when, say, a new species has been discovered, without relying on the notion of consensus within the relevant scientific community.

The question that immediately arises is why one should adopt different stances in the two cases. For two reasons, I think. First, because the realism debate has focused on the existence of unobservable entities, with both sides sharing a belief in the existence of observable objects and phenomena. Second, because to talk about the discovery of an unobservable entity one has to face a difficulty that does not appear in the case of observables. The discovery of an observable entity might simply involve its direct observation and does not require that all, or even most, of the discoverer's beliefs about it are true. For example, to discover "that there is a person in the ditch, . . . not every belief about that person needs to be true or known to be true."[6] This is not the case, however, when it comes to unobservable entities where direct physical access is, in principle, unattainable. The lack of independent access to such an entity makes problematic the claim that the discoverer's beliefs about it need not be true. If most, or even some, of those beliefs are not true it is not evident that the "discovered" entity is the same with its contemporary counterpart. It has to be shown, for instance, that Thomson's "corpuscles," which were conceived as classical particles and structures in the ether, can be identified with contemporary "electrons," which are endowed with quantum numbers, wave-particle duality, indeterminate position-momentum, etc. This would require, among other things, a philosophical theory of the meaning of scientific terms that would enable one to establish the referential stability of a term, despite a change of its meaning. In the philosophical literature there have been such proposals, most notably by Hilary Putnam, which are applicable to terms denoting observable objects. It is not clear, however, how these proposals would handle terms with unobservable referents.[7] Once more, one sees that an attempt to retain the category of scientific discovery with respect to unobservables leads us to philosophical deep water that a historian would rather avoid.

Let us now turn to Zeeman's discovery, which not only provided evidence for the existence of the electron but also led to a specification of two of its properties, its charge to mass ratio and the sign of its charge.

Zeeman's Discovery[8]

Pieter Zeeman (1856–1943) began to study magnetooptical phenomena in 1890, as Lorentz's assistant at the University of Leiden. The first phenomenon he investigated was the Kerr effect—the rotation of the plane of polarization of light upon reflection from a magnetized substance. The investigation of this phenomenon was also the subject of his doctoral dissertation, which he completed in 1893, under the supervision of Kamerlingh Onnes.[9] In the course of that research he made an unsuccessful attempt to detect the influence of a magnetic field on the sodium spectrum.[10] Several years later, inspired by reading "Maxwell's sketch of Faraday's life" and finding out that "Faraday thought of the possibility of the above mentioned relation [between magnetism and light]," he thought that "it might yet be worthwhile to try the experiment again with the excellent auxiliaries of the spectroscopy of the present time."[11] This time the experiment turned out to be a success.[12]

Zeeman placed the flame of a Bunsen burner between the poles of an electromagnet and held a piece of asbestos impregnated with common salt in the flame. After turning on the electromagnet, the two D-lines of the sodium spectrum, which had been previously narrow and sharply defined, were clearly widened. In shutting off the current the lines returned to their former condition. Zeeman then replaced the Bunsen burner with a flame of light gas fed with oxygen and repeated the experiment. The spectral lines were again clearly broadened. Replacing the sodium by lithium he observed the same phenomena.

Zeeman was not convinced that the observed widening was due to the action of the magnetic field directly upon the emitted light. The effect could be caused by an increase of the radiating substance's density and temperature. As noted by Zeeman, a similar phenomenon had been reported by Pringsheim in 1892.[13] Since the magnet caused an alteration of the flame's shape, a subsequent change of the flame's temperature and density was also possible. To exclude this possibility, Zeeman tried another more complicated experiment. He put a porcelain tube horizontally between the poles of the electromagnet, with the tube's axis perpendicular to the direction of the magnetic field (figure 5.1). Two transparent caps were attached to each terminal of the tube and a piece of sodium was introduced into the tube. Simultaneously the tube's temperature was raised by the Bunsen burner. At the same time the light of an electric lamp was guided by a metallic mirror to traverse the entire tube.

In the next stage of the experiment the sodium, under the action of the Bunsen flame, began to gasify. The absorption spectrum was obtained by means

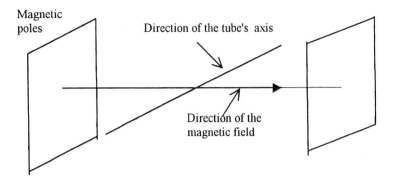

Figure 5.1

of the Rowland grating and finally the two sharp D-lines of sodium were observed. The heterogeneity of the density of the vapor at different heights of the tube produced a corresponding asymmetry in the lines' width, making them thicker at the top. By activating the electromagnet the lines became broader and darker. When it was turned off the lines recovered their initial form.

Zeeman, however, was still skeptical about whether the experiment's purpose, to demonstrate the direct effect of magnetism on light, had been accomplished. The temperature difference between the upper and lower parts of the tube was responsible for the heterogeneity of the vapor's density. The vapor was denser at the top of the tube and, since their width at a certain height depended on the number of incandescent particles at that height, the spectral lines were therefore thicker at the top. It was conceivable that the activation of the magnetic field could give rise to differences of pressure in the tube of the same order of magnitude and in the opposite direction to those produced by the differences of temperature. If this were the case, the action of magnetism would move the denser layers of vapor toward the bottom of the tube and would alter in this way the width of spectral lines without interacting directly with the light that generated the spectrum.

To exclude the possibility of these phenomena, which would undermine the experiment's aim, Zeeman performed a more refined experiment. He used a smaller tube and heated it with a blowpipe to eliminate disturbing temperature differences. Moreover, he rotated the tube around its axis and thus achieved equal densities of sodium vapor at all heights. The D-lines were now uniformly wide along their whole length. The subsequent activation of the electromagnet resulted in their uniform broadening.

Zeeman was by then nearly convinced that the outcome of his experiments was due to the influence of magnetism directly upon the light emitted or absorbed by sodium: "The different experiments ... make it more and more probable, that the absorption—and hence also the emission—lines of an incandescent vapor, are widened by the action of magnetism."[14] The sentence immediately following is instructive with respect to the theoretical significance of Zeeman's experimentation: "Hence if this is really the case, then by the action of magnetism in addition to the free vibrations of the *atoms, which are the cause of the line spectrum,* other vibrations of changed period appear"[15] (emphasis added). It is evident that Zeeman identified the origin of spectral lines with the vibration of atoms. H. A. Lorentz, Zeeman's mentor and collaborator, had developed a theory of electromagnetic phenomena that accounted for the emission of light in this way. As the above excerpt indicates, Lorentz's theory could be used to provide a theoretical understanding of Zeeman's experimental discovery. As it turned out, that theory guided Zeeman's subsequent experimental researches and was, in turn, shaped by them. Let us examine more closely the state of Lorentz's theory at that time.

Lorentz's Theory of "Ions" and its Impact on Zeeman's Investigations

In 1878 Lorentz had already suggested that the phenomenon of dispersion could be explained by assuming that molecules are composed of charged particles that may perform harmonic oscillations.[16] In 1892 he developed a unification of the continental and the British approaches to electrodynamics, which incorporated those particles. From the British approach he borrowed the notion that electromagnetic disturbances travel with the speed of light. That is, his theory was a field theory that dispensed with action-at-a-distance. From the continental approach he borrowed the conception of electric charges as ontologically distinct from the field. Whereas in Maxwell's theory charges were mere epiphenomena of the field, in Lorentz's theory they became the sources of the field.[17]

The aim of Lorentz's combined approach, in 1892, was to analyze electromagnetic phenomena in moving bodies. That analysis required a model of the interaction between matter and ether. The notion of "charged particles" provided him with a means of handling this problem.[18] The interaction in question could be understood if one reduced all "electrical phenomena to [. . . the] displacement of these particles."[19] The movement of a charged particle altered the state of the ether which, in turn, influenced the motion of other particles. Furthermore, macroscopic charges were "constituted by an excess of particles whose charges have a determined sign, [and] an electric current is a true stream of these corpuscles."[20] This proposal was similar to the familiar conception of the passage of electricity through electrolytic solutions and metals.

It is worth pointing out that in the last section of his 1892 paper Lorentz deduced a formula for the velocity of light in moving media that had been derived by Fresnel on the assumption that the ether was dragged by moving matter. Lorentz's derivation, however, discarded that assumption and capitalized on the influence of light on moving charged particles. The latter were forced to vibrate by the ethereal waves constituting light and gave rise to a complex interaction that produced the effect named after Fresnel. Lorentz's analysis enhanced considerably the credibility of his theory and facilitated the acceptance of his "charged particles" as real entities.[21]

In 1895 he explicitly associated those particles with the ions of electrolysis.[22] The transformation of "ions" to "electrons" took place as a result of Zeeman's experimental discovery, which after its initial stage was dominated by Lorentz's theory. To understand how this transformation took place it is necessary to examine Lorentz's theoretical analysis of Zeeman's initial results and its role in guiding further Zeeman's experimental research. The first form of that analysis is recorded in Zeeman's second paper on his celebrated discovery.[23] Zeeman initially thought that Lorentz's theory could provide an explanation of his experimental results. Thus, he asked Lorentz to provide a quantitative treatment of the influence of magnetism on light:

> Prof. Lorentz to whom I communicated these considerations, at once kindly informed me of the manner, in which according to his theory the motion of an ion in a magnetic field is to be calculated, and pointed out to me that, if the explanation following from his theory was true, the edges of the lines of the spectrum ought to be circularly polarized. The amount of widening might then be used to determine the ratio of charge and mass to be attributed in this theory to a particle giving out the vibrations of light.
>
> The above mentioned extremely remarkable conclusion of Prof. Lorentz relating to the state of polarization in the magnetically widened line, I have found to be fully confirmed by experiment.[24]

As I mentioned above, the emission of light, according to Lorentz, was a direct result of the vibrations of small electrically charged particles ("ions"), which exist in all material bodies. In the absence of a magnetic field an "ion" would oscillate about an equilibrium point under the action of an elastic force. The influence of a magnetic field would alter the mode of vibration of the "ion." Suppose that an "ion" is moving in the xy-plane under the action of a uniform magnetic field which is parallel to the z-axis. The equations of motion are:

$$m\frac{d^2x}{dt^2} = -k^2x + eH\frac{dy}{dt},$$

$$m\frac{d^2y}{dt^2} = -k^2y - eH\frac{dx}{dt},$$

where e and m are the charge and the mass of the "ion" respectively and H is the intensity of the magnetic field. The first term on the right side of the equations denotes the elastic force and the second term represents the force due to the magnetic field (the "Lorentz" force). Assuming that

$$x = ae^{st} \text{ and } y = \beta e^{st},$$

we get

$$ms^2 a = -k^2 a + eHs\beta,$$

$$ms^2 \beta = -k^2 \beta - eHsa.$$

In the absence of a magnetic field ($H = 0$), we can easily obtain the period of vibration of the ion:

$$s = i\frac{k}{\sqrt{m}} = i\frac{2\pi}{T} \Rightarrow T = \frac{2\pi\sqrt{m}}{k}.$$

When a magnetic field is present the period becomes

$$s \cong i\frac{k}{\sqrt{m}}\left(1 \pm \frac{eH}{2k\sqrt{m}}\right) \Rightarrow T' \cong \frac{2\pi\sqrt{m}}{k}\left(1 \pm \frac{eH}{2k\sqrt{m}}\right).$$

It follows that

$$\frac{T'-T}{T} = \frac{eH}{2k\sqrt{m}} = \frac{e}{m}\frac{HT}{4\pi}. \tag{1}$$

The physical implications of this analysis are as follows:[25] In the general case, the oscillation of the 'ion' has an arbitrary direction in space. In the absence of a magnetic field the motion of the 'ion' can be resolved into three components: a linear oscillation and two circular oscillations in a plane perpendicular to the first. All three oscillations have the same frequency, and the two circular ones have opposite directions. When a magnetic field is present, the oscillations along the direction of the field remain unaltered. But one of the circular components is accelerated, while the other is retarded. Thus, under the influence of magnetism the charged particle will yield three distinct frequencies. If the particle is observed along the direction of the field a doublet of lines will be seen. Each line represents circularly polarized light. If it is observed in a direction perpendicular to the field, a triplet of lines will be seen. The middle component represents plane-polarized light, its plane of polarization being parallel to the field. The two outer components also represent plane-polarized light, but their plane of polarization is perpendicular to the field.

All these theoretical expectations were subsequently confirmed by experiments designed specifically to detect them. In the same paper that contained Lorentz's analysis Zeeman confirmed that the polarization of the edges of the broadened lines followed the theoretical predictions. Lorentz considered the confirmation of his predictions as "direct proof for the existence of ions."[26] Furthermore, Zeeman estimated the order of magnitude of the ratio e/m. As we saw, the change in the period of vibration of an 'ion' due to the influence of a magnetic field depends on e/m (see equation 1 above). Thus, the widening of spectral lines, which is a reflection of the alteration in the mode of vibration of an 'ion,' is proportional to the 'ionic' charge to mass ratio. According to Zeeman's approximate measurements a magnetic field of 10000 Gauss produced a widening of the D-lines equal to 2.5 percent of their distance. From the observed widening of the spectral lines, Zeeman calculated (using equation 1) e/m, which turned out to be unexpectedly large (10^7 e.m.u.). As he recalled, when he announced the result of his calculation to Lorentz, the latter's response was: "That looks really bad; it does not agree at all with what is to be expected."[27]

It should be noted that this was the first estimate of the charge to mass ratio of the 'ions' that indicated that the 'ions' did not refer to the well-known ions of electrolysis, but corresponded instead to extremely minute subatomic particles. J. J. Thomson's measurement of the mass-to-charge ratio of the particles that constituted cathode rays was announced several months later and was in close agreement with Zeeman's result.[28] It is worth pointing out that the priority of Zeeman over Thomson was not always acknowledged. Oliver Lodge, for instance, claimed that Zeeman's results were obtained after Thomson's measurements.[29] Not surprisingly, Zeeman did not appreciate that remark. In a letter to Lodge, praising "your book on electrons" and thanking him for being "kind enough to send me a copy," he defended his priority over Thomson:

> May I make a remark concerning the history of the subject? On p. 112 of your book you mention that the small mass of the electron was deduced from the radiation phenomena in the magnetic field, the result "being in general conformity with J. J. Thomson's direct determination of the mass of an electron *some months previously.*" I think, my determination of e/m

being of order 10^7 has been previous to all others in this field. My paper appeared in the "Verslagen" of the Amsterdam Academy of October and November 1896. It was translated in the "Communications from the Leyden Laboratory" and then appeared in the Phil. Mag. for *March* 1897. Prof. Thomson's paper on cathode rays appeared in the Phil. Mag. for *October* 1897. [Emphasis in the original.][30]

Even though Zeeman neglected to mention that an early report of Thomson's measurements appeared in April 1897,[31] his complaint was justified. Thomson's supposed priority, however, continued to be promoted. In 1913, for instance, Norman Campbell erroneously suggested that Thomson's measurement of the charge to mass ratio of cathode ray particles preceded Zeeman's estimate of e/m.[32] Millikan also spread the same mistaken view.[33]

The splitting of lines was initially observed by Zeeman in 1897.[34] Instead of sodium he had used cadmium. Its indigo line was found to split into a doublet or triplet depending on whether the light was emitted in a direction parallel or perpendicular to the magnetic field. This stage of Zeeman's experimentation was dominated completely by the theoretical insight of Lorentz. Lorentz's theoretical anticipations led to new aspects of the novel phenomenon. The refinement of the experiment, however, soon led to theoretical advances. For instance, from the direction of polarization of the higher frequency component of the doublet Zeeman inferred that the charge of the 'ions' was negative.[35] Moreover, he gave a more accurate value of e/m and finally, by considering this unexpectedly large ratio, he was able to distinguish the 'ions' from the electrolytical ions.

As a result of Zeeman's discovery, the assumption that the radiating particles were as massive as hydrogen ions was abandoned and Lorentz's theory of ions was subsequently transformed into his theory of electrons. Zeeman's discovery had a similar effect on the transformation of the "ion's" British counterpart—the electron, as is testified to by Joseph Larmor's work.[36]

Larmor's "Electron" and its Transformation by Zeeman's Discovery

The name 'electron' was introduced by George Johnstone Stoney in 1891 to denote an elementary quantity of electricity.[37] At the Belfast meeting of the British Association in 1874 he had already suggested that "Nature presents us in the phenomenon of electrolysis, with a single definite quantity of electricity which is independent of the particular bodies acted on."[38] In 1891 he proposed that "it will be convenient to call [these elementary charges] *electrons*."[39] Stoney's electrons were permanently attached to atoms, that is, they could "not be removed from the atom," and each of them was "associated in the chemical atom with each bond." Furthermore, their oscillation within molecules gave rise to "electromagnetic stresses in the surrounding aether."[40]

In 1894 Stoney's electron was appropriated by Joseph Larmor, "at the suggestion of G. F. FitzGerald,"[41] to resolve a problem situation that had emerged in the context of the Maxwellian research tradition.[42] Larmor's adoption of the electron represented the culmination (and perhaps the abandonment) of that tradition. A central aspect of the research program initiated by Maxwell was that it avoided microscopic considerations altogether and focused instead on macroscopic variables (e.g., field intensities). This macroscopic approach ran into both conceptual and empirical problems. Its main conceptual shortcoming was that it proved unable to provide an understanding of electrical conduction. Its empirical defects were numerous: "It could not explain the low opacity of metal foils, or dispersion, or the partial dragging of light waves by moving media, or a number of puzzling magnetooptic effects."[43] It was in response to these problems that Larmor started to develop a theory whose aim was to explain the interaction between ether and matter.

The first stage in that development was completed with the publication of "A Dynamical Theory of the Electric and Luminiferous Medium. Part I" in August 1894.[44] Its initial version was submitted to the *Philosophical Transactions* on 15 November 1893 and was revised considerably in the months that preceded its publication under the critical guidance of FitzGerald. What is crucial here is that the published version concluded with a section, added on 13 August, titled "Introduction of Free Electrons."[45]

According to Larmor's representation of field processes, "the electric displacement in the medium is its absolute rotation . . . at the place, and the magnetic force is the velocity of its movement. . . ."[46] For a medium to be able to sustain electric displacement it must have rotational elasticity. In the original formulation of his theory conductors were conceived as regions in the ether with zero elasticity, since Larmor had "assumed that the electrostatic energy is null inside a conductor."[47] Conduction currents were regarded, in Maxwellian fashion, as mere epiphenomena of underlying field processes and were represented by the circulation of the magnetic field in the medium encompassing the conductor.

To explain electromagnetic induction, Larmor had to find a way in which a changing electric displacement would change that circulation. If conductors were totally inelastic, a changing displacement in their vicinity could not affect them.[48] Therefore, Larmor had to endow conductors with the following peculiar feature: they were supposed to contain elastic zones that were affected by displacement currents and were the vehicle of electromagnetic induction. This implied that in conductors the ether had to be ruptured, a consequence strongly disliked by Larmor. This problem could be circumvented, however, if one assumed that the process of conduction amounted to charge convection.[49] As he remarked,

> If you make up the world out of monads, electropositive and electronegative, you get rid of any need for such a barbarous makeshift as rupture of the aether . . . . A monad or an atom is what a geometer would call a "singular point" in my aether, i.e., it is a singularity naturally arising out of its constitution, and not something foreign to it from outside.[50]

There was another conceptual problem related to the phenomenon of electromagnetic induction. Larmor had initially appropriated William Thomson's conception of atoms as vortices in the ether, and he suggested that magnetism was due to closed currents within those atoms (already postulated by Ampère).[51] FitzGerald pointed out, however, that currents of this kind would not be affected by electromagnetic induction, since the ether could not get a hold on them. To solve this problem, Larmor suggested that the currents in question were unclosed. In connection with this issue FitzGerald sent a letter to Larmor which provided the inspiration for the introduction of the electron:[52]

> I don't see where you *require* a discrete structure except that you *say* that it is required in order to make the electric currents unclosed, yet I think that electrolytic and other phenomena prove that there is this discrete structure and you *do* require it, where you *don't* call attention to it, namely where you speak of a rotational strain near an atom. You *say* that electric currents are unclosed vortices but I can't see that this *necessitates* a *molecular* structure because in the matter the unclosedness might be a continuous peculiarity so far as I can see. That it is molecular is due to the molecular constitution of matter and not to any necessity in your theory of the ether.[53]

FitzGerald's point was that the discrete structure of electricity was an independently established fact that did not follow from Larmor's theory, but had to be added to it.

In a few months Larmor reconstructed his theory on the basis of FitzGerald's suggestion. Currents were now identified with the transfer of free charges ("monads"), which were also the cause of magnetic phenomena. Those charges had the ontological status of independent entities and ceased to be epiphenomena of the field. Furthermore, material atoms were represented as stable configurations of electrons. In Larmor's words, the core of the vortex ring [constituting an atom] . . . [is] made up of discrete electric nuclei or centres of radial twist in the medium. The circulation of these nuclei along the circuit of the core would constitute a vortex . . . its strength is now subject to variation owing to elastic action, so that the motion is no longer purely cyclic. A magnetic atom, constructed after this type, would behave like an ordinary electric current in a nondissipative circuit. It would, for instance, be subject to alteration of strength by induction when under the influence of other changing currents, and to recovery when that influence is removed.[54]

Thus, the problem that FitzGerald had brought up disappeared, since the ether could now get a hold on the core of the vortex ring and the atomic currents could be influenced by electromagnetic induction.

In July 1894 FitzGerald suggested the word "electron" to Larmor, as a substitute for the familiar "ion." In FitzGerald's words, Stoney "was rather horrified at calling these ionic charges 'ions.' He or somebody has called them 'electrons' and the ion is the atom not the electric charge."[55] This was the first hint of the need for a distinction between the entities introduced by Larmor and the well-known electrolytical ions. This distinction was obscured, however, by the fact that the effective mass of Larmor's electrons was of the same order of magnitude with the mass of the hydrogen ion. In this respect the subsequent discovery of the Zeeman effect was crucial, since it indicated that the electron's mass was three orders of magnitude smaller than the ionic mass (see below for details).

Larmor's "electrons" were conceived as permanent structures in the ether with the following characteristics:

An electron has a vacuous core round which the radial twist is distributed. . . . It may be set in radial vibration, say pulsation, and this vibrational energy will be permanent, cannot possibly be radiated away. All electrons being alike have the same period: if the amplitudes and phases are also equal for all at any one instant, they must remain so . . . Thus an electron has the following properties, which are by their nature permanent

 (i) its strength [= electric charge]
 (ii) its amplitude of pulsation
(iii) the phase of its pulsation.

These are the same for all electrons. . . . The equality of (ii) and (iii) for all electrons may be part of the pre-established harmony which made them all alike at first,—or may, very possibly, be achieved in the lapse of aeons by the same kind of averaging as makes the equalities in the kinetic theory of gases.[56]

Furthermore, he suggested that they were universal constituents of matter. He had two arguments to that effect. First, spectroscopic observations in astronomy indicated that matter "is most probably always made up of the same limited number of elements."[57] This would receive a straightforward explanation if "the atoms of all the chemical elements [were] to be built up of combinations of a single type of primordial atom."[58] Second, the fact that the gravitational constant was the same in all interactions between the chemical elements indicated that "they have somehow a common underlying origin, and are not merely independent self-subsisting systems."[59]

Larmor's electronic theory of matter received strong support from experimental evidence. First, it could explain the Michelson-Morley experiment. Inspired by Lorentz, Larmor managed to derive the so-called FitzGerald contraction hypothesis, which had been put forward to accommodate the null result of that experiment.[60] As he mentioned in a letter to Lodge, "I have just found, developing a suggestion that I found in Lorentz, that if there is nothing else than electrons—i.e., pure singular points of simple definite type, the only one possible, in the aether—then movement of a body, *transparent* or *opaque,* through the aether *does actually* change its dimensions, just in such way as to verify Michelson's second order experiment."[61] Second, Fresnel had suggested that the ether was dragged by moving matter and had derived from this hypothesis a formula for the velocity of light in moving media. Larmor's theory was able to reproduce Fresnel's result: "The application [of electrons] to the optical properties of moving media leads to Fresnel's well known formula."[62]

The introduction of the electron initiated a revolution that resulted in the abandonment of central features of Maxwellian electrodynamics. Although in Larmor's theory, as in Maxwell's, the concept of charge was explicated in terms of the concept of the ether, there were significant differences between the two electromagnetic theories. In contrast to Maxwellian theory which did not attribute independent existence to charges, in Larmor's theory the electron acquired an independent reality. Furthermore, the macroscopic approach to electromagnetism was jettisoned and microphysics was launched. Conduction currents were represented as streams of electrons and dielectric polarization was attributed to the polarizing effect of an electric field on the constituents of molecules.

Larmor's "electron" was transformed as a result of Zeeman's discovery. Before that discovery, Larmor thought that a magnetic widening of spectral lines would be beyond experimental detection. The widening in question was proportional to the charge-to-mass ratio of the electron and, on the assumption that "electrons were of mass comparable to atoms," he was led to "the improbability of an observable effect."[63] Larmor's reaction to an announcement of Zeeman's discovery in *Nature*[64] shows that he immediately realized its far-reaching implications with respect to the characteristics of the electron. In a letter to Lodge, asking him to confirm Zeeman's results, he writes:

> There is an experiment of Zeeman's . . . which is fundamental + ought to be verified. . . . It demonstrates that a magnetic field can alter the free period of sodium vapor by a measurable amount. I have had the fact as I believe it is (on my views) before my mind for months . . . [but] it never occurred to me that it could be great enough to observe: and it needs a lot of proof that it is so.[65]

Several days later he was even more skeptical about the possibility of observing the effect: "I don't expect you will find the effect all the same. The only theory I have about it is that it must be extremely small."[66] Lodge managed to reproduce Zeeman's results and informed Larmor of his success several weeks after Larmor's initial request: "Did I tell you that I had verified Zeeman's result, to the extent of seeing the broadening of a Na line from a flame between magnetic poles. It is a *small* effect though."[67]

The implications of Zeeman's discovery were clear for Larmor:

> in an ideal simple molecule consisting of one positive and one negative electron revolving round each other, the inertia of the molecule would have to be considerably less than the chemical masses of ordinary molecules, in order to lead to an influence on the period, of the order observed by Dr. Zeeman.[68]

Furthermore, Zeeman's result and his subsequent estimate of e/m enabled Larmor and Lodge to determine a property of the electron that had been left unspecified in Larmor's theory, the electron's size. The value of e/m obtained by Zeeman together with the concept of electromagnetic mass made possible an estimate of the electron's size. The concept of electromagnetic mass was introduced by J. J. Thomson in 1881. A charged spherical body would possess, besides its material mass, an additional inertia due to its charge. The value of that inertia would depend on $\mu e^2/a$, where μ was the magnetic permeability of the ether and a the radius of the sphere.[69] Now assuming that the electron's mass was purely electromagnetic, one could calculate its size. Lodge performed the calculation and asked Larmor whether the result that he obtained was acceptable: "Zeeman's $e/m = 10^7$ means if $m = 2\mu e^2/3a$ that $a = 10^{-14}$ . . . is this too small for an electron?"[70]

Larmor's reply is very revealing with respect to the process that led to the construction of the concept of the electron:

> I don't profess to know à priori anything about the size or constitution of an electron except what the spectroscope may reveal. I do assert that a logical aether theory must drive you back on these electrons as the things whose mutual actions the aether transmits : but for that general purpose each of them is a point charge just as a planet is an attracting point in gravitational astronomy. But as regards their constitution am inclining to the view that an atom of 10^{-8} cm is a complicated sort of solar system of revolving electrons, so that the single electron is very much smaller, 10^{-14} would do very well—is in fact the sort of number I should have guessed.[71]

So, originally the concept of the electron was arrived in an a priori fashion, that is, as a solution to a theoretical problem. The remaining task was to construct its properties so as to accommodate the available empirical evidence. The size of the electron, for instance, was calculated by Lodge so as to "attain Zeeman's quantitative result."[72]

Larmor's detailed analysis of the Zeeman effect was completed by November 8, 1897.[73] Larmor considered "a single ion e, of effective mass M, describing an elliptic orbit under an attraction to a fixed centre proportional to the distance therefrom."[74] If a magnetic field was introduced, Larmor proved, by solving the corresponding equations of motion, that instead of the original frequency of vibration three distinct ones would appear: one of them would coincide with the original, whereas the other two would be shifted by an amount equal to $\pm e\mathrm{H}/4\pi \mathrm{M}c^2$. A "striking feature" of Larmor's analysis was "that the modification thus produced is the same whatever be the orientation of the orbit with respect to the magnetic field."[75] This feature resulted from a general theorem that he had managed to prove a few weeks before he submitted his paper to the *Philosophical Magazine*. In his words,

> the following math prop is true:—Consider any system of (say) *negative* ions, with charges proportional to their effective masses, attracting each other according to some laws & attracted to fixed centres anywhere on the axes of the magnetic field: then their motion when the magnetic field is turned on relative to an observer fixed is the same as when it was off relative to an observer attached to a frame rotating round the axe of the field H with ang. velocity $e\mathrm{H}/\mathrm{M}c^2$ where e/M is the constant charge/mass and c is the velocity of radiation.[76]

In this respect Larmor's analysis was superior to Lorentz's less general explanation of the results obtained by Zeeman. In other respects, such as the polarization of the emitted spectral lines, Larmor reached identical conclusions to those obtained by Lorentz (see above). Larmor's analysis, in conjunction with Zeeman's experiments, enabled the approximate estimate of e/M. As it turned out, "the effective mass of a revolving ion, supposed to have the full unitary charge or electron, is about 10^{-3} of the mass of the atom."[77]

As a result of Larmor's work and the support that it received by Zeeman's experiments, by 1898 the electron had become an essential ingredient of British scientific practice in the domain of electromagnetism.[78]

To summarize here, Zeeman's discovery was crucial with respect to the "discovery of the electron" in three respects. First, it provided direct empirical support for Lorentz's and Larmor's postulation of the ion-electron. As Zeeman remarked, it "furnishes, as it occurs to me, direct experimental evidence for the existence of electrified ponderable particles (electrons) in a flame."[79] Second, it led to an approximately correct value of a central property of the electron, namely its charge to mass ratio. The small value of that ratio indicated that Lorentz's "ions" were different from the ions of electrolysis and, thus, led to a revision of the taxonomy of the unobservable realm. Whereas before Zeeman's experiments the term "ions" denoted the ions of electrolysis as well as the entities producing electromagnetic phenomena, after those experiments the extension of the term was restricted to the ions of electrolysis. That is why Lorentz started using the expression "light-ions" to refer to the entities of his electromagnetic theory,[80] and later adopted the term "electrons."[81] Third, Zeeman's results in conjunction with Lorentz's analysis of optical dispersion led to an estimate of the light-ion's mass. In particular, using his equations for dispersion Lorentz expressed the light-ion's mass as a function of e/m. By substituting Zeeman's estimate of that ratio, he obtained a value of the mass in question that was approximately 350 times smaller than the mass of the hydrogen atom.[82]

The significant contributions of Zeeman, Lorentz, and Larmor to the acceptance of the electron as a subatomic constituent of matter might (mis)lead us to the opinion that they should be given credit for the "discovery" of the electron. In fact, some have adopted this view. As early as 1901, Walter Kaufmann suggested that the existence of the electron had been established by Zeeman's discovery.[83] More recently, according to "the opinion of Leiden physicists, as told to me by H. B. G. Casimir, . . . Lorentz was the "discoverer" of the electron."[84] This view is subject to all the historiographical and philosophical problems that I have pointed out elsewhere in connection with the attribution of the electron's discovery to J. J. Thomson.[85] To begin with, we have no adequate philosophical theory of scientific discovery that could be used to justify the attribution of the electron's discovery to Zeeman. Furthermore, and more importantly, from the point of view of the physics community at that time Zeeman's experimental discovery did not establish, beyond doubt, belief in the existence of electrons.[86]

It should be clear that the purpose of the preceding narrative was not to settle a priority question and suggest that it was Zeeman, as opposed to Thomson, who discovered the electron. On the contrary, this narrative in conjunction with narratives about Thomson can help us to reconsider the historiographical issues related to the "discovery of the electron." What these narratives tell depends on the philosophical perspective adopted with respect to scientific realism and scientific discovery. One thing is, however, clear. The electron was not discovered by any particular scientist. The concept of the electron was introduced in physics in the early 1890s and was gradually transformed as a result of various theoretical and experimental developments in the context of electromagnetic theory and in the study of the discharge of electricity in gases. Several physicists, theoreticians and experimentalists provided evidence that supported the electron hypothesis. The most that can be said about one of those, say Zeeman, is that his contribution to the acceptance of the electron hypothesis was significant.

ACKNOWLEDGMENTS

I wish to thank Jed Buchwald and Kostas Gavroglu for their helpful comments. In what appears here I have drawn on my previous publications: Arabatzis (1992, 1996). Excerpts from the Joseph Larmor letters are reproduced by kind permission of the President and Council of the Royal Society of London. I would also like to thank University College London Library for permission to quote from Oliver Lodge's correspondence.

NOTES

1. For an elaboration and criticism of this view see T. Arabatzis, "Rethinking the 'Discovery' of the Electron," *Studies in History and Philosophy of Modern Physics* 27 (1996): 405–435.

2. Cf. B. C. van Fraassen, *The Scientific Image* (New York: Oxford University Press, 1980).

3. This does not imply that no such philosophical justification is possible.

4. A similar approach, with respect to the construction of the concept of the electron, has been followed by O. Darrigol, "Aux Confins de l'Électrodynamique Maxwelliene: Ions et Électrons vers 1897," *Rev. Hist. Sci.* 51/1 (1998): 5–34.

5. Only in this weak sense can the term "discovery" be used with respect to unobservable entities. In its stronger form (i.e., as implying existence) two further conditions are required. First, the consensus with respect to the reality of the entity in question should be maintained to this very day. Second, one should propose some criteria that enable us to identify the original entity with its present counterpart; but more on this below.

6. P. Achinstein, "Who Really Discovered the Electron?" (this volume, 416). Nevertheless, *some* beliefs about the discovered entity have to be true. For example, in the case of the

discovery of a man in the ditch we have to know that what we have discovered is a person and not, say, a stone (or, in general, an I-know-not-what).

7. For a detailed argument, along with the relevant literature, see T. Arabatzis, *The Electron: A Biographical Sketch of a Theoretical Entity* (Princeton University: Ph.D. Dissertation, 1995).

8. For a more detailed account of Zeeman's path to his discovery see T. Arabatzis, "The Discovery of the Zeeman Effect: A Case Study of the Interplay Between Theory and Experiment," *Studies in History and Philosophy of Science* 23 (1992): 365–388. Cf. also J. B. Spencer, *An Historical Investigation of the Zeeman Effect (1896–1913)* (Ph.D. Dissertation, The University of Wisconsin, 1964). A more recent study, based on an examination of Zeeman's laboratory notebooks, is A. J. Kox, "The Discovery of the Electron: II. The Zeeman Effect," *European Journal of Physics* 18 (1997): 139–144.

9. See J. B. Spencer, "Zeeman, Pieter," in C. C. Gillispie (ed.), *Dictionary of Scientific Biography*, 16 vols. (New York: Charles Scribner's Sons, 1970–1980), vol. 14, 597–599.

10. See P. Zeeman, "On the Influence of Magnetism on the Nature of the Light Emitted by a Substance. (part I.)" *Communications from the Physical Laboratory at the University of Leiden* 33 (1896): 1–8, 1; translated from *Verslagen van de Afdeeling Natuurkunde der Kon. Akademie van Wetenschappen te Amsterdam,* 31 October 1896, 181. There is no unpublished record of this early attempt. See Kox, "The Discovery of the Electron," 139–140.

11. Zeeman, "On the Influence of Magnetism . . . (part I)," 3. Those "excellent auxiliaries" included a concave Rowland grating that had recently been acquired by the Physical Laboratory of the University of Leiden.

12. There is a record of this experiment in Zeeman's laboratory notebook, dated September 2, 1896. See Kox, "The Discovery of the Electron," 140. The description that follows is from Zeeman's published paper, which appeared on 31 October 1896. Kox's analysis of Zeeman's notebooks shows that Zeeman's publications gave a faithful account of his research.

13. See E. Pringsheim, "Kirchhoff'sches Gesetz und die Strahlung der Gase," *Wiedmannsche Annalen der Physik* 45 (1892): 428–459, 455–457.

14. Zeeman, "On the Influence of Magnetism . . . (part I)," 8.

15. Ibid.

16. See H. A. Lorentz, "Concerning the Relation Between the Velocity of Propagation of Light and the Density and Composition of Media," in P. Zeeman and A. D. Fokker (eds.), *H. A. Lorentz: Collected Papers,* 9 vols. (The Hague: Martinus Nijhoff, 1935–1939), vol. 2, 1–119.

17. H. A. Lorentz, "La Théorie Électromagnétique de Maxwell et son Application aux Corps Mouvants," in his *Collected Papers,* vol. 2, 164–343, esp. 229. Cf. R. McCormmach, "Einstein, Lorentz, and the Electron Theory," *Historical Studies in the Physical Sciences* 2 (1970): 41–87.

18. See T. Hirosige, "Origins of Lorentz' Theory of Electrons and the Concept of the Electromagnetic Field," *Historical Studies in the Physical Sciences* 1 (1969): 151–209,

178–179, 198; and N.J. Nersessian, *Faraday to Einstein: Constructing Meaning in Scientific Theories* (Dordrecht: Martinus Nijhoff, 1984), 98.

19. "[L]es phénomènes électriques sont produits par le déplacement de ces particules." Lorentz, "La Théorie Électromagnétique de Maxwell," 228.

20. "[U]ne charge électrique est constituée par un excès de particules dont les charges ont un signe déterminé, un courant électrique est un véritable courant de ces corpuscules." Lorentz, "La Théorie Électromagnétique de Maxwell," 228–229. Cf. J. L. Heilbron, *A History of the Problem of Atomic Structure from the Discovery of the Electron to the Beginning of Quantum Mechanics* (University of California, Berkeley, Ph.D. dissertation, 1964), 98.

21. Cf. N.J. Nersessian, "Hendrik Antoon Lorentz," in *The Nobel Prize Winners: Physics* (Pasadena, CA: Salem Press, 1989), 35–42, esp. 39.

22. H. A. Lorentz, "Versuch einer Theorie der elektrischen und optischen Erscheinungen in bewegten Körpern," in his *Collected Papers,* vol. 5, 1–137, esp. 5.

23. See P. Zeeman, "On the Influence of Magnetism on the Nature of the Light Emitted by a Substance. (part II.)" *Communications from the Physical Laboratory at the University of Leiden* 33 (1896): 9–19; translated from *Verslagen van de Afdeeling Natuurkunde der Kon. Akademie van Wetenschappen te Amsterdam,* 28 November 1896, 242.

24. Ibid., 12.

25. Ibid., 16.

26. We know that from an entry in Zeeman's diary, dated 23 November 1896. See Kox, "The Discovery of the Electron," 142.

27. Cited in Zeeman, "Faraday's Researches on Magneto-Optics and their Development," *Nature* 128 (1931): 365–368, on 367. Even though Lorentz associated his 'ions' with the ions of electrolysis, he did not specify, to the best of my knowledge, their charge to mass ratio. Thus, there is no conclusive evidence that he assumed his 'ions' to be of an order of magnitude comparable with an atom's. Further indirect evidence is found in O. Lodge, "The History of Zeeman's Discovery and its Reception in England," *Nature* 109 (1922): 66–69, on 67. There we read that Larmor "*like everyone else at that time, . . .* considered that the radiating body must be an atom or part of an atom with an $e/m = 10^{4}$" (emphasis added). Moreover, Zeeman, while discussing various aspects of his discovery, remarked that "[t]he value found [for the charge to mass ratio of the 'ion'] is about 1500 times that of the corresponding value which can be derived for hydrogen from the phenomena of electrolysis. *This was something entirely new in 1896.*" (Emphasis added.) P. Zeeman, *Researches in Magneto-Optics: With Special Reference to the Magnetic Resolution of Spectrum Lines* (London: Macmillan, 1913), 39–40.

28. See J. J. Thomson, "Cathode Rays," *Proceedings of the Royal Institution* 15 (1897): 419–432.

29. O. Lodge, *Electrons, or the Nature and Properties of Negative Electricity* (London: George Bell and Sons, 1906), 112.

30. Zeeman to Lodge, 3 August 1907, University College Library, Lodge Collection, MS. Add 89/116.

31. Thomson, "Cathode Rays."

32. See N. R. Campbell, *Modern Electrical Theory,* 2nd edn (Cambridge: Cambridge University Press, 1913), 148–149.

33. See R. A. Millikan, *The Electron: Its Isolation and Measurement and the Determination of some of its Properties,* 2nd edn (Chicago: University of Chicago Press, 1924), 42–43.

34. P. Zeeman, "Doublets and Triplets in the Spectrum produced by External Magnetic Forces," *Philosophical Magazine,* 5th series, 44 (1897): 55–60, 255–259.

35. It should be noted that Zeeman initially reported that these polarization results led to the conclusion that the 'ions' were positively charged. See Zeeman, "On the Influence of Magnetism . . . (part II)," 18. He soon corrected his erroneous statement, however, in his following paper. See Zeeman, "Doublets and Triplets," 58. My attention was drawn to Zeeman's mistake by S. Endo and S. Saito, "Zeeman Effect and the Theory of Electron of H. A. Lorentz," *Japanese Studies in History of Science* 6 (1967): 1–18.

36. It is worth noting that Larmor acknowledged "Lorentz's priority about electrons which he introduced in 1892 very candidly." Larmor to Lodge, 7 February 1897, Univ. College London, Lodge Collection, MS. Add 89/65 (ii).

37. G. J. Stoney, "On the Cause of Double lines and of Equidistant Satellites in the Spectra of Gases," *The Scientific Transactions of the Royal Dublin Society,* 2nd series, 4 (1888–1892): 563–608, on 583. Cf. J. G. O'Hara, "George Johnstone Stoney, F.R.S., and the concept of the electron," *Notes and Records of the Royal Society of London* 29 (1975): 265–276; and J. G. O'Hara, "George Johnstone Stoney and the Conceptual Discovery of the Electron," *Stoney and the Electron–Papers from a Seminar held on November 20, 1991, to commemorate the centenary of the naming of the electron* (Dublin: Royal Dublin Society, 1993), 5–28.

38. Stoney's paper was first published in 1881. See G. J. Stoney, "On the Physical Units of Nature," *The Scientific Proceedings of the Royal Dublin Society,* new series, 3 (1881–1883): 51–60, on 54.

39. Stoney, "On the Cause of Double lines," 583.

40. Ibid.

41. J. Larmor, *Mathematical and Physical Papers,* 2 vols. (Cambridge: Cambridge University Press, 1929), vol. 1, 536. (Footnote added in the 1929 edition; not in the original paper.)

42. This tradition has been thoroughly studied by Jed Buchwald, Bruce Hunt, and Andrew Warwick. See J. Z. Buchwald, *From Maxwell to Microphysics: Aspects of Electromagnetic Theory in the Last Quarter of the Nineteenth Century* (Chicago: University of Chicago Press, 1985); B. J. Hunt, *The Maxwellians* (Ithaca: Cornell University Press, 1991); and A. Warwick, 'On the Role of the FitzGerald-Lorentz Contraction Hypothesis in the Development of Joseph

Larmor's Electronic Theory of Matter', *Archive for History of Exact Sciences* 43:1 (1991): 29–91. For what follows, I am indebted to their analysis.

43. Hunt, *The Maxwellians*, 210.

44. J. Larmor, *Philosophical Transactions of the Royal Society,* vol. 185, 719–822; repr. in his *Math. and Phys. Papers,* vol. 1, 414–535.

45. Ibid., 514–535.

46. Ibid., 447.

47. Ibid., 448.

48. Ibid., 462.

49. Cf. Buchwald, *From Maxwell,* 161.

50. Larmor to Lodge, 30 April 1894, University College Library, Oliver Lodge Collection, MS. Add 89/65(i); also quoted in Buchwald, *From Maxwell,* 152–153.

51. See Larmor, *Math. and Phys. Papers,* vol. 1, 467. Cf. Hunt, *The Maxwellians,* 218.

52. Cf. Buchwald, *From Maxwell,* 163–164.

53. FitzGerald to Larmor, 30 March 1894, Royal Society Library, Joseph Larmor Collection, 448. This excerpt is also reproduced in Buchwald, *From Maxwell,* 166.

54. Larmor, *Math. and Phys. Papers,* vol. 1, 515.

55. FitzGerald to Larmor, 19 July 1894; quoted in Hunt, *The Maxwellians,* 220.

56. Larmor to Lodge, 29 May 1895, Univ. College London, Lodge Collection, MS. Add 89/65 (i). Larmor's suggestion of a pulsating electron appeared in print. See J. Larmor, "A Dynamical Theory of the Electric and Luminiferous Medium—Part III: Relations with Material Media," *Philosophical Transactions of the Royal Society,* Febr. 9, 1898, repr. in his *Math. and Phys. Papers,* vol. 2, 11–132, on 25.

57. Larmor, *Math. and Phys. Papers,* vol. 1, 475.

58. Ibid. Cf. O. Darrigol, "The Electron Theories of Larmor and Lorentz: A Comparative Study," *Historical Studies in the Physical and Biological Sciences* 24 (1994): 265–336, 312.

59. Ibid.

60. See Warwick, "On the Role of the FitzGerald-Lorentz Contraction."

61. Larmor to Lodge, 29 May 1895, Univ. College Library, Lodge Collection, MS. Add 89/65 (i). This excerpt from Larmor's letter is reproduced in Warwick, "On the Role of the FitzGerald-Lorentz Contraction," 56.

62. J. Larmor, "A Dynamical Theory of the Electric and Luminiferous Medium. Part II: Theory of Electrons," *Philosophical Transactions of the Royal Society of London A* 186 (1895): 695–743; repr. in his *Math. and Phys. Papers,* vol. 1, 543–597, the quote is from p. 544. Cf. Darrigol, "The electron theories of Larmor and Lorentz," 315–316.

63. Larmor, *Math. and Phys. Papers,* vol. 1, 622. (Footnote added in the 1929 edition; not in the original paper.) The accuracy of Larmor's retrospective remark is confirmed by contemporary evidence. See below.

64. *Nature* 55 (24 December 1896), 192; cf. N. Robotti and F. Pastorino, "Zeeman's Discovery and the Mass of the Electron," *Annals of Science* 55 (1998): 161–183, 172.

65. Larmor to Lodge, 28 December 1896, University College Library, Oliver Lodge Collection, MS. Add 89/65 (ii).

66. Larmor to Lodge, 6 January 1897, *ibid.*

67. Lodge to Larmor, 6 February 1897, Royal Society Library, Larmor Collection, 1244. The repetition of Zeeman's experiment was reported by Lodge on 11 February 1897 in the *Proceedings of the Royal Society.* See O. Lodge, "The Influence of a Magnetic Field on Radiation Frequency," *Proceedings of the Royal Society of London,* 60 (1897): 513–514.

68. J. Larmor, "The Influence of a Magnetic Field on Radiation Frequency," *Proceedings of the Royal Society of London,* 60 (1897): 514–515, on 515.

69. See J. J. Thomson, "On the Electric and Magnetic Effects produced by the Motion of Electrified Bodies," *Philosophical Magazine,* 5th series 11 (1881): 229–249, on 234.

70. Lodge to Larmor, 8 March 1897, Royal Society Library, Larmor Collection, 1247. Lodge's calculation was probably prompted by a letter that he had received from FitzGerald (FitzGerald to Lodge, 6 March 1897, Univ. College London, Lodge Collection, MS. Add 89/35 (iii)). This calculation appeared a few days later in *The Electrician.* See O. Lodge, "A few notes on Zeeman's discovery," *The Electrician,* 12 March 1897: 643–644, on 644.

71. Larmor to Lodge, 8 May 1897, Univ. College London, Lodge Collection, MS. Add 89/65 (ii). Well before he wrote this letter, Larmor had found out that Lodge "had verified Zeeman's result." Lodge to Larmor, 6 February 1897, Royal Society Library, Larmor Collection, 1244. Thus, he had no reason to doubt the validity of that result. Furthermore, he had realized early on its implications with respect to the magnitude of the electron's mass. Therefore, his allusion to Lodge's estimate of the size of the electron as "the sort of number I should have guessed" is not surprising and does not contradict my previous claim that, *prior to Zeeman's discovery,* Larmor had attributed to the electron a mass comparable to the mass of the hydrogen ion.

72. Lodge, "A Few Notes," 644.

73. J. Larmor, "On the Theory of the Magnetic Influence on Spectra; and on the Radiation from moving Ions," *Philosophical Magazine,* 5th series, 44 (1897): 503–512.

74. Ibid., 503.

75. Ibid., 504.

76. Larmor to Lodge, 12 October 1897, University College Library, Lodge Collection, MS. Add. 89/65 (ii).

77. Larmor, "On the Theory of the Magnetic Influence on Spectra," 506.

78. Cf. Buchwald, *From Maxwell*, 172; and Hunt, *The Maxwellians*, 220–221.

79. Zeeman to Lodge, 24 January 1897, University College Library, Lodge Collection, MS. Add. 89/116. In a subsequent letter he clarified his previous remark: "I have called electrons ponderable particles; I wished to express that they must possess inertia." Zeeman to Lodge, 28 January 1897, *ibid*.

80. See, for instance, H. A. Lorentz (1898), "Optical Phenomena Connected with the Charge and Mass of the Ions (I and II)," in his *Collected Papers*, vol. 3, 17–39, on 24.

81. He began using this term in 1899. See C. Jungnickel and R. McCormmach, *Intellectual Mastery of Nature: Theoretical Physics from Ohm to Einstein*, 2 vols. (Chicago: The University of Chicago Press, 1986), vol. 2, 233.

82. Lorentz, *Collected Papers*, vol. 3, 24–25. Cf. C. L. Maier, *The Role of Spectroscopy in the Acceptance of an Internally Structured Atom, 1860–1920* (University of Wisconsin, Ph.D. thesis, 1964), 298.

83. See Isobel Falconer, "Corpuscles to Electrons," (this volume, 82).

84. Nersessian, "'Why Wasn't Lorentz Einstein?,' An Examination of the Scientific Method of H. A. Lorentz," *Centaurus* 29 (1986): 205–242, 209. Cf. also A. Romer, "Zeeman's Discovery of the Electron," *American Journal of Physics* 16 (1948): 216–223.

85. See Arabatzis, "Rethinking the 'Discovery' of the Electron."

86. Ibid.

6

The Electron, the Protyle, and the Unity of Matter
Helge Kragh

In 1930, in an account of his recently proposed theory of holes, Paul Dirac emphasized the attractiveness of his belief that "the electron and proton are really not independent, but are just two manifestations of one elementary kind of particle." As he wrote, "It has always been the dream of philosophers to have all matter built up from one fundamental kind of particle."[1] Dirac believed that the antielectron—a hole in the sea of negative-energy electrons— could be identified with the proton and that quantum mechanics could in this way provide an affirmative answer to the age-old question of the unity of matter. Although it was soon realized that Dirac's optimism was unwarranted, it is historically important that he appealed to the principle of unity of matter and that he considered the electron to be *the* fundamental particle. Dirac's shortlived hypothesis was the last attempt to build up matter of electrons alone and it can be regarded as a chapter in the search for the ultimate constituent of matter that has its roots back in the nineteenth century.

The earliest history of the electron, from about 1880 to 1910, was characterized by the different roles that the particle played in physical and chemical phenomena. One may speak of the electrochemical (or Stoney-Helmholtz) electron, the electrodynamical (or Larmor-Lorentz) electron, the cathode rays (or Thomson-Wiechert) electron, and the magnetooptical (or Zeeman-Lorentz) electron.[2] Only about 1900 did these conceptions converge into a unified picture of the electron. In addition to these "versions" of the electron, and not quite separable from them, there was what one might call the Proutean electron, that is, the conception of the electron as the fundamental building block of matter. The tentative identification of the electron with the *protyle* served as an important stimulus to J. J. Thomson and other researchers in their conceptualization of the particle. For a decade or so, believers in the unity of matter hoped to have found in the electron the protyle that earlier physicists and chemists had speculated about. This theme in the history of the electron was mainly restricted to Britain, and was characteristic of Thomson in particular. It is around this theme that this present chapter is structured. I suggest that two lines of development may be usefully distinguished in this area. One line, cultivated by Thomson in particular, focused on the empirical, negatively charged electron that was first identified in cathode rays and luminous bodies. The aim here was to build up an electron theory of matter that could account for physical as well as chemical phenomena. Another line of development, frequently intersecting with the first one, was more concerned with the electron as a structure in the electromagnetic field. I shall deal in some detail with the first line, but only cursorily with the second.

The Proutean Tradition

In his *Elements of Chemical Philosophy,* Humphry Davy speculated that "Matter may ultimately be found to be the same in essence, differing only in the arrangements of its particles; or two or three *simple* substances may produce all the varieties of compound bodies."[3] He even suggested that perhaps hydrogen was the true element common to all matter. According to William Prout's slightly later hypothesis, first suggested in 1815–16, the chemical elements consisted of multiples of hydrogen atoms. The empirical basis of Prout's suggestion was the atomic weights of the elements, claimed to be multiples of that of hydrogen, but it soon turned out that atomic weight determinations did not favor the hypothesis. Increasingly precise measurements, made by Edward Turner, Jöns Berzelius, Jean-Servais Stas, and many others showed convincingly that the simple version of Prout's law was untenable. But experiments were unable to refute certain modifications of Prout's idea of the unity of matter, such as the hypothesis that the primary element was a hypothetical substance with a weight a fraction of that of hydrogen. This was what Prout had speculated in 1831 when he suggested that there was no sufficient reason "why bodies still lower in the scale than hydrogen . . . may not exist, of which other bodies may be multiples without being actually multiples of the intermediate hydrogen."[4] This way of protecting the principle of the unity of matter was supported by several chemists, including Jean-Baptiste Dumas and Jean Marignac. In the 1870s the general idea received support from two new sources of knowledge—spectroscopic investigations of the stars and the periodic table. Although Dmitri Mendeleev firmly denied that the periodic system had any bearing on the principle of the unity of matter, many of his colleagues disagreed and tended to see in the classification a key to understand the unity they believed must exist among the chemical elements. Norman Lockyer's astrochemical work led him to suggest that chemical atoms decomposed into smaller parts at the extreme stellar temperatures. Both sources greatly stimulated the imagination of

William Crookes, whose presidential address to the British Association in 1886 was an eloquent defense of a modernized Prouteanism. He referred approvingly to "the well-known hypothesis of Prout" and suggested that helium—"all analogy points to its atomic weight being below that of hydrogen"—was the real protyle.[5] To Crookes and many others, evidence from spectroscopy, electrical discharges in gases, the periodic system, and the chemistry of rare earths indicated that the atoms were composite and possibly consisting of combinations of atoms of some primary matter. Only a minority of chemists supported subatomism à la Crookes, but the twin ideas of the complex atom and the unity of matter were considered respectable hypotheses by many chemists around 1890. The Danish thermochemist Julius Thomsen, a typical representative of neo-Prouteanism, emphasized in 1887 that "the atoms of our so-called elements are generated by combination of the uniform, minimal atoms of a primeval substance."[6] According to Victor Meyer, in an address of 1895: "The complex nature of the elements, though unproved at the present time, must today be counted as a well-founded hypothesis which we are justified to choose as starting point for further research."[7]

It was mostly chemists who were fascinated by Prout's hypothesis, but there were also physicists and astronomers who speculated about the unity and evolution of matter and the possible connections with the nature of electricity. Such ideas resonated with the Zeitgeist and found their way even to the earth sciences.[8] In March 1897, shortly before J. J. Thomson announced his discovery of the electron, Arthur Schuster declared that "most of us [physical scientists] are convinced in our innermost hearts that matter is ultimately of one kind, whatever ideas we may have formed as to the nature of the primordial substance. That opinion is not under discussion."[9] The Proutean dream was shared by many scientists all over Europe and North America, yet, as Crookes said in 1903, "This dream has been essentially a British dream, and we have become speculative and imaginative to an audacious extent, almost belying our character of a purely practical nation." According to Crookes:[10]

> The notion of impenetrable mysteries has been dismissed. A mystery is a thing to be solved—'and man alone can master the impossible.' There has been a vivid new start. Our physicists have remodeled their views as to the constitution of matter and as to the complexity if not the actual decomposibility of the chemical elements. To show how far we have been propelled on the strange new road, how dazzling are the wonders that waylay the researcher, we have but to recall—matter in a fourth state, the genesis of the elements, the dissociation of the chemical elements, the existence of

bodies smaller than atoms, the atomic nature of electricity, the perception of electrons, not to mention other dawning marvels far removed from the lines of thought usually associated with English chemistry.

The results obtained by J. J. Thomson about the turn of the century were evidently an important source of Crookes' dream. And Thomson was evidently one of Crookes' speculative and imaginative scientists.

According to the vortex atomic theory, introduced by William Thomson (Lord Kelvin) in 1867 and developed by many British mathematical physicists until the end of the century, the atoms were vortical modes of motion of a primitive and perfect fluid, the ether. This kind of theory became quite popular and formed the basis of *The Unseen Universe,* Balfour Stewart's and Peter Guthrie Tait's influential account of Victorian philosophy of nature.[11] The theory of vortices and knots was mathematically complex, but also physically appealing. Tait, who in 1876–77 investigated the vortex theory in great detail, believed that it might prove as useful for chemistry as for physics and mathematics.[12] Young J. J. Thomson agreed. In his Adams Prize essay of 1882, he greatly developed Kelvin's vortex theory and related his results to such chemical problems as affinity and chemical combination and affinity.[13]

It has often been pointed out that Thomson's early work with the vortex atom provided him with a framework of thinking that was of direct importance to his later interpretation of the cathode rays experiments in terms of streams of electrons. To Thomson, vortex atoms and electron atoms were more than mere analogies. For one thing, much of the mathematical analysis underlying Thomson's complicated calculations in 1882 was taken over almost directly in his model of the electron atom in the early years of the twentieth century. For another thing, the vortex theory functioned as an exemplar both in a methodological and an ontological sense. It was a highly attractive theory because it built on minimum assumptions, avoided ad hoc hypotheses, and operated with only one kind of primeval substance, the same that filled empty space and made up atoms of matter. The methodological advantage of the vortex atom theory had been highlighted already by Maxwell.[14] Although the theory "cannot be said to explain what matter is, since it postulates the existence of a fluid possessing inertia," Thomson considered the vortex atom to be "evidently of a very much more fundamental character than any theory hitherto started," and the one that "enables us to form much the clearest mental representation of what goes on when one atom influences another."[15]

In his analysis of several interacting vortex rings, Thomson examined theoretically the question of stability of vortices arranged at equal intervals round the circumference of a circle. Using standard perturbation theory adopted from celestial mechanics, he found after lengthy calculations a general formula that expressed the condition of stability. His general method was to express the perturbed coordinates as exp(bt) and then to determine the b-coefficients. If the coefficients were imaginary, the equilibrium system would have periods of vibration and be stable; for real coefficients a disturbance would lead to instability. Thomson found in this way that configurations with $n = 2, 3, 4, 5$, and 6 vortices would be stable, but that 7 vortices on the same ring could not form a stable system. For larger n he relied on the analogy with Alfred Mayer's floating magnets experiment already pointed out by Kelvin in 1878.[16] Thomson realized that there was no reason why the vortices should be of equal strength (the product of the velocity of rotation and the section area), but for reasons of simplicity he assumed that "the atoms of the different chemical elements are made up of vortex rings of the same strengths." This assumption facilitated the calculations and it also agreed with Thomson's inclination toward a unified theory of matter. The methods he used twenty-one years later, in his electron model of the atom, were similar to those used in 1883. Both from a methodological and an ontological point of view, the analogies between the two models are striking.[17]

The vortex atom approach greatly influenced Thomson's thinking about the complexity of atoms. For example, in 1890 he pointed out the suggestive similarity between Mayer's configurations of magnetized needles, the arrangements of columnar vortices, and the periodicity of the chemical elements: "If we imagine the molecules of all elements to be made up of the same primordial atom, and interpret increasing atomic weight to indicate an increase in the number of such atoms, then, on this view, as the number of atoms is continually increased, certain peculiarities will recur."[18] This was very much the same view he had held earlier, in connection with the vortex atom theory; and the same view turned up in his works from 1897 onward, only now with the primordial atom identified as an electron.[19] Thomson's high appreciation of the vortex atom theory continued even after he had abandoned the theory and replaced it with a theory of atomic structure based on electrons. In 1898, shortly after having suggested that atoms consist of electrons, he praised the vortex theory in a letter to the American physicist Silas Holman:

> I do not know of any phenomenon which is manifestly incapable of being explained by it [the vortex atom theory]; and personally I generally endeavour (often without success) to picture to myself some kind of vortex-ring mechanism to account for the phenomenon with which I am

dealing. . . . I regard . . . the vortex-atom explanation as the goal at which to aim, though I am afraid we know enough about the properties of molecules to feel sure that the distribution of vortex motion concerned is very complex.[20]

Again, in 1907, in a comprehensive account of the mature version of the electron atom, Thomson admitted that his new atomic theory "is not nearly so fundamental as the vortex theory of matter, . . . [where] the difference between matter and non-matter and between one kind of matter and another is a difference between the kinds of motion in the incompressible liquid at various places, matter being those portions of the liquid in which there is vortex motion." But Thomson, although attracted by fundamental theories of everything, was also a pragmatist: "The simplicity of the assumptions of the vortex atom theory are, however, somewhat dearly purchased at the cost of the mathematical difficulties which are met with in its development; and for many purposes a theory whose consequences are easily followed is preferable to one which is more fundamental but also more unwieldy."[21]

The important thing to note is that Thomson, from the early 1880s onward, was convinced that the atom had a complex constitution and that he was predisposed toward a Proutean unity of matter. The vortex atom theory can be seen as an extreme case of the Proutean ideal and, although Thomson and most other researchers abandoned the theory before 1890, the idea of unity continued to play an important role in his thinking.[22] This is further illustrated by Thomson's "gyrostatic" model of 1895 in which it was supposed that "atoms have a structure possessing similar properties to those which the atoms would possess if they contained a number of gyrostats all spinning in one way round the outwardly drawn normals to their surface."[23] This model, or analogy, included the idea that atoms are composite and that their energy and charge are determined by the number and configurations of the components. In 1883 the components were vortex rings, in 1895 gyrostats, and in 1897 corpuscles. The Proutean theme in Thomson's thinking is further illustrated by his work on x-rays shortly before he turned to cathode rays. In his attempt to understand the nature of Röntgen's rays, Thomson once again returned to the idea of atoms composed of identical and primordial particles. In his Rede Lecture of 10 June 1896, he discussed briefly the absorption of x-rays: "This appears to favour Prout's idea that the different elements are compounds of some primordial element, and that the density of a substance is proportional to the number of primordial atoms; for if each of these primordial atoms did its share in stopping the Röntgen rays, we should have that intimate connection between density and opacity which is so marked a feature for these rays."[24] Even before Thomson made his celebrated 1897 cathode rays experiments he tried to understand Philipp Lenard's discovery that the distance traversed by cathode rays is inversely proportional to the density of the gas. This suggested to Thomson that the carriers of electricity were Proutean elements much smaller than hydrogen atoms.[25]

In his important 1894 memoir on *A Dynamical Theory of The Electric and Luminiferous Medium,* Joseph Larmor adopted Johnstone Stoney's term "electron" to signify a singularity in the electromagnetic ether. He concluded that the vortex theory had to be replaced by an electron theory, although his electrons had in fact many features in common with the vortex atoms. In the theory of Larmor, the electrons were introduced to explain electromagnetic and optical phenomena, and not primarily as constituents of matter. But the role of electrons as building blocks of chemical atoms was not ignored. Larmor described electrons as "the sole ultimate and unchanging singularities in the uniform all-pervading medium" and conceived them as primordial units of matter.[26] Before explicitly introducing the electrons, Larmor referred to "monads" in a manner clearly reminding of Crookes' "protyles." How to explain the fact that matter is always made up of a small number of the same chemical elements? Larmor was well acquainted with the Proutean tradition and referred to the Scottish chemist Thomas Graham, a firm believer in the unity of matter.[27] Larmor's suggestion was this:[28]

> It would seem that we are almost driven to explain this by supposing the atoms of all the chemical elements to be built up of combinations of a single type of primordial atom, which itself may represent or be evolved from some homogeneous structural property of the aether. . . . We may assume that it is these ultimate atoms, or let us say monads, that form the simple singular points in the aether; and the chemical atoms will be points of higher singularity formed by combinations of them. These monads must be taken to be all quantitatively alike, the one set being, in their dynamical features, simply perversions or optical images of the other set.

This was a view with which Thomson fully agreed. The following year, 1895, Larmor went a step further and suggested "a molecule to be made up of, or to involve, a steady configuration of revolving electrons." He noted that it would then "follow that every disturbance of this steady motion will involve radiation and consequently loss of energy."[29] Larmor used Stoney's name electron, but understood the concept quite differently from that of Stoney, who had first referred to "electron" in 1891 as a quantum of electric charge associated with a chemical bond. Stoney's electrons were not subatomic particles, were not assigned any specific mass or sign of charge (nor were Larmor's), and were confined to the interior of the atom.[30] He believed that electrons were parts of the atom and that their oscillations were responsible for the emitted frequencies of light, but not that the atom was made up of electrons. "These charges, which it will be convenient to call electrons," he wrote in 1891, "cannot be removed from the atom, but they become disguised when atoms chemically unite."[31] It was only in 1893 that George FitzGerald pointed out that the electronic oscillators would have to be of subatomic dimensions.[32] Arthur Schuster was another early convert to Stoney's electron. He pictured the molecule as including one or more electrons in equilibrium positions and argued that the number of degrees of freedom must be much smaller than the number of spectral lines. In January 1895 Schuster wrote: "In the existence of the 'electron' I firmly believe; and this necessarily implies a very restricted number of variables."[33] Like Stoney, however he was convinced that the electrons resided safely within the atom. As he wrote in 1911: "The separate existence of a detached atom of electricity never occurred to me as possible; and if it had, and had I openly expressed such heterodox opinions, I should hardly have been considered a serious physicist, for the limits to allowable heterodoxy in science are soon reached."[34]

The electron became a more physical and definite particle in the fall of 1896 when Pieter Zeeman discovered the magnetic influence on the frequency of light and Lorentz explained the phenomenon in terms of electron theory.[35] The Zeeman effect, and also studies of electrical conduction and the optical properties of metals, led Lorentz and others to consider the electron as a subatomic, negatively charged particle with a mass-to-charge ratio some 1000 times smaller than the electrolytically determined value of hydrogen; moreover, Lorentz's negative electron—or "ion" as he called it until 1899—was capable of existing in a free state. Until 1896, Lorentz and most others had thought of electrons as corresponding to electrolytic ions, that is, of both signs of charge and with a mass perhaps equal to that of the hydrogen atom. Under the impact of Zeeman's surprisingly small mass-to-charge ratio Larmor reconstructed his electron and his picture of the atom. In May 1897 he wrote to Lodge that he was inclined "to the view that an atom of 10^{-8} cm is a complicated sort of solar system of revolving electrons, so that the single electron is very much smaller, 10^{-14} would do very well—is in fact the sort of number I should have guessed."[36]

Electrical atoms were not new in the 1890s. In publications between 1838 and 1851 Richard Laming, a British chemist and industrialist, hypothesized the existence of subatomic, unit-charged particles and pictured the atom as made up of a material core surrounded by an "electrosphere" of concentric shells of electrical particles.[37] Laming's speculations had similarities with the approach to electrical theory followed by Rudolf Clausius, Wilhelm

Weber, Robert Grassmann, Carl von Neumann, Bernhard Riemann, Friedrich Zöllner, and other German physicists. Weber considered the ether to consist of positive and negative particles of equal numerical charge orbiting around each other, a picture that resembles Larmor's ideas from the late 1890s.[38] Robert Grassmann, an amateur physicist, developed an elaborate and highly speculative atomic system based on ether particles consisting of electrical particles.[39] Both Weber and Grassmann thought of their electrical particles as subatomic constituents not only of ether but also of matter. Their particles had no definite mass or charge, however, and thus were not elementary particles in the sense of the later electron. The speculations of Weber, Grassmann, and Zöllner have their place in the prehistory of the electron, but it is an isolated place. They received little attention outside Germany and appear not to have influenced the scientists whose work led to the discovery of the electron. They did have an influence on Lorentz, however, whose theory of electrons included the conception of electrical particles distinct from the field.

THE PROTYLE MATERIALIZED

With Thomson's cathode-ray experiments of 1897 the electron became a material reality, an elementary particle, and the basis of a theory of matter.[40] This was an important element in the discovery of the electron, and one that distinguished Thomson's discovery from the mass-to-charge ratio (m/e) measurements made in Germany by Emil Wiechert and Walther Kaufmann. According to Thomson, the electron was the universal building block of matter in the strong sense that all matter consisted of electrons and only electrons. Of course, he called the particles "corpuscles" rather than electrons and had his reasons for it. In 1897 Thomson believed that there was "some evidence that the charges carried by the corpuscles in the atom are large compared with those carried by the ions of an electrolyte" and hence that the mass of the corpuscle might not be quite as small as indicated by Zeeman's experiment. He found it "interesting" that the m/e value of cathode rays was of the same magnitude as Zeeman's value of m/e, but did not follow up the remark.[41] From Lenard's data he argued that the corpuscles were of subatomic dimensions, indeed the primordial elements long sought. In his Royal Institution lecture of 30 April 1897 he was explicit about the Proutean theme:[42]

> The assumption of a state of matter more finely divided than the atom of an element is a somewhat startling one; but a hypothesis that would involve somewhat similar consequences—viz. that the so-called elements are compounds of some primordial element—has been put forward from time

> to time by various chemists. Thus, Prout believed that the atoms of all the elements were built up of atoms of hydrogen, and Mr. Norman Lockyer has advanced weighty arguments, founded on spectroscopic consideration, in favour of the composite nature of the elements. Let us trace the consequences of supposing that the atoms of the elements are aggregations of very small particles, all similar to each other; we shall call such particles corpuscles, so that the atoms of the ordinary elements are made up of corpuscles and holes, the holes being predominant.

Thomson's reference to Lockyer's "weighty arguments" was to the dissociation hypothesis that Lockyer had recently published and according to which atoms in hot stars were completely dissociated into "protohydrogen." Interestingly, two years later Lockyer derived by means of highly speculative arguments that the mass of an atom of protohydrogen was about 1/600 of that of an ordinary hydrogen atom.[43] No wonder that Thomson was struck by the similarity between the hypothetical protohydrogen and the real electron. As he wrote to Lockyer, "I get for the mass of the small particles with which I have been dealing values which in different experiments have varied between 1/500 and 1/700 of that of the ordinary atom, so that the two lines of enquiry lead to very concordant results."[44]

But back to 1897. The identification of the cathode-ray corpuscle with the free electron in the sense of Larmor and Lorentz was first suggested by FitzGerald immediately after Thomson had announced his discovery. FitzGerald argued that the interpretation had the advantage "that it does not assume the electron to be a constituent part of an atom, nor that we are dissociating atoms, nor consequently that we are on the track of the alchemists."[45] Also Larmor was quick to suggest that Thomson's corpuscles might "be simply electrons," in which case "there would be about 10^3 electrons in the molecule."[46] Thomson, however, considered his corpuscles to be charged material particles and not etherial charges without matter. This was undoubtedly a main reason why he resisted using the name electron. His initial resistance to the electron theory appeared in a note of 1899, a reply to a paper by the Australian physicist William Sutherland: "As far as I can see the only advantage of the electron view is that it avoids the necessity of supposing the atoms to be split up; . . . it supposes that a charge of electricity can exist apart from matter, of which there is as little evidence as of the divisibility of the atom; and it leads to the view that cathode rays can be produced without the interposition of matter at all by splitting up neutrons into electrons."[47] In 1901 Thomson still rejected the idea that the mass of the corpuscle was wholly or mainly electrical in origin. He found the idea "fascinating," but contradicted by experiments.[48] According to Thomson, the atom was not merely made up of corpuscles, it could also be broken down into corpuscles. His atoms were not indivisible entities. In fact, this was what he believed took place near the cathode in the discharge tube. Rather than considering the corpuscles to be liberated from the cathode metal, he thought that they resulted from dissociation of the molecules of the gas in the intense electric field near the cathode. That is, the cathode-ray tube acted as an atomic smasher and for this reason a remnant gas was necessary.

In his more elaborate version of October 1897, Thomson repeated his equation between the cathode-ray corpuscles and the primordial subatomic particles, including a reference to the views of Prout, Lockyer, and "many chemists."[49] The materiality of Thomson's corpuscles is further underlined by his initial conception of them as a kind of chemical element. He found that "the quantity of matter produced by means of the dissociation at the cathode is so small as to almost preclude the possibility of any direct chemical investigation of its properties," but note that he dismissed chemical analysis for practical reasons and not for reasons of principle. Compared with his April address, Thomson went a step further, now including a sketch of a more quantitative atomic theory based on the equilibrium states of a large number of corpuscles:[50]

> If we regard the chemical atom as an aggregation of a number of primordial atoms, the problem of finding the configurations of stable equilibrium for a number of equal particles acting on each other according to some law of force . . . whether that of Boscovich, where the force between them is a repulsion when they are separated by less than a certain critical distance, and an attraction when they are separated by a greater distance, or even the simpler case of a number of mutually repellent particles held together by a central force—is of great interest in connexion with the relation between the properties of an element and its atomic weight.

From his earlier work with vortex atoms, Thomson knew the kind of complex calculations that were necessary to determine the stability of the equilibrium systems. As he had done in his 1883 essay, he now referred to Mayer's experiment as a substitute for the abstruse calculations and cited Mayer's polygonal arrangements for up to forty-two magnets as a striking analogy to the periodic table. Thomson's daring hypothesis of corpuscles as constituents of atoms was more controversial than his conclusion about the nature of cathode rays. It was only after 1899, when Thomson and his research students at the Cavendish succeeded in determining the charge of the corpuscle (and then also its mass) that the hypothesis received solid empirical confirmation. The result, that the hydrogen atom was about 700 times as heavy as the corpuscle, was not very precise but it did show that "we have something smaller even than the atom, something which involves the splitting up of the atom."[51] The near equality of the charge of the corpuscle and that of the electron was an important factor in the merging of the two concepts that occurred about 1900. For simplicity I shall hereafter refer to Thomson's corpuscle as an electron, although Thomson continued to speak and write of corpuscles up to about 1915. Among the few scientists who adopted Thomson's terminology were A. C. Jessup and A. E. Jessup, who in a speculative paper of 1908 suggested that the atoms were formed as a central assemblage of corpuscles surrounded by a number of satellite corpuscles. "For the sake of distinction from the other corpuscles we will apply the term 'electron' to them," they wrote.[52]

In his mature atomic model developed between 1903 and 1907, Thomson introduced point-like electrons configured in dynamic equilibrium positions in a massless positive fluid.[53] The roots of his concept of electrons in a Proutean tradition are particularly clear from his *Electricity and Matter,* based on the Silliman lectures of 1903. For example, he used his model to discuss why the hydrogen atom is the lightest known atom and why there is only a limited number of chemical elements. In accordance with Crookes, Sterry Hunt, Thomas Carnelley, and other chemists of a Proutean inclination, he sought to illuminate these questions by referring to the inorganic evolution that had supposedly formed the elements during the long cosmic history—"the theory that the different chemical elements have been gradually evolved by the aggregation of primordial units."[54] Thomson's theory was ambitious and monistic, a worthy follower of the vortex atom theory. It aimed at reducing matter to a manifestation of electrons in motion. But Thomson's electrons were negatively charged and matter is electrically neutral, so the scheme appeared to necessitate that the negative electrons somehow produced effects corresponding to an atomic sphere of positive electricity. The nature of the positive electricity was a serious problem in Thomson's atomic model—in fact, it was its Achilles' heel.

In the earliest theories of the electron, the particle could be both negatively and positively charged and "positive electron" often referred to any kind of elementary positive charge, not necessarily a mirror particle of the negative Zeeman-Thomson electron.[55] This terminology, used by Lorentz, Wilhelm Wien, Johannes Stark, and others, was an additional reason that Thomson preferred to speak of corpuscles rather than electrons. True positive electrons, of the same mass as the empirically known negative electron, were frequently discussed from about 1898 to 1906 and they entered some of the atomic models of the period. Oliver Lodge and James Jeans, among others, considered the atom to consist of a multitude of interacting positive and negative electrons. Experimental evidence for the positive electron was missing, however, and early claims that positive electrons had been discovered were not accepted by the majority of physicists. By 1907 Norman Campbell summarized the standard view, namely "if there is one thing which recent research in electricity has established, it is the fundamental difference between positive and negative electricity."[56] The charge dissymmetry was built into Thomson's atomic theory, but in the sense that the positive charge, far from being massive, was considered a ghost-like entity whose only function was to keep the electrons together. In April 1904 Thomson wrote to Lodge about his problems and hopes for the sphere of positive electricity:[57]

> With regard to positive electrification I have been in the habit of using the crude analogy of a liquid with a certain amount of cohesion, enough to keep it from flying to bits under its own repulsion. I have however always tried to keep the physical conception of the positive electrification in the background because I have always had hopes (not yet realised) of being able to do without positive electrification as a separate entity, and to replace it by some property of the corpuscles. When one considers that all the positive electricity does, on the corpuscular theory, is to provide an attractive force to keep the corpuscles together, while all the observable properties of the atom are determined by the corpuscles, one feels, I think, that the positive electrification will ultimately prove superfluous and it will be possible to get the effects we now attribute to it, from some property of the corpuscles.

Thomson never succeeded in explaining the positive electricity as an epiphenomenon. On the contrary, his continued research showed that "the number of corpuscles must be of the same order as the atomic weight."[58] It followed that Thomson's original belief in the mass of the atom being made up of the masses of the electrons was unjustified. Lodge was acutely aware of the problem and considered it the main weakness of Thomson's otherwise attractive theory. In 1906 Lodge sketched five different possibilities for the structure of atoms, of which he found Thomson's model the best offer. Referring to Thomson's recent estimate of the number of atomic electrons, however, Lodge concluded that the Thomson atom had been reduced "to a state of exaggerated uncertainty" and now constituted "the most serious blow yet dealt at the electric theory of matter."[59] The reason was that the positive electricity had now become ponderable and seemingly defied explanation in terms of electromagnetic theory. According to this theory, electromagnetic inertia varied inversely with the radius of the charge (as n^2e^2/r), meaning that it would be negligible for the positive sphere as compared with that of a single electron.

In 1904 Harold Wilson at the Cavendish Laboratory suggested that the alpha particle might be a "positive electron exactly similar in character to an ordinary negative electron," a view which could be defended if the alpha particle was supposed to be much smaller than the electron.[60] But Thomson's positive sphere had atomic dimensions and thus practically no electromagnetic mass.

In 1907 Thomson sketched a modified version of his atomic model. Characteristically, he illustrated it with "an example taken from vortex motion through a fluid," because this "may make this idea clearer." From the analogy he concluded that "the system of the positive and negative units of electricity is analogous to a large sphere connected with vortex filaments with a very small one, the large sphere corresponding to the positive electrification, the small one to the negative."[61] In this way he explained to his own satisfaction the large mass of the positive charge. By at the latest 1909 Thomson had succumbed to the electromagnetic electron. At the meeting of the British Association that year he referred to the experiments of Walter Kaufmann and Alfred Bucherer on the magnetic deflection of rays of electrons. These experiments, Thomson said, "have shown [that] the whole of the mass of the corpuscle arises from its charge." At that time, under the impact of his and others' experiments with positive rays, Thomson was ready to abandon his original atomic model based on electrons alone. He now suggested that "the atom of the different chemical elements contain definite units of positive as well as of negative electricity, and that the positive electricity, like the negative, is molecular in structure."[62] This meant a farewell to the pure version of Prout's hypothesis.

Independent of Thomson's determination of the number of electrons, Lorentz felt it necessary to assume that practically the entire mass of the atom was made up of the positive sphere, and not vice versa. If the positive electricity was homogeneous this would preclude an electromagnetic interpretation of the mass. Not ready to accept such a conclusion, Lorentz mentioned the possibility "that part of the charge is concentrated in a large number of small particles whose mutual distances are invariable; in this case the total electromagnetic mass of the positive charge could have a considerable value."[63] Eight years later, in 1914, Owen Richardson suggested a similar idea of the positive electricity being composed of positively charged subelectrons (of which "there is no experimental evidence") with a numerical charge small compared with that of the negative electron. By adding the assumption of ad hoc non-Coulombian forces, Richardson sketched an electromagnetic, Thomson-like atomic model in which positive subelectrons "would be regularly distributed inside [the atom] so that such clusters would behave much like a continuous distribution of positive electrification."[64] Richardson's suggestion was not taken seriously and was perhaps not meant to be. At that time the Bohr-Rutherford atomic model was quickly on its way to gaining general acceptance and with it the dream of basing all matter on electrons vanished.

Thomson's electron was part of a theory of matter and as closely linked to chemical as to physical concerns. Reception among chemists differed considerably, from those who welcomed the electron to those who denied its legitimacy. Mendeleev belonged to the latter category. He dismissed the "to me, scarcely conceivable hypothesis of electrons," primarily because it was a subatomic, Proutean particle that made possible the transmutation of elements. "It appears to me that the whole question of a primary matter belongs to the province of fancy and not of science," he warned.[65] According to the American chemist Henry Bolton, Mendeleev was all wrong in his evaluation. In 1898, Bolton reported a "growing belief among advanced chemists in the theory that the elementary bodies as known to us are compounds of a unique primary matter (*protyle*), and that transformation of one kind into a similar one is not beyond the bounds of possibility."[66] Ida Freund, too, noted the revival that Prout's hypothesis had experienced with the discovery of the electron: "The primary matter, the πρφτη θλη, has been shifted down the scale, and hydrogen itself appears as a highly condensed form of matter with each of its atoms containing about 1000 of the truly elemental corpuscles (or electrons) of which there is one kind only."[67] Not surprisingly, Crookes supported enthusiastically the electron and emphasized his own role in the path to the discovery: "What I then [1879] called 'radiant matter' now passes as 'electrons,' . . . The electrons are the same as the 'satellites' of Lord Kelvin and the 'corpuscles' of J. J. Thomson."[68] The electron was useful to the chemists, but what Crookes found most appealing was that Thomson's particle justified his long held belief in the Proutean protyle. On the view that matter is "merely congeries of electrons," he wrote, "the electron would be the 'protyle' of 1886, whose different groupings cause the genesis of the elements."[69]

Although the Proutean conception of the electron was mostly cultivated by British researchers, the theme was well known also to scientists on the Continent. Kaufmann had not originally thought along this line, but in an address of 1901 he speculated about atomic structure in a manner strikingly similar to that of Thomson. According to Kaufmann, the electrons might well be "the long sought-for 'primordial atoms' whose different groupings would form the chemical elements had been formed; in that case the alchemists' old dream of the transmutation of the elements would be brought a good deal nearer realisation." Moreover, if the atoms of the chemical elements consisted of stable configurations of electrons, "perhaps a mathematical treatment will one day succeed in presenting the relative frequency of the elements as a function of their atomic weights and perhaps also in solving many other of the puzzles of the periodic system of the elements."[70]

Thomson's original belief that electrons were a sort of chemical protosubstance was taken to its extreme by a few chemists who suggested that the electron was a new element, not differing qualitatively from sodium or mercury. Janne Rydberg, the Swedish chemist and physicist, proposed in 1906 that the electron was an atom of a chemical element for which he assigned the symbol E. He placed it in the periodic system in the same group as oxygen and with atomic weight zero. Rydberg's "discovery" received wide notice in the press but was ridiculed by Arrhenius and other Swedish chemists.[71] Several other scientists speculated at the time about elements with atomic weights less than hydrogen's. The Russian Nikolai Beketov, the Yugoslavian Sima Losanitsch, and the American Benjamin Emerson all found a place for the electron in the periodic table.[72] Nor were such speculations restricted to obscure scientists. William Ramsay, the eminent British chemist and Nobel laureate of 1904, argued in 1908 that "Electrons are atoms of the chemical element, electricity; they possess mass; they form compounds with other elements; they are known in the free state, that is, as molecules; . . . the electron may be assigned the symbol 'E'."[73] And the following year: "Recent researches make it probable that what used to be called negative electricity is really a substance."[74] Ramsay's ideas belonged to the same Victorian tradition as those of Crookes and Thomson, but it was a tradition that was no longer in vogue at the time Ramsay took it up.

The Protyle Dematerialized

The electron could be seen as a material particle, an atom of a kind, but during the first decade of the twentieth century it was far from obvious what materiality meant. At about the time that Thomson's electron became accepted as the primary particle of matter, it began to lose its material attributes as a result of the popularity of the electromagnetic world view. With the theories of Wiechert, Wien, Max Abraham, Paul Langevin, and others there appeared a new picture of the electron, a purely electromagnetic particle. By 1906 the cautious Lorentz was also "quite willing to adopt an electromagnetic theory of matter and of the forces between material particles." Concerning the ultimate electrical particles of matter he wrote that "We should introduce what seems to me an unnecessary dualism, if we considered these charges and what else there might be in the particles as wholly distinct from each other."[75]

Many physicists conceived the electron (of whatever sign of charge) to be a kind of concentrated ether, the ether itself being scarcely distinguishable from the electromagnetic field. According to this view, ultimately the world consisted of one substratum only, the electromagnetic ether, and the electrical particles of matter were merely material manifestations of this underlying substratum. This was a truly unified picture of matter, with the protyle being the continuous ether rather than some elementary particle. The picture found its way even into chemistry, a science dealing with ponderable matter. Could chemical matter be just an epiphenomenon, a special version of the ether or of the electromagnetic field? Some chemists, inspired by the new electromagnetic world view, suggested that the electron might be a link between the ether (hence physics) and ponderable matter (hence chemistry). Richard Ehrenfeld, a German chemist and historian of science, concluded with Thomson that "Electrons are the final realities of matter, electricity then the material of which the atoms of our elements are constructed." He then went on: "But what is electricity itself? Light ether in a certain state . . . the light ether is thus the universal primary matter."[76] Another German chemist, H. Strache, attempted to explain the periodic system on a similar view, namely, electrons as tiny particles of ether. "Electrons," he wrote in 1908, "are identical with the smallest parts of the world ether . . . the smallest parts of which the atoms can be conceived to consist (the corpuscles) can be regarded as identical with ether particles and electrons."[77] The degree of success of Strache's electron-ether chemistry may be judged from his prediction of four new elements, with atomic weights 99, 176, 233, and 235.

The unitary ether-based view was popular, but not accepted by all physicists. Max Planck, for one, favored a dualistic theory. In a letter of 1909, he wrote:[78]

> According to the modern theory of electrons, the ether is, first of all, completely different from the electrons. The presently customary view of the ether is an absolute continuum, . . . both inside and outside the electrons. . . . It is precisely the modern theory of electrons which has established the fundamental doctrine that matter and electricity are atomistic, whereas the ether is constituted continuously; in this respect the electrons, that is, the atoms of electricity, are therefore much closer to the ponderable matter than to the ether.

On the whole, the relations among electrons, the ether, and the electromagnetic field were not clear.[79] The advent of relativity did nothing to make the relations clearer. Although Einstein's theory was often believed to be an electron theory, a variant of Lorentz's, in fact it was neutral with regard to whether matter consisted of electrons or not. Einstein indicated his distance from contemporary electron theory by writing that his result of mass variation with velocity, although derived from the Maxwell-Lorentz theory, was "also valid for ponderable material points, because a ponderable material point can be made into an electron (in our sense of the word) by the addition of an *arbitrarily small* electric charge."[80] This was a strange kind of electron. Although the ether was considered indispensable by most electron theorists, it was, in the view of many physicists, a highly abstract ether devoid of material attributes. Lorentz's ether was "the receptacle of electromagnetic energy," and he saw "no reason to speak of its mass or of forces that are applied to it."[81] Planck equated "ether" with "vacuum" and added that "I regard the view that does not ascribe any physical properties to the absolute vacuum as the only consistent one."[82] From this position there is but a small step to declare the ether nonexisting, a step that Emil Cohn had already taken in his version of electron theory.[83] Physicists could consistently deny the principle of relativity and the ether, as Cohn did; or deny the relativity principle and accept the ether, as Abraham did; or accept both the relativity principle and the ether, as Lorentz did; or deny the ether and accept the principle of relativity, as Einstein did. No wonder some physicists were confused.

The dematerialized ether was more popular on the continent than in Britain. Although Thomson came to accept the electron as an electromagnetic particle, his view was different from that held by Lorentz and the German electrodynamicists. In a little known work of 1907 he pictured the ether as an "etherial astral body" glued to electrical particles and thought that these were "connected by some invisible universal something which we call the ether . . . [and that] this ether must possess mass . . . when the electrified body is brought into motion." Thomson concluded his 1907 discourse on matter and ether with a formulation that illustrates how little his thoughts had changed since the 1870s when he first encountered *The Unseen Universe:* "We are then led to the conclusion that the invisible universe—the ether—to a large extent is the workshop of the material universe, and that the natural phenomena that we observe are pictures woven on the looms of this invisible universe."[84]

The electromagnetic program received its most sophisticated and ambitious formulation with the theory developed by the German physicist Gustav Mie between 1912 and 1914. Like his predecessors, Mie believed that ultimately the world consists of structures in the electromagnetic ether. Basing his theory on a generalization of the Maxwell-Lorentz equations, but otherwise in accordance with the views of Larmor, Wien, and Abraham, he wrote:[85]

> Elementary material particles . . . are simply singular places in the ether at which lines of electric stress of the ether converge; briefly, they are "knots" of the electric field in the ether. It is very noteworthy that these knots are always confined within close limits, namely, at places filled with elementary particles . . . The entire diversity of the sensible world, at first glance only a brightly colored and disordered show, evidently reduces to processes that take place in a single world substance—the ether. And the processes themselves, for all their incredible complexity, satisfy a harmonious system of a few simple and mathematically transparent laws.

Mie believed that the electron was a tiny portion of the ether in "a particular singular state" and pictured it as consisting of "a core that goes over continuously into an atmosphere of electric charge which extends to infinity."[86] That is, strictly speaking the electron did not have a definite radius. From Mie's fundamental equations it was possible to calculate the charge and mass of the elementary particles as expressed by a "world function." This was a notable advance and the first time that a field model of particles was developed in a mathematically precise way. The advance was limited to the mathematical program, however, and the grandiose theory was conspicuously barren when it came to real physics, not to mention chemistry. In 1913, two elementary particles were known, the electron and the hydrogen nucleus, and their properties could in principle be derived from the theory. Alas, in principle only, for the form of the world function that entered Mie's formulas was unknown. It was the spirit and aim of the theory, rather than its details, that appealed to mathematical physicists. As Hermann Weyl expressed it in 1919: "These [Mie's] laws of nature, then, enable us to calculate the mass and charge of the electrons, and the atomic weights and atomic charges of the individual elements whereas, hitherto, we have always accepted these ultimate constituents of matter as things given with their numerical properties."[87]

The aim of a unified theory is to understand the richness and diversity of the world in terms of a single theoretical scheme. The mass and charge of the electron, for example, are usually considered to be contingent properties ("things given"), that is, quantities that just happen to be what they are; they do not follow uniquely from any law of physics and could, therefore, presumably be different from what they are. According to the view of the unificationists, the mass and charge of the electron (and, generally, the properties of all elementary particles) must ultimately follow from theory—they must be turned from contingent into law-governed quantities. Not only that, the number and kinds of elementary particles must follow from theory too; not merely those particles that happen to be known at the time but also particles not yet discovered. In other words, a truly successful unified theory should be able to predict the existence of elementary particles; no more than exist in nature and no less. This is a formidable task, especially because physical theories cannot avoid relying upon what is known empirically and thus must reflect the state of art of experimental physics. In 1913, the electron and the proton were known, and thus Mie and his contemporaries designed their unified theories in accordance with the existence of these particles. But the impressive theories of the electromagnetic program had no real predictive power. For all its grandeur and advanced mathematical machinery, Mie's theory was a child of its age and totally unprepared for the avalanche of particle discoveries that occurred in the 1930s. When Mie died in 1957, the world of particles and fields was radically different from what it had been in 1912. It was much more complex and much less inviting to the kind of grand unified theories that he had pioneered in the early part of the century.

As shown by Daniel Siegel, the electrical theory of matter survived the decline of the electromagnetic world view that had already started when Mie began his work.[88] The mathematically complex theories of Abraham and Mie were only theories of matter in a rather abstract sense and were of limited interest to the physicists and chemists who were trying to understand atomic and molecular structure. What appealed to these scientists were not so much the particular theories of the electron as the general idea, common to all the theories, that matter consisted of particles with a mass of electromagnetic origin. Rutherford, Richardson, Harkins, Nicholson, and others identified the hydrogen nucleus with the positive electron and argued that it was a heavier and tinier counterpart of the negative electron. William Harkins and E. Wilson, two American chemists, noted the revived interest in the idea "that all matter is composed of some primordial substance" and welcomed the recent insight that "energy, or some form of it, electricity, might be this primordial substance."[89] Rutherford concluded that "the electron is to be regarded as a condensed charge of negative electricity existing independently of matter as ordinarily understood." He furthermore found it probable that "the hydrogen nucleus of unit charge may prove to be the positive electron, and that its large mass compared with the negative electron may be due to the minuteness of the volume over which the charge is distributed . . . It would be natural on this view to suppose that the positive and negative electrons are the two fundamental units of which all the elements are composed."[90] This was a form of Prout's hypothesis, although one operating with two different protyles and therefore not a pure form of the principle of the unity of matter. It was the beginning of the two-particle paradigm that reigned supreme in atomic physics until 1932 when the neutron and the positron were discovered and made it even more difficult to believe in any simple form of the principle of unity of matter.

The transition from a unitary to a dualistic conception of electricity is further illustrated by the views of Arnold Sommerfeld. Originally an electron theorist in the style of Abraham, by the 1910s Sommerfeld had become a leading relativist and an authority in quantum theory. He was no longer concerned with the ether or models of electron structure. In the first edition of his influential *Atombau und Spektrallinien* he advocated the then standard view that electricity is essentially negative electrons and that positive electricity is the absence of electrons. But he stressed that a purely unitary view of electricity was untenable because the two kinds of electricity were different in their essence: whereas the electron was the true representative of negative electricity, the positive electricity or hydrogen nucleus was material in nature. "Is [electricity] substantial or energetical, matter or force?" he asked. "Surely," he answered, "negative electricity is . . . something materially, . . . one of the universal elements which stands on an equal footing with the other element, the positively charged matter."[91] In 1922, after Francis Aston's work on isotopes, Sommerfeld was more inclined toward a dualistic conception of electricity and the idea of the unity of matter: "In the same way that negative electricity consists of ordinary negative electrons, according to Prout's old hypothesis and Aston's most recent experiments . . . matter decomposes probably into positive hydrogen ions. As a fundamental constituent of matter and positive electricity, the positive hydrogen ion deserves the name positive electron."[92]

CONCLUSION: TOWARD THE QUANTUM ELECTRON

The classical electromagnetic electron was an extended particle with an internal structure. The shape of the moving electron, the distribution of its charge, its deformability, and the partition between mechanical and electromagnetic mass were central questions for a period of twenty years, eagerly discussed by electron theorists such as Abraham, Poincaré, Lorentz, and Langevin. With the demise of the electromagnetic world view, however, the structured electron gradually disappeared from physics. Niels Bohr simply ignored the question in his theory of 1913 where he treated the atomic electrons as points. So did Sommerfeld in his relativistic extension of Bohr's theory. On the other hand, some scientists outside mainstream atomic physics were less restricted in their view of the electron and pictured it in whatever shape they found suitable. To indicate this kind of speculation, consider two examples. The American chemist Alfred Parson in 1915 developed a theory of valency in which the electron had the shape of a thin ring rotating on its axis.[93] Five years later Harkins suggested a model of the nucleus with electrons in "the form of rings, or disks, or spheres flattened into ellipsoids."[94]

Such pictures of the electron might serve specific purposes, but they had no physical justification and were not taken seriously by the physicists.

The complex, extended electron was not declared dead, but there appeared to be no use for it in the quantum theory of atoms. When Yakov Frenkel in 1925 demanded that electron models be abandoned, he merely gave methodological sanction to an already established practice. Frenkel's argument against the old picture of the electron was philosophical in tone and harmonized with the antimodel attitude that characterized atomic physics even before quantum mechanics:[95]

> The inner equilibrium of an extended electron becomes . . . an insoluble puzzle from the point of view of electrodynamics. I hold this puzzle (and the questions related to it) to be a scholastic problem. . . . The electrons are not only indivisible physically, but also geometrically. They have no extension in space at all. Inner forces between the elements of an electron do not exist because such elements are not available. The electromagnetic interpretation of the mass is thus eliminated.

According to Abraham Pais, this was the first time that a physicist explicitly argued for the point electron.[96] Several years earlier, however, in relation to continuum theories in general, young Wolfgang Pauli had made a similar argument, that the extended electron was methodologically illegitimate. In 1921 he insisted that it was meaningless to talk about an electromagnetic field in the interior of the electron. For the operational meaning of a field is the force acting on a test particle, and "since there are no test particles smaller than an electron or hydrogen nucleus, the field strength at a given point in the interior of such a particle would seem to be unobservable by definition, and thus be fictitious and without physical meaning."[97] Dirac's argument was more direct. He believed that "the electron is too simple a thing for the question of the laws governing its structure to arise."[98] Whatever the validity of these arguments, the point electron became accepted as a natural part of quantum mechanics and the once so advanced concept of electromagnetic mass was effectively dismissed.[99] The new quantum picture (or rather nonpicture) was a major reconceptualization of the electron and indirectly it helped refute the Proutean dream of unity in its pure form. Although one could still speak of the proton as a "positive electron"—and some continued to do so until the 1930s[100]—it could no longer be considered a true mirror particle of the negative electron. Not only was the proton much heavier, contrary to the electron, it was also an extended particle. When Dirac made his heroic attempt of 1930 to resuscitate the one and only protyle, he failed. And not only that, for ironically his attempt resulted in the prediction of three new elementary particles, in drastic conflict with Prout's hypothesis.[101]

NOTES

1. P. A. M. Dirac, "The Proton," *Nature 126* (1930); 605–606, on 605. On Dirac's hypothesis, see Helge Kragh, *Dirac: A Scientific Biography* (Cambridge: Cambridge University Press, 1990), 87–117. Also in 1930, Heisenberg sought to produce a theory in which the electron could metamorphose into a proton and thus be considered the fundamental particle of all matter. Heisenberg abandoned the theory, however, before it was published. For details, see Bruno Carazza and H. Kragh, "Heisenberg's Lattice World: The 1930 Theory," *American Journal of Physics 63* (1995): 595–605.

2. The different origins of the modern electron are discussed in L. Marton and C. Marton, "Evolution of the Concept of the Elementary Charge," *Advances in Electronics and Electron Physics 50* (1980): 449–472, David L. Anderson, *The Discovery of the Electron: The Development of the Atomic Concept of Electricity* (Princeton: Van Nostrand, 1964), and Theodore Arabatzis, "Rethinking the 'Discovery' of the Electron," *Studies in the History and Philosophy of Modern Physics 27* (1996): 405–435. See also the excellent review of one of the electron pioneers, Walther Kaufmann, "Die Entwicklung des Elektronenbegriffs," *Physikalische Zeitschrift 3* (1901): 9–15, translated in *The Electrician 48* (1901): 95–97. The most recent survey of the different routes leading to the construction of the electron is presented in Olivier Darrigol, "Aux Confins de l'Électrodynamique Maxwellienne," *Revue d'Histoire des Sciences 51* (1998): 5–34.

3. H. Davy, *Elements of Chemical Philosophy* (London, 1812), 132 in John Davy, ed., *The Collected Works of Sir Humphry Davy*, vol. IV (London, 1840; Johnson reprint, 1972).

4. Quoted in William H. Brock, *From Protyle to Proton: William Prout and the Nature of Matter, 1785–1985* (Bristol: Adam Hilger, 1985), 169. On Prout's hypothesis and other ideas of the complexity of elements, see also W. V. Farrar, "Nineteenth-century Speculations on the Complexity of the Chemical Elements," *British Journal for the History of Science 2* (1965): 297–323, and Arthur J. Meadows, *Science and Controversy: A Biography of Sir Norman Lockyer* (London: Macmillan, 1972). The principle of the unity of matter goes back in time much before Prout. For a selection of early views, starting about 1200 B.C., see Ida Freund, *The Study of Chemical Composition: An Account of its Method and Historical Development* (Cambridge: Cambridge University Press, 1904; Dover reprint 1968), 226–284.

5. W. Crookes, presidential address to Section B, *Report of the British Association for the Advancement of Science 1886,* 558–576, on 563. Crookes assumed that the elementary protyle was not itself an atom, but a "formless mist," some kind of potential matter existing before the evolution of chemical elements. Valuable background is given in Robert K. DeKosky, "Spectroscopy and the Elements in the Late Nineteenth Century," *British Journal for the History of Science 6* (1973): 400–423, Richard F. Hirsh, "The Riddle of the Gaseous Nebulae," *Isis 70* (1979): 197–212, and W. H. Brock, "Lockyer and the Chemists: The First Dissociation Hypothesis," *Ambix 16* (1969); 81–99.

6. Quoted in H. Kragh, "Julius Thomsen and 19th-century Speculations on the Complexity of Atoms," *Annals of Science 39* (1982): 37–60, on 45.

7. V. Meyer, "Probleme der Atomistik," *Verhandlungen der Gesellschaft deutscher Naturforscher und Ärtzte 67* (1895): 95–110, on 110.

8. Frank W. Clarke, "Evolution and the Spectroscope," *Popular Science Monthly* 2 (1873): 320–326. Thomas Sterry Hunt, *Chemical and Geological Essays* (Boston: James Osgood & Co., 1875). George F. Becker, "Relations of Radioactivity to Cosmogony and Geology," *Bulletin of the Geological Society of America* 19 (1908): 113–146.

9. A. Schuster, "On the chemical constitution of the stars," *Proceedings of the Royal Society* 61 (1897): 209–213, on 213.

10. W. Crookes, "Modern Views on Matter: The Realization of a Dream," *Annual Report of the Smithsonian Institution 1903*, 229–245, on 229. Crookes' address was delivered to the Congress of Applied Chemistry held in Berlin in June 1903.

11. Peter M. Heimann, "The Unseen Universe: Physics and Philosophy of Nature in Victorian Britain," *British Journal for the Philosophy of Science* 6 (1972): 73–79. *The Unseen Universe* was published anonymously in 1875. For natural philosophy and the Victorian Zeitgeist, see also Brian Wynne, "Physics and Psychics: Science, Symbolic Action, and Social Control in late Victorian England," 167–189 in Barry Barnes and Steven Shapin, eds., *Natural Order: Historical Studies of Scientific Culture* (Beverly Hills: Sage, 1979). Michael Chayut suggests that Stewart, who was Thomson's professor at Owens' College, introduced Thomson to the vortex atom theory. M. Chayut, "J. J. Thomson: The Discovery of the Electron and the Chemists," *Annals of Science* 48 (1991): 527–544. Whatever the validity of this suggestion, it appears likely that Stewart influenced Thomson's general view of physics and chemistry. For example, Thomson assisted Stewart in experiments attempting to detect a possible change in weight as a result of chemical combination. See J. J. Thomson, *Recollections and Reflections* (London: G. Bell and Sons, 1936), 20.

12. See the detailed analysis in Moritz Epple, "Topology, Matter, and Space, I: Topological Notions in 19th-century Natural Philosophy," *Archive for History of Exact Sciences* 52 (1998): 297–392 and Epple, *Die Entstehung der Knotentheorie: Kontexte und Konstruktionen einer modernen mathematischen Theorie* (Braunschweig: Vieweg, 1999), 94–148.

13. J. J. Thomson, *A Treatise on the Motion of Vortex Rings* (London: Macmillan, 1883; reprinted 1968 by Dawsons of Pall Mall). Thomson's work was reviewed by Osborne Reynolds, who had much praise for it, but also expressed some scepticism toward "the vortex atoms [which] are very slippery things." O. Reynolds, "Vortex Rings," *Nature* 29 (1883): 193–194, on 194. An historical account of Kelvin's theory is given in Robert H. Silliman, "William Thomson: Smoke Rings and Nineteenth-century Atomism," *Isis* 54 (1963): 461–474. See also Steve B. Sinclair, "J. J. Thomson and the Chemical Atom: From Ether Vortex to Atomic Decay," *Ambix* 34 (1987): 89–116, where the continuity in Thomson's ideas of atomic constitution is emphasized. Sinclair rightly points out that Thomson's "half-century interest in valency and its association with atomic structure thus begins here [in 1882] and not in 1897." (93).

14. J. C. Maxwell, "Atom," *Encyclopedia Britannica 1875*, reprinted in W. D. Niven, ed., *The Scientific Papers of James Clerk Maxwell* (New York: Dover Publications, 1965), vol. II, 445–484. According to Maxwell: "The disciple of Lucretius may cut and carve his solid atoms in the hope of getting them to combine into worlds; the follower of Boscovich may imagine new laws of force to meet the requirements of each new phenomenon; but he who dares to plant his feet in the path opened up by Helmholtz and Thomson has no such resources." 471.

15. Thomson, *Motion of Vortex Rings* (ref. 13), 1–2. The emphasis on "mental representation" was a typical feature of Thomson's methodology, as examined in David R. Topper, "'To Reason by Means of Images': J. J. Thomson and the Mechanical Picture of Nature," *Annals of Science 37* (1980): 31–57.

16. W. Thomson, "Floating Magnets," *Nature 18* (1878): 13. On Mayer's experiments, see Harry A. M. Snelders, "A. M. Mayer's Experiments with Floating Magnets and their Use in the Atomic Theories of Matter," *Annals of Science 33* (1976): 67–80.

17. The American physical chemist Harry Jones provides an interesting example of the continuity in Thomson's ideas. In the first edition of his *Elements of Physical Chemistry* (New York: Macmillan, 1902), Jones included a treatment of Thomson's ether vortex model; in the third edition of 1907 the ether vortices had disappeared, now replaced by Thomson's electron atom which was treated much in the same manner as earlier the vortex atom.

18. J. J. Thomson, "Molecular Constitution of Bodies, Theory of," 410–417 in H. F. Morley and Matthew M. Pattison Muir, eds., *Watt's Dictionary of Chemistry,* vol. 3 (London, 1892), on 410. Isobel Falconer has called attention to some manuscript notes of 1890 which apparently were drafts to the article in Watt's *Dictionary* and in which Thomson outlined "A theory of the structure of the molecule which may possibly explain the periodic law." See I. Falconer, "Corpuscles, Electrons and Cathode Rays: J. J. Thomson and the 'Discovery of the Electron'," *British Journal for the History of Science 20* (1987): 241–276, on 262.

19. In his autobiography of 1936, Thomson recalled that his early work on vortex atoms yielded "some interesting results and ideas which I afterwards found valuable in connection with the theory of the structure of the atom." Thomson, *Recollections and Reflections* (ref. 11), 95.

20. Thomson to Holman, 20 June 1898, as quoted in S. W. Holman, *Matter, Energy, Force and Work. A Plain Presentation of Fundamental Physical Concepts and of the Vortex Atom and Other Theories* (New York: Macmillan, 1898), 226. As late as 1931, Thomson thought of electric force in terms of the vortex model. See the extract from his letter to Lodge, quoted in E. A. Davis and I. J. Falconer, *J. J. Thomson and the Discovery of the Electron* (London: Taylor & Francis, 1997), 6.

21. J. J. Thomson, *The Corpuscular Theory of Matter* (London: Constable & Co., 1907), 2.

22. The vortex atom continued to be discussed by a few chemists and physicists even after 1900, but at that time it was realized that the theory was unable to produce real progress in the understanding of atoms and molecules. For a brief discussion of the late phase of the vortex atom and other ether-based atomic models, see H. Kragh, "The Aether in Late Nineteenth Century Chemistry," *Ambix 36* (1989): 49–65.

23. J. J. Thomson, "The Relation between the Atom and the Charge of Electricity Carried by It," *Philosophical Magazine 40* (1895): 511–544, on 513. This paper, together with several other of Thomson's important papers from 1881 to 1905, are reproduced in Davis and Falconer, *Thomson and the Discovery of the Electron* (ref. 20).

24. J. J. Thomson, "The Röntgen Rays," *Nature 54* (1896): 302–306, on 304.

25. J. J. Thomson, *The Discharge of Electricity Through Gases* (New York: Charles Scribner's Sons, 1898), 197–198. The book contains lectures given at Princeton University in October 1896. The preface, dated August 1897, informs that Thomson added "some results which have been published between the delivery and the printing of the lectures."

26. J. Larmor, "A Dynamical Theory of the Electric and Luminiferous Medium. Part I," *Philosophical Transactions of the Royal Society 75* (1894): 719–822, and in Larmor, *Mathematical and Physical Papers,* vol. 1 (Cambridge: Cambridge University Press, 1927), on 517. The description appeared in a section added 13 August 1894 in which the electron was first introduced.

27. The reference was to 299 of T. Graham, *Chemical and Physical Researches* (Edinburgh: T. and A. Constable, 1876), where Graham confirmed the Proutean doctrine that "It is conceivable that the various kinds of matter, now recognised as different elementary substances, may possess one and the same ultimate or atomic molecule existing in different conditions of movement."

28. Larmor, *Mathematical and Physical Papers* (ref. 26), 475.

29. J. Larmor, "A Dynamical Theory of the Electric and Luminiferous Medium. Part II. Theory of Electrons," *Philosophical Transactions of the Royal Society 186* (1895): 695–743, on 741. (In *Mathematical and Physical Papers,* 543–597.) It is not quite clear if Larmor thought of the electron as an entity smaller than the simplest atom. But a note added by Larmor in 1927 indicates that this might not have been the case: "[A]s far as was then known, the electron would be chemically an atom of very simple type such as, e.g., that of hydrogen." Larmor, *Mathematical and Physical Papers* (ref. 26), 519. This is also what can be inferred from Larmor's initial response to the Zeeman effect, according to Oliver Lodge, "The latest discovery in physics," *The Electrician 38* (1897): 568–570. On the other hand, Sinclair, "Thomson and the chemical atom" (ref. 13), 106, indicates that Larmor may have privately speculated on subatomic electrons. In the spring of 1897 Larmor had definitely accepted the subatomic electron, as discussed in Nadia Robotti and Francesca Pastorino, "Zeeman's Discovery and the Mass of the Electron," *Annals of Science 55* (1998): 161–183.

30. See James O'Hara, "George Johnstone Stoney and the Conceptual Discovery of the Electron," *Occasional Papers in Irish Science and Technology* no. 8 (1993): 5–28, and N. Robotti, "L'elettrone di Stoney," *Physis 21* (1979): 103–143.

31. G. J. Stoney, "On the Cause of Double Lines and of Equidistant Satellites in the Spectra of Gases," *Scientific Transactions of the Dublin Society 4* (1891): 563–608, on 582.

32. G. F. FitzGerald, "Note on Professor Ebert's Estimate of the Radiating Power of an Atom, with Remarks on Vibrating Systems Giving Special Series of Overtones Like those Given Out by Molecules," *Report of British Association for the Advancement of Science 1893,* 689–690. FitzGerald did not refer to electrons.

33. A. Schuster, "The Kinetic Theory of Gases," *Nature 51* (1895): 293.

34. Quoted in Stuart M. Feffer, "Arthur Schuster, J. J. Thomson, and the Discovery of the Electron," *Historical Studies of the Physical and Biological Sciences 20* (1989): 33–61, on 34.

35. On the history of Zeeman's discovery, see T. Arabatzis, "The Discovery of the Zeeman Effect: A Case Study of the Interplay between Theory and Experiment," *Studies in the History and Philosophy of Science 23* (1992): 365–388, A. J. Kox, "The Discovery of the Electron: II. The Zeeman Effect," *European Journal of Physics 18* (1997): 139–144, and Robotti and Pastorino, "Zeeman's discovery" (ref. 29).

36. Quoted from Arabatzis, "Rethinking the 'Discovery' of the Electron" (ref. 2), 419. Larmor's letter was possibly a reply to a letter from Lodge of 8 March 1897 in which Lodge suggested to combine Zeeman's $e/m = 107$ with Thomson's 1881 expression for the electromagnetic mass and in this way estimate the electron's linear dimension. See Robotti and Pastorino, "Zeeman's discovery" (ref. 29), 175.

37. R. Laming, "Observations on a Paper by Prof. Faraday Concerning Electrical Conduction and the Nature of Matter," *Philosophical Magazine 27* (1845): 420–423. W. F. Farrar, "Richard Laming and the Coal-Gas Industry, with His Views on the Structure of Matter," *Annals of Science 25* (1969): 243–253, where further references to Laming's papers can be found.

38. J. C. F. Zöllner, *Principien einer Elektrodynamischen Theorie der Materie* (Leipzig: Engelmann, 1876) includes on 5–290 Weber's "Abhandlungen zur Atomistischen Theorie der Elektrodynamik." See also M. Norton Wise, "German Concepts of Force, Energy, and the Electromagnetic Ether: 1845–1880," 269–307 in Geoffrey N. Cantor and M. J. S. Hodge, eds., *Conceptions of Ether: Studies in the History of Ether Theories 1740–1900* (Cambridge: Cambridge University Press, 1981).

39. R. Grassmann, *Die Lebenslehre oder die Biologie* (Stettin, 1882). A brief description is included in H. Kragh, "Concept and Controversy: Jean Becquerel and the Positive Electron," *Centaurus 32* (1989): 203–240.

40. J. J. Thomson, "Cathode Rays," *Proceedings of the Royal Institution* (1897): 1–14. Thomson, "Cathode Rays," *Philosophical Magazine 44* (1897): 293–316. (Hereafter "Cathode Rays I" and "Cathode Rays II").

41. Thomson's suggestion of the corpuscles having relatively large charges appeared in "Cathode Rays II," 309, his brief comment on Zeeman in the earlier "Cathode Rays I," 14 (where m/e is mistakenly stated as e/m). Thomson did not refer to Zeeman in "Cathode rays II," and he did not mention Stoney, Larmor, or Lorentz in either of the two 1897 papers.

42. Thomson, "Cathode Rays I," 13. It is a curious coincidence that Thomson wrote of "holes" and thus used the same term as Dirac thirty-three years later. Thomson's holes meant vacuum or the absence of corpuscles.

43. N. Lockyer, "The Methods of Inorganic Evolution," *Nature 61* (1899): 129–131. N. Robotti and Matteo Leone, "Lockyer's 'Proto-elements' and the Discovery of the Electron," forthcoming in H. Kragh, Geert Vanpaemel and Pierre Marage, eds., *Modern Physics. Proceedings of the XXth International Congress of History of Science* (Turnhout: Brepols, 2000).

44. Thomson to Lockyer, 15 November 1899, as quoted in Meadows, *Science and Controversy* (ref. 4), 170.

45. G. FitzGerald, "Dissociation of atoms," *The Electrician* 39 (1897): 103–104, on 104.

46. J. Larmor, "On the theory of the magnetic influence on spectra; and on the radiation from moving ions," *Philosophical Magazine* 44 (1897): 503–512, on 506. Larmor's term "molecule" should be understood as "atom."

47. J. J. Thomson, "Notes on Mr. Sutherland's Paper on the Cathode Rays," *Philosophical Magazine* 47 (1899): 415–416, referring to W. Sutherland, "Cathode, Lenard and Röntgen Rays," *ibid.*, 269–284. Sutherland assumed the ether to consist of doublets of positive and negative electrons for which he coined the term "neutron," a name and concept taken over by Walther Nernst in his *Theoretische Chemie* (Stuttgart: F. Enke, 1907), 392. For the difference between electrons and corpuscles, see also O. Lodge, "Modern Views of Matter," *Annual Report of the Smithsonian Institution 1903*, 215–228, on 220. The subject is analyzed in Falconer, "J. J. Thomson and the 'Discovery' of the Electron" (ref. 18) and Arabatzis, "Rethinking the 'Discovery' of the Electron" (ref. 2).

48. J. J. Thomson, "The Existence of Bodies Smaller than Atoms," *Proceedings of the Royal Institution* 16 (1901): 138–150.

49. According to Chayut, "The Discovery of the Electron" (ref. 11), Thomson's speculative theory of matter and revival of Prout's hypothesis met "universal disbelief" and "alienated Thomson from the physicists." I do not agree with this evaluation. Thomson's atomic model and general approach to atomic theory were well received and highly influential both among physicists and (somewhat later) among chemists. This is seen, for example, in the many attempts of the period to understand radioactivity on a subatomic basis. See H. Kragh, "The Origin of Radioactivity: From Solvable Problem to Unsolved Non-problem," *Archive for History of Exact Sciences* 50 (1997): 331–358.

50. Thomson, "Cathode Rays II" (ref. 40), on 313.

51. J. J. Thomson, "On the Masses of the Ions at Low Pressures," *Philosophical Magazine* 48 (1899): 547–567, on 548. See also N. Robotti, "J. J. Thomson and the Cavendish Laboratory: The history of an Electric Charge Measurement," *Annals of Science* 52 (1995): 265–284.

52. A. C. Jessup and A. E. Jessup, "The Evolution and Devolution of the Elements," *Philosophical Magazine* 15 (1908): 21–55, on 31.

53. On Thomson's atomic model, see John L. Heilbron, "J. J. Thomson and the Bohr Atom," *Physics Today* 30 (April 1977): 23–33, and H. Kragh, "J. J. Thomson, the Electron, and Atomic Architecture," *The Physics Teacher* 35 (1997): 328–332. Thomson's electron was tiny, but not of zero extension. In the later versions, when he accepted the mass of the electron to be electromagnetic, he gave its radius as 10^{-13} cm. "The volume of a corpuscle bears to that of the atom about the same relation as that of a speck of dust to the volume of this room," he mentioned at the Winnipeg meeting of the British Association. J. J. Thomson, presidential address, *Report of the British Association for the Advancement of Science 1909*, 3–29, on 12.

54. J. J. Thomson, *Electricity and Matter* (London: Constable & Co., 1904), 101. Thomson suggested that atoms much lighter than hydrogen had once existed, but that they had now formed aggregations the smallest of which happened to be hydrogen. It followed

from his argument that in the far future there would be no hydrogen left. The following year he discussed the opposite possibility, that the final stage of the universe would consist of the most simple atoms, containing only one electron. J. J. Thomson, "The Structure of the Atom," *Proceedings of the Royal Institution* 18 (1905): 1–15. For a geologist's appreciation of Thomson's speculations, see Becker, "Relations of Radioactivity" (ref. 8).

55. The early history of the positive electron is detailed in Kragh, "Concept and Controversy" (ref. 39). See also Per F. Dahl, *Flash of the Cathode Rays: A History of J. J. Thomson's Electron* (Bristol: Institute of Physics Publishing, 1997), 251–264.

56. N. R. Campbell, *Modern Electrical Theory* (Cambridge: Cambridge University Press, 1907), 130.

57. Quoted in Davis and Falconer, *Thomson and the Discovery of the Electron* (ref. 20), 195.

58. Thomson, *Corpuscular Theory of Matter* (ref. 21), 151.

59. O. Lodge, *Electrons, or the Nature and Properties of Negative Electricity* (London: Bell and Sons, 1906), 151 and 194. For a survey of atomic models between 1870 and 1920, see H. Kragh, *Quantum Generations: A History of Physics in the Twentieth Century* (Princeton: Princeton University Press, 1999), 44–57.

60. H. A. Wilson, "The Nature of the α Rays Emitted by Radio-active Substances," *Nature* 70 (1904): 101. If of electromagnetic origin, the alpha particle would have a radius of about 5×10^{-17} cm.

61. Thomson, *Corpuscular Theory of Matter* (ref. 21), 151.

62. Thomson, "Presidential Address" (ref. 53), 13.

63. H. A. Lorentz, *The Theory of Electrons and its Applications to the Phenomena of Light and Radiant Heat* (New York: Dover, 1952; original 1909), 119.

64. O. W. Richardson, *The Electron Theory of Matter* (2nd edn., Cambridge: Cambridge University Press, 1916), 586–588.

65. D. M. Mendeleev, *The Principles of Chemistry* (London: Longmans, Green & Co., 1905), vol. I, 33 and 218. See also Kragh, "The Aether" (ref. 22) and Michael D. Gordin, "Making Newtons: Mendeleev, Metrology, and the Chemical Ether," *Ambix* 45 (1998): 96–115. For the chemists' reception of the electron, see Chayut, "The Discovery of the Electron" (ref. 11), Anthony N. Stranges, *Electrons and Valence: Development of the Theory, 1900–1925* (College Station, Texas: Texas A&M University Press, 1982), and T. Arabatzis and Kostas Gavroglu, "The Chemists' Electron," *European Journal of Physics* 18 (1997): 150–163.

66. Quoted in Thaddeus J. Trenn, "The Justification of Transmutation: Speculations of Ramsay and Experiments of Rutherford," *Ambix* 21 (1974): 53–77, 55.

67. Freund, *Chemical Composition* (ref. 4), 617.

68. W. Crookes, "The Stratifications of Hydrogen," *Proceedings of the Royal Society* 69 (1902): 399–413. The reference to Kelvin's satellites was to the atomic model proposed in W. Thomson, "Aepinus Atomized," *Philosophical Magazine* 3 (1902): 257–283.

69. Crookes, "Modern Views of Matter" (ref. 10), 243.

70. Kaufmann, "Die Entwicklung" (ref. 2), 15.

71. J. R. Rydberg, *Elektron, der erste Grundstoff* (Lund: Håkan Ohlsson, 1906). Examples of press coverage are "Swedish Scientist's Reported Discovery," *Evening Standard,* 6 October 1906, and "New Element Found; it Will Reveal Marvels," *New York American,* 6 October 1906.

72. J. W. van Spronsen, *The Periodic System of Chemical Elements: A History of the First Hundred Years* (Amsterdam: Elsevier, 1969).

73. W. Ramsay, "The Electron as an Element," *Journal of Chemical Society* 93 (1908): 774–788, on 778. See also Stranges, *Electrons and Valence* (ref. 65), 93–96.

74. W. Ramsay, *Essays, Biographical and Chemical* (London: Constable & Co., 1909), 196. For Ramsay as a late follower of the Crookes-Thomson tradition, see Trenn, "Justification of Transmutation" (ref. 66).

75. Lorentz, *Theory of Electrons* (ref. 63), 45.

76. R. Ehrenfeld, *Grundriss einer Entwicklungsgeschichte der chemischen Atomistik* (Heidelberg: Carl Winter's Universitätsbuchhandlung, 1906), 269.

77. H. Strache, "Die Erklärung des periodischen Systems der Elemente mit Hilfe der Elektronentheorie," *Verhandlungen der deutschen physikalischen Gesellschaft* 10 (1908), 798–803. At that time the term "corpuscle" was rarely used by chemists or physicists other than Thomson himself. When it was used, as Strache did, it was associated with atomic structure whereas "electron" referred to the ether and electromagnetic theory.

78. Letter from Planck to Friedrich Kuntze, quoted in F. Kuntze, "Die Elektronentheorie der Brüder Hermann und Robert Grassmann," *Vierteljahrschrift für Wissenschaftliche Philosophie und Soziologie* 33 (1909): 273–298, on 283.

79. See the review in József Illy, "Revolutions in a Revolution," *Studies in the History and Philosophy of Science* 12 (1981): 173–210.

80. A. Einstein, "Zur Elektrodynamik bewegter Körper," *Annalen der Physik* 17 (1905): 891–821, on 919. For a comment on Einstein's somewhat obscure passage, see Arthur I. Miller, *Albert Einstein's Special Theory of Relativity* (Reading, Mass.: Addison-Wesley, 1981), 330.

81. Lorentz, *Theory of Electrons* (ref. 63), 31.

82. Quoted in Vladimir P. Vizgin, *Unified Field Theories in the First Third of the 20th Century* (Basel: Birkhäuser, 1994), 17.

83. Olivier Darrigol, "Emil Cohn's Electrodynamics of Moving Bodies," *American Journal of Physics* 63 (1995): 908–915.

84. J. J. Thomson, "Die Beziehung zwischen Materie und Äther im Lichte der neueren Forschungen auf dem Gebiete der Elektrizität," *Physikalische Zeitschrift* 9 (1908): 543–550, on 544 and 550. The paper was the Adamson lecture of 4 November 1907.

85. G. Mie, *Moleküle, Atome, Weltäther* (Leipzig: Teubner, 1911). Here quoted from Vizgin, *Unified Field Theories* (ref. 82), 18 and 28. On Mie's electron theory, see also Leo Corry, "From Mie's Electromagnetic Theory of Matter to Hilbert's Unified Foundations of Physics," *Studies in the History and Philosophy of Modern Physics 30* (1999): 159–183.

86. G. Mie, "Die Grundlagen einer Theorie der Materie," *Annalen der Physik 37* (1912): 511–534, on 511–512.

87. H. Weyl, *Raum, Zeit, Materie: Vorlesungen über allgemeine Relativitätstheorie* (Berlin: Julius Springer, 1919), 172. Weyl's "electrons" included atomic nuclei.

88. D. M. Siegel, "Classical-Electromagnetic and Relativistic Approaches to the Problem of Nonintegral Atomic Masses," *Historical Studies in the Physical Sciences 9* (1978): 323–360.

89. W. D. Harkins and E. D. Wilson, "Energy Relations Involved in the Formation of Complex Atoms," *Philosophical Magazine 30* (1915): 723–734, on 723. John W. Nicholson, "A Structural Theory of Chemical Elements," *Philosophical Magazine 22* (1911): 864–889.

90. E. Rutherford, "The Structure of the Atom," *Scientia 16* (1914): 337–351, and in *The Collected Papers of Lord Rutherford of Nelson,* vol. 2 (London: Allen and Unwin, 1963), 445–455. Quotations on 445 and 451.

91. A. Sommerfeld, *Atombau und Spektrallinien* (Braunschweig: Vieweg & Sohn, 1919), 24. Sommerfeld held that electricity was even more substantial than matter because the electrical charge, contrary to the mass, does not vary with the velocity.

92. Ibid., third ed 1922, 25–26.

93. A. L. Parson, "A Magneton Theory of the Atom," *Smithsonian Miscellaneous Publication 65* (1915): 1–80. Parson's theory has been considered a precursor of the spinning electron, see Max Jammer, *The Conceptual Development of Quantum Mechanics* (New York: McGraw-Hill, 1966), 148 and Edvige Schettino, "Il Modello Ring-electron di A. L. Parson," *Physis 26* (1984): 361–371. On the differences between physicists' and chemists' conceptions of the electron, see Arabatzis and Gavroglu, "The Chemists' Electron" (ref. 65), H. Kragh, "Historiography of Electronic Valency Theory," *Annals of Science 40* (1983): 289–295, and H. Kragh, "Bohr's Atomic Theory and the Chemists," *Rivista di Storia della Scienza 2* (1985): 463–486.

94. W. D. Harkins, "The Nuclei of Atoms and the New Periodic System," *Physical Review 15* (1920): 73–94. See also Roger H. Stuewer, "The Nuclear Electron Hypothesis," 19–67 in William R. Shea, ed., *Otto Hahn and the Rise of Nuclear Physics* (Dordrecht: Reidel, 1983).

95. Y. Frenkel, "Zur Elektrodynamik punktförmiger Elektronen," *Zeitschrift für Physik 32* (1925): 518–534, on 518.

96. A. Pais, *Inward Bound: Of Matter and Forces in the Physical World* (Oxford: Clarendon Press, 1986), 372.

97. W. Pauli, *Theory of Relativity* (Oxford: Pergamon Press, 1958; first published 1921), 206. See also John Hendry, *The Creation of Quantum Mechanics and the Bohr-Pauli Dialogue*

(Dordrecht: Reidel, 1984), 43–44. Pauli made the same point at the German Physical Society's meeting on relativity in Nauheim in 1920. See discussion remark in *Physikalische Zeitschrift 21* (1920), 650.

98. P. A. M. Dirac, "Classical theory of radiating electrons," *Proceedings of the Royal Society A 167* (1938): 148–169, on 149.

99. Work on classical electromagnetic electrons continued after 1925 but increasingly focused on point electrons. For a survey, see Fritz Rohrlich, "The Electron: Development of the First Elementary Particle Theory," 331–369 in Jagdish Mehra, ed., *The Physicist's Conception of Nature* (Dordrecht: Reidel, 1973).

100. Robert Millikan continued to call protons "positive electrons" as late as 1935, after the positron had been confirmed and its name accepted. R. A. Millikan, *Electrons (+ and –), Protons, Photons, Neutrons, and Cosmic Rays* (Cambridge: Cambridge University Press, 1935). Curiously, Millikan used the word proton only once, on the title page.

101. P. A. M. Dirac, "Quantised Singularities in the electromagnetic Field," *Proceedings of the Royal Society A 133* (1931): 60–72. The new particles were the positron, the negative proton, and the magnetic monopole. Although the positron was only discovered in 1932, the negative proton (or antiproton) in 1955, and the magnetic monopole has still defied all attempts of detection, the years 1931–1933 definitely marked the end of the Proutean dream. In addition to Dirac's predictions, the neutron was discovered in 1932, and the neutrino was predicted in 1930 and detected in 1956; the antineutron was suggested in 1933 and detected in 1957.

7

O. W. RICHARDSON AND THE ELECTRON THEORY OF MATTER, 1901–1916

Ole Knudsen

To the generation of physicists who began their careers around the turn of the century the existence and characteristics of the electron were facts, the establishment of which constituted the latest triumph of physical science. The various determinations of the electron's charge-to-mass ratio, coming from optical measurements of the Zeeman effect and direct measurements on cathode rays agreed within ever diminishing margins of error, and in 1900 Planck could infer a very precise value for the electronic charge from his new theory of blackbody radiation. The electron was soon established as a universal constituent of all matter, some 2,000 times lighter than the smallest atom, and it lent a new sense of reality to microphysical theories and models of all kinds.

One important part of the early history of the electron, the development of models of the electron as part of the structure of atoms and molecules, ending with Niels Bohr's famous theory of 1913, has been well described in the historical literature.[1] In this chapter I study a different aspect of the history of physics before 1916, one in which atomic structure was of little significance but in which physicists nevertheless relied heavily on the electron for the explanation of the macroscopic properties of matter. I do this by focusing on the career of one such physicist, Owen Willans Richardson. By following his career to about 1916, I hope to present some characteristic features of the electron theory of matter in the period when it was still dominated by classical dynamics and electrodynamics, even though the quantum theory was, so to speak, lurking in the background.

Richardson entered Trinity College, Cambridge, in 1897 and soon became one of a lively group of students, among them such luminaries as Ernest Rutherford, C. T. R. Wilson, and Paul Langevin, who were working at the Cavendish Laboratory to explore under J. J. Thomson's leadership the exciting new fields opened up by the discoveries of x-rays, radioactivity, and the electron, known at the Cavendish as Thomson's subatomic corpuscle. Richardson's education followed the pattern that had become standard at Cambridge since the reform instigated by Thomson around 1890. The essential new element was that, instead of taking the Mathematical Tripos, physics students could now study for the Natural Sciences Tripos which had become a proper physics education, combining a solid grounding in differential calculus and theoretical physics with practical laboratory training at the Cavendish.[2] Richardson passed the Tripos with first class honors in 1900 and thereafter worked full time at the Cavendish until 1906 when he was appointed professor of physics at Princeton. He was elected fellow of Trinity in 1902 and won a Maxwell Scholarship and a D.Sc. (London) in 1904.[3]

Richardson clearly belonged to the generation that grew up with the electron and whose scientific career centered on the physics of this new constituent of matter. He began his physics education in the year of the discovery of the electron and was trained as a researcher at the world's leading center for experimental and theoretical work on the new physics. Moreover he stayed with the electron and did not stray much into such fields as radioactivity or x-rays; his research during the period dealt with here was mostly concentrated on one aspect of the electron theory of matter: the thermal emission of electrons. Furthermore, after about fifteen years of research which brought him international fame (and eventually a Nobel Prize), he took time off to write a textbook that gave a comprehensive and critical survey of electron physics up to the time of writing, thus providing us with an insider's view of what the electron had meant for the development of microphysics during this period.

Richardson's Research, 1901–1916

Richardson's first piece of work as a new research student was a typical example of "Cavendish physics" as characterized by I. Falconer.[4] It consisted in an attempt to look for a new effect, the existence of which had been suggested to him by Thomson. The idea, inspired by Thomson's new conception of electric currents being carried by corpuscles[5] was that, since in a wire carrying an alternating current of high frequency the moving corpuscles would be confined to a thin layer near the surface, it would be reasonable to expect the wire to emit some kind of radiation, either in the form of emitted ions or corpuscles (like radioactive radiation) or of Röntgen radiation produced when the rapidly oscillating corpuscles collided with the atoms in the wire. Hence Richardson tried to detect such radiation, first by means of a photographic plate (which indeed showed a line of fogging when brought near to the wire; unfortunately this turned out to be due to a luminous discharge round the wire) and then by using a sensitive electrometer to look for ionization in the gas near the wire in a pressure range from one atmosphere down to .01 mm of mercury. The results of numerous experiments of successively finer sensitivity with wires of different metals were all negative, but the young man learned to handle vacuum equipment and delicate measuring instruments, skills that he would soon put to good use.[6]

Richardson's next work, which marked the beginning of his long-lasting research on the thermal emission of electrons, was an almost direct continuation of his first: If rapidly alternating currents did not cause a wire to emit radiation, heating a wire was known to make it emit electricity. The researches of Elster and Geitel had shown that a heated wire would leak either positive or negative electricity, depending on the temperature and the nature of the surrounding gas, and McClelland had shown that at high temperatures a platinum wire would emit negative electricity and that the amount emitted increased with the temperature.[7] Richardson decided to concentrate on the effect at very low pressures where the influence of the surrounding gas could be assumed negligible, and to investigate the temperature dependence of the saturation current from the heated wire, that is, of the number of electrons emitted per unit time.

The apparatus that he constructed for this investigation (figure 7.2) was a modified version of one he had used in his earlier work (figure 7.1) so that with respect to equipment and experimental technique his study of the thermal emission of electrons was closely related to his failed attempt to detect radiation from alternating currents.

Richardson published his results in a paper read to the Cambridge Philosophical Society in November 1901.[8] He began with a short theoretical consideration in which he reviewed the "corpuscular theory of conduction in metals" and used the Maxwell distribution of velocities to calculate the number of free corpuscles hitting unit area of the metal surface in unit time. Assuming that to penetrate the surface the corpuscles had to overcome a potential discontinuity w, he could then derive the first version of the expression that would later be known as Richardson's law for the number N of electrons emitted from unit area of the metal surface in unit time:

$$N = n\sqrt{\frac{kT}{2m\pi}}e^{-w/kT}. \qquad (1)$$

Here n is the number of free electrons in unit volume of the metal, k Boltzmann's constant, T the absolute temperature, and m the mass of an electron.[9]

In the experimental part of the paper Richardson first established that the current between the wire and the cylinder was indeed caused by negative particles emitted from the wire; with a positive potential of 400 volts on the wire he obtained no current, while negative potentials resulted in quite large

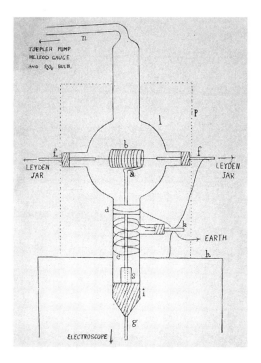

Figure 7.1
Richardson's apparatus for detecting ionization in the gas round a wire ff_1 carrying an alternating current. The spiral b was charged positively or negatively and connected to a sensitive electrometer. Source: O. W. Richardson: "On an Attempt to Detect Radiation from the Surface of Wires Carrying Alternating Currents of High Frequency," *Proceedings of the Cambridge Philosophical Society* 11 (1902): 175.

currents. Having made sure that the current reached saturation for a negative potential well below 80 volts, he proceeded to measure the saturation current as a function of temperature using a fixed potential of −120 volts on the wire and determining the temperature of the wire by measuring its resistance. To compare his results with eq. (1) he rewrote it in the form

$$n = i/\varepsilon S = AT^{1/2}e^{-b/T}, \qquad (2)$$

where i is the saturation current, ε the electronic charge, and S the surface area of the wire. In the temperature interval from 1300 K to 1600 K where the current increased three orders of magnitude from 2.5×10^{-9} to 4.0×10^{-6} ampères, he found a very good agreement between his measured values and eq. (2). Furthermore, from his measurements he could determine the constants A and b leading to the following values for n, the density of electrons in platinum, and the discontinuity in the electric potential at the surface:

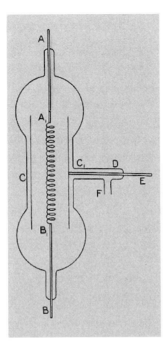

Figure 7.2
Apparatus for measuring the saturation current from the hot platinum wire A_1B_1. The wire was surrounded by a metal cylinder C put to earth through a Thomson galvanometer measuring the current between the cylinder and the wire. (O. W. Richardson, "The Electrical Conductivity Imparted to a Vacuum by Hot Conductors," *Philosophical Transactions of the Royal Society* 202A (1903): 507.) The purely verbal description in O. W. Richardson: "On the Negative Radiation from Hot Platinum," *Proceedings of the Cambridge Philosophical Society* 11 (1902): 287–288, appears to indicate that originally the wire was straight.

$$n = \left(\frac{2m\pi}{R}\right)^{1/2} A = 1.3 \times 10^{21} \text{cm}^{-3} \quad (3)$$

and

$$\delta\phi = \frac{w}{\varepsilon} = \frac{b}{\varepsilon k} = 4.1 \text{ volts.} \quad (4)$$

(The value of n was taken at a temperature of 1542 K while that of $\delta\phi$ represented an average between 1378 K and 1571 K.) Both values appeared to be of the right order as compared, respectively, to the value $n = 1.37 \times 10^{22}$ cm^{-3} obtained from Patterson's measurements of the change of resistance of platinum in a magnetic field using Thomson's theory of conduction,[10] and to contact emf's between metals.

Fifteen months after reading this paper to the Cambridge Philosophical Society, Richardson submitted a lengthy article to the Royal Society in which he brought his theoretical and experimental research on the thermal emission of electrons to a temporary completion.[11] In it he repeated the derivation of eq. (1) in a slightly more detailed manner and supplemented it by an alternative derivation using the ideal gas law and the first law of thermodynamics on unit mass of electrons passing from the inside to the outside of the metal. He ended the theoretical part with a veiled remark on the analogy between the emission of corpuscles and evaporation, thus supplementing the concept of the electron gas by that of the electron vapor.[12] In the experimental part a report of his results on platinum, taken almost verbatim from the earlier paper, was followed by descriptions of new experiments on carbon filaments and sodium. Because of its volatility and high photoelectric activity, sodium posed particular difficulties that required the construction of a completely different type of apparatus and even then made it impossible to achieve saturation, so that it was necessary to use the current at a fixed voltage to measure the number of emitted electrons per unit time as a function of temperature. The new results provided further confirmation of Richardson's law, but the values of n for both carbon and sodium turned out to be several orders of magnitude too high both from a theoretical point of view (they would correspond to pressures of millions of atmospheres) and compared to available experimental results. Richardson put this down to a slight temperature variation of w; this assumption also helped to improve the fit between differences in his values of $\delta\phi$ and known values of contact potentials among the three substances.

Richardson's early work on thermal emission established his reputation as a physicist and was instrumental for his appointment at Princeton in 1906. The phenomenon turned out to be complicated, however, and it continued to take up a large proportion of his efforts, both during his time in America and after his return to England in 1914 as a newly elected FRS and Wheatstone professor of physics at King's College, London. Before the outbreak of World War I he had published some thirty papers on many different aspects of "thermionics" as he dubbed the phenomenon in 1909,[13] and the subject had developed into a flourishing research field with many contributors, as is amply demonstrated by Richardson's monograph of 1916.[14] The interest in this field was due partly to its relevance for the important progress of radio technology[15], partly to its theoretical implications for the electron theory of metals, in particular with respect to their thermoelectric properties.[16]

To follow Richardson's later work in detail would take us too far afield, but a few points may be mentioned here.[17]

First, in collaboration with his student F. C. Brown, Richardson demonstrated experimentally that the electrons emitted by a hot strip of platinum have velocities that agree with the Maxwell distribution, thus providing the first direct experimental verification of that distribution for any gas. The authors interpreted this result as a support for Richardson's original theory of the electronic emission, in particular for the assumption that the Maxwell distribution held for the conduction electrons inside the metal as well. Thus they saw their work as confirming the electron gas theory of metals.[18]

Second, Richardson's original derivation of his law relied on the kinetic theory of the electron gas in metals. As the difficulties of this theory about specific heats and in relation to radiation theory became more and more evident,[19] Richardson came to rely more on purely thermodynamical arguments. In two theoretical papers in 1912 he developed a new theory that brought electron emission into relation with thermoelectric phenomena and also led to a modified form of his law. Instead of the formula

$$i = AT^{1/2}e^{-w/kt}, \tag{5}$$

which follows from eq. (1), he now found

$$i = AT^{2}e^{-w/kT}, \tag{6}$$

where i is the saturation current, w is the work function (the work an electron has to perform to pass from inside the metal to the outside), and A denotes different constants in the two formulas. Because the variation with temperature was dominated by the exponential function, the difference between the two expressions could not be detected experimentally. A further elaboration of the thermodynamical relations between the work function and thermoelectric properties showed that the relation

$$w = w_0 + \frac{3}{2}kT \tag{7}$$

would be a good approximation for substances having a small Thomson effect, that is, for most metals.[20]

Third, the basic assumption underlying the whole of Richardson's work was that the thermionic currents consisted of conduction electrons evaporating out through the surface of the metal. The currents were greatly influenced, however, by residual gases, gases occluded in the metal, impurities in the hot filament, and so forth. Already in 1903 H. A. Wilson had found results indicating that the main part of the current from platinum was due to occluded hydrogen,[21] and by 1912 Richardson's results on carbon and sodium also had been cast in doubt, so that there was a real possibility that the emission of electrons was in all cases a secondary effect accompanying some chemical or other process at the surface of the hot filament. In 1913 I. Langmuir of General Electric provided Richardson with specimens of ductile tungsten and taught him the best way of removing gas from his apparatus; this enabled him to prove conclusively that at least in the case of tungsten the overwhelming part of the current came from the conduction electrons. The tungsten filaments could stand a high temperature for a long time, so Richardson could show that in some of his experiments the total number of electrons emitted was 10^4–10^8 times the number of gas molecules liberated from the filament or impinging on it. In another experiment the total mass of emitted electrons was close to three times the mass of tungsten lost by evaporation or sputtering from the hot filament. Thus, the only possible source for the thermionic electrons was that they must have flowed into the filament from outside the tube, that is, by conduction.[22]

Fourth, another disturbing influence was the photoelectric effect. Already in his 1903 experiments on sodium, Richardson had had to take special precautions to protect the emitting surface from light to remove the photoelectric emission that for this very electropositive metal was considerable even at ordinary temperatures.[23] In 1912 he began a thorough study of the photoelectric effect both theoretically and experimentally, the latter in collaboration with his student Karl T. Compton. In his theory Richardson used statistical and thermodynamical arguments, similar to those used in his thermionic theory, to establish equilibrium conditions for the electron vapor near a metal surface subject to blackbody radiation described by Wien's radiation law (the high frequency limit of Planck's law). The result was an integral equation for the number $F(v)$ of electrons emitted by a unit of incoming radiative energy of frequency v, and a second integral equation containing $F(v)$ and the maximum kinetic energy T_v of the electrons emitted by radiation of frequency v. A solution to the first integral equation was

$$F(v) = 0 \qquad \text{when } 0 < hv < w_0,$$

$$F(v) = A_1 \frac{h}{k^2 v^2}\left(1 - \frac{w_0}{hv}\right) \qquad \text{when } w_0 < hv < \infty, \qquad (8)$$

where A_1 is a constant characteristic of the metal, h is Planck's constant and w_0 is the constant part of the work function, cf. eq. (3). With $F(v)$ given by eq. (4) the second integral equation had the solution

$$T_v = hv - w_0. \qquad (9)$$

These results were not surprising since Einstein had derived eq. (9) already in 1905, but Richardson emphasized that in his derivation he had used only Planck's radiation law, not Einstein's lightquantum hypothesis or, as he also called it, the "unitary" hypothesis. Hence experimental confirmation of eqs. (8) and (9) did not constitute compelling evidence for Einstein's theory.[24]

Compton's and Richardson's experiments confirmed the general features of the theory, particularly the linear relation between T_ν and ν as well as the existence of a threshold frequency below which photoelectric emission ceased, and they provided two independent methods of determining the value of Planck's constant: from the slope of the experimental (T_ν, ν)-curves or from the experimental values of the threshold frequency $\nu_0 = w_0/h$, using values of w_0 from thermionic and thermoelectric data. The first method yielded a value some 20 percent smaller than the well-established radiation value $h = 6.55 \times 10^{-27}$ erg sec, while the second gave a value almost as much in excess. A fairly thorough discussion of possible sources of error did not enable the authors to reach a firm conclusion about the reasons for these deviations.[25]

The problem that originally had turned Richardson's attention to the photoelectric effect was that of distinguishing it from the thermionic emission, or to prove that the latter was a genuine effect and not simply photo-emission due to the ubiquitous blackbody radiation. His theory showed, however, that the temperature variation of photoemission caused by blackbody radiation at the temperature of the hot filament was given by the same expression that governed thermionic emission (eq. (2)), and his theoretical expression (8) for the number of emitted electrons as a function of the frequency of the radiation turned out to agree rather badly with experimental results. It was only in 1916 that Richardson had enough theoretical results and experimental data to be able to conclude with some certainty that blackbody photoemission could account for only an insignificant fraction (less than 1/5,000 in the worst case) of the observed thermionic current from platinum at 2,000 K.[26]

Richardson's impressive research activity during the fifteen years we have been considering clearly showed the influence of his Cavendish education. The most characteristic feature of his work was the integration of experiment with theory. Almost all his experimental papers were introduced by a theoretical section in which fundamental theory, usually statistical mechanics or thermodynamics, was combined with microphysical assumptions to yield results that were then tested in experiments in which ever improved vacuum technique went hand in hand with manipulation of electrons or ions by electric fields and measurements of currents by sensitive galvanometers or electrometers. An example of Cavendish ingenuity in solving an experimental problem by inexpensive means may be seen in his and Compton's method of obtaining a fresh sodium surface free of oxidation by furnishing their vacuum tube with an additional bulb containing a small electrically heated furnace by means of which they could evaporate sodium on to their target. An external magnet acting on a piece of soft iron connected with the target allowed them to then hoist the target back into position in the measuring bulb without breaking the vacuum.[27] On the other hand the influence of Richardson's exposure to a different American laboratory practice with ties to the affluent industrial laboratories may perhaps be discerned in the apparatus he built for his 1908 determination of the specific charge of thermionic particles.[28] With its precisely machined moving parts this complicated piece of equipment was a long way from the Cavendish "string and sealing wax" approach, strikingly illustrated by Richardson himself during his Cavendish days in a paper on electroscopes.[29]

When Richardson in 1916 published his monograph *The Emission of Electricity from Hot Bodies,* summing up his own and others' work on thermionics, he was hailed by an anonymous reviewer in *Nature* as the acknowledged master of his field:

> The author was one of the first workers in this new field of work . . . A large part of our knowledge of this subject is due to his investigations.
>
> As a consequence, we have a first-hand account of this interesting subject, written by one who has a full appreciation of the experimental difficulties and the adequacy of the theories proposed.[30]

RICHARDSON'S TEXTBOOK

In 1914 Richardson published a textbook called *The Electron Theory of Matter* (*ETM*), a second revised edition of which appeared in 1916.[31] Based on a course of lectures he had been giving to his graduate students at Princeton, this was an advanced textbook aiming at bringing the students rapidly up to the research front in electron physics in general. Thus it had a much wider scope than his later monograph and it was praised for this by Niels Bohr in a review in *Nature:*

> It will be seen that the book covers a very extensive field. To give an adequate representation of the entire electron theory is naturally a task of the greatest difficulty, but the author appears to have done this in an admirable manner.[32]

The value of *ETM* as a historical source for the first phase of the electron theory lies not only in its being probably the most comprehensive survey available, but also in the fact that it was published at a particularly interesting point in time. Again it is appropriate to quote *Nature,* this time from an editorial note reporting the award of the Nobel Prize to Richardson:

> Richardson's "Electron Theory of Matter" is also well known to students of electricity and atomic physics, and although published between the advent of the Bohr and the Wilson-Sommerfeld theories of the atom and with a strong classical bias, is still much used.[33]

There is indeed a world of difference between ETM^2 and, say, Arnold Sommerfeld's *Atombau und Spektrallinien,* published only three years later. A comparison makes the former stand out as perhaps the last important book on the constitution of matter written "with a strong classical bias" by an acknowledged master of electron physics in general. In the following sections I describe some characteristic features of the book on three main points: electromagnetic principles, electrons in matter, and quantum theory.

Fundamental Principles of Electromagnetism

It is interesting to compare ETM with two other surveys of the electron theory, H. A. Lorentz's *Theory of Electrons* and Niels Bohr's *Metallernes Elektrontheori,*[34] both strongly theoretical with little concern for experiment; not surprisingly, Richardson showed more appreciation of experimental work and reported much more fully on experimental results. Both works were written for specialists who were tacitly assumed to accept a standard version (Lorentz's) of the general principles of the electromagnetic theory. Bohr plunged directly into his investigation of the extent to which the behavior of electrons in metals could be accounted for by these principles combined with statistical mechanics and reasonable assumptions about the interactions between electrons and metal atoms, while Lorentz began his book with a succinct account of his own theory, including the concept of electromagnetic mass and the electromagnetic world view.

By contrast, ETM was composed as a textbook for beginning graduate students who could be assumed to have from the outset only an elementary knowledge of electromagnetic phenomena. Thus its account of electromagnetic principles was much more detailed than Lorentz's; but, more significantly, it presented different theoretical points of view without explicitly preferring one over the other. An example is the discussion of the Ampère law and the Faraday law (which Richardson called the First and Second Law of Electrodynamics) where a Maxwellian conception of the ether was compared to the more axiomatic view inherent in the electromagnetic world view:

> The Second Law of Electromagnetism may be looked upon from two different standpoints according to the attitude we take towards electrical science. If we regard electrodynamics as more fundamental than dynamics proper, then we must regard the Second Law as a fundamental law of nature empirically given. We may however take the standpoint that the aether, which we postulate as a medium in which all electrical actions occur, will in the last analysis prove to be a mechanical system subject to the basic laws of dynamics. . . . The view that electrical actions are ultimately dynamical is one whose development in the hands of Maxwell led to notable advances in the science, and it is the view towards which, at any rate until quite recently, most authorities have leaned. Nevertheless it is equally logical to accept the Second Law as an ultimate fact and then, later on, to consider what we can make of the laws of dynamics from the standpoint thus adopted. (ETM^2, 102)

This comment introduced seven pages explaining the analytical dynamics of the ether and the deduction of the second law from the first, so that the late Maxwellian views with which Richardson must have been thoroughly familiar from his Cambridge education were faithfully transmitted to his readers.[35]

The electromagnetic world view was no less fully represented. In a chapter headed "The Fundamental Equations," the four microscopic Maxwell equations and the Lorentz force expression were presented as the equations "associated with the name of Lorentz," and a later chapter entitled "The Aether" gave a detailed account of Lorentz's theory of the electrodynamics of moving media. Furthermore the concept of electromagnetic mass was introduced as a foundation for the electromagnetic world view:

> The idea of electromagnetic inertia, which is due to J. J. Thomson, is fundamental to the electron theory of matter. For it opens up the possibility that the mass of all matter is nothing else than the electromagnetic mass of the electrons which certainly form part, and perhaps form the whole, of its structure. It obviously opens up the possibility of an electrical foundation for dynamics. (ETM^2, 229)

In the beginning of the book Richardson defined the electron as a particle consisting "of a geometrical configuration of electricity and nothing else, whose mass, that is, is all electromagnetic" (ETM^2, 8). This was followed up at the end of the book by a detailed discussion of the unsuccessful attempts to account for gravitation on an electromagnetic basis, concluding that "the electron theory is not in a position to make very definite assertions about the nature of gravitational attraction" (ETM^2, 619).

Having mastered the technical and conceptual complications of understanding first the Maxwellian derivation of electrodynamics from ether mechanics, then Lorentz's electrodynamics of systems of charged particles moving through the ether and its associated hope of an electrical foundation for dynamics, Richardson's students must have felt slightly dizzy upon encountering, immediately after the chapter on "The Aether," yet a third world view that denied the existence of the ether. Richardson's chapter on "The Principle of Relativity" was taken almost verbatim (with due reference) from Einstein's 1907 review in Stark's *Jahrbuch*. Einstein had taken great pains to make his review as clear as possible and Richardson must have appreciated this for he made no substantial changes in the presentation.[36] He showed his sympathy with the reader's difficulties, however, (perhaps his own as well) in accepting the consequences of the theory by a remark following the derivation of the Lorentz contraction, the time dilation, and the addition of velocities:

> Some of the preceding results differ so considerably from those which follow from the generally accepted notions of space and time that many readers will probably regard them as serious objections to the views here developed. If, however, the principle of relativity is accepted they appear to follow with logical certainty. (ETM^2, 303)

The chapter ended with a section entitled "The Principle of Relativity and the Aether" which stated that Lorentz's theory could explain all the known facts, but that the principle of relativity "describes them in a simpler and more symmetrical manner," and that, by denying the possibility of determining the motion of the ether it "finds the aether a superfluous hypothesis" (ETM^2, 323–325).

Clearly, Richardson's approach to the fundamental principles must be characterized as pluralistic. He made an effort to present each of the three paradigms in its own terms and on its own premises, emphasizing the strength of each and to a great extent letting his readers form their own judgments on their relative merits. This attitude may of course be put down to the pragmatism of a working physicist to whom debates on fundamental theory mattered little, either for his own research or for the main chapters of his textbook; this will be discussed more fully in the conclusion.

The Classical Electron Theory of Matter

The main part of Richardson's book consisted of a number of chapters on electron-theoretic explanations of specific groups of physical phenomena. These chapters usually began with a short description of the main features of the phenomena in question. Then a microphysical model involving the electron was sketched and subjected to a detailed and rigorous treatment using electromagnetic theory, classical dynamics and, if necessary, statistical mechanics or thermodynamics. The resulting formulas and constants were compared with the best established empirical laws and experimental measurements, and the free parameters of the model were adjusted so as to give the best possible fit between theory and experiment. Sometimes the original model proved unable to accommodate all the data; an attempt was then made to generalize the model by dropping some too specific assumptions. As an example consider the chapter on "Dispersion, Absorption and Selective Reflection." First "an ideal substance" was described "which in all probability is somewhat simpler in its constitution than any real substance occurring in nature" (ETM^2, 142). It consisted of molecules, each of which contained a number of electrons with fixed positions of equilibrium around which they could oscillate under the combined influence of forces of restitution obeying Hooke's law, velocity-proportional damping forces, and external electric and magnetic fields. The description included a comment to the effect that the damping forces were badly understood—attempted explanations in terms of radiation damping or molecular collisions that resulted in absorption coefficients much smaller than those actually observed—and that generally the absorption process was still mysterious. Next the consequences of the model were worked out and compared with an extensive amount of optical data, and finally it was shown how by means of Lagrangian dynamics a generalized theory might be obtained that was free of unfounded assumptions about the unknown details of atomic structure and at the same time could be adapted to fit a wider range of experimental data, but at the cost of formulas so unwieldy as to be of limited practical use.

Richardson devoted two chapters entitled "The Kinetic Theory of Electronic Conduction" and "The Equilibrium Theory of Electronic Conductors" to the electron gas theory of metals, including his own specialty of thermionics. In the first of these he applied Boltzmann's kinetic theory, taken from Jeans's textbook,[37] as a foundation for a thorough theoretical and experimental discussion of electrical and thermal conductivities, and for the theory of thermoelectricity, deriving relations among the thermoelectromotive force, the Peltier coefficient, and the Thomson coefficient. He also discussed galvanomagnetic phenomena emphasizing the difficulties with understanding the negative Hall effect. In the second chapter he gave a full account of his own 1912 thermodynamical treatment of thermionic emission and its relations to thermoelectricity and repeated his theory of the photoelectric effect.

In these two chapters, as in several others, much discussion was given to various models of the structure of atoms and the behavior of electrons in matter. We have already met one such: electrons oscillating around fixed positions of equilibrium. In his chapter on magnetism Richardson needed atoms to possess magnetic dipole moments and so introduced orbiting electrons. In the kinetic theory of the electron gas, electrons were originally regarded as free, except for brief hard-sphere collisions with atoms, but Richardson showed that if the electrons were supposed to move under the influence of a force from the rest of the atom, proportional to the inverse cube of the distance to the center of the atom, this would make the constant in the so-called Wiedemann-Franz law for the ratio between the thermal and electrical conductivities of a metal agree better with experimental measurements (ETM^2, 413–422).[38] At this point Richardson gave his readers a glimpse of a possible unification: an atom might have a core containing strongly bound, oscillating electrons. This core would constitute an atomic dipole with an oscillating electric moment. In the r^{-3}-field from these dipoles other, more loosely bound electrons might describe orbits and thus give the atom a net magnetic moment. Some of these orbiting electrons might be so weakly bound that they could become "free" from time to time in the sense that they might be able to move from one atom to the next under the influence of an external field and thus produce an electric current. Richardson listed a number of uses of this dipole model, as I shall call it (ETM^2, 422–425; 461–468):

1. A model of this type had been used by J. J. Thomson to explain the photoelectric effect.[39]

2. The magnetic moment of the orbiting electrons would be proportional to the electrical moments of the cores and since one could infer from the "universality of the law connecting radiation and temperature" that the latter were matter-independent one would have an explanation of Weiss's magnetons.

3. One might identify the oscillating cores with the vibrators in Einstein's theory of specific heats. The amplitudes of their dipole moments would then tend exponentially to zero as the temperature approached absolute zero; by making the "free" electrons "more free" this would explain the increase of conductivity at low temperatures. Even superconductivity, recently discovered by Kamerlingh Onnes, had been explained by J. J. Thomson using the dipole model.[40]

4. On the simple kinetic theory of the electron gas, the conductivity of a metal would be proportional to the number of free electrons per unit volume, while the Peltier effect at a junction between two metals would be proportional to the logarithm of the ratio between these numbers for the two metals. Hence there ought to be a large Peltier effect at the junction between a good and a bad conductor. This was not borne out by experiments; in some cases the Peltier effect even went in the direction opposite to the expected. The dipole model removed this difficulty by making the conductivity proportional to the mean number of

electrons free at a given moment, while making the Peltier effect depend on the mean potential energy of the electrons that might become free from time to time. In a bad conductor there might be many more of the latter than of the former type of electrons. The same feature of the model might also help explain the relation between conductivity and thermoelectric power in alloys.

The dipole model was partly inspired by J. J. Thomson and was in general agreement with Thomson's views on atomic structure. Though Richardson was impressed by the number of phenomena for which this model furnished a qualitative (and sometimes quantitative) explanation, he was aware of the recent doubts cast on the Thomson atom. In a discussion of the scattering of α and β rays, for instance, he compared Thomson's theory of multiple scattering with Rutherford's theory of single scattering against atomic nuclei. "Reviewing the whole evidence broadly" he concluded that the phenomena were "quite decisively in favor of Rutherford's view" (ETM^2, 490–496), and he then gave a veiled reference to Bohr's theory of the atom, but he did not take the opportunity to discuss the consequences of this new theory for the dipole model.

The Quantum Theory

On page 347 of ETM^2, more than halfway through the book, Richardson's readers made their first acquaintance with Planck's quantum. This happened in the middle of a chapter on "Radiation and Temperature" after a twenty page discussion of blackbody radiation ending with a demonstration that any theory in which the emission and absorption of radiation by matter was assumed to be a continuous process, subject to the laws of dynamics and electrodynamics, would inevitably lead to the radiation formula of Rayleigh and Jeans, a formula that, except for long wavelengths, went against all experimental evidence. "Although it may appear revolutionary to some," Richardson continued, "it seems to the writer that the only logical way out is to deny the adequacy of dynamics and electrodynamics for the explanation of the emission and absorption of radiation of energy by matter." He then went on to recount Planck's latest (1912) version of his theory in which the only discontinuous feature was that an atomic oscillator was assumed to emit all its energy, with a certain probability dependent on the radiation density, whenever its energy reached an integral multiple of Planck's constant h times its frequency ν, while absorption was treated as a completely continuous and classical process.

After deriving Planck's radiation formula and emphasizing that the experimental determination of Boltzmann's constant from measurements on blackbody radiation had led to values for Loschmidt's number and the electronic charge in excellent agreement with those found independently by more direct methods, Richardson stated that Planck's theory had recently received "unexpected support" in other directions. One of these was Einstein's theory of specific heats (1907) and Debye's modification of it (1912), another was Richardson's own theory of the photoelectric effect, and yet another was Bohr's theory of atoms and molecules. All of these he described not as essentially new theories but rather as natural extensions of Planck's quantum theory of the interaction between radiation and matter.

On the whole, despite his use of the word "revolutionary" in the quotation above, Richardson appears to have regarded the quantum as signifying a new, mysterious property of atoms rather than as the beginning of a new fundamental theory. Thus he justified his preference for Planck's latest theory by saying:

> In his earlier papers the assumptions made were equivalent to postulating that the energy itself had a discontinuous structure, but Planck has now shown that equivalent results may be obtained by merely supposing that the radiant energy is emitted by jumps, the absorption taking place continuously. As the emission of radiant energy might be expected to be conditioned by the breaking up of some structure present in the matter, this seems a very natural hypothesis. (ETM^2, 350)

Two pages earlier he had characterized the theory as "free from self-contradiction and from assumptions, such as that of the discontinuous nature of energy, which appear to do violence to the fundamental ideas of physics. So much could not be said of the earlier forms." One of the violators against the "fundamental ideas of physics" was no doubt Einstein's lightquantum hypothesis. In their first report in the 17 May 1912 issue of *Science* on their photoelectric experiments Richardson and Compton said that their results were "favorable to a theory . . . of the type of Einstein's,"[41] but in the 12 July issue Richardson published a note in which he argued that Planck's new derivation of the blackbody radiation law and his own derivation of the photoelectric equation (9) had shown that "the unitary theory of light" was unnecessary to account for either of these laws.[42] In *ETM,* after repeating his derivation of eq. (9), he said that a similar equation "was first given by Einstein as a consequence of the view that the energy of light waves was distributed in discrete quanta" (ETM^2, 473–474). This laconic statement is the only direct mention of the light quantum in the whole book.

In his book on the wave-particle dualism Wheaton has shown both the widespread use of the "triggering hypothesis" as an explanation for the observed particle features of x-rays and visible light and the relation between this hypothesis and Planck's 1912 theory, clearly exemplified in Richardson's "breaking up" hypothesis in the above quotation.[43] Wheaton has also demonstrated the growing recognition after 1911 of the similarity between the photoelectric effect and the emission of secondary electrons by x-rays, and its significance for the discussions of the nature of x-rays and light.[44] Richardson was clearly aware of these problems. In his discussion of x-rays he pointed out the similarity between W. H. Bragg's experiments showing a preponderance in the forward direction of secondary electrons produced by x-rays and the similar experiments of O. Stuhlmann and R. D. Kleeman on photoelectrons from ultraviolet light.[45] These experiments, he said, could not be reproduced by the simple view that the kinetic energy of the electrons derived from the work done as the electromagnetic pulse passed over them; this view would in any case lead to far too small values of the kinetic energy. He then referred to Planck's 1912 theory and his own photoelectric theory as having led to the view that "when radiant energy causes the disruption of an electron from a material system, the electron acquires an amount $h\nu$ of energy" and then gave a long and involved statistical argument about the exchange of momentum between radiation and the electrons in a thin slab of material to prove that the Bragg and Stuhlmann effects could be brought out "without supposing the primary radiations which exhibit them to be of a material nature" (ETM^2, 478–481). After surveying many more phenomena relating to x-, γ-, and β-rays, among them the fact that the maximum energy of secondary electrons produced by a characteristic x-ray was equal to the minimum energy of the primary electrons required to produce that ray, he admitted that these facts "receive a simple and obvious explanation on the view that the x-rays and light consist of showers of material particles or of bundles of energy." He immediately rejected this view, however, as unable to account for interference phenomena and deemed it "a little safer" to adopt the triggering hypothesis that he now described as a condition "of a very general character and necessarily inherent in all types of matter." This condition would determine the disruption of matter under the stimulus of a given radiation in such a way that the energy of the disrupted electrons would be equal to $h\nu$, or an integral multiple of this quantity. He ended this discussion with an aside, put between square brackets, which shows a very clear appreciation of the paradoxical, dualistic character of the evidence that existed on the nature of light and x-rays, that is of radiation in general:

> [It is difficult, in fact it is not too much to say that at present it appears impossible, to reconcile the divergent claims of the photoelectric and the interference groups of phenomena. The energy of the radiation behaves as though it possessed at the same time the opposite properties of extension and localization. At present there seems no obvious escape from the con-

clusion that the ordinary formulation of the geometrical propagation involves a logical contradiction, and it may be that it is impossible consistently to describe the spatial distribution of radiation in terms of three dimensional geometry.] (ETM^2, 507–508)

Although Richardson left no doubt of his preference for an approach that, like Planck's and his own, combined a vaguely expressed version of the triggering hypothesis with the wave theory of light and x-rays, using general statistical and thermodynamical methods, the final period of this quote appears to express an uneasiness that this might not suffice and that more radical measures might turn out to be required. Hence perhaps the square brackets.

Richardson's penultimate chapter was headed "The Structure of the Atom." Two-thirds of its pages were devoted to J. J. Thomson's atom, chiefly emphasizing Thomson's explanation of the periodic system of the elements and his theory of chemical combination. In keeping with the pluralistic character of the book, this was followed by a brief review of Rutherford's arguments for the nuclear atom, and then by a full treatment of Bohr's 1913 theory of the atom, one of the earliest, if not the earliest such treatment in a regular textbook.[46] In the preface to the second edition Richardson said that this treatment was considerably expanded relative to the first edition and wrote about the "remarkable successes of this theory" (ETM^2, vii). In the chapter he detailed Bohr's explanation of spectra and Moseley's and Kossel's work on x-rays as instances of these successes. Interestingly he described Bohr's postulates as "closely related to those underlying Planck's theory of Radiation," and he ended the chapter by emphasizing that although Bohr's theory was "non-mechanistic" it preserved "continuity with the ordinary dynamics in the region of slow vibrations" (ETM^2, 606). Thus Bohr's theory was characterized rather as a natural extension of Planck's than as a radical departure from previous theory. In view of his express preference for the triggering hypothesis and Planck's second theory, it is surprising that Richardson did not mention the fact that Bohr diverged from Planck in making the absorption of radiation as discontinuous a process as emission. Neither did he discuss the implications of Bohr's theory for the dipole model or other models of atomic structure used in previous parts of the book. He did remark, however, that the kinetic energy of an electron liberated by radiation of frequency ν from a Bohr orbit of energy $-W_D$ would be given by

$$\frac{1}{2}mv^2 = h\nu - W_D, \tag{10}$$

and that this result agreed with that of his own photoelectric theory.

Conclusion

In a recent work A. Warwick has introduced the term theoretical technology to distinguish "the pieces of theoretical work . . . which are used to solve particular problems" from "the idealized conceptual schema of a general theory," and he has employed this concept to analyze the reception of the theory of relativity by physicists at Cambridge.[47] Although Warwick in his introduction of this term appears to define it as a sociological concept characterizing a theoretical school or group of physicists, I would suggest that it might be useful also to apply it to the case of an individual physicist. As an example it is evident that even if Richardson in his textbook gave an excellent account of the fundamental concepts and theorems of the theory of relativity (taken almost verbatim from Einstein himself) that theory was not a part of the theoretical technology that he employed either in his research papers or elsewhere in his book.

The example indicates that by adopting Warwick's point of view we might get behind the apparent pluralism of Richardson's book and obtain a deeper understanding of his view of physics. I have characterized Richardson's account of the fundamentals of electromagnetic theory as pluralistic, one might say indifferent, with respect to the three world views: the mechanistic, the electromagnetic, and the relativistic. His research papers, however, as well as the "applied" chapters of his book (dealing with optical effects, radiation, magnetism, properties of metals, etc.) give a different impression. The theoretical technology that Richardson applied came from electromagnetic theory, statistical mechanics and thermodynamics, classical mechanics, and Planck's quantum theory of radiation. I have already noted the absence of any use of relativistic concepts or arguments. Likewise I find no use of ideas that can be referred to the mechanistic view of the ether. On the other hand, the Lorentz force expression was used without comment whenever appropriate, but even more revealing is Richardson's explicit definition of the electron as "a geometrical configuration of electricity and nothing else" as well as a remark in his discussion of the Thomson atom to the effect that the electromagnetic inertia of its positive sphere "is negligible compared with that of a single electron, so that the greater part of the mass is entirely unaccounted for by this theory" (ETM^2, 586–587), a remark that only makes sense within the electromagnetic world view. We may safely conclude that despite the disinterestedness displayed in his chapters on the three paradigms, Richardson thought and worked within the electromagnetic world view.

About Richardson's views on atomic structure it should first be noted that in his research he had not had much use for detailed models of atomic structure, and he had never taken active part in the work of the "atom builders," to use J. L. Heilbron's phrase for the constructors of atomic models before 1913.[48] He had used few specific properties of the dipole model, such as the r^{-3}-field, otherwise he had needed only more general features, such as the distinction between bound and free electrons in a metal and the existence of a characteristic potential energy jump or work function for an electron passing out through the surface of a metal. In his work on the photoelectric effect he had used the quantum theory, but only in the form of Planck's radiation law as a condition for statistical equilibrium; in his derivation of the photoelectric equation he had completely bypassed Einstein's light quantum as indeed any detailed consideration of the process by which a single electron was forced out of a metal by radiation.

In his textbook one finds more extensive use of models of specific atomic properties. At first glance it appears as if Richardson postulated different such properties according to the phenomenon under discussion: fixed electrons for dispersion, orbiting electrons for magnetism, free electrons for conduction. As I have emphasized above, however, the dipole model appeared to him to unify these structures. It was precise enough to serve as a basis for calculations, on the other hand it was sufficiently flexible to allow for the different types of electronic motions that were needed for the explanations of the many and varied types of properties of matter. It is worth noting that the atomic models of Thomson and Bohr both contained orbiting electrons of which some were strongly bound, while others, for example the valency electrons in the alkali metals, were easily removed. For this reason either of them may have appeared to Richardson as a particular version of the dipole model. There is certainly no indication of his being aware that Bohr's theory might require profound revisions of the theories propounded in earlier parts of the book, though he did note that it could furnish an explanation of Weiss's magneton (ETM^2, 395 and 592).

To grasp Richardson's understanding of Bohr's theory it is instructive to consider the dispersion theory that was put forward by Debye and Sommerfeld in 1915. Using Bohr's models for the molecules of hydrogen, oxygen, and nitrogen, they evaluated the perturbations caused by an electromagnetic wave in the orbit of a molecular electron, calculated the mean dipole moments corresponding to the proper vibrations into which the perturbations could be resolved, and then used these oscillating dipole moments in the formulas of Lorentz's dispersion theory. Thus the interaction between radiation and the orbiting electron was described in completely classical terms. The resulting dispersion formula contained the frequency of revolution in the unperturbed orbit as a free parameter, and the whole exercise consisted in determining this frequency from the best dispersion measurements and comparing it with that determined by Bohr's quantization of angular momentum in the ground state of the molecule. In other words, they accepted Bohr's quantum condition for the ground state, but rejected his quantum postulate for the emission and absorption of radiation in favor of a classical treatment of the interaction between light and the orbiting electron. Not surprisingly, this hybrid theory found little favor in Copenhagen and was criticized in public by C. W. Oseen; nevertheless it lived on in the literature until about 1919.[49] The point here is to suggest that Richardson's conception of Bohr's theory had some similarity to Debye's and Sommerfeld's and so was not untypical. He, too, had no difficulty in accepting Bohr's quantum postulate for the stationary states, in fact he saw it as "closely connected with the quantum hypothesis of Planck," probably because like Planck's hypothesis it allowed one to think of the quantum exclusively as reflecting a property of the structure of atoms. Bohr's frequency postulate, on the other hand, he just repeated without comment, hence one can only guess as to how he conceived of it; my conjecture is that he regarded it as just a version of the triggering hypothesis. What is certain is that he gave no indication that he saw a fundamental conflict between this postulate and the many applications of Lorentz's electrodynamics in the electron theory of matter.

What was the status of the electron theory of matter twenty years after the discovery of the electron? In Richardson's opinion it was quite good. From a modern perspective, informed by the extensive historical literature on the quantum revolution with its emphasis on the failure of classical physics in accounting for the structure of matter and radiation, the tone of Richardson's book may appear surprisingly optimistic. He reported judiciously on difficulties as well as on successes, but usually as problems not yet solved rather than as insuperable obstacles. Typical of the general attitude of the book is a passage in the preface to the first edition in which he remarked that recent developments

> lead one to think that the difficulties which beset the electron theory of metallic conduction in its usual form may be overcome by the application of the ideas underlying Planck's theory of radiation. In any event the theories of Chapters XVII and XVIII should be valid at sufficiently high temperatures when the results of the quantum theory coalesce with those of the continuous theory. Many other branches of the subject are in a similar, though possibly less aggravated, situation; amongst these the questions of atomic structure, spectroscopic emission, x-rays and the magnetic properties of bodies are conspicuous examples. At the present time this field is unquestionably a very fruitful one both for the experimental and the theoretical physicist. (ETM^2, vi)

The message to Richardson's students would clearly have been that although many problems still remained to be solved, the numerous successes of the electron theory showed that electron physics was on the right track. In the same vein, after having laid out the theoretical difficulties involved in accounting for the details of the Hall effect and the change of resistance in a magnetic field, Richardson remarked: "These effects are unquestionably very complicated, and so far the electron theory has not been able to furnish an adequate quantitative explanation of them. On the other hand it is the only theory which has been able to account for them qualitatively" (ETM^2, 409).

These passages give Richardson's general verdict on the achievements of the classical electron theory of matter up to 1916: it had been extremely successful and it still offered many possibilities for further exploration. The quantum theory did not pose a threat to the theory; on the contrary, it formed one of the most promising of these possibilities.

Notes

1. The best overall treatment is J. L. Heilbron: *A History of the Problem of Atomic Structure from the Discovery of the Electron to the Beginning of Quantum Mechanics* (Ph.D. dissertation, University of California, Berkeley, 1964, University Microfilms, Inc., Ann Arbor, Michigan).

2. For Thomson's views on physics education, and his role in the reform at Cambridge, see David B. Wilson: "Experimentalists among the mathematicians: Physics in the Cambridge Natural Sciences Tripos, 1851–1900," *Historical Studies in the Physical Sciences 12* (1982): 325–371. Cf. also Andrew Warwick: "Cambridge Mathematics and Cavendish Physics: Cunningham, Campbell and Einstein's Relativity 1905–1911, Part II: Comparing Traditions in Cambridge Physics" *Studies in the History and Philosophy of Science 24* (1993): 1–25, 2–3.

3. E. W. Foster: "Richardson, Sir Owen Willans," *The Compact Edition of the Dictionary of National Biography,* vol. II, Oxford University Press 1975, 2856; W. Wilson: "Owen Willans Richardson 1879–1959," *Biographical Memoirs of Fellows of the Royal Society 5* (1960): 206–215; Loyd S. Swenson, Jr.: "Richardson, Owen Willans," in: C. C. Gillispie, ed.: *Dictionary of Scientific Biography,* vol. XI (New York: Charles Scribner's Sons, 1975), 419–423.

4. Isobel Falconer: "J. J. Thomson and 'Cavendish Physics,'" in Frank A. J. L. James, ed.: *The Development of the Laboratory: Essays on the Place of Experiment in Industrial Civilisation* (Macmillan Press, 1989), 104–117.

5. This conception was put forward in J. J. Thomson: "Indications relatives à la constitution de la matière fournis par les recherches récentes sur le passage de l'électricité a travers les gaz," in: Ch.-Éd. Guillaume & L. Poincaré, eds.: *Rapports présentés au congrès international*

de physique réuni à Paris en 1900 (Paris: Gauthier-Villars, 1900), vol. III, 138–151. This report was undoubtedly Richardson's first introduction to the electron gas concept or, in Thomson's phrase, "the corpuscular state of matter."

6. O. W. Richardson: "On an Attempt to Detect Radiation from the Surface of Wires Carrying Alternating Currents of High Frequency," *Proceedings of the Cambridge Philosophical Society 11* (1902): 168–178.

7. For a brief survey of these early experiments on thermal emission of electricity, see O. W. Richardson: *The Emission of Electricity from Hot Bodies* (1916), 2. ed. (London: Longmans, Green and Co., 1921), 2–4.

8. O. W. Richardson: "On the Negative Radiation from Hot Platinum," *Proceedings of the Cambridge Philosophical Society 11* (1902): 286–295.

9. I have changed Richardson's notation a little. Like Jeans, he had the confusing (to a modern reader) habit of denoting Boltzmann's constant by R and calling it "the gas constant for a single corpuscle (or molecule)." Cf. J. H. Jeans: *The Dynamical Theory of Gases* (Cambridge: At the University Press), 1904.

10. J. Patterson: "On the Change of the Electrical Resistance of Metals when Placed in a Magnetic Field," *Philosophical Magazine 3* (1902): 643–656; Thomson, "Indications relatives à la constitution."

11. O. W. Richardson: "The Electrical Conductivity Imparted to a Vacuum by Hot Conductors," *Philosophical Transactions of the Royal Society 202A* (1903): 497–549. On p. 498 he referred to Drude's papers as well as Thomson's Paris report for "the hypothesis of the conduction in metals by corpuscles" and on p. 499 to Drude's application of the kinetic theory of gases to the free electrons in a metal.

12. For the importance of the electron vapor concept, see W. Kaiser: "Early Theories of the Electron Gas," *Historical Studies in the Physical and Biological Sciences 17* (1987): 271–297, on 280–282.

13. O. W. Richardson: "Thermionics," *Philosophical Magazine 17* (1909): 813–833.

14. O. W. Richardson, *The Emission of Electricity;* for a list of sixteen contributors to the field, see 60–61.

15. Cf. Leonard S. Reich: "Irving Langmuir and the Pursuit of Science and Technology in the Corporate Environment," *Technology and Culture 24* (1983): 199–221, on 211; and Hugh G. J. Aitken: *The Continuous Wave: Technology and American Radio, 1900–1932* (Princeton, N.J.: Princeton University Press, 1985), 231.

16. See, *e. g.,* Lorentz's 1924 Solvay Report, H. A. Lorentz: "Application de la théorie des électrons aux propriétés des métaux," in H. A. Lorentz: *Collected Papers,* vol. 8 (The Hague: Martinus Nijhoff, 1935), 263–306, on 283–294; and R. Seeliger: "Elektronentheorie der Metalle" (1921), in: A. Sommerfeld, ed.: *Enzyklopädie der mathematischen Wissenschaften mit Einschluss ihrer Anwendungen* vol. 5, part 2 (Leipzig: B. G. Teubner, 1904–1922), 778–878, on 835–851.

17. These points were all touched upon in Richardson's Nobel lecture. See O. W. Richardson: "Thermionic Phenomena and the Laws which Govern Them" (1928), in: *Nobel Lectures. Physics 1922–1941* (Amsterdam: Elsevier, 1965), 224–236

18. O. W. Richardson and F. C. Brown: "The Kinetic Energy of the Negative Electrons Emitted by Hot Bodies," *Philosophical Magazine 16* (1908): 353–376; see particularly the discussion on 374–376.

19. Cf. Kaiser, "Early Theories," 285–292.

20. O. W. Richardson: "The Electron Theory of Contact Electromotive Force and Thermo-electricity," *Philosophical Magazine 23* (1912): 263–278; and *id.*: "Some Applications of the Electron Theory of Matter," *ibid.*, 594–627.

21. H. A. Wilson: "On the Discharge of Electricity from Hot Platinum," *Philosophical Transactions of the Royal Society 202 A* (1904): 243–275; see 273–275 for his criticism of Richardson's views.

22. O. W. Richardson: "The Emission of Electrons from Tungsten at High Temperatures: An Experimental Proof that the Electric Current in Metals Is Carried by Electrons," *Philosophical Magazine 26* (1913): 345–350.

23. O. W. Richardson, "The Electrical Conductivity," 535.

24. O. W. Richardson, "The Electron Theory" ("Some Applications . . ."), 615–627; *id.*: "The Theory of Photoelectric Action," *Philosophical Magazine 24* (1912): 570–574; *id.*: "The Laws of Photoelectric Action and the Unitary Theory of Light (Lichtquanten Theorie)," *Science 36* (1912): 57–58.

25. O. W. Richardson and Karl T. Compton: "The Photoelectric Effect," *Philosophical Magazine 24* (1912): 575–594; Karl T. Compton and O. W. Richardson: "The Photoelectric Effect.-II." *Philosophical Magazine 26* (1913): 549–565.

26. O. W. Richardson: "The Complete Photoelectric Emission," *Philosophical Magazine 31* (1916): 149–155.

27. Compton and Richardson, "The Photoelectric Effect," 557–558.

28. O. W. Richardson: "The Specific Charge of the Ions emitted by Hot Bodies," *Philosophical Magazine 16* (1908): 740–767. The apparatus is described on 743–750.

29. O. W. Richardson: "The Construction of Simple Electroscopes for Experiments on Radioactivity," *Nature 71* (1905): 274–276.

30. Anonymous: "Escape of Electrons from Hot Bodies," *Nature 98* (1916): 146.

31. O. W. Richardson: *The Electron Theory of Matter* (Cambridge: At the University Press, 1914; Second Edition 1916). In the following these two editions will be denoted *ETM*[1] and *ETM*[2].

32. N. B.: "Modern Electrical Theory," *Nature 95* (1915): 420–421. Although the initials do not identify the author with complete certainty there can be little doubt that the

review was written by Niels Bohr. On the exchanges and correspondence between Bohr and Richardson on the electron theory, see Niels Bohr: *Collected Works,* vol. 1 (Amsterdam: North-Holland Publishing Company, 1972), 111–117 and 482–491.

33. "News and Views," *Nature 124* (1929): 814.

34. H. A. Lorentz: *The Theory of Electrons,* 1st ed. 1909; Reprint of 2d ed. (1915) (New York: Dover Publications, 1952); Niels Bohr: *Studier over Metallernes Elektrontheori: Afhandling for den filosofiske Doktorgrad,* København (V. Thaning & Appel) 1911; English Translation in Bohr, *Collected Works,* 291–395.

35. This section was modeled on the treatments by Larmor and Jeans, cf. Joseph Larmor: *Aether and Matter* (Cambridge: At the University Press, 1900), ch. VI; and J. H. Jeans: *The Mathematical Theory of Electricity and Magnetism* (Cambridge: At the University Press, 1908), ch XVI.

36. A. Einstein: "Über das Relativitätsprinzip und die aus demselben gezogenen Folgerungen," *Jahrbuch der Radioaktivität und Elektronik 4* (1907): 411–462, reprinted in: *The Collected Papers of Albert Einstein,* vol. 2, ed. by John Stachel (Princeton, N.J.: Princeton University Press, 1989), 433–484; cf. 272 for Einstein's pedagogical efforts.

37. Jeans, *The Dynamical Theory.*

38. Richardson had developed this theory of conductivity in his 1912 paper, "The Electron Theory" ("Some Applications . . ."), 594–601, and had noted that Niels Bohr had obtained identical formulae in his dissertation, cf. Bohr, *The Studier over Metallernes* 46–63; English translation, 335–345.

39. J. J. Thomson: "On the Theory of Radiation," *Philosophical Magazine 20* (1910): 238–247. For Thomson's discussion of the photoelectric effect and his attempt to avoid "the unitary theory of light," see 243–246.

40. J. J. Thomson: "Conduction of Electricity through Metals," *Phil. Mag. 30* (1915): 192–202. In Thomson's model, atomic dipoles will show, below a certain critical temperature, complete alignment in an external electric field combined with the internal field from the dipoles themselves; this alignment will be maintained by the internal field even after the external field has vanished, just like the alignment of magnetic dipoles in the case of permanent magnetization. In the internal field from the dipoles, loosely bound electrons will then be able to pass "along the chain of atoms like a company in single file passing over a series of stepping stones" (195), thus producing an electric current even in the absence of an external electromotive force.

41. O. W. Richardson and Karl T. Compton: "The Photoelectric Effect," *Science 35* (1912): 783–784.

42. *Cf.* p. xxx above.

43. Bruce R. Wheaton: *The Tiger and the Shark: Empirical Roots of Wave-Particle Dualism,* (Cambridge University Press, 1983), 178–180.

44. *Ibid.* ch. 8. Richardson's theory of the photoeffect and his attitude toward Einstein's light quantum are discussed on 236–238 and 260 where his attitude is characterized as typical for experimentalists at the time.

45. For a description of these experiments, see *ibid.*, 97–98 (Bragg) and 235 (Stuhlmann and Kleeman).

46. The preface is dated 11 January 1916, so work on the book must presumably have been finished by the end of 1915.

47. Andrew Warwick: "Cambridge Mathematics and Cavendish Physics: Cunningham, Campbell and Einstein's Relativity 1905–1911. Part I: The Uses of Theory," *Studies in the History and Philosophy of Science 23* (1992): 625–656, 633.

48. Heilbron, *A History of the Problem of Atomic Structure* 296.

49. The Debye-Sommerfeld theory is discussed in detail in an unpublished M. Sc. dissertation, Kirsten Kragbak: *Udviklingen af teorien for optisk dispersion i den gamle kvantemekanik, med særligt henblik på perioden 1913–21,* University of Aarhus, History of Science Department, 1989, 28–73. For the reactions of Bohr and Oseen, see Ulrich Hoyer's introduction in Niels Bohr: *Collected Works,* vol. 2 (Amsterdam: North-Holland Publishing Company, 1981), 336–341.

8

Electron Gas Theory of Metals: Free Electrons in Bulk Matter
Walter Kaiser

In 1887 Heinrich Hertz discovered electromagnetic waves, and Maxwell's electrodynamics were seemingly confirmed by experiment. A surprising number of questions, however, remained unanswered. Issues related to bulk matter remained vague,[1] including the conception of an electric current, of electric resistance, and the reaction of matter to high frequency fields (that is, dispersion of light in dielectric substances and absorption of light by metallic surfaces). Physicists such as Oliver Heaviside and Hertz were well aware of this precarious situation, notwithstanding how clearly they otherwise recognized the outstanding achievement of Maxwell's theory.

Typical discussions of this situation appear in the correspondence of Heaviside and Hertz, where Heaviside speaks of the "embryonic state"[2] of Maxwell's theory, of a "sort of skeleton-framework,"[3] and in which he suggests that electrodynamics has to be a molecular theory as well.[4] A letter by Hertz to Paul Drude confirms the recognition of difficulties. Hertz admitted that with metal optics he had been running up against the same "rock," obviously now struck by Drude.[5] Quite naturally, however, electrodynamics in Britain was still strongly influenced by Maxwell's theory. It is not at all surprising, therefore, that British physicists were somewhat reluctant to abandon the idea of a continuously distributed "electricity." Joseph Larmor's long and complicated path toward electron theory is an outstanding example of this difficulty.[6]

On the other hand, it is easy to comprehend how Hendrik Antoon Lorentz, with his understanding of Wilhelm E. Weber's electrodynamics of moving charges, was able to develop his electron theory from 1892 in a straightforward manner.[7] In fact Weber had already developed a fairly elaborate microphysical model to explain electric resistance,[8] which he ascribed to the paths traced by his charges. More influential at first, however, was the general model of an electric current on which Weber's law of electrodynamics was based. In this model (due originally to Fechner) an electric current consisted of opposite electric charges moving with equal but opposite velocities.[9] Evidence from experiment came rather late. In 1879 the U.S. physicist

Edwin Herbert Hall discovered that a transverse current develops in a solid material when it carries an electric current and is placed in a magnetic field that is perpendicular to the current. In 1886, in his basic theory of the Hall effect, Ludwig Boltzmann, noting that the Hall effect could be understood as a result of the force that a perpendicular magnetic field exerts on moving positive or negative particles presumed to constitute Weber's electric current, pointed out that Fechner's model of equal but opposite speeds had to be abandoned. Moreover, Boltzmann was able to explain the opposite signs of the Hall coefficient in different conductors when he assumed that both types of carriers have different velocities.[10]

THE LEGITIMACY OF THE ELECTRON GAS MODEL

The Theories of Riecke and Drude

In 1898 Eduard Riecke in Göttingen attempted to amalgamate the qualitative attempts toward a theory of conductivity made since Ampère and Weber (who was Riecke's teacher) and to build a mathematical theory. Most important for the justification of his new approach, which was designed to supersede Maxwell's electrodynamics, was the model offered by the theory of electrolysis as well as experimental results in the field of gas discharge.[11]

Although the intellectual climate in physics at the end of the nineteenth century certainly was strongly influenced by positivism, in practice it was often important to strive for a tentative unification of different fields with the help of a common theoretical model. In 1887 Riecke himself discussed—somewhat behind the times—hydrodynamical analogies in electrodynamics. This discussion was explicitly based on the idea that those analogies allow for a transfer of laws valid for one group of phenomena to a completely different field.[12] He used this concept to legitimize his theory of electric conductivity as well.[13]

In his explicit formulation of the theory of conductivity Riecke combined Weber's microphysical explanation of resistance with the conduction mechanism associated with Weber's force law. Thus he assumed a dual conduction mechanism, in which "in the space between the ponderable molecules [the metal atoms] positive as well as negative electrical particles move."[14] The phenomena of electrolysis and experiments with "cathode rays" and "canal rays" presumably furnished evidence for the assumption of a dual conduction mechanism.[15] Riecke's calculations were based on a rather small concentration of charge carriers. Thus he only needed to consider the interaction of carriers and metal atoms. Consequently, Riecke's theory was not yet an electron gas theory. It was, rather, a "geometry of particle trajectories,"[16] in the sense of providing linear mean free paths and curved molecular paths, which one could consider as resulting from collisions. For the important relation between energy and temperature, Riecke assumed only that the mean kinetic energy of the charge carriers is proportional to the temperature. Obviously the constant of proportionality was not yet numerically determined, since that would require the equipartition theorem of the kinetic theory of gases that Riecke did not use. About experiment, Riecke's theoretical expressions remained much too general. This is true for what later turned out to be important in the electron gas theory of metals, namely the ratio of the expressions for conductivity of heat and for electrical conductivity—the famous law of Gustav Wiedemann, Rudolph Franz, and Ludvig Valentin Lorenz.[17]

The electron gas theory of metals came into being two years later when Paul Drude, still in Leipzig, published two lengthy papers on the subject in the Annalen der Physik.[18] The method of using a successful theoretical model developed for one field in a different one was quite evident in the work of Drude, in two ways: first, despite his beliefs, Drude eventually accepted Maxwell's merger of electromagnetic theory and of the theory of light;[19] second, in his electron gas theory of metals he applied significant elements of the kinetic theory of gases.

Both steps were remarkable not only with regard to their tendency toward a unified science[20] but were also significant in the difficulties they approached and the questions of legitimacy they raised. The previously mentioned letter by Hertz to Drude tackled one of these problems. Applying Maxwell's theory to metal optics, Drude struggled with one of the puzzles Maxwell's continuity theory had left unsolved. Drude eventually tried to explain the dispersion of light and the reflection of metal surfaces in terms of a response of free ("conducting") electrons and of bound ("isolating") electrons to the fast changing electromagnetic field.[21] More specifically, he wanted to investigate the ratio of free electrons to bound electrons and thus to get more information about the reaction of bound electrons to light.[22] Out of this approach his free electron gas theory of metals almost immediately emerged. In turn, the emerging theory of the electron gas of metals reflected one of the uncertainties typical for the time immediately before and after the publication of the fully developed special theory of relativity. It remained unclear, for example, whether the carriers of metallic conduction possess only apparent electrodynamical mass or true mass as well.[23]

What was important, however, was the problem of applying a thermodynamical model in electrodynamics at a time when statistical mechanics was by no means universally accepted. Around 1900 the memory of the controversies in Oxford on specific heats, on the equipartition theorem, and in general on mechanics and irreversibility, the debate on energetics in Lübeck and the fierce polemic between Ernst Zermelo and Ludwig Boltzmann was still too fresh. Even in 1911, Paul and Tatiana Ehrenfest were quite surprised that the theory was so widespread, despite the remaining "severe incompleteness" of statistical mechanics "with regard to logic."[24]

Drude relied on Henricus van't Hoff's kinetic theory of osmotic[25] pressure and Walther Nernst's theory of concentration cells in electrolysis, which itself depended on van't Hoff's theory of dilute solutions. Drude thereby conceived the possibility of applying the model of the kinetic theory to the carriers of metallic conduction. The crucial point was his adoption of Boltzmann's equipartition theorem, including the numerical value of the "universal constant" a, assigning each degree of freedom a kinetic energy of $1/2\ mv^2$, that is, $1/2\ m_1 v_1^2 + 1/2\ m_2 v_2^2 + \ldots + 1/2\ m_n v_n^2 = aT$ (where T is the absolute temperature, indices $1, 2, \ldots, n$ denote the degrees of freedom). He remarked:

> These laws for gases obviously have now proven to be valid for osmotic pressure as well, shown by ions in electrolytes, and not only in a formal sense but also with use of the same numerical constants. If a metal is now immersed in an electrolyte in the case of 'temperature-equilibrium' [that is, thermodynamic equilibrium] the free electrons ['kernels'] in the metal would have the same kinetic energy as the ions in the electrolyte. Therefore, we must also derive for our equation (1) [for the equipartition theorem] the constant a occurring in the gas laws [that is, assuming spherical particles as well as translational states, $2/3\ a = R$, R = gas constant].[26]

For the physicists of the time this was probably the decisive core of Drude's theory. A prime example is Arnold Sommerfeld's lecture notes for a chapter on "Electron theory of metals and some related statistical questions" in the summer semester 1912. Drude's "transfer of the numerical value furnished by the gas theory" appears here as the "salient point [springender Punkt]" of Drude's electron gas theory of metals.[27]

Drude's own calculations were based on a general dual mechanism of conduction. In contrast to Riecke, Drude admitted arbitrary integer multiples of the elementary charge. Starting from a diffusion equation, which he took from Boltzmann's "Lectures on gas theory,"[28] Drude derived an expression for the flow of heat carried by the movement of electrons through a temperature gradient. From collisions between electrons and metal atoms he calculated the constant drift velocity of electrons in an electric field. For single charged positive or negative carriers the ratio of the expressions for the conduction of heat and of electrical conductivity proved to be independent of typical "kinetic" properties (mean free path, concentration or number of carriers, velocity of carriers). Thus, Drude was able to derive a theoretical expression for the famous empirical Wiedemann-Franz law:[29]

$$\frac{k}{c} = \frac{4}{3}\left(\frac{a}{e}\right)^2 T;$$

where k is the conductivity for heat, c is the electrical conductivity, a equals 3/2 R, R is the gas constant, and e is the elementary charge.

Since Wiedemann, Franz, Lorenz, Weber, and Kohlrausch, the close relationship between the conductivity of heat and electrical conductivity appear beyond any reasonable doubt. Drude's theory offered the possibility of calculating theoretically the constant of the experimentally established Wiedemann-Franz law through its connection to the gas constant and the value of the elementary charge. Thus Drude's theory explained an important property of solid matter with assistance from a basic thermodynamic constant, which had a distinct microphysical meaning, and with help of a microphysical quantity in electricity. Vice versa, a physicist could deduce the value of a constant in the theory of gases from the electric behavior of metals.[30]

The agreement of Drude's expression for the Wiedemann-Franz law with the measurements of Wilhelm Jaeger and Hermann Diesselhorst at the Physikalische Technische Reichsanstalt in Berlin was particularly impressive.[31] Furthermore, what often happens in physics also happened here: the experimental confirmation of the theoretically derived Wiedemann-Franz law in turn was a new indication of the outstanding importance of the gas (or the Boltzmann) constant. Therefore it is not surprising that Max Planck—who always thought in terms of universal constants—called the constant a (= 3/2 R) the "Boltzmann-Drude constant."[32] Indicative of Planck's approbation, is the proposal for the election of Paul Drude as a full member ("ordentliches Mitglied") of the Prussian Academy of Sciences. The nomination was written by Emil Warburg and was—among others—signed by Max Planck. In this proposal, Drude's "fundamental result" was stressed according to which "the ratio of thermal conductivity and electrical conductivity is a universal constant proportional to the absolute temperature."[33]

As mentioned before, Drude assumed the existence of negative and positive carriers in an electric current. Thus, the further development of the electron gas theory of metals was also influenced by questions concerning the nature of positive electricity. With help of deflection in crossed electromagnetic fields, among others, J. J. Thomson and Walter Kaufmann were able to measure increasingly precise values of the large specific charge e/m of negative electrons. Another source for the value of the specific charge was Lorentz's theory of the Zeeman effect. On the other hand, positive electricity occurring in gas discharge appeared to be exclusively attached to masses of an order of magnitude of the atoms.[34]

Some scientists expressed remarkable uncertainty about positive electricity, including Julius Edgar Lilienfeld in Leipzig[35] and Willy Wien.[36] These doubts were expressed more clearly in letters from Planck to Wien, in which Planck encouraged Wien's experimental research, because he felt there was "bloody little (blutwenig)" to be done in theory.[37] Deriving Planck's radiation formula with help of the electron gas theory, Alfred Bucherer in letters to Lorentz expressed his chagrin about the lack of a "clear insight in the nature of positive electricity."[38] As a consequence of doubtful experimental findings in the field of gas discharge as well as measurements of the Faraday effect,[39] for a short time around 1905, physicists could in fact conceive of material positive carriers which had the specific charge of the electron, or they could imagine positive carriers with only "apparent mass."[40] Even J. J. Thomson, the "discoverer" of the electron, would ask in 1909 whether there was "anything analogous" to the extraction of negative electrons from different substances "in the case of positive electricity." According to Thomson, however, such a particle would probably have the specific charge of the hydrogen ion.[41]

Lorentz's Electron Gas Theory of Metals
The Hall effect with its different signs of the observed electric fields would have provided good arguments for maintaining the model of two oppositely charged elementary carriers in the electron gas theory of metallic conduction. Nevertheless, development aimed at a greater simplicity of the electron gas theory. Arthur Schuster, for one, deliberately abandoned the dual conductivity model.[42] Most influential of all, Hendrik Antoon Lorentz envisioned metallic conduction exclusively in terms of negative electrons, despite his own doubts and Drude's objections concerning galvanomagnetic effects.[43]

In his most influential paper, "The movement of electrons in metals," Lorentz introduced his new conduction concept rather timidly: "We start with the assumption that the metal only contains a single kind of free electron which possesses the same charge e and the same mass m; . . . ."[44] Upon closer examination of the paper, we can see that the problem of the charge carrier in metallic conduction is treated even more cautiously. Engendered by difficulties surrounding the sign of the Hall effect in different metals,

Lorentz's rather complicated approach again reflects widespread uncertainty about the nature of positive electricity.[45] What he believed barely understandable[46] was the creation of energy due to the recombination of material positive and negative particles in contact phenomena. Such a system, in his view, violated the second law of thermodynamics.

Not long after this article was published, Lorentz cut off the debate. His decision to assume negative electrons as the only carriers of electric conduction was considered to be the crucial step in the development of the electron gas theory of metals. Riecke admitted that this contributed a "simplification" of the theory.[47] Despite his reservations concerning the different signs of the Hall coefficients, Drude also conceded the advantages of the unitary theory. Rudolf Seeliger in his article on the electron gas theory of metals in the influential "Encyklopädie der mathematischen Wissenschaften" was willing to justify Lorentz's theory as consistent with the general tendency toward a unified physical science, which in 1921 had become a faint echo of the electromagnetic world view at the beginning of the century:

> The above mentioned exception [from a certain stagnancy[48] in the electron gas theory] is to be seen in the bold idea to assume only one single kind of carrier and to identify this carrier with the otherwise already known and studied electrons. (This idea actually does not come from Lorentz. But he was the first to have thoroughly worked it out with all its consequences). Also with respect to the desirable objective of a simplification and a unification of the whole physical world view one should consider this, in general, a great success.[49]

The choice of the charge carrier was one important point in Lorentz's electron gas theory of metals. Another achievement with long lasting influence was Lorentz's more elaborate statistical approach. For reasons of simplicity Riecke and Drude had only considered mean velocities in their calculations. Lorentz assumed that the velocities of the particles in his electron gas are distributed according to Maxwell-Boltzmann statistics. With the addition of a small term, he was able to introduce the influence of the electric field and the temperature gradient (and eventually the influence of abrupt shifts in the case of contact phenomena) into the distribution function. The solution of the one-dimensional Boltzmann transport equation

$$\frac{df}{dv_x} X + v_x \frac{df}{dx} + \frac{df}{dt} = b - a$$

for the modified distribution function in the steady state case furnished the equations for heat conduction and for electrical conductivity.[50] (X is a component of the electric field, f is the velocity distribution function, v_x is the velocity component in the direction of the x-axis, $b - a$ is the collision term, "collisions" are encounters only of electrons and metal atoms.)

The result was a new expression for the Wiedemann-Franz law. Lorentz's calculations, however, led to a numerical factor (8/9) that was different from Drude's (4/3). Therefore, interestingly enough, apparent improvement in the foundations of the theory accompanied less precise agreement with experimental results. Lorentz, himself, was not without restraint concerning the validity of his theory. During a talk in Berlin in 1904 on his revision of Drude's calculations and his inquiries into the movement of single particles in the "electron swarm," Lorentz called the deviations in his theory from the experimental Wiedemann-Franz law "not inconsiderable"—"nicht unbeträchtlich."[51] More than Riecke and Drude, he was clear about the problematic basic assumptions of the model. Therefore Lorentz was willing to discuss whether it was legitimate to treat electrons in solid matter according to the model of the kinetic theory of gases. He also asked whether a complete separation of the equipartition theorem from the gaseous state of matter was justified. Eventually he questioned whether it was possible to imagine the enormously high velocities of charged particles in the densely packed matter of a metal.[52] In 1905 in his elaborate paper entitled "Le mouvement des électrons dans les métaux," however, Lorentz still emphasized the surprising accordance of numbers in the entirely distinct regimes of the ideal gas and metallic conduction in bulk matter.[53]

A crucial question is how Lorentz convinced himself of the electron gas theory of metals. Almost certainly the answer can be found in his continuous struggle to extend the kinetic theory of gases in which the most important step, as in the case of Drude, was the theory of dilute solutions.[54] In 1909, in a lecture in Leyden, Lorentz explicitly discussed the transfer of the gas model, the equipartition theorem included, to liquids and to the solid state form of matter.[55] Lorentz's argument here was that the kinetic approach worked well in practice. It had worked in van't Hoff's theory of osmotic pressure and in Nernst's theory of concentration cells. Furthermore, it was developing into a "successful first approximation"[56] in the theory of metallic conduction. In addition, Lorentz referred to the recently published experiments of Jean Perrin. Using aqueous suspensions of microscopic small raisin particles, Perrin had been able to test Einstein's and Smoluchowsky's statistical theories of Brownian movement.

How close this interaction of the kinetic theory of dilute solutions was to electron gas theory can be spied in the details of terminology. In his influential theory of the Hall effect, which was based on Lorentz's transport theory, Richard Gans explained the limitation of the transverse flow of electricity due to the growing concentration of carriers precisely in terms of an "osmotic force" acting on the metal electrons.[57] An exchange of letters between Lorentz and Paul Hertz further illustrates the cross-fertilization of models:[58] Paul Hertz suggested exploiting Lorentz's electron gas theory of metals to explain transport phenomena in electrolytes; obviously, he was trying to move in the other direction—from the solid state to the liquid.

The argument that the kinetic model simply worked well does not encompass the whole story. Behind this belief in the heuristic value of the model there may have been a deeply rooted philosophical conviction. Henri Poincaré stated that the perfect parallelism and the numerical coincidences among the kinetic theory of gases, the kinetic theory of solutions, and the electron gas theory of metals had changed their individual status from ingenious but doubtful hypotheses to a set of three possible theories that yield those numerical coincidences by necessity.[59] Another indication is a copy of a letter of nomination for the Nobel Prize which can be found in the Lorentz correspondence. In this letter Hermann Haga had nominated Jean Perrin for the physics prize for the year 1918. In Haga's view, Jean Perrin's achievement in the experimental test of the statistical theories of Brownian movement was inseparably connected with his belief in the principle of continuity in nature, which was so obvious in the success of kinetic theories of ideal gases and dilute solutions.[60]

A New Element: The Electron Vapor
Around 1910, physicists involved in electron gas theory were able to add a new constituent to their understanding of the electron theory of metals, which had impact both in theory and in experiment. Due to this new perspective the concept of an electron and the picture of an electron gas in metals acquired new aspects. Progress in theory was represented by the new idea of an electron vapor building up above a metal surface. The development in experiment was related to thermoelectricity and to the emission of electrons by hot metals.

The basic idea of the electron vapor theory originated with Harold A. Wilson, Owen W. Richardson, and J. J. Thomson.[61] It was worked out independently by Friedrich Krüger (in Göttingen), and Karl Baedeker (in Jena). Thomson, Krüger,[62] and Baedeker[63] dealt with the transition of the electron vapor from one solid phase to another. Thus they abandoned calculating the diffusion effects of electrons. Baedeker considered thermoelectric work as equivalent to the work that is necessary to transfer an electron vapor from one metal in a thermocouple (with a characteristic vapor pressure) to the other metal (with a different vapor pressure) and vice versa to the starting point of the thermodynamic cycle.

Wilson and Richardson concentrated at first on the thermodynamics of a system composed of solid phase and a truly "free" electron vapor. Richardson pictured the emission of electrons from hot metal surfaces precisely as "an electron gas evaporating from the hot source."[64] With the help of a kinetic approach he derived the well-known exponential law for the thermionic current.[65] Wilson, on the other hand, restricted his considerations from the beginning to a purely "thermodynamical" treatment, applying the Clausius-Clapeyron equation for the vapor pressure above the surface of a liquid to the vapor pressure of electrons above the surface of the metal.[66]

The idea of an electron vapor proved to be extraordinarily fruitful. In retrospect, Richardson felt that in the years between 1900 and 1913 physicists were able to learn much more about the behavior of electrons in the electron gas than about molecules in an ordinary gas.[67] One was able, for instance, to heat cold bodies by absorption of electrons or to cool hot matter by emission of electrons.[68] But doubts about "the legitimacy" of applying the equation of a "perfect gas . . . [$pV = nRT$] to the electrons" remained. Richardson raised this question when he ventured to derive an expression for the Peltier effect. The small size of the electrons (compared with gas molecules) and the low concentration of external electrons (to which the gas law was applied) however, appeared to make the electron gas an even "closer dynamical approximation to the ideal gas" than "any real gas under comparable conditions."[69]

Richardson's measurements of the velocity distribution of electrons emitted by hot metals were most important for the analogy of the kinetic theory of gases with the electron gas theory. Although not beyond doubt, these data were evidence enough for the assertion that the velocity distribution among the electrons outside the source was the Maxwell-Boltzmann distribution. From this distribution outside the hot metal, Richardson concluded that the velocities of the electrons inside the metal also obey the Maxwell-Boltzmann distribution function.[70] Interestingly enough, he changed his mind in the case of the interior of the metal some years later, as a consequence of the degenerate gas theory of Willem Hendrik Keesom as well as the more general feeling that one had to apply quantum theory to the metallic electron gas.[71] Those thoughts clearly foreshadowed Fermi-Dirac statistics, which indeed apply to electrons at high densities.

No doubt the electron vapor theory was rather promising from the beginning. Remarks and judgments of Planck and Lorentz[72] are indications of the role this idea was to play. On the other hand, the origin of the electrons in thermionic processes was not beyond doubt. Chemical reactions of the remaining gas in the vacuum tube or in the filament were discussed as a possible source for the emitted electrons as well.[73] In the meantime, due to the development of metal filament lamps new materials had become available. Willis R. Whitney and Irving Langmuir of the General Electric Company supplied Richardson with "specimens of ductile tungsten."[74] With the help of improved vacuum and especially by using this new ductile tungsten, which withstands very high temperatures and furnishes large currents, Richardson was able to rule out this objection quantitatively in 1913.[75]

Another problem, which was sometimes rather polemically discussed among Richardson, Bohr, and Walter Schottky, was the precise definition of the thermionic work function. In turn, the knowledge of the work function had some relevance in the application of electron gas theory to the technology of vacuum tubes. One question was whether the energy to liberate the already free electron gas from the lattice and the metal surface—or rather the energy to liberate electrons still bound to metal atoms—should be considered as the true thermionic work function.[76] Bohr and Richardson also discussed the influence of a surface charge that may depend on the temperature. Obviously, both the complex process of liberating an electron from an atom and the occurrence of a temperature dependent surface layer, contradicted the assumption of a constant specific heat of the electrons in the metal. Therefore, these questions did touch upon the validity of the basic representation of electrons evaporating from a hot liquid source.[77] (For more on Richardson, see Knudsen, this book.)

Bohr's Attempt To Save the Phenomena
Although Niels Bohr's dissertation of 1911 was rather like a final step in the development of the classical electron gas theory, it does not fit exactly into the picture discussed above. Bohr's "Studies on the electron theory of metals" did not truly reflect the need to justify the transfer of the gas model to the solid state. Although Bohr's theory was still based on Lorentz's statistical approach and on the transport equation,[78] the central focus of Bohr's dissertation was a much more general treatment of the interaction of electrons and metal atoms. Therefore, Bohr's theory showed a peculiar structural similarity in the development of kinetic theory of gases to the electron gas theory of metals.

James Clerk Maxwell had tried to reconcile the kinetic theory of gases with the experimental findings on the temperature dependence of the viscosity of gases. Maxwell abandoned the analogy of hard spheres and assumed "softer," repulsive forces between gas molecules of the form $1/r^5$ (where r is the distance).[79] Ludwig Boltzmann proceeded in a similar manner. He postulated attractive forces proportional to $1/r^5$ and, in addition, assumed short range repulsive forces to furnish a plausible physical model of collisions.[80] To improve the agreement between Lorentz's theoretically derived expressions of the Wiedemann-Franz law and the measurements of Diesselhorst and Jaeger, Bohr formulated a special law for the repulsive force between electrons and metal atoms in collisions—similarly in his case a force proportional to $1/r^3$.[81] This mere mathematical analogy is even more impressive if one looks at the justifications for the different force laws. Boltzmann was already quite clear about the conventionalist character of his approach.[82] Although Bohr's dissertation does not contain a discussion of philosophical repercussions, his assumption that the "ratio between the thermal and electric conductivity depends, to a certain extent, upon the special assumptions introduced concerning the forces acting between the metal atoms and the electrons" is reminiscent of Boltzmann's remarks on to the kinetic theory.[83]

Bohr must have been nearly forced to adopt a conventionalist view when he received a letter from Peter Debye in Zurich. Debye pointed out that one had probably to assume two different kinds of force laws. But even then, according to Debye, it would be impossible to fit the theoretically derived expression for the Wiedemann-Franz law with experimental results and, at the same time, to fit the two separate expressions for conductivity of heat and for conductivity of electricity with experiment.[84] In a more desperate manner, Bohr continued struggling with different force laws to explain, for example, black body radiation with the help of electron gas theory. Moreover, James H. Jeans had assumed special force laws "to reconcile experiment with electron-theory." For Jeans, this meant reconciling the phenomena of electric conductivity of metals and the radiation curve—which falls off for high frequencies—with the equipartition theorem; as well as eventually reconciling it with the assumption of a nonequilibrium in the interaction of matter and radiation.[85]

The Tolman-Stewart Experiment

The basic material assumption of the electron gas model is the free electron; that is, a charged particle with a certain mass moving freely within the metal, whatever the origin of the mass may be. The more general question of whether moving electricity possesses mass at all, and whether this mass is observable or not, is rather old. One can trace this problem back to Maxwell's "Treatise . . . ," to H. Hertz's early research, and to Boltzmann's "Vorlesungen über Maxwell's Theorie . . ."[86] But Lorentz's calculations for the reaction of electrons suffering centrifugal accelerations predicted only a very small effect, possibly beyond the reach of experiment.[87] This prediction appeared to be confirmed by earlier measurements of Ernest Fox Nichols in 1906.[88] Although Nichols was encouraged by H. A. Lorentz and by Walther Nernst (both had lectured at Columbia University, New York), he left his "whirling disk experiments" without definitive results.[89]

The first successful experiments were performed in 1916 by Richard Chase Tolman and Thomas Dale Stewart (who later concentrated on chemistry) in Berkeley. In a quite imaginative way, Tolman and Stewart wanted to show the inertia of metal electrons with help from massive acceleration processes applied to conductors. Therefore, they tried to measure potential differences by abruptly slowing down pieces of conducting material, for example, rotating metal coils. Tolman and Stewart did not perform their experiments solely using solid conductors, for they began by measuring potential differences in accelerated electrolytes.[90] This experiment was designed as early as 1888 by Walther Nernst.[91]

Although these experiments on the inertia of metal electrons are quite striking, they played only a marginal historical role. Rudolf Seeliger in his article on electron gas theory of metals in the "Encyklopädie der mathematischen Wissenschaften" gave for the time (1921) a precise account of the possibilities inherent in the Tolman-Stewart experiment:

> [Lorentz's] assumption [of a single carrier of electric conduction in metals] was in fact very bold. It has been confirmed by results proceeding from a consequently developed theory and by results indirectly provable by observation. It is, however, remarkable that this assumption has also experienced a direct confirmation by experiment. Obviously R. C. Tolman and T. D. Stewart have succeeded in determining the mass of the charge carriers and in finding a value which agrees well with values originating from different experiments, namely 1/1.900 to 1/1.940 of the hydrogen atom. Strange enough [the physicists] did not pay much attention to these experiments.[92]

Not surprisingly, Hendrik Antoon Lorentz, with his sensitivity to the needs of a justification of the electron gas theory, was among those[93] who did pay attention to the Tolman-Stewart experiment. Moreover, the great difficulties bearing upon the electron gas theory of metals, he considered the Tolman-Stewart experiment to be a kind of consolation.[94] But in 1935–36 the eminent theorist, Charles G. Darwin, with help of the modern band model, demonstrated the limits of the Tolman-Stewart experiment. According to Darwin's authoritative paper,[95] these acceleration experiments measure only the "normal mass of free electrons."[96] In particular, they do not furnish any knowledge about the Fermi surfaces of a given metal.

Severe Problems in Electron Gas Theory

The Equipartition Theorem

In this discussion of the electron gas theory of metals, we have already mentioned difficulties posed by different kinds of statistics, the equipartition theorem, and the Hall effect. The following section looks more closely at these problems and deal with the logical inconsistencies of the theory, the enormous experimental difficulties, and the thoughts of the physicists who struggled with these problems.

From the beginning there was much skepticism about even the basic theoretical idea. Marcel Brillouin for instance was outspoken on the shortcomings ("des lacunes") of Riecke's and Drude's theories.[97] The notion of random collisions in electron gas theory apparently contradicted a basic equation of movement of charged particles in metal optics. Brillouin probably had in mind Drude's equation: $eE = m\partial^2 x/\partial^2 t - 1/v \partial x/\partial t$, because this equation contains a resistance term ($1/v\partial x/\partial t$) which at each moment could be expressed as a term proportional to a velocity (e is the elementary charge, E is the electric field, m is the mass, x is the elongation, and v is the mobility of the particle). Perhaps with these objections in mind, Brillouin argued in a letter to Lorentz, that he felt it necessary to restrict the "crude application of thermodynamical formulae."[98]

The major problem of the application of thermodynamic formulae was, no doubt, the equipartition theorem. In retrospect, this appears obvious:[99] if one assigns a mean kinetic energy of $3/2\ RT$ per mole to the electrons, then the electron gas should contribute the corresponding amount of $3/2\ R$ to the specific heat of a metal. A look at the specific heats of metals and of insulators will, however, immediately show that the metals by no means have a specific heat that consists of the Dulong-Petit value for solid elements (namely $3R = 6$ cal/degree and mole) plus the contribution of the electron gas (namely $3/2\ R$).

But this deviation was certainly not a problem of the complex nature of solid matter. For example, in contrast to the Hall effect, it has nothing to do with the structure of the energy band of the metals. It was more of a homemade problem that emerged almost necessarily from the application of the gas model to electrons in solid matter.

Max Reinganum, who worked at that time with Heike Kamerlingh Onnes in Leyden, responded early to this contradiction. In fact he published his remarks before Drude had published the second part of his fundamental paper in the "Annalen der Physik" in 1900.[100] Although Reinganum's paper is essentially in favor of Drude's electron gas theory of metals, Reinganum's objections indicated why physicists were not necessarily ready to answer this question and why the answer was not simple:

> Due to the experiments of Mr. Kaufmann on cathode rays and due to Zeeman's splendid confirmation of Lorentz's theory, which assigns a certain number of degrees of freedom to the bound electrons in luminous gases, it seems not too bold to assume completely free electrons in metals. Nevertheless, the extended theory of Giese [which pictured a direct transfer of charge from one atom to another] deserves to be taken into consideration as well . . . For, if one assumes free electrons and provided the number of free electrons are comparable with the number of metal atoms, the theoretical explanation of the Dulong-Petit law for the specific heats has to be replaced by a completely different theory of this empirical law. That means replacing a theory which according to the basic ideas of Boltzmann was developed by Richarz . . . and which has, up to now, not led to contradictions.[101]

Surrounding the question of the specific heats, a rather precarious situation developed,—both theoretical and experimental. The young Einstein was seriously concerned with the problem emerging from Drude's electron gas theory of metals. He felt that it indicated a deep-rooted inconsistency in classical statistical mechanics.[102] The prevailing opinion, which was most decisively expressed by Walther Nernst, was that either the application of the equipartition theorem in the electron gas theory, or the assumed number of electrons, does not agree with the experimentally observable contribution of the electrons to the specific heat of the metals.[103]

A typical way out of the dilemma was the construction by, among others, J. J. Thomson and Niels Bohr, of complicated conduction mechanisms to reduce the number of truly free electrons.[104] Lorentz utilized it more in the sense of a "reductio ad absurdam" because reducing the number of free electrons meant increasing the mean free path; otherwise, one would not have been able to save the explanation of electric conduction by free electrons (since conduction depends on the number of carriers and on the mean free path).

The enormous mean free path was already a severe problem for the physical basis of the electron gas theory of metals. Calculations based on the densities of metals and on x-ray diffraction had furnished figures for the mean distance between the centers of the atoms in the order of magnitude of small multiples of the radius of Bohr's atom.[105] Calculations based on the electron gas theory, however, had furnished mean free paths of metal electrons in the range of 10 to 100 atomic radii.[106] The dilemma was even worse if one included superconductivity, which was discovered by Heike Kamerlingh

Onnes in 1911.[107] The discovery of the Ramsauer effect at the beginning of the 1920s provided no relief. At first the confirmation of the kinetic effective cross-section in the case of slow electrons moving in "normal" gases was much more impressive.[108] The discovery of a drastic drop in the cross-section in the case of argon was still no indication for a solution. The minimum of the cross-section occurs at much higher energies than the mean kinetic energy of electrons in metals.[109] Troubling as well was the notorious gap between a gas and solid matter.

Blackbody Radiation

The application of the equipartition theorem in electron gas theory led to problems beyond that of specific heats. The equipartition theorem also created a rather puzzling situation in the theory of blackbody radiation. The situation must have been even more puzzling to physicists if one considers Willy Wien's[110] and Paul Drude's[111] expectations that the kinetic theory of gases and the theory of blackbody radiation should confirm or at least support one another.

Based on Drude's electron gas theory Lorentz was able to calculate emission and absorption of electromagnetic radiation due to free electrons in metals as early as 1903. He obtained an expression which—for long wavelengths—agreed with Planck's radiation law, which seemed to be experimentally confirmed.[112] Despite the problems, Riecke and Baedeker felt that this was an important success of the electron gas theory.[113] Further investigations by, among others, Harold Albert Wilson, James Jeans, and Niels Bohr stabilized the result.[114] But with Bohr's dissertation, the perspective changed. In 1911 Samuel Bruce McLaren finally demonstrated that calculating the emission and absorption based on the statistics of electron gas theory and aiming at the whole range of frequencies necessarily yields the result now known as the Rayleigh-Jeans law.[115] Contrary to experiment, this law indicated that—in the words of Planck—there may be no equilibrium in the distribution of energy between radiation and matter.[116]

Basing a theory of blackbody radiation on the electron gas theory of metals made much sense. Since blackbody radiation was experimentally produced by hot solid matter—mostly emitted by heated metallic cavities[117]—it simply meant taking advantage of a rather successful theory of the solid state. The carriers of metallic conduction were a reasonable choice in the search for a conceivable mediator between matter and heat radiation. Although Bohr felt that the radiation law may have revealed the limitations of classical electrodynamics (in the case of solid matter and of high frequencies) there was ample hope of finding an appropriate interaction mechanism of electrons and metal atoms to explain the fall off in the experimentally established radiation curve for high frequencies.[118]

The complex interaction of the theories of blackbody radiation and of the electron gas theories also reflects and explains the painful process of realizing the break that had occurred in Planck's and to a greater extent in Paul Ehrenfest's and Albert Einstein's derivations of the radiation law. A most telling example of this difficult transition appears in a letter from Planck to Wien in which Planck complains about Jeans' stubborn behavior, comparing Jeans to "Hegel in philosophy."[119] On the other hand, Planck himself restricted discontinuity to the process of absorption of radiation (by his ideal "resonators") rather than allowing for discontinuities in the ether.[120] In his "second theory," he ascribed discontinuity to the mere process of emission of radiation by the so-called oscillators. Planck's idea that colliding electrons in metals also obey his "new quantum hypothesis" rather than being simply "reflected" belongs to this "second theory." Due to the hypothesis that the emission energy of metal electrons is discrete, these electrons might possibly exchange only a part of their kinetic energy and thus add only a small fraction to the specific heat of solid matter. About Drude's free electron gas theory (and the assumption of the equipartition theorem), however, which had "partially" led to results in "obviously good agreement with experience," Planck was cautious with his own contribution to electron gas theory.[121]

Willy Wien was one of the physicists who felt quite uncomfortable about this apparent link between the equipartition theorem in electron gas theory and the Rayleigh-Jeans law in the theory of blackbody radiation. In 1913 he was willing to question the crucial role played by the equipartition theorem. Therefore, he left the theoretical explanation of heat conductivity aside and abandoned the explanation of the empirical Wiedemann-Franz law.[122] Due to the application of the quantum hypothesis of Planck's second theory of blackbody radiation to the collisions of atoms and electrons, which led to the derivation of the temperature dependence of the resistance (namely proportional to the absolute temperature), his attempt to solve the dilemma was influential, but it also met with criticism. In 1913, the "Wolfskehlstiftung" of the "Königliche Gesellschaft der Wissenschaften" at Göttingen invited, among others, M. Planck, W. Nernst, A. Sommerfeld, and H. A. Lorentz, to give talks about the "Kinetic Theory of Matter and Electricity." Lorentz, in his Wolfskehl lecture, complained about Wien's abandonment of the free electron gas theory. Lorentz found it unpleasant that Wien "even went so far as to give up completely the picture of a real heat movement, for he assigned only a velocity to the electrons which is independent of the temperature."[123]

Lorentz based his objections to Wien's assumption of an electron velocity that is independent of the temperature on solid arguments. He referred once more to the successful kinetic approach in the theory of electrolytes. In addition, he used one of the most striking experimental arguments within the reach of the electron gas theory of metals: the result of Richardson's measurements of the Maxwell-Boltzmann distribution for the velocities of electrons emitted by hot metals.[124] As mentioned above, Richardson had concluded that the electrons within the metal must possess the Maxwell-Boltzmann distribution of velocities, meaning that the electron velocities depend on temperature, which in turn implies that the equipartition theorem does hold. Neither was Max Planck pleased with Wien's theory, probably because of his own attempt to reconcile the quantum theory of radiation (or at least the emission of radiation) with the Maxwell-Boltzmann distribution among free electrons and with thermodynamic equilibrium ("Massenwirkungsgesetz") between oscillators and electrons.[125]

The assumption of a "Maxwell-Boltzmann distribution among the emitted electrons" had already been used by James Jeans and in S. B. McLaren's paper entitled "The emission and absorption of energy by electrons" which had destroyed the hope of a reconciliation of the theory of the classical free electron gas and of Planck's radiation law.[126] Richardson's measurement of thermionic currents was also mentioned by Paul and Tatiana Ehrenfest in their famous encyclopedia article entitled "Begriffliche Grundlagen der statistischen Mechanik." The Ehrenfests felt that the glow discharge was a "direct and almost complete confirmation" of the applicability of the kinetic theory to electron gas theory without including the ether or the radiation, despite the certainly close connection between the "heat movement of electrons" and the ether in this field.[127] Charles G. Darwin's considerations in a manuscript entitled "The theory of radiation," in which he went so far as to discuss the contribution of the inner degrees of freedom of the electrons, are similar to the arguments made by the Ehrenfests.[128]

Obviously the paradox of Planck's quantum theory of the radiation law with its recourse to classical statistics[129] as well as the impact of Richardson's conclusions derived from his thermionic measurement sometimes obscured the need of quantum theory in electron gas theory. Therefore, Lorentz's suggestion in his Wolfskehl lecture appears to be rather vague. He referred to the importance of collisions in relation to the radiation puzzle, in terms of a future possibility, conceding that the collision mechanism problem might become a subject of quantum theory.[130] This was, perhaps, a faint reflection of the letters he exchanged with Planck. In fact, Planck had stressed that in the case of the more bound states of electrons that suffer collisions, quantum theory might account for the oscillations that take place.[131]

Bohr's successful quantum theoretical treatment of Rutherford's atom might have been even more of a temptation to concentrate on the mechanism of the collisions. The echo of Bohr's atomic theory, however, in electron theory is not substantial though it can be found, for example, in papers of Fritz Haber[132] and Owen W. Richardson.[133] Richardson and Haber treated superconductivity in terms of electrons occupying Bohr orbits with common tangents. Thus, they pictured a movement of electrons without resistance. Later Paul Ehrenfest in a letter to Lorentz discussed these penetrating orbits in Bohr's elaborate theory as a possible explanation of the behavior of electrons in metals.[134] On the other hand, from at least the time of the Wolfskehl conference in Göttingen, the possibilities of quantum statistics in the electron gas theory—the "degenerate gas"—were up for discussion.[135] An early and important response to the degenerate gas theory can be seen in letters and papers of O. W. Richardson, who almost immediately abandoned his earlier conclusion that inside the metal the Maxwell-Boltzmann distribution holds.[136]

The Hall Effect

Although Edwin Herbert Hall was not willing to accept precisely Willy Wien's approach, he was also inclined to abandon the theory of a free electron gas. This was discussed in a letter to Owen W. Richardson in 1913 when Richardson still worked in Princeton. In spite of Richardson's results in the field of thermionics, Hall focused on the problems of equipartition and specific heat. As a consequence he questioned the basic notion of a free electron gas:

> I got much satisfaction from your treatment of the equilibrium between the external and the internal free electrons, though I thought your machinery for handling the electrons in this thermo-electric cycle was sometimes more elaborate than it need be. Now, however, I am a good deal impressed by the arguments, especially the specific heat argument, against the assumption that the electrons within a metal are really free, to the extent of having the same heat energy as true gas molecules would have. On the other hand, the theory of metallic conduction put forth by Stark qualitatively and by Wien quantitatively does not seem to be a probable one. Do you feel that the "gaskinetic" theory of metallic conduction, the theory of free electrons within the metal, is really done for? I am just now feeling of the proposition [?] that the course of the electrons may be through the metal atoms rather than around them.[137]

Obviously through his critical evaluation of the electron gas theory, Hall was urged to resume his thinking about an effect he had discovered in 1879 while he was working with Henry Rowland at Johns Hopkins University in Baltimore.[138] Hall had found a transverse electric field in a solid material when it carries an electric current and is placed in a magnetic field that is perpendicular to the current. Once the effect was separated from Maxwell's suggestion of a rotation of the primary electric field, the explanation did not appear to be an insurmountable obstacle.[139] Lorentz[140] and Boltzmann[141] in their elementary theories tried to explain the changing sign of the effect (which was already measured by Hall in the case of gold and iron[142]) by opposite forces that act on oppositely charged moving electricity in the transverse magnetic field, that is, with help of the dual mechanisms of conductivity.

The first electron gas theories of metals were still flexible. Riecke and Drude assumed carriers with different signs, with different "drift velocities" (Riecke)[143] or "mobilities" (Drude).[144] Drude even considered different temperature dependencies of the concentration of the different carriers.[145] Thus, the Hall effect appeared to fit rather nicely into the framework of electron gas theory. The remaining deviations appeared to be mainly experimental as well as numerical problems.

But when Lorentz restricted electron gas theory to a single carrier, instantly the Hall effect became more than a numerical problem. The electron gas theory of metals appeared to be unable to account for the different signs, which was a crucial experimental result. Lorentz was quite aware of the situation. On the other hand, he always believed that the assumption of different types of carriers would create more problems, for example in explaining contact phenomena.[146] From a more general point of view these problems have to do mainly with the difficulties of physicists around 1905 (and after) in picturing positive carriers as anything other than charged particles with a certain mass.

Although Lorentz's research is full of self-critical overtones, its impact was primarily a successful simplification of the theory in the sense of a restriction to an electrical conductivity exclusively due to negative electrons. With the help of an historical episode, in which the Hall effect figures prominently, one may illustrate the claims that Lorentz's electron gas theory had staked out. It is an irony of history that this episode also provides an example of the takeoff in semiconductor research. It is to be found in the work of Karl Baedeker in Leipzig and Jena.

In 1907 Baedeker wrote a Habilitationsschrift entitled "Über die elektrische Leitfähigkeit und die thermoelektrische Kraft einiger Schwermetallverbindungen"—"On the electrical conductivity and the thermoelectric power of some compounds of heavy metals." Specially interesting is

Baedeker's investigation of cuprous iodide (CuI). In the case of fresh preparations, Baedeker obtained a small value for the specific resistance[147] which was rather "strange for a transparent substance."[148] Further investigations with samples that were still in equilibrium with iodine vapor showed a marked drop in resistance. Varying the concentration of iodine in cuprous iodide, Baedeker eventually was able to maintain a variable and comparatively high electrical conductivity in cuprous iodide. Thus Baedeker had, for the first time, deliberately doped a semiconductor.[149] His earlier findings were confirmed by experiments that indicated the temperature range of a purely metallic conductivity, that is, a conduction without a considerable contribution of electrolytic conduction.[150] Baedeker proceeded to combine his variable semiconductor with platinum to perform thermoelectric measurements. He explained the variable thermoelectric power in terms of the above mentioned electron vapor theory.[151]

The most interesting experimental step, by far, was the measurement of the Hall effect in the doped cuprous iodide: "The sign of the effect was positive according to the accepted nomenclature, that is, it was opposite to the effect in bismuth."[152] Additionally, what appears to be quite modern is the measurement of the Hall effect as a function of the iodide concentration. But at the same time, this experiment—conducted by Baedeker's graduate student, Karl Steinberg—again veiled the problem of the type of charge carrier involved in the metallic conduction of cuprous iodide. Steinberg's data showed that with increasing iodine concentration (and decreasing resistance) the absolute value of the Hall coefficient decreases. This tendency alone—that is, if one left aside the sign of the Hall coefficient—fit well into the picture of the electron gas theory.[153] The alleged increase in the electron concentration (due to doping with iodine) would have led to a decrease in the absolute value of the Hall coefficient. But Baedeker and Steinberg were also somewhat lucky. They only measured below the degeneracy limit; that is, where the mobility of the carriers still decreases with increasing concentration of the carriers.[154] In a lecture at the Deutsche Physikertagung in 1912, entitled "Artificial metallic conductors," Baedeker still explained the variation of the absolute value of the Hall coefficient with the model of a varying electron concentration.[155] It is obvious that the electron vapor theory of thermoelectric phenomena, the emission of electrons from hot metals and—in general—the simplicity of a single carrier of electric conduction were persuasive arguments in favor of Lorentz's unitary approach.[156]

It was not possible, however, to turn away from the problem of the changing signs of the Hall coefficients. Steinberg was clear on this subject: ". . . with help of negative particles alone in the elementary theory it is only possible to derive an effect in one direction—namely opposite to [the effect of] cuprous iodide...."[157] At the time, physicists already knew of a rather impressive number of conductors with positive Hall coefficients. About this number, cuprous iodide included, Baedeker was also forced to state that the electron gas theory of metals is "not sufficient"["unzureichend"].[158]

The rather complicated physics of the doped cuprous iodide was clarified late and, even then, only to some extent. In 1951 Robert J. Maurer and Benjamin H. Vine investigated again the electrical properties of the semiconductor. Their paper, which was dedicated to Walter Schottky's 65th birthday, was based completely on modern solid state physics with its concept of different carriers. Although it remained a somewhat qualitative approach, it was now beyond any doubt that the doped cuprous iodide was a p-conductor. Different from Baedeker and Steinberg, Maurer and Vine also crossed the degeneracy limit and obtained an increasing mobility of the charge carriers (of the holes) with increasing concentration of iodine.[159]

The analysis of the research of Baedeker and of related investigations shows, once more, the possibilities of the electron gas theory of metals, but it also demonstrates the limitations of the theory and, above all, the consciousness of those limitations. One can certainly sense this realization in the papers and talks of Lorentz. It appears to be a major feature of Bohr's dissertation. He even added the problem of diamagnetism; that is, the classical free electron gas did not allow for diamagnetic behavior of metals.[160]

But it was also Eduard Riecke, one of the founders of electron gas theory, who clearly marked the limitations. At the "Hauptversammlung der Deutschen Bunsengesellschaft" in Aachen, in 1909, he discussed the problems of a unitary electron gas theory about an unsolved Hall effect problem.[161] In 1913 Riecke wrote an article for the "Handbuch der Radiologie" entitled "Electron theory of the galvanic properties of metals" ["Elektronentheorie galvanischer Eigenschaften der Metalle"]. In this article Riecke argued against simply ignoring the Hall effect and its related phenomena to proceed with Lorentz's unitary theory. He even warned "that as long as it excludes these phenomena from its sphere, the electron [gas] theory will remain a fragment."[162]

But where physicists really tried to include galvanomagnetic effects, they almost necessarily disturbed the picture of a free electron gas. They concentrated more on the interaction of electrons and metal atoms. This is true for the so-called directive field theories—"Richtfeldtheorien." Johann Koenigsberger[163] and Niels Bohr,[164] among others, assumed an interplay of macroscopic and oppositely directed microscopic fields that might eventually be responsible for the changing signs of the Hall coefficients. Hall himself went so far as to split the electron theory of metals. He wanted to describe thermionics and thermoelectricity in terms of the free electron gas theory. About metallic conductivity and the Hall effect, he assumed a much more direct transfer of electrons from one atom to the other. He assigned a certain mobility to the resultant positive metal ions. The positive metal ions would, therefore, also be deflected in magnetic fields. Thus an explanation of positive Hall coefficients appeared to be possible. The Hall effect nevertheless remained an unsolved problem.[165]

The Solution to the Problems
Around 1913–14, physicists had also opened a way, starting with the solution of the thermodynamical puzzle, out of the difficulties of the electron gas theory of metals. From 1900 the young Einstein was deeply concerned with the problem emerging from Drude's electron gas theory of metals. He felt that the specific heat dilemma indicated a deep-rooted inconsistency in classical statistical mechanics.[166] So it is not at all surprising that the decisive stimulation came from Einstein's quantum theory of the specific heat of solid matter, which was published in 1907, followed by Peter Debye's work in 1912.[167] The kind of quantum statistics implicit in Einstein's and Debye's quantum theories of lattice vibrations soon became a standard method in electron gas theory. To explain electrical resistance, even for low temperatures, the starting point was usually the vibration of the metal atoms or the vibration of the lattice. The decisive idea was that, due to a variety of interaction mechanisms, vibrating metal atoms in the lattice could limit the mean free path of the electrons. Thus the vibrating metal atoms forced the colliding electrons to behave according to the statistics of the lattice. The theories of Heike Kamerlingh Onnes, along with those of Nernst's pupil Frederick Alexander Lindemann, Willy Wien, and Percy William Bridgman belong within this context.[168]

Although Walther Nernst had the feeling that "at the time [1911] the quantum hypothesis is essentially a calculation method" ["Rechnungsregel"] . . . of a very strange or even grotesque character," he clearly saw the need for a modification of the theory of gases with the help of the quantum hypothesis.[169] The direct transfer of Debye's statistical model to the theory of gases to explain deviations from the behavior of an ideal gas, was even more important for the further revision of electron gas theory. In 1912 Debye had calculated the energy distribution among the vibration modes (below a maximum frequency) of a solid body, according to Planck's theory of the radiation law. In 1913[170] and 1914[171] Hermann Tetrode and Willem Hendrik Keesom adopted this idea to calculate the energy distribution among the acoustic vibrations in a volume of a gas or liquid. Thus they were able to explain the vanishing of the specific heats for $T = 0$, which had been stated earlier by Walther Nernst.[172] At the same time, the theory represented the behavior of an ideal gas, which is "degenerate" ["entartet"] at "lowest" ["tiefste"] temperature. This "degeneracy behavior" appeared to occur in the electron gas even for considerably higher temperatures due to the smaller masses of the electrons.[173] Despite the somewhat tentative character of this "most radical" application of quantum theory, which added to the inconsistency of Planck's quantum theory of blackbody radiation with its recourse to classical statistics,[174] it was the first glimmer of hope in solving the notorious problem of the small contributions of the electron gas to the specific heats of the metals at normal temperatures.[175]

What followed, however, was a long period of no real progress. H. A. Lorentz almost ignored the concept of a degenerate gas. In general, there was no development of the principles of the electron gas theory which may depend, to some extent, on the decline of the electromagnetic world view starting in 1910. It was, rather, a time of critical overtones and of details: "Thermionics,"[176] behavior of conductors under high pressure,[177] conductivity of alloys,[178] and discussion of mechanisms to explain the Hall effect as well as the rise of Bohr's atom. Furthermore, it was war time, which was, perhaps, more a drain of energy than a stimulation. On the one hand, there was a considerable development of valves and amplifiers for wireless telegraphy which depended, at least partially, on the construction of vacuum tubes and eventually on electron gas theory.[179] On the other hand, there were bitter battles between leading physicists on German and allied warfare which fill, for example, the correspondence of Lorentz.[180] It was also a time of summarizing, of overviews, and of articles in encyclopedias.[181] No doubt, it was a period of blind alleys as well. Fritz Haber's, Johannes Stark's, and Frederick Alexander Lindemann's idea of an electron space-lattice, which under the influence of an electric field moves through the atomic lattice, proved to be nothing like an observable entity.[182]

It was not until the mid-1920s that leading physicists, Erwin Schrödinger and Albert Einstein, dealt once again with the degenerate gas and with the possible consequences in the electron gas theory of metals. But the breakthrough occurred with the formulation of the new quantum statistics in early 1926, by Enrico Fermi (and shortly afterwards by Dirac), which obeyed Pauli's exclusion principle.[183] With Fermi statistics, there was an enormous acceleration of development. Ralph Howard Fowler applied the new quantum statistics to stellar matter.[184] With help of Fermi statistics, Wolfgang Pauli was able to explain the (surprisingly small) paramagnetism of alkali metals which was, to some extent, independent of the temperature.[185] Stimulated by galley proofs of Pauli's paper on the paramagnetism of the alkalis, Arnold Sommerfeld resumed research on the electron gas theory of metals, using virtually the whole institute in Munich as a kind of thinktank.[186] Based on Fermi statistics, Sommerfeld's revision of Lorentz's transport theory led to an improved version of the free electron gas theory. The results were mostly in better agreement with experimental data. This is true for the contribution of the electron gas to the specific heats of the conductors as well as for the expression for the Wiedemann-Franz law and for thermionic phenomena.[187]

For a short period of time, Sommerfeld's even appeared to be the correct electron theory of metals,[188] but severe problems remained. What remained difficult to understand was the mean free path of an electron in such a crowded metal lattice. Furthermore, contrary to the expression of the Wiedemann-Franz law—where the mean free path simply does not appear—the temperature dependence of the electric conductivity remained a notorious stumbling block.[189] The signs of the Hall coefficients were, almost necessarily, beyond the reach of the theory of a free electron gas.[190]

The problem of the mean free path, or of metallic conduction in general, was solved by Felix Bloch who worked with Werner Heisenberg in Leipzig. Sommerfeld's guest, William Houston, had first derived an expression for the mean free path from the interference of electron waves in a lattice of atoms with extended electron "atmospheres."[191] Bloch demonstrated, however, that the wave function of the conduction electron is only periodically modulated by the lattice. The ideal lattice, therefore, did not cause any diffraction. Only perturbations in the lattice, that is, chemical and structural anomalies, and the heat vibration of the lattice, lead to diffraction and eventually to electrical resistance.[192]

Progress in the field of galvanomagnetic effects, on the other hand, developed somewhat more slowly. The starting point was Pauli's paper on the exclusion principle in 1925 which indicated that not only existent but also missing electrons in a shell may have a physical meaning. About the number of terms (energy levels), atoms with n electrons resemble those atoms that lack n electrons to complete a shell. Referring to the quantum numbers of the missing electrons, Pauli had a notion of "hole values," of "Löckenwerte."[193] In 1929 Rudolf Peierls, who was also working with Heisenberg, adopted Pauli's idea. His paper indicated that, according to the occupation of the shells, one may expect negative or positive Hall coefficients.[194]

In 1931 Heisenberg himself invented the now well-known concept of an electrical conduction that is carried by "holes" (by "Löcher"), contrary to—or in addition to—a conduction carried by the material negative electrons. He demonstrated that one can approximately replace the solution of a Schrödinger equation for n electrons by the solution of a Schrödinger equation for N − n holes, where N is the number of electrons of a fully occupied shell. A citation from Heisenberg's paper reads:

> If one compares . . . [the wave equation for holes] with the wave equation . . . [for electrons] one recognizes that under the influence of perturbative external fields the holes behave exactly like electrons with positive charge. On the other hand, the holes also contribute to the current and to the charge density just as electrons with positive charge. Electrical conduction in metals, which is carried by a small number of holes, can therefore be described in every respect, like conduction in metals carried by a small number of positive conduction electrons. From this immediately follows the anomalous Hall effect for those metals.[195]

In 1931 Heisenberg's result, which is valid for the energy levels of a single atom, was translated into the language of band theory, which refers to the energy levels of a substance in a whole lattice.[196] From the point of view of modern solid state physics the Hall effect is certainly much more important than specific heats, heat radiation, or conductivity. Only in connection with Hall effect measurements, are we able to obtain data on the densities of carriers, on the types of carriers, on mobilities, and, eventually, on band structures.

CONCLUSION

This chapter has attempted to analyze the first mathematically formulated microphysical theory of the electrical properties of metals. What at first appeared to be a bold step—namely, to apply the model of the physics of gases to the solid state—is intelligible from the point of view of the historical development, the prehistory, and the subsequent results. If one is not inclined to consider the influence of the fluid model in classical electrodynamics, the early justification of the electron gas theory of metals depended widely on the paradigm of van't Hoff's theory of the osmotic pressure and on Nernst's theory of concentration cells.

Like the kinetic theory of gases in the field of viscosity and diffusion, the electron gas theory celebrated an early success in the theoretical explanation of the Wiedemann-Franz law, which was experimentally established long before and which had almost urged the physicists to think of a close relationship of conductivity to heat and electricity. Moreover, the explanation of thermoelectricity in terms of a thermodynamic cycle performed by an electron vapor of varying density and, above all, the almost visible appearance of an electron vapor in the emission of electrons by hot metals, contributed to the feeling of the reality of the electron gas model. The somewhat underestimated electron inertia experiments of Tolman and Stewart also belong in this context.

Due to the inherent weakness of the electron gas model and the complexity of solid matter, a growing number of problems appeared which met with growing consciousness of these problems. The transfer of the equipartition theorem, the small contribution of the electrons to the specific heat of the metals, the radiation puzzle, and the different signs of the Hall effect were increasingly recognized as severe obstacles for the development of the free electron gas theory.

Although Lorentz was an authoritative person in early electron gas theory, he did not, like Sommerfeld and his school in later years,[197] truly dominate the field. Lorentz's open-hearted exchange of letters with almost every physicist dealing with electron gas theory is evidence enough. The result was a long-lasting dialogue with a three dimensional network of theoretical and experimental arguments gradually shifting toward criticism and even disillusion. Perhaps the most fascinating feature of this dialogue is that, despite the inherent problems and notwithstanding the tendency toward applicability in technology, which was obvious since World War I, the electron gas theory was foremost a crossroad for the most important ways of theoretical thinking at the beginning of the twentieth century.

To some extent, the electron gas theory carried further the feeling of the dignity of a mechanical explanation, which was one of the driving wheels of the kinetic theory of gases. On the other hand, there is a close interaction of electron theory and the electromagnetic world view. Thus, the electron gas theory of metals is also a part of the early history of the special theory of relativity. In the field of statistics, it was important for a long time and was probably the most impressive application of a mathematical structure emerging from the thermodynamics of gases. In turn, this made the electron gas theory problematic for a kind of radiation theory that wanted to include the state of the art in solid state physics. The electron gas theory of metals was, at first, more of a hindrance for the reception of the quantum concept. On the transfer of the degenerate gas theory to the electron gas, however, it paved the way for the new quantum theory to some extent. Seen from the perspective of the concept of elementary particles, the changing history of the electron gas eventually indicates how the electron was burdened—or even overburdened—with an evergrowing number of theoretical and experimental aspects.

Acknowledgments

Valuable information and criticism were given by Jed Z. Buchwald, John L. Heilbron, David Cahan, Michael Eckert, Kai Handel, Helge Kragh, Jürgen Renn, Bruce R. Wheaton, and Norton Wise.

Notes

Abbreviations used in the notes

AHES	Archive for History of Exact Sciences
AHQP	Archive for History of Quantum Physics, University of California, Berkeley, and elsewhere
AP	Annalen der Physik
BCW	Niels Bohr, Collected Works, vol. 1 (Amsterdam and New York, 1972)
BSC	Bohr Scientific Correspondence
DM	Deutsches Museum, Munich
EMW	Felix Klein (ed.), Encyklopädie der mathematischen Wissenschaften
HSPS	Historical Studies in the Physical Sciences
JRE	Jahrbuch der Radioaktivität und Elektronik
LCP	Hendrik Antoon Lorentz, Collected Papers, 9 vols. (The Hague, 1934–39)
LTZ	Hendrik Antoon Lorentz, Correspondence, Algemeen Rijksarchief, Den Haag, Microfilm at the Office for History of Science (OHST) at the University of California, Berkeley
PM	The Philosophical Magazine
PR	The Physical Review
PZ	Physikalische Zeitschrift
RDN	The University of Texas [Austin] Richardson Collection, Microfilm at the OHST
SB	Sitzungsberichte der Königlich Preussischen Akademie der Wissenschaften, Berlin
SW	Sitzungsberichte der Kaiserlichen Akademie der Wissenschaften zu Wien, Mathematisch-Naturwissenschaftliche Classe, 2.Abt.
WWN	Wilhelm Wien-Nachlaß, Berlin, Staatsbibliothek Preussischer Kulturbesitz, Microfilm at the OHST
ZPC	Zeitschrift für Physikalische Chemie
ZSP	Zeitschrift für Physik.

1. See, e.g., J. C. Maxwell, *A Treatise on Electricity and Magnetism,* vol. 1–2 (3rd. ed., Oxford, 1904), in vol. 1, no. 241, on 362–363 (nature of electric resistance); in vol. 2, no. 574, on 216–218 (nature of electric current); ibid., no. 789, on 437–438 (dispersion of light), ibid., no. 800, on 446–447 (metal optics). An interesting comment on Maxwell's notion of an electric current can be found in the German translation of (the 2nd. ed. of) Maxwell's "Treatise. . . ." In a footnote, the translator, B. Weinstein, refers to the Hall effect, which furnished a measuring method for "both the absolute velocity and the direction of a current." See J.Cl. Maxwell, *Lehrbuch der Electricität und des Magnetismus,* Authorized translation by B. Weinstein, vol. 1–2 (Berlin, 1883), in vol. 2, no. 569, on 257.

2. O. Heaviside to H. Hertz, 13 July 1889, DM, call. no. 2924.

3. O. Heaviside to H. Hertz, 14 August 1889, DM, call. no. 2925.

4. O. Heaviside to H. Hertz, 13 July 1889, DM, call. no. 2924.

5. P. Drude to H. Hertz, 15 May 1892, DM, call. no. 3209.

6. Jed Z. Buchwald, "The Abandonment of Maxwellian Electrodynamics : Joseph Larmor's Theory of the Electron," Archives internationales d'histoire des sciences 31 (1981); 135–180, 373–438; *From Maxwell to Microphysics* (Chicago: Chicago University Press, 1985), 141–173.

7. H. A. Lorentz, "La théorie électromagnétique de Maxwell et son application aux corps mouvants," Archives Néerlandaises des Sciences exactes et naturelles, [1] 25 (1892); 363–553, on 432–551.

8. W. Weber, "Ueber die Bewegung der Elektricität in Körpern von molecularer Constitution," AP, 156 (1875); 1–61, on 34–41, 49–55.

9. W. Weber, "Elektrodynamische Maassbestimmungen. Ueber ein allgemeines Grundgesetz der elektrischen Wirkung [1846]," *Werke*, vol. 3 (Berlin, 1893); 25–214, on 132–136.

10. L. Boltzmann, "Zur Theorie des von Hall entdeckten elektromagnetischen Phänomens," SW, 94 (1886); 644–669, on 645–647; also in L. Boltzmann, *Wissenschaftliche Abhandlungen*, 3 vols. [1909] (Reprint, New York, 1968), in vol. 3, 187–211, on 188–189.

11. E. Riecke, "Zur Theorie des Galvanismus und der Wärme," AP, 66 (1898); 353–389, 545–581, on 352–356, 569–572. For a qualitative explanation of electric conductivity, based on the analogy with electrolytic conduction, see W. Giese, "Grundzüge einer einheitlichen Theorie der Electricitätsleitung," AP, 37 (1889); 576–609, on 592. E. Riecke had already given a kinetic treatment of conductivity and diffusion in electrolytes. See E. Riecke, "Molekulartheorie der Diffusion und Elektrolyse, ZPC, 6 (1890); 564–572. W. Nernst's theory was qualitatively based on a kinetic model. Similar to van't Hoff's theory of osmosis, it contains, however, no explicitly kinetic calculations. See W. Nernst, "Zur Kinetik der in Lösung befindlichen Körper," ZPC, 2 (1888); 613–637, and "Die elektromotorische Wirksamkeit der Ionen, ZPC, 4 (1889); 129–181. For Riecke's belief that he went beyond Maxwell's theory, see also E. Riecke, "Über die Elektronentheorie des Galvanismus und der Wärme," JRE, 3 (1906); 24–47, on 24.

12. E. Riecke, "Ueber einige Beziehungen zwischen hydrodynamischen und electrischen Erscheinungen," Nachrichten von der Königlichen Gesellschaft der Wissenschaften . . . zu Göttingen (1887); 10–28, on 11.

13. E. Riecke, "Zur Theorie des Galvanismus und der Wärme," AP, 66 (1898); 353–389, 545–581, on 355.

14. E. Riecke, "Zur Theorie des Galvanismus und der Wärme," AP, 66 (1898); 353–389, 545–581, on 356.

15. Riecke identified the negative particles as particles of the "cathode rays." The positive carriers he identified with metal ions. See E. Riecke, "Zur Theorie des Galvanismus und der Wärme," AP, 66 (1898); 353–389, 545–581, on 569–571. The latter assumption was

soon disproved. See E. Riecke, "Ist die metallische Leitung verbunden mit einem Transport von Metallionen?[1901]," PZ, 2 (1900–1901); 639.

16. Rudolf Seeliger, "Elektronentheorie der Metalle [1921]," EMW, 5,2 (Leipzig, 1904–1922); 777–878, on 782.

17. E. Riecke, "Zur Theorie des Galvanismus und der Wärme," AP, 66 (1898); 353–389, 545–581, on 379–381. For an evaluation see E. Riecke, Über die Elektronentheorie des Galvanismus und der Wärme," JRE, 3 (1906); 24–47, on 32–34; and Rudolf Seeliger, "Elektronentheorie der Metalle [1921]," EMW, 5,2 (Leipzig, 1904–1922); 777–878, on 784–785.

18. P. Drude, "Zur Elektronentheorie der Metalle," AP, 1 (1900); 566–613; AP, 3 (1900), 369–402.

19. P. Drude, Physik des Aethers auf elektromagnetischer Grundlage (Stuttgart,1894).

20. A unique physical world view as an argument in favor of the electron gas theory of metals is obvious in Rudolf Seeliger, "Elektronentheorie der Metalle [1921]," EMW, 5,2 (Leipzig, 1904–1922); 777–878, on 778–781.

21. P. Drude, "Zur Ionentheorie der Metalle," PZ, 1 (1900); 161–165.

22. P. Drude, "Zur Elektronentheorie der Metalle," AP, 1 (1900); 566–613; AP, 3 (1900); 369–402, on 568, 584. Deriving the number of free electrons from the optical properties of metals however, led only to a rough approximation. See P. Drude, "Optische Eigenschaften und Elektronentheorie," AP, 14 (1904); 677–725, 936–961, on 960. For Bohr's objections see N. Bohr to C. W. Oseen, 1 December 1911," BCW, [426]–[431], on [427]. In 1915, Bohr felt that he had "pulled" the calculations of the number of electrons from optical properties "to pieces." See N. Bohr to G. H. Livens, 14 February 1915, BCW, [479]–[480], on [480]. Interestingly enough, from the point of view of modern semiconductor physics, G. H. Livens referred to the Hall effect and to the related galvanomagnetic effects as an "apparently better and more direct means of obtaining an estimate of the number of free electrons taking part in the conduction phenomena." See G. H. Livens, "On the Number of Electrons Concerned in Metallic Conduction," PM, [6] 30 (1915), 105–112, on 111–112.

23. P. Drude, "Zur Elektronentheorie der Metalle," AP, 1 (1900), 566–613; AP, 3 (1900); 369–402, on 566–567, 570, 572; H. A. Lorentz, "Ergebnisse und Probleme der Elektronentheorie [1904]," LCP, 8 (1935); 76–124, on p. 101.

24. Paul and Tatiana Ehrenfest, "Begriffliche Grundlagen der statistischen Auffassung in der Mechanik [1911]," EMW, 4,4 (Leipzig, 1907–1914), article no. 32; 3–90, on 74.

25. J. H. van't Hoff was clear about the microphysical analogy of the pressure of a gas with the osmotic pressure. See J. H. van't Hoff, "Die Rolle des osmotischen Druckes in der Analogie zwischen Lösungen und Gasen," ZPC, 1 (1887); 481–508, on 482–483. An explicitly formulated kinetic theory of the osmotic pressure is due to Boltzmann. See L. Boltzmann, "Die Hypothese van't Hoffs über den osmotischen Druck vom Standpunkt der kinetischen Gastheorie," ZPC, 6 (1890); 474–480. Max Planck, and, from the point of view of energetics, Wilhelm Ostwald, questioned, however, the need for a kinetic approach in the theory of the dilute solutions. See M. Planck, "Über den osmotischen Druck," ibid.,

187–189. See also M. Planck, "Allgemeines zur neueren Entwicklung der Wärmetheorie" [Vortrag . . . 64. Versammlung der Deutschen Naturforscher und Ärzte in Halle . . . 1891], ZPC, 8 (1891);647–656, on 648. For the theory of electrolytes, see W. Nernst, "Zur Kinetik der in Lösung befindlichen Körper," ZPC, 2 (1888); 613–637, and "Die elektromotorische Wirksamkeit der Ionen, ZPC, 4 (1889); 129–181. See also E. Riecke, "Zur Theorie des Galvanismus und der Wärme," AP, 66 (1898); 353–389, 545–581, on 352–356, 569–572.

26. P. Drude, "Zur Elektronentheorie der Metalle," AP, 1 (1900); 566–613; AP, 3 (1900); 369–402, on 572.

27. A. Sommerfeld, [chapter of a ?] Lecture course entitled "Elektronenth[eorie] d[er] Metalle und verwandte statistische Fragen," Sommerfeld-Nachlaß, reproduced in AHQP, Microfilm 21, 8. The date given in AHQP is summer semester 1908. Michael Eckert, has pointed out, however, (see Michael Eckert, "Propaganda in science: Sommerfeld and the spread of the electron theory of metals," in HSPS, 17 (1986); 191–233, in Eckert's ref. 29) that the correct date must be summer semester 1912. For the importance of the transfer of the gas constant to the electron gas, see also E. Riecke, "Ueber das Verhältnis der Leitfähigkeiten der Metalle für Wärme und für Elektricität," AP, 2(1900); 835–842, on 835; and E. Riecke, "Die jetzigen Anschauungen über das Wesen des metallischen Zustandes [Lecture . . . 1908]," PZ, 10 (1909); 508–518, on 509.

28. P. Drude, "Zur Elektronentheorie der Metalle," AP, 1 (1900); 566–613; AP, 3 (1900); 369–402, on 573. Interestingly enough, the section in Boltzmann's lectures cited by Drude is almost directly followed by a chapter on the electric conductivity as well as the viscosity of gases. See L. Boltzmann, *Vorlesungen über Gastheorie,* 2 parts (Leipzig, 1896–1898); a reprint is Stephen G. Brush (ed.) and Roman U. Sexl (ed. in chief), *Ludwig Boltzmann Gesamtausgabe,* vol. 1 (Braunschweig, Wiesbaden, Graz, 1981), in part 1, on 77–80.

29. P. Drude, "Zur Elektronentheorie der Metalle," AP, 1 (1900); 566–613; AP, 3 (1900); 369–402, on 577–578.

30. See, e.g., H. A. Lorentz, "The Methods of the Theory of Gases Extended to Other Fields [1909]," LCP, 8 (The Hague, 1935); 159–182, on 179.

31. W. Jaeger and H. Diesselhorst, " Wärmeleitung, Electricitätsleitung, Wärmecapacität und Thermokraft einiger Metalle (Mitteilung aus der Physikalisch-Technischen Reichsanstalt . . .)," SB (1899); 719–726.

32. M. Planck, "Über die Elementarquanta der Materie und der Elektrizität," AP, 4 (1901); 564–566. For a demonstration of the guiding role of Planck's constant h in the development of quantum theory see Armin Hermann, *Frühgeschichte der Quantentheorie (1899–1913)* (Mosbach, 1969), 28–35; see also A. Hermann, *Max Planck. In Selbstzeugnissen und Bilddokumenten* (Reinbek bei Hamburg, 1973), 26–31. A different interpretation is in Thomas S. Kuhn, *Black-Body Theory and Quantum Discontinuity, 1894–1912* (Chicago and London: University of Chicago Press, 1987), 278, in Kuhn's ref. 30.

33. Emil Warburg, "Wahlvorschlag für Paul Drude . . . ," see Christa Kirsten, Hans-Günther Körber, Hans-Jürgen Treder (eds.), *Physiker über Physiker, Wahlvorschläge zur Aufnahme von Physikern in die Berliner Akademie 1870 bis 1929 . . .* vol. 1 (Berlin, GDR, 1975); 163–164, on 164.

34. See, e.g., J. J. Thomson, *Die Korpuskulartheorie der Materie* (Braunschweig, 1908); 25–26.

35. J. E. Lilienfeld, "Über neuartige Erscheinungen in der positiven Lichtsäule der Glimmentladung. (Zweite Mitteilung. Positive Elektronen.)," Berichte der Deutschen Physikalischen Gesellschaft, 5 (1907), containing Verhandlungen der Deutschen Physikalischen Gesellschaft, 9 (1907); 125–135, on 132; J. E. Lilienfeld, "Die Elektrizitätsleitung im extremen Vakuum (Leipziger Habilitationsschrift), AP, 32 (1910); 673–738, on 675, 736–737.

36. W. Wien, "Untersuchungen über die elektrische Entladung in verdünnten Gasen," AP, 8 (1902); 244–266; Ueber die Natur der positiven Elektronen, AP, 9 (1902); 660–664, especially on 660.

37. M. Planck to W. Wien, 16 March 1903, WWN, no. 5. In another letter, Planck asserted his interest in Lilienfeld's research on "positive electrons," see M. Planck to W. Wien, 24 May 1907, WWN, no. 21.

38. A. Bucherer to H. A. Lorentz, 31 March 1912, LTZ.

39. The opposite rotation of polarized light in a magnetic field, in the region of the absorption bands, was explained by a response of negative or of positive electrons to light frequencies. See R. W. Wood, "On the existence of positive electrons in the sodium atom," PM, [6] 15 (1908); 274–279. See also Jean Becquerel, "On the Dispersion of Magnetic Rotatory Power in the Neighborhood of Bands of Absorption in the Case of Rare Earths . . . , ibid., 16 (1908); 153–161, on 155. J. Becquerel, who had worked with H. A. Lorentz in Leyden, wrote to Lorentz about "some new results which . . . give the existence of a positive electron a degree of probability close to certainty [quelques résultats nouveaux qui . . . donnent a l'existence des électrons positifs un degré de probabilité bien voisin de la certitude]." See J. Becquerel to H. A. Lorentz, 11 July 1908, LTZ. An overview is in Helge Kragh, "Concept and Controversy: Jean Becquerel and the Positive Electron," Centaurus, 1989, 32 (1989); 203–240.

40. J. E. Lilienfeld, "Über neuartige Erscheinungen in der positiven Lichtsäule der Glimmentladung. (Zweite Mitteilung. Positive Elektronen.)," Berichte der Deutschen Physikalischen Gesellschaft, 5 (1907), containing Verhandlungen der Deutschen Physikalischen Gesellschaft, 9 (1907); 125–135, on 132.

41. See among a sequel of papers J. J. Thomson, "Positive Electricity," PM, [6] 18 (1909), 821–845, on 821. For objections against J. J. Thomson's views see W. Wien, "On rays of positive electricity," ibid., 14 (1907); 212–213; O. W. Richardson, "The Specific Charge of the Ions Emitted by Hot Bodies," ibid., 16 (1908); 740–767.

42. See J. J. Thomson, "Some Speculations as to the Part Played by Corpuscles in the Physical Phenomena," Nature, 62 (1901); 31–32; see also J. J. Thomson, "Indications relatives a la constitution de la matière fournis par les recherches récentes sur le passage de l'électricité a travers les gaz," Rapports du Congres de Physique de 1900, Paris; vol. 3, 138–150. See further A. Schuster, "On the Number of Electrons Conveying the Conduction Currents in Metals," PM, [6] 7 (1904); 151–157, on 157.

43. P. Drude, "Optische Eigenschaften und Elektronentheorie," AP, 14 (1904); 677–725, 936–961, on 679 and on 939. In 1884 H. A. Lorentz had pointed out that any theory of

the Hall effect, which is based on the movement of "electric particles" in the magnetic field, must assume some kind of asymmetry in the behavior of "positive and negative electricity." See H. A. Lorentz, "Le phénomène découverte par Hall et la rotation électromagnétique du plan de polarisation de la lumière," LCP, 2 (The Hague, 1936); 136–163, on 141–142. In 1903 Lorentz still assumed different kinds of carriers in order to explain the different signs of the Hall effect. See H. A. Lorentz, "Weiterbildung der Maxwellschen Theorie. Elektronentheorie [1903]," EMW, 5, 2 (Leipzig,1904–1922); 145–280, on 222.

44. H. A. Lorentz, "Le mouvement des électrons dans les métaux," Archives Néerlandaises des Sciences exactes et naturelles, [2] 10 (1905) ; 336–371; reprinted in LCP, 3 (The Hague, 1936); 180–214, on 180 (subsequently only the LCP will be cited).

45. H. A. Lorentz, "Le mouvement des électrons dans les métaux," Archives Néerlandaises des Sciences exactes et naturelles, [2] 10 (1905) ; 336–371; reprinted in LCP, 3 (The Hague, 1936); 180–214, on 214.

46. H. A. Lorentz, "Le mouvement des électrons dans les métaux," Archives Néerlandaises des Sciences exactes et naturelles, [2] 10 (1905) ; 336–371; reprinted in LCP, 3 (The Hague, 1936); 180–214, on 206–208. See also H. A. Lorentz, "Ergebnisse und Probleme der Elektronentheorie [1904]," LCP, 8 (1935); 76–124, on 117; H. A. Lorentz, "Anwendung der kinetischen Theorien auf Elektronenbewegung [1913]," LCP, 8 (The Hague, 1935); 214–243, on 217.

47. E. Riecke, "Die jetzigen Anschauungen über das Wesen des metallischen Zustandes [Lecture . . . 1908]," PZ, 10 (1909); 508–518, on 509.

48. Karl Baedeker complained about how the electron gas theory of metals became "discredited" among a number of physicists. See K. Baedeker to N. Bohr, 6 May 1911, BCW, [398]. For Bohr's "skeptical" view see N. Bohr to O. W. Richardson, 23 November 1914, RDN, R–000 139. O. W. Richardson considered the lack of quantum theory as one "reason why the electron theory of metallic conduction has so far been to a great extent a failure"; see O. W. Richardson to Edwin Herbert Hall, 13 October 1914, RDN, L–000 121. James H. Jeans did simply not "believe in the Electron theory of conduction anyhow," see J. H. Jeans to O. W. Richardson, 30 [?] April [1914?], RDN, R–000 754. See also E. H. Hall to O. W. Richardson, 8 December 1913, RDN, R–000 537. Carl Benedicks in Stockholm felt disillusion with electron gas theory. He found it not at all strange, however, ("wenig befremdend") that ideas, which are derived from the kinetic theory of gases did not fit into a different field, namely into the phenomena of metallic conduction. See C. Benedicks, "Beiträge zur Kenntnis der Elektrizitätsleitung in Metallen und Legierungen," JRE, 13 (1916); 351–395, on 351.

49. Rudolf Seeliger, "Elektronentheorie der Metalle [1921]," EMW, 5, 2 (Leipzig, 1904–1922); 777–878, on 779.

50. H. A. Lorentz, "Le mouvement des électrons dans les métaux," Archives Néerlandaises des Sciences exactes et naturelles, [2] 10 (1905) ; 336–371; reprinted in LCP, 3 (The Hague, 1936); 180–214, on 184–192.

51. H. A. Lorentz, "Ergebnisse und Probleme der Elektronentheorie [1904]," LCP, 8 (1935); 76–124, on 109.

52. H. A. Lorentz, "Ergebnisse und Probleme der Elektronentheorie [1904]," LCP, 8 (1935); 76–124, on 103–104.

53. H. A. Lorentz, "Le mouvement des électrons dans les métaux," Archives Néerlandaises des Sciences exactes et naturelles, [2] 10 (1905) ; 336–371; reprinted in LCP, 3 (The Hague, 1936); 80–214, on 194.

54. H. A. Lorentz, "On the Molecular Motion of Dissolved Substances [1889]," LCP, 6 (The Hague, 1938); 112–113; "Sur la théorie moléculaire des dissolutions diluées [1892]," ibid., 114–133.

55. H. A. Lorentz, "The Methods of the Theory of Gases Extended to Other Fields [1909]," LCP, 8 (The Hague, 1935); 159–182. See also H. A. Lorentz, *The Theory of Electrons . . . A course of lectures delivered in Columbia University, New York, in March and April 1906* (Leipzig, 1909); on 10.

56. H. A. Lorentz, "The Methods of the Theory of Gases Extended to Other Fields [1909]," LCP, 8 (The Hague, 1935); 159–182, on 176.

57. R. Gans, "Zur Elektronenbewegung in Metallen," AP, 20 (1906); 293–326, on 310.

58. P. Hertz to H. A. Lorentz, 4 March 1908, 20 March 1908, 7 April 1911; H. A. Lorentz to P. Hertz [undated fragment], on P. Hertz's letter of 4 March 1908; LTZ.

59. H. Poincaré, "Les rapports de la matière et de l'éther [1912]," Œuvres, 9 (Paris, 1954) ; 669–682, on 670.

60. H. Haga to Comité Nobel pour la physique, Stockholm, 2 January 1918, LTZ. In the Berkeley microfilm the copy is filed under "Lan [last name], Perrin," instead of "Perrin, Jean." Jean Perrin, in his later Nobel lecture, referred precisely to van't Hoff's theory of the osmotic pressure as an extension of the gas laws to the dilute solutions and in turn to his own idea, according to which the theory of emulsions is analogous to the theory of solutions. See J. Perrin, "Discontinuous Structure of Matter. Nobel lecture . . . 1926," Nobel Lectures . . . Physics 1922–1941 (Amsterdam, London, New York, 1965); 138–164, on 145.

61. See J. J. Thomson, *Die Korpuskulartheorie der Materie* (Braunschweig, 1908); 29, 71–75.

62. M. Planck considered the electron vapor theory and its analogy with Nernst's theory of electrolytes a promising idea. His criticism of how Krüger actually had worked out the theory, however, prevented a publication in AP. See M. Planck to W. Wien, 18 March 1908, WWN, no. 27.

63. K. Baedeker, "Zur Elektronentheorie der Thermoelektrizität," PZ, 11 (1910); 809–810. Similar to Krüger, Baedeker found it "useful" to refer to the "complete analogy" of the electron vapor theory with W. Nernst's notion of an electrolytic "solvent tension [Lösungsdruck]" of metals as well as with the concept of an osmotic pressure of ions. See K. Baedeker, *Die elektrischen Erscheinungen in metallischen Leitern* (Braunschweig, 1911), on 91. For a critical review see O. W. Richardson, "The Electron Theory of Contact Electromotive Force and Thermoelectricity," PM, [6] 23 (1912); 263–278, on 277. Baedeker defended his approach, in a letter to O. W. Richardson [then at Princeton], 11 February 1912, RDN, R–000 060.

64. O. W. Richardson, "Thermionic Phenomena and the Laws which Govern Them. Nobel lecture, December 12, 1929"; Nobel lectures . . . Physics 1922–1941 (Amsterdam, London, New York, 1965); 224–236, on 226.

65. O. W. Richardson, "The Electrical Conductivity Imparted to a Vacuum by Hot Conductors," Philosophical Transactions of the Royal Society of London, [A] 201 (1903); 497–549, on 499–503. See also O. W. Richardson, "Die Abgabe negativer Elektrizität von heißen Körpern," JRE, 1 (1904); 300–315. The notion of "thermionics" was coined by O. W. Richardson in 1909, see "Thermionics," PM, [6] 17 (1909); 813–833; see also O. W. Richardson, "Notes on the Kinetic Theory of Matter," PM, [6] 18 (1909); 695–698.

66. H. A. Wilson, "On the Discharge of Electricity from Hot Platinum," Philosophical Transactions of the Royal Society of London, [A] 202 (1904); 243–275, on 258–259.

67. O. W. Richardson, "Thermionic Phenomena and the Laws which Govern Them. Nobel Lecture, December 12, 1929"; Nobel Lectures . . . Physics 1922–1941 (Amsterdam, London, New York, 1965); 224–236, on 226.

68. O. W. Richardson, "Thermionic Phenomena and the Laws which Govern Them. Nobel lecture, December 12, 1929"; Nobel lectures . . . Physics 1922–1941 (Amsterdam, London, New York, 1965); 224–236, 227; see also O. W. Richardson and H. L. Crooke, "The Heat Developed During the Absorption of Electrons by Platinum," PM, [6] 20 (1910); 173–206; "The Heat Liberated During the Absorption of Electrons by Different Metals," ibid., 21 (1911); 404–410; "The Absorption of Heat Produced by the Emission of Ions from Hot Bodies," ibid., 25 (1913); 624–643; 26 (1913); 472–476.

69. O. W. Richardson, "The Electron Theory of Contact Electromotive Force and Thermoelectricity," PM, [6] 23 (1912); 263–278, on 273.

70. See O. W. Richardson and F. C. Brown, "The Kinetic Energy of the Negative Electrons Emitted by Hot Bodies," PM, [6] 16 (1908); 353–376. See also O. W. Richardson, "Thermionic Phenomena and the Laws which Govern Them. Nobel lecture, December 12, 1929"; Nobel lectures . . . Physics 1922–1941 (Amsterdam, London, New York, 1965); 224–236, on 226–227. Interesting here is how Richardson claimed, like Otto Stern in the case of his measurements of fast moving silver atoms, that, for the first time, he had demonstrated "Maxwell's law for any gas, although the law was enunciated by Maxwell in 1859." See also O. W. Richardson, "Glühelektroden," in E. Marx (ed.), Handbuch der Radiologie, vol. IV (Leipzig,1917); 445–602, on 517. See further O. W. Richardson, "The kinetic energy of the ions emitted by hot bodies-II," PM, [6] 18 (1909); 681–695, on 688, fig. 3, a diagram comparing the calculated and the observed velocity distribution. See also O. W. Richardson, "The Electron Theory of Contact Electromotive Force and Thermoelectricity," PM, [6] 23 (1912); 263–278, on 263–264.

71. O. W. Richardson, "The Distribution of the Molecules of a Gas in a Field of Force, with Applications to the Theory of Electrons," PM, [6] 28 (1914); 633–647, on 633–635, and on 638. Most likely due to the war, Richardson's abandonment of the Maxwell-Boltzmann distribution inside the metal did not enter into his German "Handbuch" article. See O. W. Richardson, "Glühelektroden," in E. Marx (ed.), Handbuch der Radiologie, vol. IV (Leipzig,1917); 445–602O, on 517, and on 591–602.

72. For Planck's judgment see M. Planck to W. Wien, 18 March 1908, WWN, no. 27. For Lorentz's application of the electron vapor theory see H. A. Lorentz, "Anwendung der kinetischen Theorien auf Elektronenbewegung [1913]," LCP, 8 (The Hague, 1935); 214–243, on 224–225; on 216 a discussion of Richardson's thermionic measurements. Lorentz's paper was first published in David Hilbert (ed.), *Vorträge über die kinetische Theorie der Materie und der Elektrizität. Gehalten auf Einladung . . . der Wolfskehlstiftung* [1913]. = Mathematische Vorlesungen an der Universität Göttingen: VI (Leipzig, Berlin, 1914); 169–193, here on 170–171, and on 174–178. See also H. A. Lorentz, "Ergebnisse und Probleme der Elektronentheorie [1904]," LCP, 8 (1935); 76–124, on 110.

73. For a discussion of these problems see e.g. W. Schottky, "Bericht über thermische Elektronenemission," JRE, 12 (1915); 147–205, on 150–167.

74. See O. W. Richardson, "The Emission of Electrons from Tungsten at High Temperatures: An Experimental Proof that the Electric Current in Metals is Carried by Electrons," PM, [6] 26 (1913); 345–350, on 350. For the influence of the development of incandescent light on the thermionic measurements, see also O. W. Richardson, "The Specific Charge of the Ions Emitted by Hot Bodies.-II," PM, [6] 20 (1910); 545–559, on 555.

75. See O. W. Richardson, "Thermionic Phenomena and the Laws which Govern Them. Nobel lecture, December 12, 1929"; Nobel lectures . . . Physics 1922–1941 (Amsterdam, London, New York, 1965); 224–236, on 227. See also O. W. Richardson, "The Emission of Electrons from Tungsten at High Temperatures: An Experimental Proof that the Electric Current in Metals is Carried by Electrons," PM, [6] 26 (1913); 345–350. The paper was immediately published in a German translation: "Die Emission von Elektronen seitens des Wolframs bei hohen Temperaturen; ein experimenteller Beweis dafür, daß der elektrische Strom in Metallen von Elektronen getragen wird," PZ, 14 (1913); 793–796.

76. See N. Bohr to O. W. Richardson, 29 September 1915; O. W. Richardson to N. Bohr, 9 October 1915, BCW, [482]–[488]. See also W. Schottky, "Bericht über thermische Elektronenemission," JRE, 12 (1915); 147–205, on 150–167, on 170–185.

77. N. Bohr to O. W. Richardson, 15 October 1915, BCW, [489]–[491], on [489]. See also F. A. Lindemann to O. W. Richardson, 30 June 1915, RDN, R 000 753.

78. N. Bohr, Studier over metallernes elektrontheori [1911], translated by J. Rud Nielsen as "Studies on the electron theory of metals," BCW, [291]–[393], on [302]–[317].

79. James Clerk Maxwell, "On the Dynamical Theory of Gases [1866]," The Scientific Papers [1890], 2 vols. (Reprint, New York, 1965), in vol. 2, 26–78, on 29, and on 40–43. Maxwell's contribution to the problem of force laws, however, is only a small fraction of a complex history. See Stephen G. Brush, *The Kind of Motion We Call Heat*, 2 vols. (Amsterdam and elsewhere, 1976), in vol. 2, on 386–418; Karl von Meyenn, "Dispersion und mechanische Äthertheorien im 19. Jahrhundert," in P. L. Butzer and F. Feher (eds.), E. B. Christoffel. The influence of his work on mathematics and the physical sciences (Basel, Boston, Stuttgart, 1981); 680–703; K. von Meyenn, "Engpässe in der Atomtheorie des frühen 19. Jahrhunderts," in Charlotte Schönbeck (Ed.), Atomvorstellungen im 19. Jahrhundert (Paderborn, 1982); 35–55; C. C. Gillispie (R. Fox, I. Grattan-Guinness), "Laplace," DSB, 15 (= Supplement I) (New York, 1978); 273–403, on 358–360.

80. L. Boltzmann, "Über die Möglichkeit der Begründung einer kinetischen Gastheorie auf anziehende Kräfte allein [1884]," AP, 24 (1885); 37–44; see also L. Boltzmann, Wissenschaftliche Abhandlungen [1909], 3 vols. (reprint, New York, 1968), in vol. 3, on 101–109.

81. N. Bohr, "Studier over metallernes elektrontheori [1911]," translated by J. Rud Nielsen as "Studies on the electron theory of metals," BCW, [291]–[393], on [341].

82. L. Boltzmann, "Über die Möglichkeit der Begründung einer kinetischen Gastheorie auf anziehende Kräfte allein [1884]," AP, 24 (1885); 37–44, on 37–38.

83. N. Bohr, "Studier over metallernes elektrontheori [1911]," translated by J. Rud Nielsen as "Studies on the electron theory of metals," BCW, [291]–[393], on [341].

84. P. Debye to N. Bohr, 30 May 1911, BCW, [400]–[402]; N. Bohr to P. Debye, undated, ibid., [402]–[403].

85. N. Bohr, "Studier over metallernes elektrontheori [1911]," translated by J. Rud Nielsen as "Studies on the electron theory of metals," BCW, [291]–[393], on [364]–[365], [379]. For Jeans's attempts see J. H. Jeans, "The Motion of Electrons in Solids. Part II," PM, [6] 20 (1910); 642–651; see also J. H. Jeans, "On the Analysis of the Radiation from Electron Orbits," PM, [6] 20 (1910); 642–651.

86. J. C. Maxwell, *A Treatise on Electricity and Magnetism,* vol. 1–2 (3rd. ed., Oxford, 1904), in no. 574, on 216–218; H. Hertz, "Versuche zur Feststellung einer oberen Grenze für die kinetische Energie der elektrischen Strömung," AP, 10 (1880); 414–448; "Obere Grenze für die kinetische Energie der bewegten Elektricität, AP, 14 (1881); 581–590; L. Boltzmann, *Vorlesungen über Maxwells Theorie der Elektricität und des Lichtes,* 2 parts [Leipzig, 1891–1893], reprinted as W. Kaiser (ed.) and R. U. Sexl (Ed. in chief), Ludwig Boltzmann, Gesamtausgabe, vol. 2 (Graz, 1982), in part 1, on 18.

87. H. A. Lorentz, "Anwendung der kinetischen Theorien auf Elektronenbewegung [1913]," LCP, 8 (The Hague, 1935); 214–243, on 238–240.

88. E. F. Nichols, "Die Möglichkeit einer durch zentrifugale Beschleunigung erzeugten elektromotorischen Kraft," PZ, 7 (1906), 640–642. The problem of measuring may have been that these small potential differences of the order of magnitude of 10^{-8}V are masked by thermoelectric phenomena caused by sliding contacts.

89. E. F. Nichols to H. A. Lorentz, 21 October 1906, 17 July 1909, LTZ. For H. A. Lorentz's lecture course at Columbia see H. A. Lorentz, *The Theory of Electrons . . . A course of lectures delivered in Columbia University, New York, in March and April 1906* (Leipzig, 1909).

90. R. C. Tolman and T. D. Stewart, "The Electromotive Force Produced by the Acceleration of Metals," PR, 8 (1916); 97–116, on 97–98 the experiments with accelerated electrolytes. See also R. C. Tolman, Sebastian Karrer, and Ernest W. Guernsey, "Further experiments on the mass of the electric carrier in metals," PR, 21 (1923); 525–539; R. C. Tolman and Lewis M. Mott-Smith, "A Further Study of the Inertia of the Electric Carrier in Copper," PR, 28 (1926); 794–832.

91. W. Nernst, "Zur Kinetik der in Lösung befindlichen Körper," ZPC, 2 (1888); 613–637, on 636–637; "Die elektromotorische Wirksamkeit der Ionen," ZPC, 4 (1889); 129–181, on 130–131.

92. Rudolf Seeliger, "Elektronentheorie der Metalle [1921]," EMW, 5,2 (Leipzig, 1904–1922); 777–878, on 788, Seeliger's ref. 18.

93. See E. Riecke, "Elektronentheorie galvanischer Eigenschaften der Metalle [1913]," in Erich Marx (ed.), Handbuch der Radiologie, vol. VI (Leipzig,1925), = Die Theorien der Radiologie; 281–494, on 286, note of the article's editor Max von Laue. See also Walther F. Meißner, "Der Stand der Forschung über das Wesen der Elektrizitätsleitung," Elektrotechnische Zeitschrift, 60 (1939); 333–336, on 333; D. I. Blochinzew, *Grundlagen der Quantenmechanik,* 5th ed. (Frankfurt, Zürich, 1966), on 214. Christian Gerthsen, Hans O. Kneser, *Physik, Ein Lehrbuch . . .* , 9th ed. (Berlin, Heidelberg, New York, 1966), on p. 204.

94. H. A. Lorentz, "Physics in the new and the old world [1926]," LCP, 8 (The Hague, 1935); 405–417, on 409. See also "Positive and Negative Electricity [1920]," ibid., 48–75, on 63–66; "Application de la théorie des électrons aux propriétés des métaux [Solvay lecture, 1924]," ibid., 263–306, on 282–283; "The motion of electricity in metals [1925]," ibid., 307–332, on 312–316. For a discussion of experimental problems, see J. A. Fleming to H. A. Lorentz, 1 December 1925, 22 January 1926, and 4 February 1926, LTZ.

95. See, e.g., Norman Rostoker, "Interpretation of the Electron-Inertia Experiment for Metals with Positive Hall Coefficients," PR, 88 (1952); 952–953; William Shockley, "Interpretation of e/m Values for Electrons in Crystals," ibid., 953.

96. C. G. Darwin, "The Inertia of Electrons in Metals," Proceedings of the Royal Society of London, [A], 154 (1936); 61–66.

97. M. Brillouin to H. A. Lorentz, 12 February 1901, LTZ.

98. M. Brillouin to H. A. Lorentz, 25 (28?) March 1901, LTZ. In fact, P. Drude was well aware that, seen from the perspective of electron gas theory, his assumption in metal optics only refer to mean values. See P. Drude, "Zur Ionentheorie der Metalle," PZ, 1 (1900); 161–165, on 162.

99. See, e.g., A. Sommerfeld and H. Bethe, "Elektronentheorie der Metalle," in H. Geiger and K. Scheel (eds.), Handbuch der Physik, 2nd. ed., vol. 24,2 (Berlin, 1933), 333–622, on 334; Charles Kittel, *Introduction to Solid State Physics,* 3rd. ed. [1966], translated as *"Einführung in die Festkörperphysik"* (München, 1968); on 258–259.

100. P. Drude, "Zur Elektronentheorie der Metalle," AP, 1 (1900); 566–613; AP, 3 (1900), 369–402.

101. M. Reinganum, "Theoretische Bestimmung des Verhältnisses von Wärme-und Elektricitätsleitung der Metalle aus der Drude'schen Elektronentheorie," AP, 2 (1900); 398–403, on 401. For Giese's theory see W. Giese, "Grundzüge einer einheitlichen Theorie der Electricitätsleitung," AP, 37 (1889); 576–609. For Richarz's contribution see Franz Richarz, "Über das Gesetz von Dulong und Petit," AP, 48 (1893), 708–716.

102. Jürgen Renn, "Einstein's Controversy with Drude and the Origin of Statistical Mechanics: A New Glimpse from the 'Love Letters'," Preprint 55, Max Planck Institute for

the History of Science, Berlin 1997, 10–16, 20–22. Jürgen Renn, "Einstein's Controversy with Drude and the Origin of Statistical Mechanics: A New Glimpse from the 'Love Letters'," Archive for History of Exact Sciences, 51 (1997), 315–354. An abbreviated version is Jürgen Renn, "Die Geburt der Statistischen Mechanik aus dem Geist der Elektronentheorie der Metalle," Physikalische Blätter, 53 (1997); 860–862.

103. See, e.g., J. J. Thomson, *Die Korpuskulartheorie der Materie* (Braunschweig, 1908), 83; Arnold Eucken, "Neuere Untersuchungen über den Temperaturverlauf der spezifischen Wärme," JRE, 8 (1911); 489–534, on 509–511; W. Nernst, "Untersuchungen über die spezifische Wärme bei tiefen Temperaturen. III," SB (1911); 306–315, on 310; Walther F. Meißner, "Thermische und elektrische Leitfähigkeit der Metalle," JRE, 17 (1920); 229–273, on 263; Rudolf Seeliger, "Elektronentheorie der Metalle [1921]," EMW, 5,2 (Leipzig, 1904–1922); 777–878, on 853–854.

104. J. J. Thomson introduced his concept of electric duplets in metals, which, under the influence of an electric field, line up. The negative charge in the duplet, that is, the carrier of the current, is pictured to move from one duplet to the other. So the Wiedemann-Franz ratio depends on the distance of the charges in the duplet as well as on the distance of the centers of the dipoles. See J. J. Thomson, *Die Korpuskulartheorie der Materie* (Braunschweig,1908); 84–87. See also N. Bohr, "Studier over metallernes elektrontheori [1911]," translated by J. Rud Nielsen as "Studies on the electron theory of metals," BCW, [291]–[393], on [325]; N. Bohr to C. W. Oseen, 1 December 1911, BCW, [426]–[431], on [427]–[428]; on [428]–[429] Bohr argues against J. J. Thomson's theory; N. Bohr to S. B. McLaren, 17 December 1911, ibid., [432]–[434], on [432]. See also H. A. Lorentz, "Anwendung der kinetischen Theorien auf Elektronenbewegung [1913]," LCP, 8 (The Hague, 1935), 214–243, on 216; "Application de la théorie des électrons aux propriétés des métaux [Solvay lecture, 1924]," ibid., 263–306, on 275; "The Motion of Electricity in Metals [1925]," ibid., 307–332, on 312–316, on 323–325. See further J. H. Jeans, "The Motion of Electrons in Solids. Part I. Electric conductivity, Kirchhoff's Law and Radiation of Great Wave-Length," PM, [6] 17 (1909); 773–794. On 794 he discusses a number of free electrons that depend on the temperature and on the energy to free an electron in a certain metal.

105. See Rudolf Seeliger, "Elektronentheorie der Metalle [1921]," EMW, 5,2 (Leipzig, 1904–1922); 777–878, on 852.

106. See, e.g., Frederick Alexander Lindemann, "Untersuchungen über die spezifische Wärme bei tiefen Temperaturen. IV.," SB (1911); 316–321, on 317–318; K. Baedeker, *Die elektrischen Erscheinungen in metallischen Leitern* (Braunschweig, 1911), on 18. See also the letters N. Bohr to C. W. Oseen, 1 December 1911, BCW, [426]–[431], and N. Bohr to S. B. McLaren, 17 December 1911, ibid., [432]–[434]. See further Rudolf Seeliger, "Elektronentheorie der Metalle [1921]," EMW, 5,2 (Leipzig, 1904–1922); 777–878, on 853.

107. Heike Kamerlingh Onnes, "Investigations into the Properties of Substances at Low Temperatures, which Have Led, amongst other Things, to the Preparation of Liquid Helium," Nobel lecture, December 11, 1913, Nobel lectures . . . Physics 1901–1921 (Amsterdam, London, New York, 1967); 306–336, on 335; J. J. Thomson, "Conduction of Electricity through Metals," PM, [6] 30 (1915); 192–202, on 193.

108. Carl Ramsauer, "Über den Wirkungsquerschnitt der Gasmoleküle gegenüber langsamen Elektronen," PZ, 21 (1920); 576–578; Hans Ferdinand Mayer, "Über das Verhalten von Molekülen gegenüber freien langsamen Elektronen," AP, 64 (1921); 451–480; C. Ramsauer, "Über den Wirkungsquerschnitt der Gasmoleküle gegenüber langsamen Elektronen," ibid., 513–540. See also Rudolf Seeliger, "Elektronentheorie der Metalle [1921]," EMW, 5,2 (Leipzig, 1904–1922); 777–878, on 852.

109. In terms of electron gas theory, the mean kinetic energy of metal electrons is 0.04 eV. A minimum of the effective cross-section of argon for electrons of 0.39 eV was reported in J. S. Townsend and V. A. Bailey, "The motion of electrons in argon and in Hydrogen," PM, [6] 44 (1922); 1033–1052, on 1051. See also Friedrich Hund, "Theoretische Betrachtungen über die Ablenkung von freien langsamen Elektronen in Atomen," ZSP, 13 (1923); 241–263, on 242.

110. W. Wien, "Ueber die Energievertheilung im Emissionsspectrum eines schwarzen Körpers," AP, 58 (1896); 662–669, on 664.

111. P. Drude, *Lehrbuch der Optik* (Leipzig, 1900), preface.

112. H. A. Lorentz, "On the Emission and Absorption by Metals of Rays of heat of Great Wave-Lengths [1903]," LCP,3 (The Hague, 1936); 155–176.

113. E. Riecke, "Über die Elektronentheorie des Galvanismus und der Wärme," Vortrag gehalten auf der Versammlung deutscher Naturforscher in Meran, 25. September 1905, JRE, 3 (1906); 24–47, on 47; K. Baedeker, *Die elektrischen Erscheinungen in metallischen Leitern* (Braunschweig, 1911); 135.

114. H. A. Wilson, "The Electron Theory of Optical Properties of Metals," PM, [6] 20 (1910); 835–844, on 841–844; J. H. Jeans, "The Motion of Electrons in Solids. Part I. Electric conductivity, Kirchhoff's law and radiation of great wave-length," PM, [6] 17 (1909); 773–794. In 1910 Jeans compared an easily intelligible theory of radiation, which depends on continuous motion and on equipartition of energy, with a quantum theory "to many still unthinkable," while different from Planck's explicit statements assumes not only a system of vibrators with "definite multiples of a fixed unit of energy" but also indivisible "atoms" of energy in the aether. See J. H. Jeans, "On Non-Newtonian Mechanical Systems, and Planck's Theory of Radiation," PM, [6] 20 (1910); 943–954, on 943, 953. See also N. Bohr, "Studier over metallernes elektrontheori [1911]," translated by J. Rud Nielsen as "Studies on the Electron Theory of Metals," BCW, [291]–[393], on [357]–[379].

115. S. B. McLaren, "The Emission and Absorption of Energy by Electrons," PM, [6] 22 (1911); 66–83, on 66–72. Although McLaren was inclined to abandon "classical dynamics" he was not "prepared", however, to accept "Einstein's atomism" for "radiation." See S. B. McLaren, "The Theory of Radiation," PM, [6] 25 (1913); 43–56, on 44.

116. M. Planck to H. A. Lorentz, 1 April 1908, LTZ; a part of the letter is published with altered spelling in Armin Hermann, *Frühgeschichte der Quantentheorie (1899–1913)* (Mosbach, 1969); on 47–48. See also J. H. Jeans, "On the partition of energy between matter and ether," PM, [6] 10 (1905); 91–97.

117. See Hans Kangro, *Early History of Planck's Radiation Law* (1976), 75–77, 159, 168, 171, 175.

118. H. A. Lorentz to W. Wien, 6 June 1908, see Armin Hermann, *Frühgeschichte der Quantentheorie (1899–1913)* (Mosbach, 1969), on 46. See also W. Wien, "Theorie der Strahlung [1909]," EMW, 5,3 (Leipzig, 1909–1926); 282–357, on 326–333. See also M. Reinganum to H. A. Lorentz, 16 January 1910; G. H. Livens to H. A. Lorentz, 9 July 1914; LTZ. See O. Sackur, "Die universelle Bedeutung des sog.[nannten] elementaren Wirkungsquantums," AP, 40 (1913); 67–86, on 85. See C. W. Oseen, "Zur Kritik der Elektronentheorie der Metalle," AP, 49 (1916); 71–84; and C. W. Oseen to H. A. Lorentz, 13 January 1916, LTZ. For the attempts of Cambridge physicists, among others J. J. Thomson, to reconcile electron gas theory with the radiation law see John L. Heilbron and Thomas S. Kuhn, "The Genesis of the Bohr Atom," HSPS, 1 (1969); 211–290, on 217.

119. M. Planck to W. Wien, 27 February 1909, WWN, no. 34. In 1905 Jeans demonstrated that in a system consisting of a gas and of radiation, which is "enclosed by an ideal perfectly reflecting boundary," the "energy accumulates in the ether," that is, that there is no radiation equilibrium. See J. H. Jeans, "On the Partition of Energy between Matter and Ether," PM, [6] 10 (1905); 91–97, on 97. For Jeans's "stubborn behavior" see also J. H. Jeans, "The Motion of Electrons in Solids. Part I. Electric conductivity, Kirchhoff's Law and Radiation of Great Wave-Length," PM, [6] 17 (1909); 773–794; J. H. Jeans, "On Non-Newtonian Mechanical Systems, and Planck's Theory of Radiation," PM, [6] 20 (1910); 943–954.

120. See Armin Hermann, *Frühgeschichte der Quantentheorie (1899–1913)* (Mosbach, 1969); Thomas S. Kuhn, *Black-Body Theory and Quantum Discontinuity, 1894–1912*, (Chicago and London: University of Chicago Press, 1987), 235–244.

121. At first, Planck considered only the stronger bound electrons. Probably their characteristic oscillations, which could be described in terms of quantum theory, would guarantee an equilibrium between matter and radiation. See M. Planck to H. A. Lorentz, 21 November 1908, 10 July 1909, LTZ. See also M. Planck to W. Wien, 14 January 1911, WWN, no. 53. For Planck's second theory see M. Planck, "Eine neue Strahlungshypothese," Verhandlungen der Deutschen Physikalischen Gesellschaft, 13 (1911); 138–148; "Zur Hypothese der Quantenemission," SB (1911); 723–731. Thomas S. Kuhn, *Black-Body Theory and Quantum Discontinuity, 1894–1912,* (Chicago and London: University of Chicago Press, 1987), 235–244. In 1913 Planck discussed a thermodynamic equilibrium between "oscillators" and radiation that is mediated by free electrons. See M. Planck, "Über das Gleichgewicht zwischen Oszillatoren, freien Elektronen und strahlender Wärme," SB (1913); 350–363.

122. W. Wien, "Zur Theorie der elektrischen Leitung in Metallen," SB (1913); 184–200, on 186.

123. H. A. Lorentz, "Anwendung der kinetischen Theorien auf Elektronenbewegung [1913]," LCP, 8 (The Hague, 1935); 214–243, on 214. For the Wolfskehl lectures see David Hilbert (ed.), *Vorträge über die kinetische Theorie der Materie und der Elektrizität. Gehal-*

ten auf Einladung . . . der Wolfskehlstiftung [1913]. = Mathematische Vorlesungen an der Universität Göttingen: VI (Leipzig, Berlin, 1914).

124. H. A. Lorentz, "Anwendung der kinetischen Theorien auf Elektronenbewegung [1913]," LCP, 8 (The Hague, 1935); 214–243, on 216.

125. See M. Planck, Über das Gleichgewicht zwischen Oszillatoren, freien Elektronen und strahlender Wärme," SB (1913); 350–363, on 352. Arguing against W. Wien's abandonment of free electrons, Planck referred to super conductivity which indicated enormous mean free paths of metal electrons. See M. Planck to W. Wien, 10 January 1913, WWN, no. 65.

126. See J. H. Jeans, "The Motion of Electrons in Solids. Part I. Electric conductivity, Kirchhoff's Law and Radiation of Great Wave-Length," PM, [6] 17 (1909); 773–794, on 792; S. B. McLaren, "The Emission and Absorption of Energy by Electrons," PM, [6] 22 (1911); 66–83, on 68.

127. Paul and Tatiana Ehrenfest, "Begriffliche Grundlagen der statistischen Auffassung in der Mechanik [1911]," EMW, 4,4 (Leipzig, 1907–1914), article no. 32, 3–90, on 75. Preparing the final version of the article, P. Ehrenfest had asked O. W. Richardson for reprints of his papers on thermionics. See P. Ehrenfest to O. W. Richardson, 24 May 1911 [probably old style], RDN, R–000 402.

128. C. G. Darwin, "The Theory of Radiation [August 1912]," AHQP, Microf. 36, 2; in no. 13.

129. Kuhn, *Black-Body,* 184–185.

130. H. A. Lorentz, "Anwendung der kinetischen Theorien auf Elektronenbewegung [1913]," LCP, 8 (The Hague, 1935); 214–243, on 215.

131. See M. Planck to H. A. Lorentz, 21 November 1908, 10 July 1909, LTZ. See also M. Planck to W. Wien, 14 January 1911, WWN, no. 53.

132. See Rudolf Seeliger, "Elektronentheorie der Metalle [1921]," EMW, 5,2 (Leipzig, 1904–1922); 777–878, on 862–863. See also Per F. Dahl, "Kamerlingh Onnes and the Discovery of Superconductivity: The Leyden years, 1911–1914," HSPS, 15 (1984), 1–37, on 36.

133. See Per F. Dahl, "Superconductivity after World War I and Circumstances Surrounding the Discovery of a State B = 0," HSPS, 16 (1986); 1–58, on 6.

134. P. Ehrenfest to H. A. Lorentz, 5 February 1922 [the letter is erroneously dated 1912], LTZ. For similar ideas see P. W. Bridgman, "The Electrical Resistance of Metals," PR, 17 (1921); 161–194. E. Kretschmann discusses in this context also theories of Kr. Höjendahl, J. Frenkel, and of A. Wolf. See Erich Kretschmann, "Kritischer Bericht über neue Elektronentheorien der Elektrizitäts- und Wärmeleitung in Metallen," PZ, 28 (1927); 565–592, on 572–576.

135. Michael Eckert, "Propaganda in Science: Sommerfeld and the Spread of the Electron Theory of Metals," in HSPS, 17 (1986); 191–233.

136. See O. W. Richardson to Edwin Herbert Hall, 13 October 1914, RDN, L–000 121; see also O. W. Richardson, "The Distribution of the Molecules of a Gas in a Field of Force, with Applications to the Theory of Electrons," PM, [6] 28 (1914); 633–647.

137. E. H. Hall to O. W. Richardson, 8 December 1913, RDN, R–000 537. For Stark's theory of a combined atom and electron space lattice, see J. Stark, "Folgerungen aus einer Valenzhypothese. II. Metallische Leitung der Elektrizität," JRE, 9 (1912); 188–203. See also Rudolf Seeliger, "Elektronentheorie der Metalle [1921]," EMW, 5,2 (Leipzig, 1904–1922); 777–878, on 859–860. See further W. Wien, "Zur Theorie der elektrischen Leitung in Metallen," SB (1913); 184–200.

138. For an analysis of the early history of the Hall effect, see Jed Z. Buchwald, "The Hall effect and Maxwellian electrodynamics in the 1880's," Centaurus, 23 (1979); 51–99, 118–162; see also Jed Z. Buchwald, *From Maxwell to Microphysics* (Chicago: Chicago University Press, 1985), 141–173, on 73–108.

139. For Maxwell's suggestion, see J.Cl. Maxwell, *A Treatise on Electricity and Magnetism*, vol. 1–2 (3rd. ed., Oxford, 1904, no. 297–303, 418–423. See also Jed Z. Buchwald, "The Hall Effect and Maxwellian Electrodynamics in the 1880's," Centaurus, 23 (1979); 51–99, 118–162, on 91–93.

140. See H. A. Lorentz, "Le phénomène découverte par Hall et la rotation électromagnétique du plan de polarisation de la lumière," LCP, 2 (The Hague, 1936); 136–163.

141. L. Boltzmann, "Zur Theorie des von Hall entdeckten elektromagnetischen Phänomens," SW, 94 (1886); 644–669, on 645–647; also in L. Boltzmann, *Wissenschaftliche Abhandlungen,* 3 vols. [1909] (reprint, New York,1968), in vol. 3, 187–211.

142. E. H. Hall, "On Boltzmann's Method for Determining the Velocity of an Electric Current," PM, [5] 10 (1880); 136–138, on 137–138. Hall's second objection that, contrary to the primary electric field, the Hall electric field gives rise to ponderomotive forces (that is, to "technical" magnetic forces) was clarified only in terms of F. Bloch's lattice theory of conduction. See Norman Rostoker, "Hall Effect and Ponderomotive Force in Simple Metals," American Journal of Physics, 20 (1952): 100–107.

143. E. Riecke, "Zur Theorie des Galvanismus und der Wärme," AP, 66 (1898); 353–389, 545–581, on 368, 563; see also E. Riecke, "Nachtrag zu der Abhandlung: 'Zur Theorie des Galvanismus und der Wärme'," AP, 66 (1898); 1199–1200.

144. P. Drude, "Zur Elektronentheorie der Metalle," AP, 1 (1900); 566–613; AP, 3 (1900); 369–402, on 576; 374.

145. P. Drude, "Zur Elektronentheorie der Metalle," AP, 1 (1900); 566–613; AP, 3 (1900); 369–402, on 579; 374.

146. H. A. Lorentz, "Le mouvement des électrons dans les métaux," Archives Néerlandaises des Sciences exactes et naturelles, [2] 10 (1905) ; 336–371; reprinted in LCP, 3 (The Hague, 1936); 180–214, on 206–208. See also H. A. Lorentz, "Ergebnisse und Probleme der Elektronentheorie [1904]," LCP, 8 (1935); 76–124, on 117; H. A. Lorentz, "Anwendung der kinetischen Theorien auf Elektronenbewegung [1913]," LCP, 8 (The Hague, 1935); 214–243, on 217.

147. K. Baedeker, "Über die elektrische Leitfähigkeit und die thermoelektrische Kraft einiger Schwermetallverbindungen," AP, 22 (1907); 749–766, on 756–758, 765–766.

148. K. Baedeker and E. Pauli, "Das elektrische Leitvermögen von festem Kupferjodür," PZ, 9 (1908); 431.

149. See K. Baedeker, "Über eine eigentümliche Form elektrischen Leitvermögens," PZ, 9 (1908); 431–433.

150. See K. Baedeker, "Über eine eigentümliche Form elektrischen Leitvermögens bei festen Körpern," AP, 29 (1909); 566–584, on 566, 574–575.

151. See K. Baedeker, "Über eine eigentümliche Form elektrischen Leitvermögens bei festen Körpern," AP, 29 (1909); 566–584, on 581–583. See also K. Baedeker, "Eine direkte Prüfung der Elektronentheorie der Thermoelektrizität," Festschrift. Walther Nernst zu seinem fünfundzwanzigsten Doktorjubiläum gewidmet von seinen Schülern (Halle, 1912); 62–67.

152. See See K. Baedeker, "Über eine eigentümliche Form elektrischen Leitvermögens bei festen Körpern," AP, 29 (1909); 566–584, on 580.

153. K. Steinberg, "Über den Halleffekt bei jodhaltigem Kupferjodür," AP, 35 (1911); 1009–1033, on 1030.

154. K. Steinberg, "Über den Halleffekt bei jodhaltigem Kupferjodür," AP, 35 (1911); 1009–1033, on 1026–1030.

155. K. Baedeker, "Künstliche metallische Leiter," PZ, 13 (1912), 1080–1082, on 1081.

156. See K. Baedeker, "Eine direkte Prüfung der Elektronentheorie der Thermoelektrizität," Festschrift. Walther Nernst zu seinem fünfundzwanzigsten Doktorjubiläum gewidmet von seinen Schülern (Halle, 1912); 62–67. See also K. Baedeker, *Die elektrischen Erscheinungen in metallischen Leitern* (Braunschweig, 1911); on 10–14, 77–94, 123.

157. K. Steinberg, "Über den Halleffekt bei jodhaltigem Kupferjodür," AP, 35 (1911); 1009–1033, on 1030.

158. See K. Baedeker, *Die elektrischen Erscheinungen in metallischen Leitern* (Braunschweig, 1911); on 122–123, citation on 122.

159. B. H. Vine and R. J. Maurer, "The Electrical Properties of Cuprous Iodide," ZPC, 198 (1951); 147–156, on 155.

160. N. Bohr, "Studier over metallernes elektrontheori [1911]," translated by J. Rud Nielsen as "Studies on the Electron Theory of Metals," BCW, [291]–[393], on [380]–[383]. See John L. Heilbron and Thomas S. Kuhn, "The genesis of the Bohr atom," HSPS, 1 (1969), 211–290, on 218–223. See also Ulrich Hoyer, *Die Geschichte der Bohrschen Atomtheorie* (Weinheim, 1974); 28–32. Ch.G. Darwin struggled with Bohr's result that free electrons do not explain the magnetic properties of metals in letters to Bohr, on 15 October 1913, and on 19 October 1913, BSC 1, 4.

161. E. Riecke, "Die jetzigen Anschauungen über das Wesen des metallischen Zustandes [Lecture . . . 1908]," PZ, 10 (1909); 508–518, on 509, on 517.

162. See E. Riecke, "Elektronentheorie galvanischer Eigenschaften der Metalle [1913]," in Erich Marx (Ed.), Handbuch der Radiologie, Vol. VI (Leipzig,1925), = Die Theorien der Radiologie; 281–494, on 431, see also on 286. Riecke's article was posthumously published by Max von Laue in 1925. Therefore, it reflects, to some extent, the opinion of the publisher, Max von Laue, as well.

163. J. Koenigsberger and G. Gottstein, "Über den Halleffekt," PZ, 14 (1913); 232–237.

164. N. Bohr, "Studier over metallernes elektrontheori [1911]," translated by J. Rud Nielsen as "Studies on the electron theory of metals," BCW, [291]–[393], on [383]–[395]. See also G. H. Livens, "The Electron Theory of the Hall Effect and Allied Phenomena," PM, [6] 30 (1915); 526–548, on 545. Obviously Livens had borrowed an English translation of Bohr's dissertation, see 529.

165. E. H. Hall, "Illustrations of the Dual Theory of Metallic Conduction," PR, 28 (1926); 392–417. See also L. L. Campbell, *Galvanomagnetic and Thermomagnetic Effects. The Hall and Allied Phenomena* (New York and elsewhere, 1923), on 89–91; W. Hume-Rothery, *The Metallic State. Electrical Properties and Theories* (Oxford, 1931), on 223–229. See also E. H. Hall to H. A. Lorentz, 13 November 1922, LTZ.

166. Jürgen Renn, "Einstein's Controversy with Drude and the Origin of Statistical Mechanics: A New Glimpse from the 'Love Letters'," Preprint 55, Max Planck Institute for the History of Science, Berlin 1997, 10–16, 20–22. Jürgen Renn, "Einstein's Controversy with Drude and the Origin of Statistical Mechanics: A New Glimpse from the 'Love Letters'," Archive for History of Exact Sciences, 51 (1997); 315–354. An abbreviated version is Jürgen Renn, "Die Geburt der Statistischen Mechanik aus dem Geist der Elektronentheorie der Metalle," Physikalische Blätter, 53 (1997); 860–862.

167. A. Einstein, "Die Planck'sche Theorie der Strahlung und die Theorie der spezifischen Wärme," AP, 22 (1907); 180–190, on 800 a correction; P. Debye, "Zur Theorie der spezifischen Wärmen," AP, 39 (1912); 789–839.

168. See Rudolf Seeliger, "Elektronentheorie der Metalle [1921]," EMW, 5, 2 (Leipzig, 1904–1922), 777–878, 864–865. For Kamerlingh Onnes see Per F. Dahl, "Kamerlingh Onnes and the discovery of superconductivity: The Leyden years, 1911–1914," HSPS, 15 (1984); 1–37, on 12–13. See also W. Nernst, "Untersuchungen über die spezifische Wärme bei tiefen Temperaturen. III.," SB (1911); 306–315, on 311–315. Nernst used, Planck's radiation formula, however, only ad hoc to calculate the electric resistance for low temperatures. For an elaborate physical theory see F. A. Lindemann, "Untersuchungen über die spezifische Wärme bei tiefen Temperaturen. IV.," SB (1911); 316–321, on 318–319. A similar theory, which started with the quantum theory of the energy of the electron, was Karl F. Herzfeld, "Zur Elektronentheorie der Metalle," AP, 41 (1913); 27–52; see also K. F. Herzfeld to H. A. Lorentz, LTZ. See further W. Wien, "Zur Theorie der elektrischen Leitung in Metallen," SB (1913); 184–200; P. W. Bridgman, "The Electrical Resistance of Metals," PR, 17 (1921); 161–194, on 163.

169. W. Nernst, "Über neuere Probleme der Wärmetheorie," SB (1911); 65–90, on 86.

170. H. Tetrode, "Bemerkungen über den Energiegehalt einatomiger Gase und über die Quantentheorie für Flüssigkeiten," PZ, 14 (1913); 212–215.

171. W. H. Keesom, "Über die Zustandsgleichung eines idealen einatomigen Gases nach der Quantentheorie," PZ, 14 (1913); 665–670.

172. W. Nernst, "Der Energieinhalt fester Stoffe," AP, 36 (1911); 395–439, on 435.

173. H. Tetrode, "Bemerkungen über den Energiegehalt einatomiger Gase und über die Quantentheorie für Flüssigkeiten," PZ, 14 (1913); 212–215, on 214. Tetrode did not allow, however, for a complete analogy of an ideal gas with conduction electrons, ibid.

174. Thomas S. Kuhn, *Black-Body Theory and Quantum Discontinuity, 1894–1912,* (Chicago and London: University of Chicago Press, 1987), 184–185.

175. W. H. Keesom, "Zur Theorie der freien Elektronen in Metallen," PZ 14 (1913); 670–675, on 671. Preliminary remarks were made during the Wolfskehl conference in Göttingen. See David Hilbert (ed.), *Vorträge über die kinetische Theorie der Materie und der Elektrizität. Gehalten auf Einladung . . . der Wolfskehlstiftung* [1913]. = Mathematische Vorlesungen an der Universität Göttingen: VI (Leipzig, Berlin, 1914), on 193–196. After the Wolfskehl conference Keesom discussed his degenerate gas theory with H. A. Lorentz as well as with A. Sommerfeld (Sommerfeld contributed his coworker Wilhelm Lenz's paper on quantum theory of an ideal gas to the conference). See W. H. Keesom to Lorentz, 29 April 1913, LTZ; W. H. Keesom to A. Sommerfeld, 29 April 1913, AHQP, Microf. 31,11. Although Keesom, with his introduction of the zero-point energy, could rely both on Planck and on Einstein, Einstein himself questioned the zero point energy referring to Kamerlingh Onnes's discovery of superconductivity. See A. Einstein to H. A. Lorentz, 2 August 1915, LTZ. W. Schottky argued against Keesom's assumption of a concentration of one electron per hundred metal atoms. Schottky found it therefore difficult to restrict the consideration of the "heat movement" to the electrons instead of including the metal atoms as well. See W. Schottky, "Bericht über thermische Elektronenemission," JRE, 12 (1915); 147–205, on 188. For the "somewhat tentative character" see also W. Nernst, *Die theoretischen und experimentellen Grundlagen des neuen Wärmesatzes* (Halle, 1918); 163–171.

176. An indication are the lengthy overview articles: See, e.g., O. W. Richardson, "Glühelektroden," in E. Marx (ed.), Handbuch der Radiologie, vol. IV (Leipzig, 1917); 445–602; W. Schottky, "Bericht über thermische Elektronenemission," JRE, 12 (1915); 147–205; Irving Langmuir, "The Pure Electron Discharge and its Applications in Radio Telegraph and Telephony [1915]," The collected works, vol. 3 (Oxford and elsewhere, 1961); 38–58; E. Riecke, "Elektronentheorie galvanischer Eigenschaften der Metalle [1913], in Erich Marx (ed.), Handbuch der Radiologie, vol. VI (Leipzig, 1925), = Die Theorien der Radiologie; 281–494.

177. P. W. Bridgman, "The Electrical Resistance of Metals," PR, 17 (1921); 161–194. See also Erich Kretschmann, "Kritischer Bericht über neue Elektronentheorien der Elektrizitäts- und Wärmeleitung in Metallen," PZ, 28 (1927); 565–592, on 567.

178. In 1909 the metallurgist William M. Guertler expressed his hope that it might be possible to repeat the success of the theory of electrolysis in the theory of conductivity of metals and alloys. See W. Guertler, "Stand der Forschung über die elektrische Leitfähigkeit der kristallisierten Metalllegierungen," JRE, 5 (1908); 17–81, on 17–18. The physicist and metallurgist Carl Benedicks traced the shortcomings of the electron gas theory, however, precisely back to the transfer of a theoretical model into a completely different field. See C. Benedicks, "Beiträge zur Kenntnis der Elektrizitätsleitung in Metallen und Legierun-

gen," JRE, 13 (1916); 351–395,. See further W. Guertler, "Beiträge zur Kenntnis der Elektrizitätsleitung in Metallen und Legierungen," JRE, 17 (1920); 276–292.

179. For the interaction of electron theory and technology during World War I, see Irving Langmuir, "The Pure Electron Discharge and its Applications in Radio Telegraph and Telephony [1915]," The collected works, vol. 3 (Oxford and elsewhere, 1961), 38–58. During the war M.von Laue worked with W. Wien in Würzburg on developing electronic amplifying tubes for improving the army's communication techniques. See A. Hermann, "Max von Laue," in Ch.C. Gillispie (ed.), DSB, 8 (New York, 1973); 50–53, on 51. O. W. Richardson, then at the University of London, King's College, held close contact to the Signal School of the Admiralty at Portsmouth. See Eduard Watson to O. W. Richardson, 14 June 1918, RDN, R–000 529. H. A. Lorentz was in contact with coworkers of Philips in Eindhoven, who dealt with vacuum tubes for wireless telegraphy. See E. Oosterhuis and G. Holst to H. A. Lorentz, 13 June 1918, LTZ. For the stimulation by the war see also [Amtliche Denkschrift der Reichspost] "Das deutsche Telegraphen-, Fernsprech- und Funkwesen 1899–1924" with a supplement "Die deutsche Telegraphie im Weltkrieg" (Berlin, 1925), on 7, [supplement] on 26, 36–37.

180. See letters, among others, of Marcel Brillouin, Paul Ehrenfest, Albert Einstein, Max Planck, Eduard Riecke, Ernest Rutherford, and Woldemar Voigt, LTZ, 1914–1918. For Planck's signature under the notorious "Appeal to the Cultured People of the World," "[Aufruf] an die Kulturwelt," see John L. Heilbron, "The Dilemmas of an Upright Man: Max Planck as Spokesman of German Science" (Berkeley, 1986), 70–79. See, however, also M. Planck to H. A. Lorentz, 3 December 1923, LTZ. In this letter Planck refuted to withdraw his signature under the "[Aufruf] an die Kulturwelt."

181. See Rudolf Seeliger, "Elektronentheorie der Metalle [1921]," EMW, 5, 2 (Leipzig, 1904–1922), 777–878; Erich Kretschmann, "Kritischer Bericht über neue Elektronentheorien der Elektrizitäts- und Wärmeleitung in Metallen," PZ, 28 (1927); 565–592; E. Riecke, "Elektronentheorie galvanischer Eigenschaften der Metalle [1913]," in Erich Marx (ed.), Handbuch der Radiologie, Vol. VI (Leipzig,1925), = Die Theorien der Radiologie; 281–494; O. W. Richardson, "Glühelektroden," in E. Marx (ed.), Handbuch der Radiologie, vol. IV (Leipzig,1917); 445–602; W. Schottky, "Bericht über thermische Elektronenemission," JRE, 12 (1915); 147–205.

182. See Rudolf Seeliger, "Elektronentheorie der Metalle [1921]," EMW, 5, 2 (Leipzig, 1904–1922); 777–878. For Stark's theory see J. Stark, "Folgerungen aus einer Valenzhypothese. II. Metallische Leitung der Elektrizität," JRE, 9 (1912); 188–203. For the lack of experimental evidence see E. Riecke, "Elektronentheorie galvanischer Eigenschaften der Metalle [1913]," in Erich Marx (ed.), Handbuch der Radiologie, vol. VI (Leipzig,1925), = Die Theorien der Radiologie; 281–494; on 491–494, in a paragraph on the lattice theories of metallic conduction, the article's publisher, M. von Laue, stated that the electron space lattice does not make x-ray diffraction. F. A. Lindemann had calculated the (small) specific heat of the electron space lattice to remove the "chief stumbling-block of the old [electron gas] theory." See F. A. Lindemann, "Note on the theory of the metallic state," PM, [6] 29 (1915); 127–140. Lindemann found additional support for his electron space-lattice theory in the "very large latent heat of emission" of electrons from tungsten which "seems to preclude the idea that the free electrons form a gas inside the metal." See F. A. Lindemann to O. W. Richardson, 30 June 1915, RDN, R–000 753.

183. See Lillian H. Hoddeson and G. Baym, "The Development of the Quantum Mechanical Electron Theory of Metals: 1900–28," Proceedings of the Royal Society of London [A], 371 (1980); 8–23, on 11–20; Lillian Hoddeson, Gordon Baym, and Michael Eckert, "The Development of the Quantum Mechanical Electron Theory of Metals, 1926–1933," in: Lillian Hoddeson, Ernest Braun, Jürgen Teichmann, Spencer Weart, *Out of the Crystal Maze, Chapters from the History of Solid-State Physics* (New York, Oxford: Oxford University Press, 1992).

184. R. H. Fowler, "On Dense Matter," Monthly Notices of the Royal Astronomical Society, 87, 2 (1926), 114–122. For Fowler's chagrin, see E. A. Milne, "Ralph Howard Fowler," obituary notice, ibid., 105, 2 (1945); 80–87, on 86.

185. W. Pauli, "Über Gasentartung und Paramagnetismus," ZSP, 41 (1927); 81–102.

186. See Michael Eckert, "Propaganda in Science: Sommerfeld and the Spread of the Electron Theory of Metals," in HSPS, 17 (1986); 191–233.

187. A. Sommerfeld, "Zur Elektronentheorie der Metalle," Die Naturwissenschaften, 15 (1927) [No. 41, 14 October 1927]; 825–832; "Zur Elektronentheorie der Metalle auf Grund der Fermischen Statistik," ZSP, 47 (1928); 1–32, 43–60; "Zur Elektronentheorie der Metalle," Die Naturwissenschaften, 16 (1928); 374–381.

188. See A. Einstein to A. Sommerfeld, 9 November 1927, in A. Hermann (ed.), *Albert Einstein/Arnold Sommerfeld. Briefwechsel* (Basel, Stuttgart, 1968), 111–112; Richard Gans to A. Sommerfeld, 25 October 1927, AHQP, Microf. 31,1; A. Rubinowicz to A. Sommerfeld, 28 January 1936, AHQP, Microf. 33,6. Rubinowicz praises the "correct" electron theory of metals given by Sommerfeld. See also Adolf Smekal to Sommerfeld, 17 April 1931, AHQP, Microf. 34,4. In this letter A. Smekal suggests that Sommerfeld, who had "created [geschaffen]" the electron theory of metals, may write an article for Karl Scheel's "Handbuch der Physik." Later, the greater part of the work was done, however, by Hans Bethe. See A. Sommerfeld and H. Bethe, "Elektronentheorie der Metalle," in H. Geiger and K. Scheel (eds.), Handbuch der Physik, 2nd, ed., V=vol. 24, 2 (Berlin, 1933); 333–622.

189. See A. Sommerfeld, "Zur Elektronentheorie der Metalle," Die Naturwissenschaften, 15 (1927) [no. 41, 14 October 1927]; 825–832, on 832; A. Sommerfeld "Zur Elektronentheorie der Metalle auf Grund der Fermischen Statistik," ZSP, 47 (1928); 1–32, 43–60, on 60. See also W. Hume-Rothery, *The Metallic State. Electrical Properties and Theories* (Oxford, 1931), on 292–293.

190. See A. Sommerfeld "Zur Elektronentheorie der Metalle auf Grund der Fermischen Statistik," ZSP, 47 (1928); 1–32, 43–60, ZSP, on 55. Interestingly enough Sommerfeld refers here to K. Steinberg's and to K. Baedeker's Hall effect measurements with doped CuI. I expect that in a theory based on Fermi statistics, Sommerfeld might have discussed Steinberg's Hall effect data for high concentrations of carriers which indicated a degeneracy behavior; however, Sommerfeld only discussed the case of low concentrations. J. J. Frenkel was very outspoken that despite the success of Sommerfeld's electron gas theory it does not give quantitative results in the case of the magnetic phenomena, because it does not include the interaction of electrons and positive ions. See J. J. Frenkel to A. Sommerfeld, AHQP, Microf. 30,14.

191. See W. V. Houston to Sommerfeld, 30 May 1928, 6 July 1928, AHQP, Microf. 31,8. Sommerfeld, at first, considered Houston's theory to be the solution to the problems of the mean free path and of the temperature dependence of the resistance (namely proportional to the absolute temperature at high temperatures). See A. Sommerfeld, "Zur Elektronentheorie der Metalle," Die Naturwissenschaften, 16 (1928); 374–381.

192. F. Bloch, Über die Quantenmechanik der Elektronen in Kristallgittern," ZSP, 52 (1929); 555–600. A reformulation of Bloch's theory, which includes the explanation of electric resistance due to imperfections and admixtures, is L. Nordheim, "Zur Elektronentheorie der Metalle," AP, [5] 9 (1931); 641–678.

193. W. Pauli, "Über den Zusammenhang des Abschlusses der Elektronengruppen im Atom mit der Komplexstruktur der Spektren," ZSP, 31 (1925); 765–783, on 778–779.

194. R. Peierls, "Zur Theorie der galvanomagnetischen Effekte," ZSP, 53 (1929); 255–266. The recollections of Peierls indicate that this "first paper of any importance" he "published" was due to Heisenberg's qualitative understanding of the positive Hall effect in terms of "holes." See interview with R. E. Peierls, 17 June 1963, conducted by J. L. Heilbron, AHQP.

195. W. Heisenberg, "Zum Paulischen Ausschließungsprinzip," AP, [5] 10 (1931); 888–904.

196. A. Sommerfeld and H. Bethe, "Elektronentheorie der Metalle," in H. Geiger and K. Scheel (eds.), Handbuch der Physik, 2nd ed., vol. 24, 2 (Berlin, 1933); 333–622.

197. See Michael Eckert, "Propaganda in Science: Sommerfeld and the Spread of the Electron Theory of Metals," in HSPS, 17 (1986); 191–233.

III

ELECTRONS APPLIED AND APPROPRIATED

9

The Electron and the Nucleus
Laurie M. Brown

Nearly four decades elapsed between the discovery of the electron and the recognition that the electron is only a minor participant in nuclear interactions (at least under terrestrial conditions) and plays no role in nuclear structure. Electron beams have become extremely useful probes of nuclear and nucleon structure precisely because they have no strong nuclear interaction. Whether electrons were parts of the nucleus was not an issue for a decade and a half after 1897 because physicists did not know that the nucleus existed, let alone that it was the seat of radioactivity. When these facts were established, the emission of electrons as nuclear β-rays appeared to confirm their presence in the nucleus and for two decades that was assumed to be the case. Even after the neutron discovery, many leading physicists continued to believe that the electron, the most visible player in weak and electromagnetic interactions, also mediated the nuclear binding force; this unified picture was just too beautiful to abandon.

After the Bohr-Rutherford nuclear atom became established in 1913, together with the concept of atomic number Z, Bohr viewed the atom's external electron cloud as responsible for most of the properties of matter, but he said later that it was "evident" to him that the nucleus served as the unique source of radioactivity, including β-rays and α-particles. Rutherford identified the hydrogen nucleus, or proton, as a nuclear constituent in 1919 by observing its ejection from nitrogen and other light nuclei by α-particles. Thus until about 1933, nuclei were generally thought to be constructed of protons, α-particles, and electrons.

The neutron was discovered in 1932 by James Chadwick, who considered it to be an electron-proton compound. It added to the nuclear mix, but did not eliminate electrons and α-particles as basic nuclear constituents. In 1932–33 Werner Heisenberg wrote a path-breaking three-part paper proposing a so-called *n-p* model. His nucleus also contained electrons, however; some were bound in α-particles and neutrons, but there were "unbound" electrons as well. In late 1933 Enrico Fermi published a theory of β-decay, employing the neutrino, which had been postulated by Wolfgang Pauli.

Heisenberg used the idea of the new theory in constructing a "Fermi-field" nuclear model, involving the exchange of electron-neutrino pairs within the nucleus.

Eventually, the problems associated with nuclear electrons became so severe that they had to be entirely eliminated. This task, essentially accomplished by 1936, required many new ideas, beginning with the recognition that the neutron was just as "elementary" as the proton. It was necessary to accept the reality of Pauli's neutrino and to acknowledge the success of Fermi's theory of β-decay. This also implied the acceptance of microscopic energy, momentum and angular momentum, conservation in elementary interactions (which had been doubted by some physicists, notably Bohr). Moreover, Dirac's "hole in the vacuum," the positron was needed to understand the large probability of interaction of high energy radiation with the nucleus in the absence of nuclear electrons. This was essential for the processes of pair production, bremsstrahlung, and pair annihilation, and to account for cosmic-ray cascade shower production.

Nuclear Electrons to 1932

Until the Rutherford-Bohr-Sommerfeld atom gained acceptance, about 1916, the dominant atomic models were of the type introduced by J. J. Thomson in 1903. They were characterized by concentric rings of electrons, rotating within a sphere of diffuse positive material.[1] These models obviously had no nucleus. In 1904 Hantaro Nagaoka had proposed a so-called Saturnian atom, with coplanar rings of electrons circulating about a central positive body, but this model was soon shown to be mechanically unstable. In contrast to the gravitationally attracting particles in Saturn's rings (from which the model takes its name), the electrons would have repelled each other, producing unstable oscillations. Nagaoka's atom was mentioned in Rutherford's famous paper of 1911 in which he proposed his own nuclear atom, which had the same mechanical flaw.

Niels Bohr's quantized nuclear atom of 1913 treated the nucleus as pointlike, but Bohr felt that it was the source of radioactivity. And if nuclei emitted β-rays, this was a strong suggestion that they contained electrons. One might think that the example of photons being emitted by atoms, which certainly do not "contain" them, would suggest a scheme for a similar production of β-decay electrons, but that was not the case. For even if one *did* allow that photons existed, their status as particles was still highly problematic, so photon emission was not a suitable analogy for β-decay.

In 1919 Rutherford discovered the artificial disintegration of light elements by α-particle bombardment, demonstrating that α's from RaC knock protons out of nitrogen nuclei. This was taken as experimental proof that nuclei contain protons and initiated a new round of nuclear model-building by Rutherford and others.[2] Besides α-particles and electrons, the new models contained protons as well as neutral objects such as the "neutron" or "neutral doublet" that Rutherford proposed in 1920, a kind of collapsed hydrogen atom. Other objects, for example, an α-particle neutralized by electrons, were suggested as nuclear constituents by Lise Meitner and by W. D. Harkins. Stuewer has listed some physics textbooks of the 1920s in English and in German; they all assumed the existence of electrons in the nucleus.[3] In his article *Electron* in the XIV Edition of the Encyclopedia Britannica (1929), Robert Millikan proclaimed that all matter, including nuclear matter, is made of "positive and negative electrons." (By "positive electrons" he meant protons.)

Thus until the discovery of the neutron in 1932, nuclear electrons were taken for granted by almost all physicists and by some for several years after the neutron. Nevertheless, as early as 1926 serious doubts were raised about their theoretical treatment.[4] For example, Fritz Houtermans reviewed the quantum theory of the electron-containing nucleus in 1930 and stated: "Up to now we do not yet know whether quantum mechanics can really explain the processes in atomic nuclei and their structure, or whether again a new physics, so to say a kind of 'super-quantum mechanics' is necessary."[5]

The most serious puzzles connected with having electrons in the nucleus can be placed under four headings:

1. *Spin and statistics:* This difficulty arises acutely in the case of even A, odd Z nuclei (N^{14}, Li^6, etc.). In the nuclear electron picture, these nuclei contain an even number of protons and an odd number of electrons. Thus they must have, according to quantum mechanics, odd half-integral spin and obey Bose-Einstein statistics. Instead they possess integral spin and obey Fermi-Dirac statistics.

2. *Magnetic moments of nuclei:* As measured by the hyperfine structure of atomic spectra, nuclear moments are about three orders of magnitude smaller than those of an electron. This is especially hard to understand with an odd number of nuclear electrons, where at least one electron must be unpaired.

3. *Confinement:* This is a word that I borrowed from current usage concerning quarks to describe the difficulty of keeping electrons in the nucleus. The Heisenberg uncertainty relation requires an electron in the nucleus to be highly relativistic, and such high energies lead to the paradox, first pointed out by Oskar Klein in 1929, that relativistic electrons will tunnel through the nuclear potential (no matter how deep the well) and will escape.

4. *Energy non-conservation:* This refers to the difficulty that the β-decay energy spectrum is continuous, while the excitation spectrum of the nucleus is discrete. If only an electron emerges in β-decay, then the conservation of energy is violated.[6]

We might imagine that most, if not all, of these problems could have been solved by assuming that there were no electrons in the nucleus and instead *inventing* a neutron. In fact, Rutherford did invent one in 1920! None of the above problems, however, would have been solved by Rutherford's composite neutron, which according to the laws of quantum mechanics would be a boson with zero or integer spin and would contain an unpaired electron whose confinement was still unexplainable.

Chadwick's discovery of the neutron in 1932 had a tremendous impact on the theory of nuclear structure,[7] but as Stuewer has noted, ". . . Chadwick saw his discovery of the neutron fundamentally as vindicating Rutherford's 1920 prediction, and specific model, rather than as resolving the contradiction associated with the nuclear electron hypothesis."[8] Chadwick maintained this view at least until May 1933.

Heisenberg's N-P Nuclear Model and the Electron

At a conference on the history of nuclear physics in May 1977,[9] Hans Bethe, the Moses of the subject,[10] attributed the importance of Heisenberg's seminal work of 1932 to its exclusion of electrons as nuclear constituents, saying:[11]

> Heisenberg's paper on the theory of the nucleus was submitted in June and published in July 1932. . . . First, Heisenberg definitely made the point that only by assuming neutrons and protons to be the constituents of the nucleus is it possible to have a quantum mechanics of the nucleus. Therefore, in this essential point he started our modern ideas. . . . He said, in effect "let's describe the nucleus as if the neutron were an elementary particle, and let's worry later about the neutron."

Although he was aware that in 1932 Heisenberg had assumed the neutron to be an electron-proton compound, at the 1977 conference Bethe criticized Eugene Wigner, saying that the latter had blamed Heisenberg (in 1933) "unjustly for assuming that there are still electrons in the nucleus."[12] In a later discussion period Bethe again chided Wigner on this point and Wigner even apologized.[13] Nevertheless, Heisenberg's so-called *n-p* model did contain electrons—not only those bound in neutrons and α-particles but also "free" electrons. Bethe's selective reading of this part of the history of nuclear physics is, unfortunately, not unusual.[14]

Beginning with the picture of a neutron as a strange kind of collapsed hydrogen atom,[15] Heisenberg assumed that the *dominant* attractive nuclear force arises through the exchange of a charge between a neutron and a proton. That is, the compound neutron emits its electron and becomes a proton; the electron is captured by another proton, which transforms into a neutron. There is also an attraction between two neutrons, but with a much smaller force, occurring through exchanging their electrons. Two protons, on the other hand, being elementary charged particles in Heisenberg's model, have only their repulsive Coulomb interaction. The exchange forces are modeled upon those of molecular theory.[16] In this picture, the n-p force is taken to be dominant to explain why the stable lighter nuclei are approximately charge-symmetric, i.e., why they have equal numbers of neutrons and protons. That approximate symmetry is possible only if the n-n and p-p forces, which have opposite sign, are negligible compared to the n-p force.

Heisenberg's model has been discussed previously by historians, most recently by Catherine Carson,[17] and a few differences of emphasis or interpretation appear in these works. For example, what was exchanged in realizing Heisenberg's charge-exchange forces (i.e., only the charge or the whole electron) and what was the significance of this exchange? Another issue is the extent to which Heisenberg's model nucleus did or did not contain electrons.

This discussion begins with a translation of text near the beginning of Heisenberg's part 1:[18]

> For the following considerations it will be assumed that the neutron follows the rules of Fermi statistics and possesses the spin $(1/2)h/2\pi$. This assumption is necessary to explain the statistics of the nitrogen nucleus and agrees with the empirical results on the nuclear moments. If one wished to understand the neutron as composed of proton and electron, one would have to ascribe Bose statistics and zero spin to the electron; but it does not seem expedient to elaborate this picture. Rather, the neutron will be treated as an independent fundamental particle *which, however, under suitable circumstances can split into proton and electron, where presumably the conservation laws for energy and momentum are no longer applicable.*

In explicating this paragraph, Carson argued that Heisenberg's spinless Bose electrons "had effectively been the standard response to the earlier problems of nuclear electrons" and she went on to say: "From this it should be clear that there was nothing terribly tricky going on with Heisenberg's treatment of the electrons." Heisenberg spent more than 50 percent of parts II and III of his paper, however, on the structure of the neutron and on nuclear electrons. If nuclear electrons were "standard," why was he so concerned about them? To regard the compound neutron as an elementary particle to make quantum mechanics work in the nucleus appears *terribly tricky*. When the neutron splits "under suitable circumstances" into electron and proton, as in β-decay, these are the ordinary observed spin-particles, but it is the "spinless Bose electron" that is exchanged to make the *n-p* and *n-n* forces happen.

The interaction terms of Heisenberg's Hamiltonian give attractions for the *n-p* and *n-n* pairs. Thus $J(r_{kl})$ is the attractive interaction energy between the kth neutron and the lth proton at the distance r_{kl}; similarly $K(r_{kl})$ is the attractive energy for two neutrons.[19] In the case of the simplest nucleus, the deuteron, only the space-exchange integral *(Platzwechselintegral)* $J(r)$ is present, analogous to the case of the H_2^+ molecular ion.[20] The deuteron's single electron has a probability distribution that is *constant* in time and it is shared equally between two protons, as in the molecular ion. For this reason, Heisenberg warned against trying to visualize the process in terms of *electron motions*. A similar situation prevails for the exchange interaction of two neutrons, which is viewed as analogous to the homopolar bonding of the hydrogen molecule H_2.

The Heisenberg effective Hamiltonian does not contain any electron coordinates since all have been "integrated out" in the integrals *J* and *K*. Electrons obviously play an important *dynamical* role, but one might nevertheless assume that all the electrons have been "shoved into" neutrons, although they may reappear when sufficient excitation energy is available, for example, in β-decay or in interactions of the nucleus with high energy γ-rays or electrons. Under these "suitable circumstances [the neutron] can split into proton and electron." While this still left unsolved all the problems of nuclear electrons, such as conservation laws in β-decay, and so forth, one might still think the model was a satisfactory provisional, if not "standard," contribution to the understanding of nuclear systematics. The suppressed electrons were not the only ones, however, that Heisenberg needed; he also required a fair number of "free" electrons. In the next section we shall show that Heisenberg *did* believe they were necessary.

HEISENBERG'S *N-P* NUCLEAR MODEL AND ITS "FREE ELECTRONS"

Recall Bethe chiding Wigner in 1977 for having written four decades earlier that Heisenberg considered only protons and electrons to be elementary. Recall Joan Bromberg's pioneering study in which she wrote that to Heisenberg "the neutron seemed to be elementary as well as complex."[21] Carson stated that in the *n-p* model "protons and neutrons could handle the most troublesome nuclear phenomena [e.g., β-decay]," but added: "Heisenberg did not, however, eliminate the electrons for good, but merely buried them inside the neutrons."[22] Finally, in Brink 1965, a "selected readings" volume on the history of nuclear physics, an early sentence of part I of Heisenberg's *n-p* trilogy is translated as: "This [the neutron discovery] suggests that atomic nuclei are composed of protons and neutrons *but do not contain any electrons.*" The phrase we have emphasized is a translation of "ohne Mitwirkung von Elektronen," which means literally "without the participation of electrons." There appears to have been a tacit agreement among historians that Heisenberg's model nucleus contained electrons only *inside* neutrons and alpha particles, but did not contain any *free nuclear* electrons.

Some excerpts are now offered, one from each part of Heisenberg's three-part paper, to show that he had electrons, both bound and free; later I will explain why he needed both. Here is the first selection from the last paragraph of part I (after showing that in *nonrelativistic approximation* the neutron can be taken as a static structure):[23]

> However one must realize that there are other physical phenomena for which the neutron cannot be considered a static system (*statisches Gebild*) and of which [our Hamiltonian] can give no account. To these phenomena belong, e.g., the Meitner-Hupfeld effect, the scattering of γ-rays on nuclei. Likewise, to this class belong all experiments in which the neutrons can be split into protons and electrons; an example is the slowing of cosmic-ray electrons in passing through nuclei. For this one must investigate more precisely the fundamental difficulties that appear in the continuous β-spectra.

Part II has three sections: the first deals with nuclear systematics, the second with scattering of γ-rays on nuclei, the third with the structure of the neutron as a composite system.[24] The second section noted that the γ-ray scattering could come from motions of protons and neutrons produced by the external radiation, or else:[25]

> Secondly, an individual neutron, that is, the *negative charge bound in it,* could be excited by the incident radiation to emit Rayleigh or Raman scattered radiation . . . and on account of the small electron mass it is appreciably more intense than the scattered radiation of the first kind.

Part III discussed a new possibility—that the neutron might be an elementary particle, as Dmitri Iwanenko had suggested (Iwanenko 1932), a possibility "not excluded by experiment." In that case, said Heisenberg:[26]

> the assumption that the neutron as a heavy elementary particle cannot participate appreciably in the scattering leads to the following consequence: On account of the approximate proportionality of the scattered intensity with the square of atomic number, the α-particles in the nucleus must be built up of protons and electrons (*not protons and neutrons*) and the electrons

bound in the α-particles, in spite of their large binding energy, must contribute to the γ-ray scattering at least as much as the *free nuclear electrons* [*freien Kernelektronen*].

Thus one has this paradoxical result: treating the neutron as an elementary particle in the nucleus meant that it could not be used to construct even the most stable nuclear structure, namely the α-particle, which must be built up instead from protons and electrons; and indeed, far from banishing electrons from the nucleus, additional *free* electrons would be required to understand the cosmic ray experiments.

Why was Heisenberg not willing simply to take the neutron to be the "neutral proton," as (according to Emilio Segrè) Ettore Majorana pronounced it after reading the paper of Joliot and Curie that inspired Chadwick's discovery?[27] Was it excessive stubbornness (perhaps even obsession) on Heisenberg's part to insist on free nuclear electrons? The answer is that Heisenberg was trying to explain more than the table of nuclides with his *n-p* model—especially, the interactions of high-energy cosmic rays with nuclei. He had been working on that very subject at the time of the neutron discovery. The next section summarizes these investigations, gives Heisenberg's conclusions, and discusses their relevance to electrons in nuclei.

The Role of Cosmic-Ray Multiplicity in Heisenberg's Thinking about Nuclear Electrons

On 13 February 1932, in the same month as the neutron discovery, *Annalen der Physik* received a paper from Heisenberg on cosmic ray phenomena, the first of nine papers that he would publish on this subject in the 1930s. Entitled "Theoretical considerations on cosmic radiation" (Heisenberg 1932a), the paper analyzed five kinds of cosmic ray experiments "from the point of view of existing theories" to determine "at which points the experiments roughly agree with the theoretical expectation, and where such great deviations show up that one has to be prepared for important surprises."[28]

The primary cosmic rays entering the atmosphere were considered at this time to be high-energy photons. Referred to as γ-rays, they ostensibly produced the high-energy electrons and secondary γ-rays observed at lower altitudes by interacting with the atmospheric matter. The electrons lost energy by atomic and nuclear collisions, while the principal interaction of the γ-rays was thought to be Compton scattering.[29] This last process was described by the famous Klein-Nishina formula (K-N), which was based upon Dirac's relativistic electron theory.[30] Laboratory experiments using x-rays and γ-rays from radioactivity had confirmed the high accuracy of the K-N formula up to the energy of the ThC″ γ-ray, namely 2.6 MeV, for almost all of the elements tested. For the heaviest elements at the highest energy, however, the formula failed. This anomaly, found by several workers, became known as the Meitner-Hupfeld effect. As it appeared to arise from the nucleus, it was not considered to be a defect of the K-N formula, which had assumed free electrons.

The theory of slowing or "stopping" of charged particles had been worked out classically by Bohr in 1915 and modified quantum mechanically at the beginning of the 1930s by Hans Bethe and by Felix Bloch. The first section of Heisenberg's cosmic ray paper dealt with the slowing of electrons by ionization and excitation. After reviewing the theory, Heisenberg's analysis proceeded as follows: "In an atom of charge Z_a [i.e., atomic number Z], in which Z_k electrons sit in the nucleus, we have to distinguish practically three kinds of electrons."[31]

He then presented a table giving the number of "shell-electrons," namely Z_a, the number of electrons in α-particles, $(Z_k + Z_a)/2$, and the number of "free" electrons, $(Z_k - Z_a)/2$. The table also gave effective ionization energies, respectively: $Z_a Ry$, $30mc^2$, and $2mc^2$. He obtained the stopping distances (ranges) of electrons of various energies in water and lead by adding the contributions from the three types of electrons, and he concluded from agreement with the observations that the nuclear electrons contributed about 20 percent to the stopping.

Taken alone, that was not very compelling evidence for nuclear electrons, but the situation on the absorption and scattering of hard γ-rays was far more arresting. To understand why, consider the attenuation of a beam of γ-rays in matter. In early 1932, the fundamental interactions of high-energy electromagnetic radiation were considered to be the photoelectric effect and Compton scattering. The photoelectric effect is important at optical and x-ray frequencies but falls off rapidly with energy and is not important for cosmic-rays.

Scattering of electromagnetic radiation from charged particles can be conceptualized as a two step process: The electric field, or equivalently the changing magnetic field, of the radiation accelerates the charged particle which in turn acts as an antenna and reradiates its energy. The radiated power, given by the Larmor formula, is proportional to the square of the particle's acceleration times its charge. By Newton's Second Law, the acceleration is proportional to the particle's charge and inversely to its mass. The power radiated by a charge q of mass m is therefore proportional to $(q^2/mc^2)^2$; it follows that the radiation due to electrons is about 4×10^6 as strong as that due to either protons or α-particles, for a given electromagnetic stimulus.[32]

Although we have used classical reasoning to arrive at this result, the same conclusion holds in quantum theory.

With this in mind, we can look at Heisenberg's analysis in the section entitled "Absorption and scattering of hard γ-rays." Using the K-N formula, Heisenberg calculated an "average range" for light quanta. Applying the K-N formula to the "shell-electrons" was safe since, compared to cosmic-ray energies, they are practically free; and I quote, however, "It is otherwise for the nuclear electrons." Since the wavelength of a 100 MeV γ-ray is about the size of the nucleus, Heisenberg assumed that the free nuclear electrons scatter coherently, that is, as a single charge, and therefore as the square of their number. Adding the contribution of the shell electrons, the absorption coefficient of the γ-rays is then proportional to a quantity f, which is 0.5 for oxygen and 6.6 for lead. In this way Heisenberg tried to account for the observed large relative absorption in lead, that is, for the Meitner-Hupfeld effect.[33]

After the theoretical review, there followed about eight pages of analysis of five kinds of cosmic-ray experiments.[34] Referring to two-fold coincidence experiments with lead absorber on charged cosmic-ray particles by W. Bothe and W. Kolhörster, and newer measurements measuring three-fold coincidences by Bruno Rossi, Heisenberg concluded that, at least at sea level, the cosmic rays consist only of fast electrons and so:[35]

> If these fast electrons are produced through primary γ-rays which strike from outside the atmosphere, then according to Rossi's research the γ-rays must be almost completely absorbed in the upper part of the atmosphere. This would mean that the absorption of γ-rays calculated from the Klein-Nishina formula is at least a factor 25 too small.

In other words, the rapid absorption of γ-rays from the cosmic rays meant that they interacted far more strongly with matter and produced far more secondary electrons than theory allowed, even if one included a large contribution from nuclear electrons. (This preneutron paper of course, used the *e-p*, or *e-p-α*, model.) That is why in Heisenberg's later *n-p* model paper, he assumed that there were contributions to the γ-ray scattering from electrons in neutrons, in α-particles, and also from free nuclear electrons.

In the final summary section of his cosmic-ray paper, Heisenberg explained what might be going on:[36]

> As the physical basis for the discrepancy we could perhaps look at the radiation necessarily accompanying the scattering processes, which has been neglected in the present theory. The collision process with simultaneous emission of radiation could very likely be adjusted to raise the number of secondary electrons and thus also to increase the absorption of the primary

particles. A satisfactory estimate of the frequency of these secondary electrons seems to be hardly possible, since the failure in principle of the necessary Dirac radiation theory or of the equivalent quantum electrodynamics is already an established fact [bereits feststeht] on other grounds.

The last sentence reminded the reader that "even assuming the correctness of the physical presuppositions," the conclusions have only a qualitative validity.

Provisional Solution of the Electron-Nucleus Problem

The discovery of the neutron and Heisenberg's n-p nuclear model answered some questions but left many others unanswered. Wolfgang Pauli wrote in May 1932: "Although [the neutron's] existence does not solve the principle difficulties of nuclear structure (β-spectrum, inverted statistics), it appears very useful in many respects...."[37] Among the remaining unsolved puzzles were some features of nuclear systematics, such as the saturation of nuclear forces. Another difficulty was the anomalous scattering of γ-rays, that is, large cosmic ray radiative interactions and the Meitner-Hupfeld effect.[38] Still to be accounted for was β-decay with its apparent nonconservation of energy.

On the problem of nuclear systematics, Heisenberg's charge-exchange force combined with Pauli's exclusion principle predicted a larger binding energy per nucleon in the deuteron than in the α-particle, a serious defect referred to as "saturation of forces at the deuteron." Majorana and Wigner both advocated a more phenomenological approach, in which they introduced other types of exchange forces—exchanging spin, charge, position (or some combination) multiplied by potential functions. They postulated no fundamental mechanisms for the forces and thus could entirely forget about nuclear electrons.[39] With this extra freedom, they were able to correct the α-particle binding energy. A second type of "saturation," namely, the linear increase of nuclear volume with particle number, as in a liquid, could also be accommodated in these phenomenological approaches.

At the Solvay conference in Brussels in October 1933, Heisenberg discussed all of these approaches and laid the foundation for the subsequent development of the phenomenological theory of nuclei, while Gamow discussed his liquid drop model as well as the anomalous scattering of γ-rays.[40] The same conference was also notable as the site of Pauli's first "public" espousal of the neutrino hypothesis, which he had proposed privately as early as December 1930.[41]

In August 1932, Carl Anderson announced the discovery in the cosmic rays of a positive electron. This was confirmed at the Cavendish Laboratory by P. M. S. Blackett and G. P. S. Occhialini. Using their new counter-triggered cloud chamber, they also photographed the production of electron-positron pairs and even small electron showers containing several pairs. In contrast to Anderson, the Cambridge observers were aware that pair-production and annihilation processes were allowed by Dirac's hole theory and thought that might well explain some of the cosmic-ray puzzles. They wrote: "Perhaps the anomalous absorption of gamma radiation by heavy nuclei may be connected with the formation of positive electrons and the re-emitted radiation with their disappearance."[42]

The explanation of the large γ-ray interactions and electron multiplicities in cosmic rays, and of the Meitner-Hupfeld anomaly, was indeed to be found in the large cross-sections for electron-positron pair production and bremsstrahlung. Together with the annihilation of positrons, these three "shower phenomena" were responsible for the cascade showers for which Rossi had found preliminary evidence. The Meitner-Hupfeld anomaly faded away as a problem when it was realized just how rapidly radiative interactions with nuclei rose after crossing the electron-positron pair production threshold, about 1 MeV.

After attending the Solvay conference, Enrico Fermi returned to Rome and immediately began to work on a theory of β-decay, involving the creation and annihilation of particles and using the procedure for quantizing fermion fields.[43] In Fermi's theory, the created particles were the electron and the neutrino, produced simultaneously as a pair.[44] This theory was so successful that it eventually convinced even Niels Bohr that the usual conservation laws were respected in the quantum theory of the nucleus and elementary particles.[45]

Heisenberg liked Pauli's neutrino and Fermi's new theory. To mitigate some of the problems of his original charge-exchange force, he replaced it with a new fundamental n-p interaction. Its mechanism was the exchange of an electron-neutrino pair (called the *Fermi-field*) between nucleons, produced and absorbed by Fermi's β-decay interaction.[46] Thus nuclear electrons were no longer required to explain the neutron (as a composite), the cosmic ray interactions, or β-decay, but they *reemerged* as the hypothetical carriers, together with neutrinos, of the strong nuclear interaction.

Heisenberg's Fermi-field model was taken very seriously indeed; in Bethe's nuclear bible it was considered to be *the* fundamental theory of nuclear forces. Although it was difficult to adapt to the known nuclear phenomenology, Bethe and his collaborator Robert Bacher wrote that "the general idea of a connection between β-emission and nuclear forces is so attractive that one would be very reluctant to give it up."[47] At the 1977 Minnesota conference on the history of nuclear physics, John Wheeler spoke poignantly of the difficulty he had had in accepting what he called the "Standard Strong-Force Credo: (1) the nucleus contains no electrons; (2) the force that binds nucleons has nothing to do with electromagnetism and constitutes a new "strong force"; and (3) this force is transmitted by mesons."[48]

The first charge-independent fundamental theory of nuclear forces was a generalization of the Fermi-field theory formulated in 1937 by Nicholas Kemmer, who added to the exchanges of $e^-\bar{\nu}$ and $e^+\nu$, the "neutral currents" e^+e^- and $\bar{\nu}\nu$.[49] The next year, Kemmer adapted this procedure to make a charge-independent meson theory. After the acceptance of the meson theory of nuclear forces, electrons were finally denied any role in the strong nuclear interaction. The "heavy electron," or muon, the particle that I. I. Rabi said nobody had "ordered," however, could still be considered as a player in strong interactions as late as 1950.[50]

ARE THERE ELECTRONS IN THE NUCLEUS OR NOT?

After being told that nuclei were first thought to contain electrons, and were seen not to only after a lengthy struggle, the reader may well be exasperated at the suggestion contained in the heading of this final section. Electrons are *not* building blocks of nuclei, but that being said, was something wrong with the argument that large radiative interactions require charged particles of small mass? No—the idea behind Larmor's formula remains valid, and reradiation *is* proportional to $(q^2/m)^2$. One can gain some insight into this puzzle by examining the relevant radiative cross-sections.

First note that J. J. Thomson's scattering formula of 1906 is rigorously valid as the nonrelativistic limit of the cross-section that measures the scattering of photons off free electrons. The Thomson cross-section is:

$$\phi_0 = 8\pi r_0^2/3, \text{ with } r_0 = e^2/mc^2.$$

The length r_0, called the "classical electron radius," is $\approx 10^{-13}$ cm. The quantum and relativistic Compton cross-section (K-N formula) for the Z electrons of an atom is:

$$\phi_C = Z\phi_0 f(k_0/m),$$

where k_0 is the incident photon energy; the function $f(k_0/m)$ falls off rapidly with increasing energy. As Heisenberg inferred from cosmic ray observations, the absorption of radiation in matter actually *increases* with energy, and must therefore be due to the nuclear charge and not the outer atomic electrons.

In his famous book, *The Quantum Theory of Radiation,* Walter Heitler introduces the quantity ϕ_1, given as

$$\phi_1 = Z^2 r_0^2 / 137 \approx (Z^2/1000)\phi_0$$

as "a suitable unit in which to express the cross-section for Bremsstrahlung and similar processes."[51] For moderate energies, the Compton scattering is important; for heavier elements the nuclear charge effects may be comparable at modest energies and above about $10mc^2$ nuclear radiative effects are completely dominant. For lead, the cross-sections for bremsstrahlung and pair production at cosmic-ray energies are about $10\phi_1$ and slowly increasing.

What is notable about these cross-sections is, first of all, that they are proportional to the square of the nuclear charge Z, showing that the nucleus acts as a single coherent charge and not as a collection of nucleons, α-particles, and so forth. Secondly, the formulas contain *only the electron's mass* and not that of the nucleus or any of its constituents. That is easy to understand in the bremsstrahlung case: an incoming electron is deflected in the nuclear Coulomb field and radiates according to its acceleration. There is a nuclear recoil, but it carries little energy. The pair production case, however, is more puzzling: an incoming photon knocks out an electron-positron pair from the nuclear Coulomb field.

Does this mean that these pairs are contained in the nuclear Coulomb field? Indeed, probed by the high energy incident photon, the Coulomb field *does* contain negative and positive electrons—namely, virtual electron-positron pairs. In fact they are present even in a vacuum. Since only the nuclear charge enters the formulas, and not its mass or structure,[52] why does one need the nucleus at all? For one thing, the incoming photon only knows it has energy exceeding the pair threshold in the rest frame furnished by the nucleus. In some other rest frame it might be only an infrared photon. Also, the nuclear recoil is needed to conserve momentum—the photon cannot do that in interacting with the vacuum. Another way of viewing the process is to say that the nucleus polarizes the vacuum, separating the virtual electron and positron enough to let them interact with the incoming photon.

An even more telling example is Delbrück scattering, the scattering of photons by a nuclear Coulomb field. Again, it is the electron that sets the scale, the cross section being proportional to $(Ze^2)^4 r_0^2$, even though no electrons or positrons are present in either the initial or final state.[53] The same is true of so-called light-by-light scattering.

The question then is partly a matter of semantics (and some might regard it as a "trick question"), the answer depending upon whether unobservable virtual electrons are to be counted or not? While we are pondering this question, we might also consider what to think about unobservable quarks!

Notes

1. An excellent account of these models is contained in Heilbron 1977a, b.

2. Rutherford proposed the name *proton* for the nucleus of hydrogen.

3. Stuewer 1983, 31.

4. Kronig 1926. See also Purcell 1964, Stuewer 1983, Pais 1986, and Brown and Rechenberg 1996.

5. Houtermans 1930, 24.

6. Linear and angular momentum conservation laws would also be violated.

7. See the general references cited in note 9.

8. Stuewer 1983, 45. See 42–46 for more details on the neutron discovery.

9. Stuewer 1979.

10. Hans Bethe is responsible for the bible of nuclear physics, a Triptych: Bethe and Bacher 1936; Bethe 1937; Livingston and Bethe 1937.

11. Bethe 1979, 12.

12. Ibid., 13. Wigner 1933 states on the first page that "there are three different assumptions possible concerning the elementary particles." The first of these: "The only elementary articles are the proton and the electron. This point of view has been emphasized by Heisenberg and treated by him in a series of papers."

13. Stuewer 1979, 175.

14. In Brink 1965, an English translation of the three-part paper of 1932–1933 is given, but the parts that discuss electrons and the theory of the neutron's composition are omitted without explanation.

15. It was a strange compound not only for being collapsed but for its spin and statistics.

16. Heisenberg 1932b.

17. Carson 1996a, b. See also Purcell 1964, Bromberg 1971, Miller 1984 and 1985, and Brown and Rechenberg 1988 and 1996.

18. Heisenberg 1932b, 1 (emphasis added). Heisenberg cited Bohr 1932 for the violation of the conservation laws.

19. More precisely, if we use the molecular analogy, the distances are measured between the proton at the center of one composite neutron and another proton or "central" proton.

20. In the H2+ molecular ion there exists another small attraction, and in part III (Heisenberg 1932b, p. 217) Heisenberg introduces another attractive n-p interaction,

which "corresponds to the electrostatic part of the binding energy between H and H+ in the H2+ ion." This confirms the seriousness with which Heisenberg took the molecular analogy.

21. Bromberg 1971.

22. Carson 1996b, 102.

23. Heisenberg 1932b, part I, p.11.

24. As a bound system, the neutron mass should be somewhat less than that of proton plus electron. This matter is not dealt with here as it has been thoroughly treated in Stuewer 1993.

25. Heisenberg 1932b, part II, 160.

26. Heisenberg 1932b, part III, 595 (emphasis added).

27. Segrè 1979, 48. Majorana's conception of the proton, however, is not known.

28. Heisenberg 1932a, 430. Cosmic ray experiments were also discussed at the regular Easter gathering in Copenhagen, and in two notes added in proof Heisenberg thanks Bohr for his friendly advice to be cautious regarding the experiments.

29. Stuewer 1975.

30. Klein and Nishina 1929.

31. Heisenberg 1932a, 434.

32. It is assumed here that the α-particle scatters coherently, that is, as a charge of two units.

33. Brown and Moyer 1984.

34. Heisenberg 1932a, 44–452.

35. Ibid., 448.

36. Ibid., 452.

37. Pauli to Meitner, 29 May 1932. Letter [291] in Pauli 1985.

38. The M-H effect was ascribed by Heisenberg to nuclear electrons. The same view was adopted by Gamow (Gamow 1931, 82) and by Pauli (Pauli to Meitner, 1 Aug. 1930, letter [248] in Pauli 1985).

39. Majorana 1933; Wigner 1933. Although three hypotheses concerning nuclear constituents were cited by Wigner and all had nuclear electrons, he wrote that he could ignore them for his purposes "because the first elements, even up to Cl, do not contain any free electrons."

40. Stuewer 1994 and 1995 and Brown and Rechenberg 1996.

41. See, e.g., Brown 1978.

42. Blackett and Occhialini 1933, 716.

43. Jordan and Wigner 1928.

44. Fermi 1934.

45. Bohr 1936.

46. Brown and Rechenberg 1988, 1996.

47. Bethe and Bacher 1936, p. 203.

48. Wheeler 1979, 255.

49. Kemmer 1938.

50. Wentzel 1950.

51. Heitler 1936, 165. The second edition of 1944 is almost identical with the edition of 1936.

52. Of course, very high energy electrons are used to probe nuclear and even nucleon structure *a la* Rutherford.

53. Max Delbrück's proposed this effect in a *note in proof* attached to a paper in 1933 by Meitner and H. Kösters [*Z. f. Phys.* 84 (1933), 137] on the Meitner-Hupfeld effect. (An English translation is given in Brown and Moyer 1984, 135.) Delbrück assumed that "[Dirac's] electrons of negative energy are capable of scattering γ-rays, and in fact coherently. . . . We propose the hypothesis that the scattered radiation described in the [work of Meitner and Kösters] is identical with this radiation of the electrons of negative energy."

REFERENCES

Bethe, H. A. 1937. "Nuclear Physics. B. Nuclear Dynamics, Theoretical." *Rev. Mod. Phys.* 9, 69–244.

Bethe, H. A. 1979. "The Happy Thirties." In Stuewer 1979, 11–31.

Bethe, H. A. and Bacher, R. F. 1936. "Nuclear Physics. A. Stationary States of Nuclei." *Rev. Mod. Phys.* 8, 81–229.

Blackett, P. M. S. and Occhialini, G. P. S. 1933. "Some Photographs of the Tracks of Penetrating Radiation." *Proc. Roy. Soc. (London)* A139, 699–727.

Bohr, N. 1936. "Conservation Laws in Quantum Theory." *Nature* 138, 25–26.

Brink, D. M. 1965. *Nuclear Forces* (Oxford: Pergamon Press).

Bromberg, J. 1971. "The Impact of the Neutron: Bohr and Heisenberg." *Hist. Studies in the Physical Sciences* B3, 307–341.

Brown, L. M. 1978. "The Idea of the Neutrino." *Physics Today* 38 (11), 23–28.

Brown, L. M., and Moyer, D. F. 1984. "Lady or tiger?—the Meitner-Hupfeld Effect and Heisenberg's Neutron Theory." *Amer. Journ. Of Physics* 52, 130–136.

Brown, L. M. and Rechenberg, H. 1988. "Nuclear Structure and Beta Decay (1932–1933)." *Amer. Jo. Physics* 55, 932–938.

Brown, L. M., and Rechenberg, H. 1996. *The Origin of the Concept of Nuclear Forces* (Bristol: Institute of Physics Pub.).

Carson, C. 1996a. "The Peculiar Notion of Exchange Forces—I. origins in Quantum Mechanics, 1926–1928." *Stud. Hist. Phil. Mod. Phys. 27B(1)*, 23–43.

Carson, C. 1996b. "The Peculiar Notion of Exchange Forces—II. From Nuclear Forces to QED, 1929–1950." *Stud. Hist. Phil. Mod. Phys. 27B(2)*, 99–131.

Fermi, E. 1934. "Versuch einer Theorie der β-Strahlen." I *Zeit. Phys* 88, 161–171.

Gamow, G. 1931. *Constitution of Atomic Nuclei and Radioactivity* (Oxford).

Heilbron, J. 1977a. "Lectures on the History of Atomic Physics." in *History of Twentieth Century of Physics* (New York: Clarendon), 40–108.

Heilbron, J. 1977b. "J. J. Thomson and the Bohr Atom." *Physics Today* 30 (4), 40–48.

Heisenberg, W. 1932a. "Theoretische Überlegungen zur Höhenstrahlung." *Ann. Phys.* 13, 430–452.

Heisenberg, W. 1932b. "Über den Bau der Atomkerne." *Zeit. Phys.* 77, 1–11; 78, 156–164; 80, 587–596 (1933).

Heitler, W. 1936. *The Quantum Theory of Radiation* (New York: Clarendon)

Houtermans, F. G. 1930. "Neuere Arbeiten in der Quantentheorie des Atomkerns." *Ergebnisse der Exakten Wissenschaften* 9, 123–221.

Iwanenko, D. 1932. "The neutron hypothesis." *Nature* 129, 798.

Jordan, P. and Wigner, E. 1928. "Über das Paulische Äquivalenzverbot." *Zeit. Phys.* 47, 631–651.

Kemmer. 1938. "Quantum Theory of Einstein-Bose Particles and Nuclear Interaction." *Proc. Roy. Soc. (London)* A166, 127–153.

Klein, O. and Nishina, Y. 1929. "Über die Streuung von Strahlen durch freie Elektronen nach der neuen relativistischen Quantendynamik von Dirac." *Zeit. Phys.* 53, 853–868.

Kox, A. J., and Siegel, D. M., eds. 1995. *No Truth Except in the Details* (Boston: Kluwer Academic).

Kronig, R. 1926. "Spinning Electrons and the Structure of Spectra." *Nature* 117, 550.

Livingston, M. S. and Bethe, H. A. 1937. "Nuclear Physics. C. Nuclear Dynamics, Experimental." *Rev. Mod. Phys.* 9, 245–390.

Majorana, E. 1933. "Über die Kerntheorie." *Zeit. Phys.* 82, 137–145.

Miller, A. I. 1984. *Imagery in Scientific Thought* (Boston: Birkhauer).

Miller, A. L. 1985."Werner Heisenberg and the Beginning of Nuclear Physics." *Physics Today* 38 (11), 60–68.

Pais, A. 1986. *Inward Bound* (New York: Clarendon).

Pauli, W. 1985. *Scientific Correspondence, vol. II* (Berlin: Springer), edited by K. von Meyenn.

Purcell, E. M. 1964. "Nuclear Physics without the Neutron; Clues and Contradictions." in *Proceedings of the Tenth International Congress of the History of Science* (Ithaca, New York, 1962) (Paris), 121–132.

Segrè, E. 1979. "Nuclear Physics in Rome." in Stuewer 1979, 35–62.

Shea, W. R. 1983. *Otto Hahn and the Rise of Nuclear Science* (Dordrecht: D. Reidel).

Stuewer, R. H. 1975. *The Compton Effect* (New York: Scientific History Pub.).

Stuewer, R. H. 1979. *Nuclear Physics in Retrospect* (Minneapolis: University of Minnesota Press).

Stuewer, R. H. 1983. "The Nuclear Electron Hypothesis." in Shea 1983, 19–67.

Stuewer, R. H. 1993. "Mass-Energy and the Neutron in the Early Thirties." *Science in Context* 6, 195–238.

Stuewer, R. H. 1994. "The Origin of the Liquid Drop Model and the Interpretation of Nuclear Fission." *Perspectives on Science* 2, 76–129.

Stuewer, R. H. 1995. "The Seventh Solvay Conference: Nuclear Physics at the Crossroads." in Kox and Siegel 1995.

Wentzel, G. 1950. "μ-pair Theories and the π-meson." *Phys. Rev.* 79, 710–716.

Wheeler, J. A. 1979. "Some Men and Moments in the history of Nuclear Physics: The Interplay of Colleagues and Motivations." in Stuewer 1979, 213–322.

Wigner, E. 1933. "On the Mass Defect of Helium." *Phys. Rev.* 43, 252–257.

10

The Electron, the Hole, and the Transistor
Lillian Hoddeson and Michael Riordan

During the annual dinner at the Cavendish Laboratory, Cambridge physicists are said to have made the toast: "To the electron—may it never be of any use to anybody."[1] For seventeen years after its discovery in 1897, the electron remained a particle that was real to physicists in the hallowed halls of Cambridge, but not to scientists and engineers in industrial laboratories.

This picture changed when the particle began to work in electron devices, such as the vacuum tube amplifier. The electron stepped out into the workaday world and gained "thingness," or "operational reality," a state that extended beyond the physical reality it obtained when J. J. Thomson "discovered" it.[2] It was a reality the electron had to earn through service to industry and commerce by work in devices whose value on the street was measured in "real money."

The "hole"—the physical embodiment of the idea that a vacancy near the top of an otherwise filled electronic band behaves like a positively charged particle—made a similar leap from physical to operational reality. Born as a physical construct in a 1929 paper of Rudolf Peierls,[3] the hole gained its operational reality in 1947 with the invention of the transistor by John Bardeen and Walter Brattain at Bell Laboratories.

We recognize that difficult philosophical issues are at stake in speaking about the reality of particles. But as historians, we are at liberty to sidestep such issues and focus on the historically potent questions: Real *for whom*? Real *with what effects*? And real *as compared with what* (for example, imaginary, ideal, theoretical, trivial)? By introducing the term "operational reality," we hope to bracket off the deeper questions of reality, leaving them for the philosophers to debate, while we pursue the historian's goal of identifying the functions that the electron and the hole performed in particular contexts at various times.[4]

The Operational Reality of the Electron

One could argue that the electron had been doing useful work long before its discovery, for example, in batteries and motors. But to make devices such as these work, there was no need to conceptualize electric current as a swarm of particles. The fluid picture was adequate. Nor was the particle picture needed in the first years after the invention of vacuum tube devices. To make his original 1904 vacuum diode (or "oscillation valve") convert radio waves into the direct current signals needed to drive headphones, John Ambrose Fleming could use the electric fluid idea. Nor did Lee de Forest need electrons as particles to describe how changing the grid voltage influences current in the plate circuit of his "audion," the vacuum triode he invented in 1906 by inserting a third electrode into one of Fleming's valves.

The picture changed, however, when the audion was adapted for mass communications systems. The motivation was a financial crisis that AT&T faced in 1907. As part of his program to meet this crisis, Theodore N. Vail, AT&T's newly reinstated president, demanded that his engineers build a transcontinental telephone line to be demonstrated at the celebration of the opening of the Panama Canal, an event scheduled for 1914 and later postponed to 1915.[5] At this point amplifiers, which at AT&T were called "repeaters," were not yet part of the telephone system. For when two or more of the existing mechanical repeaters were hooked into telephone circuits, the electrical echoes that occurred drowned out the signal. Telephone lines could work without repeaters, but not when they were longer than the distance between New York and Denver (about 2,100 miles). Coast-to-coast service required repeaters.

Frank Jewett, a Ph.D. physicist from the University of Chicago, took charge of AT&T's repeater problem. He recognized that the echoes from the mechanical repeaters were the result of the sluggishness of their mechanical diaphragms. Drawing on his knowledge of atomic physics, he suggested using lightweight particles, perhaps molecules, as the vibrating element. Then, because no one on the company's staff knew enough atomic physics to apply the suggestion, Jewett called on his University of Chicago colleague, Robert Millikan, well known for his precise measurement of the electronic charge using the oil-drop method. In January 1911, Millikan sent a graduate student named Harold Arnold to Jewett. Arnold had worked on the oil-drop experiments,[6] and his group would evolve into the research division of Bell Telephone Laboratories, which was incorporated in 1925.

When de Forest demonstrated his new audion at AT&T in October 1912, Arnold recognized that this device might be the solution of the repeater problem. But he also saw that some physicist's work needed to be done on the audion. For whenever the plate voltage was high enough to amplify telephone signals, the tube would, as one observer put it, "fill with blue haze, seem to choke, and then transmit no further speech until the incoming current had been greatly reduced."[7] Arnold realized that the blue haze was due to ionized gas inside the tube. Although de Forest believed that some gas is necessary for the triode to function, Arnold understood that it had to be removed, for the gas molecules were scattering the electrons and sapping their energy as they flowed from cathode to anode.

Arnold's high-vacuum tube then became the heart of the repeaters with which AT&T engineers achieved the first transcontinental telephone line, in time to meet Vail's goal of demonstrating it at the Panama-Pacific exhibition. Millikan later reflected on this pivotal moment in the electron's transition to operational reality: "The electron—up to that time largely the plaything of the scientist—had clearly entered the field as a patent agent in the supplying of man's commercial and industrial needs."[8] During the 1920s, the term "electronics" arose to describe this area of human endeavor in which understanding of electrons *as particles* was necessary to develop useful devices, circuits, and systems.

The electron also continued to work in basic research. A patent dispute arose between AT&T and General Electric, where Irving Langmuir had also fashioned a high-vacuum amplifying tube. In the process of exploring the technical issues of this dispute, Clinton Davisson and Lester Germer working at Bell Labs found that electrons can be diffracted by crystals, thus confirming Louis de Broglie's hypothesis of the wave nature of particles. Davisson's 1937 Nobel Prize for this research marked Bell Labs as a fertile environment for fundamental electron studies as well as for the applied research that had made the electron operationally real.

THE EMERGENCE OF THE HOLE

In this industrial environment the hole also achieved its operational reality. Before explaining this process, which is inseparable from the invention of the transistor, we must first identify two earlier historical strands. One concerns the crystal set, whose central element was the "cat's whisker" detector, essentially a sharp metal point pushed down on a piece of semiconductor, such as galena or silicon.[9] Until the 1930s, one could not understand how this device helped convert radio signals into the sounds heard through headphones. The explanation required the application of quantum mechanics.

Crystal sets dominated radio between 1920 and about 1927. Then vacuum tubes took over, for it was found that when vacuum tubes are used radio receivers can better separate the signals from stations with wavelengths close to one another. But by the late 1930s, radio engineers were noticing that at very high frequencies in the gigahertz range, cat's-whisker detectors were more sensitive than their vacuum-tube counterparts (which couldn't deal with rapid pulses oscillating at billions of cycles per second because of the finite time it took electrons to transit the tube). At this point, a Bell Labs radio engineer, Russell Ohl, decided to explore systematically which materials are most sensitive in cat's whisker detectors at high frequencies. He found that silicon works best. But the behavior of silicon detectors was erratic: at the "hot spots" where the rectification occurred, the current sometimes flowed in one direction, sometimes in the other.

In an attempt to improve the behavior, Ohl and two metallurgists began purifying the silicon by melting samples in an inert helium atmosphere. In the process they accidentally produced two different samples, which they later called "n-type" or "p-type" silicon, having opposite electrical properties. Ohl did not then realize that he was observing the practical effects of excess electrons and holes, but early in 1940 he noticed that a particular region in an apparently malfunctioning silicon ingot had a strikingly large photovoltaic effects at least an order of magnitude larger than had ever been observed before. He had discovered the p-n junction. At this point, World War II intervened, delaying further study of this finding.[10]

Quantum mechanics contributed another crucial strand. During the late 1920s, Wolfgang Pauli, Arnold Sommerfeld, Felix Bloch, and several other European physicists applied quantum mechanics to problems of solids, developing what came to be known as the quantum theory of solids. It was in the context of working on this theory that Rudolf Peierls put forth the fecund concept of the "hole"—the idea that empty electron states near the top of an otherwise filled band act like positively charged particles. Heisenberg claimed in 1931 that the holes, or *Löcher,* "behave exactly like electrons with *positive* charge under the influence of a disturbing electric field."[11]

That year, Alan Wilson explained the long-standing puzzle of the difference between metals and insulators in terms of the quantum-mechanical theory of energy bands. He recognized that insulators must have completely filled bands while metals must have partially filled ones. He pictured semiconductors essentially as insulators with a gap between their filled valence band and unfilled conduction bands. Impurities in the crystal lattice led to energy levels between these bands; thermal excitations could kick electrons up from these intermediate bands to the conduction band, thus permitting current to flow through the semiconductors.[12] Wilson also used his model to forge a pioneering theory of rectification at metal-to-semiconductor junctions,[13] one superseded later in the decade by the work of Walter Schottky, Nevill Mott, and Boris Davydov.[14] Davydov's 1938 theory was the only one of that period to take the holes into account, but his highly mathematical papers received little attention. Bardeen read them with interest at the end of World War II, but their mathematical complexity masked their prescient discussion of minority carriers in semiconductors. He recognized that role during his work on the first transistor.[15]

The quantum theory of solids proved pivotal to the invention of the transistor. A flurry of review articles helped the theory find its way into graduate programs of study by 1933. The first three physics programs to offer study of this theory formed at Princeton around Eugene Wigner, MIT around John Slater, and Bristol around Nevill Mott and Harry Jones. Bardeen studied the subject under Wigner in the early 1930s, at the same time Shockley studied it under Slater. Brattain, then already working as an experimental physicist at Bell Labs, tried to learn the theory on his own, for example, by attending Arnold Sommerfeld's course on the electron theory of metals at the Michigan Summer School in 1931. But understanding the mathematically complex formalism without an instructor proved to be extremely difficult for him.

Mervin J. Kelly, another Millikan graduate student who was then head of the Vacuum Tube Department at Bell Labs, recognized the importance of the new physics of solids for many of the problems faced by his staff. He would have been happy to hire a few recently trained quantum theorists, but a hiring freeze at Bell during the Great Depression at first made that impossible. He encouraged his staff to study the new physics on their own, for example, by attending courses or participating in study groups, a phenomenon that flourished on layoff days during the Great Depression.

As soon as the freeze ended in 1936, Kelly hired Shockley and encouraged him to work on finding replacements for the electromechanical relay, which was slow, and for the vacuum tube, which although fast was costly, fragile, bulky, and unreliable. Shockley took Kelly's interest seriously but did not make much progress on these problems before the war. Dean Wooldridge, who shared a laboratory with him in the late 1930s, recalled with amusement the exceedingly crude methods with which Shockley tried to build a solid-state amplifier. Part of Shockley's apparatus looked to Wooldridge as though it had been "cut out of some very old copper back porch screen with some very dull scissors." This jagged piece of screen had "evidently been out in the elements for years and years, because it was all heavily oxidized."[16] Working by analogy with the vacuum triode, Shockley positioned two wires on opposite sides of this copper screen, so that they barely touched the oxide coating. He then tried to control the current flowing from one wire to the other by adjusting the voltage on the screen. When this crude contraption didn't amplify, Shockley asked Brattain to build a better device. The seasoned experimentalist was sure the design was not feasible but, to humor Shockley, he built it anyway. It didn't work.[17]

During World War II, defense work generally interrupted research in laboratories throughout the United States. But on the problem of developing a solid-state amplifier, wartime developments contributed greatly, because the U.S. and British governments, recognizing the crystal rectifier's crucial ability to serve as a detector of radar in the centimeter range, strongly supported research and development on that device. The work yielded marked improvements in the technology of forming point contacts and producing pure crystals of silicon and germanium doped with small amounts of impurities. It also led to major advances in the understanding of semiconductor structure and phenomena, including rectification. These wartime results would prove crucial for the invention of the transistor.[18]

INVENTION OF THE TRANSISTOR

Kelly also noticed that the wartime research on crystal rectifiers was taking place in a network of institutions that were communicating with each other almost freely. Coordinated by the MIT Radiation Laboratory, the U.S. network included—besides MIT—General Electric, Pennsylvania, Sperry, Westinghouse, Sylvania, DuPont, Purdue, and Bell Labs. Information exchange occurred at "crystal meetings" held approximately every other month.[19] Kelly realized that when the war ended, these groups would stop cooperating and begin competing again. But now they would all be armed with state-of-the-art knowledge of microwave and semiconductor technology.[20] By mid-1945, Kelly's prescription for maintaining Bell's competitive edge included an expanded program of basic research in solid-state physics.

The emphasis on research fit the spirit of the times, as captured by Vannevar Bush in his 1945 report stressing the value of research to the nation, *Science: the Endless Frontier*. While it would take the U.S. government about five years to begin investing lavishly in basic science, Kelly was almost immediately able to turn Bell Labs into what he would call an "institute of creative technology," one that approached science using the practice of team research, which had proved so fruitful in the wartime laboratories.[21]

Kelly's solid-state program appealed to John Bardeen, then completing his wartime assignment in Washington, D.C. The talented theoretical physicist chose to join Bell Labs rather than return to his academic post at Minnesota, not only to take advantage of a higher salary but because of the opportunity to work full time on solid-state research. He joined Shockley's semiconductor group at the new Murray Hill facility of Bell Labs, initially sharing an office with Brattain and another experimental physicist, Gerald Pearson.

Bardeen soon became involved in the historic work that would impart operational reality to the hole. He began by trying to understand Brattain's and Pearson's data on semiconductors. He also took an active part in the group's review of the extensive progress on semiconductors and the theory of rectification made by the wartime radar program. The group focused on silicon and germanium because they are the simplest semiconductors and also because wartime advances had made it possible to obtain consistently behaving n-type and p-type samples.

Shockley asked Bardeen to check one of his earlier calculations, for a silicon "field-effect" amplifier. Drawing on theories by Mott and Schottky, Shockley had argued in April 1945 that an externally applied electric field should draw electrons to the surface of a semiconductor. If the sample is thin, he reasoned, changes in the applied field should markedly increase the number of charge carriers and therefore amplify the current passing through. Shockley did not understand why this design did not work experimentally.[22]

Bardeen soon verified Shockley's calculation and quickly became absorbed in figuring out how the theory was flawed, or at least incomplete. By March 1946, Bardeen had an explanation: electrons at the semiconductor surface were being trapped in localized "surface states" and thus could not participate in the conduction process. The trapped electrons also shielded the interior of the semiconductor from further penetration of the electric field.[23] This idea explained the failure of Shockley's design, and it also became the basis of an extensive program of research by other members of the semiconductor group. Although Shockley continued to offer suggestions, he channeled most of his energies into research on the theory of dislocations and the flow of electrons through alkali and silver halides.

About a year and a half later, another accident put Bardeen and Brattain on a direct line to the transistor. In mid-November 1947, Brattain was bothered by a hysteresis effect he was encountering in his experimental studies of the contact potential in silicon. Recognizing that water condensation was causing the problem, he impulsively immersed his apparatus in various electrolytes and dielectric liquids in an ad hoc attempt to eliminate it. To his surprise he found that the photovoltaic effect *increased*. Bardeen realized that mobile ions in the liquids close to the surface were neutralizing the surface states and permitting an applied electric field to penetrate into the semiconductor. That meant that a field effect amplifier might be possible after all![24]

Now Bardeen and Brattain began to work quickly, with input from the other members of the group. The two friends resembled a unified research organism, in which Brattain offered skillful hands that had worked for two decades with semiconductors and electronics. He also offered his expressive words, much appreciated by the extraordinarily reticent Bardeen, who offered his creative genius and encyclopedic understanding of the existing research on solid-state physics.[25] Recognizing their ignorance of many particular features of the materials they were using, Bardeen and Brattain tried to alter only a few parameters at a time as they moved step-by-step toward the historic amplifier design that would finally work for the first time on December 16, 1947.

For example, instead of the thin film of semiconductor in Shockley's field-effect design, Bardeen suggested using an "inversion layer," a thin region that forms in certain circumstances near the surface of a semiconductor. Here, the charge carriers that are in the majority have opposite charge to those in the bulk. Such a layer would allow them to use the large change in conductivity, which the wartime radar studies had shown is possible in thin films, and they could simultaneously circumvent both the known difficulty of depositing a thin film of semiconductor and the problem of the low mobility of charge carriers in such films. Bardeen also suggested using a point-contact design, mainly because he was familiar with the wartime development of point contacts. He knew such an experiment could be set up and tested in a single day.

A second accident brought them directly to the first transistor. On 8 December, responding to a suggestion by Shockley, Bardeen had suggested replacing the silicon used in their first experiments by a piece of the "high-back-voltage germanium" developed by the Purdue group during the wartime radar project.[26] This change allowed them to observe a two-fold voltage amplification and a power gain of 330 that very afternoon. But the amplification occurred when the polarity was *opposite* what they had been observing in silicon. Bardeen recognized almost instantaneously what was going on inside the slab. "Bardeen suggests that the surface field is so strong that one is actually getting p-type conduction near the surface," wrote Brattain in his laboratory notebook, "and the negative potential on the grid is increasing this p-type or hole conduction."[27] The transistor was close to being born.

The frequency response of their device was still poor. In an attempt to improve it, they tried using an oxide film grown on the germanium surface to help modulate the current flowing through an inversion layer just under it. They thought the oxide would act as an insulator, but the layer had probably been washed off during processing. To Bardeen and Brattain's surprise, they observed the modulation of output current and voltage, but again at the opposite polarity from what they had expected. It occurred when their gold electrode was biased positively instead of negatively.

Bardeen immediately understood that holes were responsible for the modulation. They were being created at the positively biased gold electrode, which was in direct contact with the germanium, and were flowing through the inversion layer to a tungsten point contact. This critical moment is recorded in Brattain's notebook entry on December 19: "It would appear then that the modulation obtained when the grid point [the gold] is bias + is due to the grid furnishing holes to the plate point [the tungsten]." Realizing that the grid point was in effect furnishing holes, they later named the grid point the "emitter" and the output contact the "collector." Due to a fortuitous accident, they had discovered it was possible to build an amplifier according to principles entirely different from Shockley's field-effect approach.

They had yet to produce a usable amplifier in which the change in the collector current not only followed but greatly exceeded the change in the emitter current. To achieve this goal, the hole current had to markedly increase the conductivity beneath the collector, making it easier for electrons to flow in the output circuit. Bardeen suggested using two contacts placed only a few mils apart on the germanium. At this stage they believed they would get a stronger effect using line contacts. Following a suggestion of Bardeen, Brattain brought the contacts close together by skillfully wrapping a piece of gold tape around the tip of a triangle of polystyrene, cutting a slit in the tape at the apex using a razor, and filling the cut with wax. The separation was about 2 mils. This apparatus worked the first time Bardeen and Brattain tried it, on December 16, 1947. In one of the first experiments at an input frequency of 1,000 Hz, the power gain was 1.3 and the voltage gain 15. This was the birthday of the transistor.

Bardeen and Brattain spent a week checking their results before demonstrating the device to executives at Murray Hill Laboratories on December 23. They connected the input to an audio signal, so that the circuit could be spoken over. They hooked the output to an oscilloscope and also listened to it using earphones. Brattain reported the event in his notebook the next day, Christmas Eve:

> This circuit was actually spoken over and by switching the device in and out a distinct gain in speech level could be heard and seen on the scope presentation with no noticeable change in quality. By measurements at fixed frequency in, it was determined that this power gain was the order of a factor of 18 or greater.[28]

That same day, Bardeen included a fuller explanation of how the holes worked in his notebook entry:

Voltage gains up to about 100 and power gains up to about 40 have been observed. The explanation is believed to be as follows. When A [the gold electrode] is positive, holes are emitted into the semi-conductor. These spread out into the thin P-type layer. Those which come in the vicinity of B [the tungsten point] are attracted and enter the electrode. Thus A acts as a cathode and B as a plate in the analogous vacuum tube circuit. . . . The signs of the potentials are reversed from the [sic] those in a vacuum tube because conduction is by holes (positive charge) rather than by electrons (negative charge).[29]

The hole had finally achieved operational reality: it was the "thing" that made the transistor work.

The detailed understanding of how electrons and holes behave in the intimate presence of one another permitted the great advances in semiconductor technology of the past fifty years to occur. In the process, the hole—the "no-thing" the transistor had elevated into a useful entity—became a major player in the Information Age. While the work of the electron in devices such as radio, telephone, and television, created modern mass culture, the work of the hole in devices like fax machines, ATMs, cellular phones, modems, copiers, and personal computers, carried the program begun by the electron much farther, creating the postmodern culture and information-based economy we inhabit today.

ACKNOWLEDGMENT

This paper is adapted in part from Michael Riordan and Lillian Hoddeson, *Crystal Fire: The Birth of the Information Age* (New York: W. W. Norton, 1997). Hoddeson first presented the argument at the symposium organized by Alan Morton and Andrew Warwick in London to celebrate the centenary of the discovery of the electron. The meeting was held at the Royal Society of London on April 11, 1997, and sponsored by The British Society for the History of Science and the Royal Institution of London.

NOTES

1. E. N. da C. Andrade, cited in Abraham Pais, "The Discovery of the Electron," *Beam Line*, vol. 27, no. 1 (spring 1997); 4–16, quote on 5.

2. We bypass the complex issues surrounding this discovery. They are treated in other contributions to this volume. Also see E. A. Oavis and I. J Falconer, *J. J. Thomson and the Discovery of the Electron* (London: Taylor and Francis, 1997); Per F. Dahl, *Flash of the Cathode Rays* (London: Institute of Physics, 1997); and Pais, "The Discovery of the Electron."

3. The idea of the hole was almost made explicit in Peierls's paper on the positive Hall effect: R. E. Peierls, "Zur Theorie der galvanomagnetischen Effekte," *Zeitschrift fur*

Physik 53 (1929); 255–266. It was fully delineated in 1931 by Werner Heisenberg, "Zum Paulischen Ausschliessungsprinzip," *Annalen der Physik* 10 (1931); 888–904.

4. The experimental physicist Percy W. Bridgman wrote copiously in the 1920s and 1930s about a related notion known as "operationalism," or sometimes "operationism," in which physical particles are defined by the set of operations used to determine them. See Percy W. Bridgman, *The Logic of Modern Physics* (New York: McMillan, 1927), 5; *The Nature of Physical Theory* (Princeton: Princeton Univ. Press, 1936); and *Reflections of a Physicist* (New York: Arno Press, 1980). See also Maila L. Walter, *Science and Cultural Crisis: An Intellectual Biopraphy of Percy Williams Bridgman (1882–1961)* (Stanford, Stanford Univ. Press, 1990), 2–4.

5. L. Hoddeson, "The Emergence of Basic Research in the Bell Telephone System, 1875–1915," *Technology and Culture* 22 (1981); 512–544.

6. Arnold is mentioned in R. A. Millikan, *The Electron* (Chicago: University of Chicago Press, 1917), 93–96. At Chicago, he determined the validity of Stokes's Law used in Millikan's analysis.

7. John Mills, "The Line and the Laboratory," *Bell Telephone Quarterly* 19 (January 1940); 5–21, quote on 13.

8. Millikan cited in M. Riordan, "The Industrial Strength Particle," *The Beam Line,* vol. 27, no. 1 (spring 1997); 30–35. Quote on 32.

9. See, e.g., Gerald Pearson and Walter Brattain, "History of Semiconductor Research," *Proc. of IRE,* vol. 43 (December 1955); 1794–1806, and references therein.

10. Michael Riordan and Lillian Hoddeson, "The Origins of the pn Junction," *IEEE Spectrum,* vol. 34, no. 6 (June 1997); 46–51.

11. Peierls, "Zur Theorie der galvanomagnetischen Effekte"; Heisenberg, "Zum Paulischen Ausschliessungsprinzip"; L. Hoddeson, G. Baym , and M. Eckert, "The Development of the Quantum-mechanical Theory of Metals, 1928–1933," *Reviews of Modern Physics* 59/1 (1987); 287–327. Quote on 294.

12. A. H. Wilson, "The Theory of Electronic Semi-conductors," *Proceedings of the Royal Society of London* A-133 (1931); 458–491; Wilson, "The Theory of Electronic Semiconductors-II," *Proceedings of the Royal Society of London* A-134 (1931); 277–287.

13. A. H. Wilson, "A Note on the Theory of Rectification," *Proceedings of the Royal Society of London* A-136 (1932); 487–498.

14. N. F. Mott, "The Theory of Crystal Rectifiers," *Proceedings of the Royal Society of London* A-171 (1939); 27–28; W. Schottky, "Halbleitertheorie der Sperrschicht," *Naturwissenschaften* 26 (1938); 843; "Zur Halbleitertheorie der Sperrschicht- und Spitzengleichrichter," *Zeitschrift für Physik* 113 (1939); 376–414; "Vereinfachte und erweiterte Theorie der Randschichtgleichrichter," *Zeitschrift für Physik* 118 (1942); 539–592; B. Davydov, "On the Rectification of Current at the Boundary Between Two Semi-conductors," *Comptes Rendus (Doklady) de l'Académie des Sciences de l'URSS* XX:4 (1938); 279–282; "On the Theory of Solid Rectifiers," *Comptes Rendus (Doklady) de l'Académie des Sciences de l'URSS* XX:4 (1938); 283–285.

15. J. Bardeen, "Trends in Semiconductor Research," *J. Phys. Chem. Solids* 8 (1959); 2–6; M. Riordan, L. Hoddeson, and C. Herring, "The Invention of the Transistor," *Reviews of Modern Physics* 71/2 (March 1999); 5336–5345.

16. Dean Wooldridge interview by L. Hoddeson, August 21, 1976, 63.

17. Michael Riordan and Lillian Hoddeson, *Crystal Fire: The Birth of the Information Age* (New York: W. W. Norton, 1997); 86.

18. See Frederick Seitz and Norman G. Einspruch, *Electronic Genie: The Tangled History of Silicon* (Urbana, IL: University of Illinois Press, 1998); F. Seitz, "Research on Silicon and Germanium in World War II," *Physics Today* (January 1995); 22–77; Henry C. Torrey and Charles A. Whitmer, *Crystal Rectifiers* (New York: McGraw-Hill, 1948), viii. The classic work on this radar program is Henry E. Guerlac, *Radar in World War II* (New York: American Institute of Physics/Tomash, 1987); see also L. Hoddeson, "Research on Crystal Rectifiers During World War II, and the Invention of the Transistor," *History and Technology* 11(1994); 121–130.

19. Torrey to Hoddeson, 6 June 1993. The Telecommunications Research Establishment played a similar role in England, coordinated with the General Electric Company, British Thompson Houston, Ltd., and Oxford University.

20. M. J. Kelly, "A First Record of Thoughts Concerning an Important Postwar Problem of the Bell Telephone Laboratories and Western Electric Company," 1 May 1943, AT&T Archives, Warren, N.J.

21. M. J. Kelly, "The Bell Telephone Laboratories—an Example of an Institute of Creative Technology," *Proc. Roy. Soc. Lond.* 203a (1950); 287–301.

22. See Riordan and Hoddeson, *Crystal Fire,* 120–122; also L. Hoddeson, "The Discovery of the Point-Contact Transistor," *Historical Studies in the Physical Sciences* 12/1(1981); 41–76.

23. J. Bardeen, "Surface States and Rectification at a Metal-Semiconductor Contact," *Physical Review* 7l (1947), 717–727.

24. Walter Brattain, "Discovery of the Transistor Effect: One Researcher's Personal Account," *Adventures in Ezperimental Physics* 5 (1976); 3–13.

25. Lillian Hoddeson and Vicki Daitch, *Gentle Genius: the Life and Physics of John Bardeen,* in preparation.

26. Paul Henriksen, "Solid State Physics Research at Purdue," *Osiris* 2:3 (1987); 237–260.

27. Brattain, Bell Laboratories Notebook #18194 8 December, 1947, 176–177. AT&T Archives.

28. Brattain, Bell Laboratories Notebook #21780 24 December, 1947, 7–8. AT&T Archives.

29. Bardeen, Bell Laboratories Notebook #20780 24 December, 1947, 71–74. AT&T Archives.

11

Remodeling a Classic: The Electron in Organic
Chemistry, 1900–1940
Mary Jo Nye

During the nineteenth century the practice of organic chemistry constituted the mainstream of the discipline of chemistry. Experimental papers in organic chemistry, whether agricultural, biological, or industrial, filled the pages of the weekly *Berichte* of the German Chemical Society and the other chemical journals. Theories of chemical constitution and chemical reaction largely fit within a natural history tradition that employed biological paradigms of descriptive classification by orders and species as well as statements of relationships between form and function. One of the most significant theoretical developments in late nineteenth-century chemistry, the chemical valence theory, emerged out of these chemical theories of type and structure, not out of physical theories about mechanical forces.[1]

Chemists had long engaged with natural philosophers' physical hypotheses about the forces that might hold the chemical elements together and break them apart. Obvious candidates for these forces of chemical "affinity" were some kind of Newtonian gravitational force on the one hand, or some type of magnetic or electrical force on the other hand. Gravitational force differs fundamentally from chemical affinity, however, because it varies with quantity of mass, not kind of mass and magnetic or electrical force, while meaningful in studying electrolysis, plays no role among the vast number of organic materials that do not form ions. So chemical constitution was best explained in the nineteenth century by representations of chemical compounds with type formulas and structural formulas, suggesting the ways in which structure can be made to vary by addition or substitution (figure 11.1).

In a recent article, the British chemist Brian Sutcliffe discussed the reticence with which Edward Frankland introduced the term chemical "bond" in the 1860s, as Frankland noted that he could not offer a convincing theory of chemical affinity and that he wanted to guard against any naive model of material connections among the elements.[2] Sutcliffe is not surprised at Frankland's caution, but finds it "rather harder to understand "why, within forty years of [the assumption of the "bond"] being made, chemists (with some

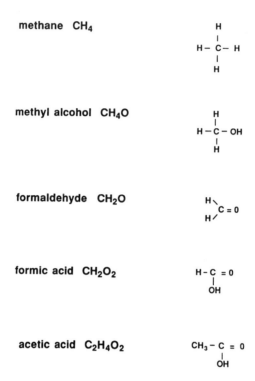

Figure 11.1
Structural formulas.

notable exceptions) were happy to throw caution to the winds and espouse the electron as the originator of the chemical bond."[3]

Also worth noting is the fact that the reconceptualization of chemistry in terms of a physical "electron theory" did not become mythologized as a "second chemical revolution," even though this chemical electron was the same electron as the one identified with a "revolutionary" new quantum physics.

In the history of chemistry, we have a "chemical revolution" associated with Lavoisier and Dalton, and we have what Alan Rocke has called the "Quiet Revolution" that produced the structural organic chemistry of Kekulé, Frankland, and Wurtz.[4] We have a number of small and large revolutions in the history of science.[5] Why was there no second chemical revolution identified by chemists or historians with the electron?

It was around 1900 that the electron was identified as a subatomic particle of matter. The electron became a wave-like distribution of electrical charge by the late 1920s. Textbooks of organic chemistry began using the electron theory by the early 1940s. The first electron-inspired organic textbook was *Organic Chemistry,* published in 1935 by the Caltech organic chemist Howard Lucas, a colleague of Linus Pauling. This volume later was characterized by another Caltech chemist, John D. Roberts, as "a pathbreaking texbook . . . which was probably the first to introduce, at that level, modern ideas of valence and thermodynamics to organic chemistry."[6]

Melvin Calvin and Gerald Branch, who were colleagues in the chemistry department at Berkeley, coauthored *The Theory of Organic Chemistry* in 1941. "Our book in effect organized all of organic chemistry in terms of electron theory," Calvin later said of the result.[7] That same year George Wheland, who had been a postdoctoral fellow at Caltech with Linus Pauling from 1932 to 1936 and who had spent the following year in England with Christopher Ingold, Nevil Sidgwick, and John E. Lennard-Jones, published *The Theory of Resonance.*[8] Two years later Edward Remick published *Electronic Interpretations of Organic Chemistry.*[9]

That there had been a battle for establishing the electron as a fundamental organizing principle of modern chemistry and that this battle was over about 1940 also can be inferred from a remark by Christopher Ingold: "The task has . . . been to show that organic chemistry is . . . provided with a framework of principles. . . . From 1925 to 1940 it was uphill work, because of initial opposition to the new approach."[10]

Ingold was not alone in what he recalled as a struggle for establishing a new theoretical foundation for organic chemistry in the 1930s, nor had Ingold been a quick convert to the view that the electron was the way to do it. The opposite was true for Linus Pauling, who instantly saw the appeal of the electron in 1920, while an undergraduate at Oregon Agricultural College, when he discovered Irving Langmuir's and G. N. Lewis's early papers on the electron theory of valence.[11]

For Pauling, too, the period around 1940 came to mark a time for the recasting of all of general chemistry by the electron theory, and in 1941 Pauling's first, lithographed, version of his *General Chemistry* textbook became available at the Caltech bookstore.[12]

But neither Ingold nor Pauling claimed to have made a "revolution" in chemistry. Indeed, in their initial remarks in the first chapter of the *Introduction to Quantum Mechanics with Applications to Chemistry,* Pauling and E. Bright Wilson Jr. write that quantum mechanics is the "most recent step in the very old search" for general laws governing the motion of matter, that the "modifications in older laws" of mechanics has had the result of "depriving physics of sole claim upon them," and that a historical approach rightly introduces "many concepts which are retained in the later theory."[13]

What can be seen in the history of the electron-in-chemistry is a history of the fulfillment of a long tradition, the remodeling, but not the disposal, of the classic theories of chemistry, particularly the framework of classical structural chemistry in organic chemistry. The electron theory became a buttress for the well-established theories and the enormous content of a flourishing discipline of organic chemistry rather than a replacement for earlier foundations.

I turn now to some of the steps by which the electron theory made its way into chemistry, by way of emphasizing three areas of investigation and debate among chemists: first, the acceptance of the electron as a material particle in the valence bond; second, the electron as a crucial participant in organic reaction mechanisms; and, third, the behavior of the electron as an explanation of the chemistry of conjugated molecules like benzene. I will conclude with a few remarks about the gradualist tradition in chemistry.

THE ELECTRON AND CHEMICAL VALENCE

In 1903 J. J. Thomson gave lectures at Yale University in which he included two proposals of particular interest here. One was his hypothesis of the "plum-pudding" model of the atom, associating the chemical properties of the groups in Mendeleev's periodic table of the elements with hypothetical numbers of electrons in concentric spheres within a material atom.

This was not the first proposal linking electrons to the periodic table. The physical chemists Richard Abegg and Guido Bödlander had published a paper in 1901 suggesting a system of positive and negative valency numbers for each element, so that the sum of the positive and negative numbers, neglecting signs, was always eight. The positive valency was taken to be the number of places occupied by electrons in a neutral atom, and the electrons were thought to be detachable, accounting for chemical instability and recombination.[14]

In his Yale lectures Thomson also postulated that electrons can be represented by Faraday tubes of force linking atoms within the chemical molecule. This conception of Faraday tubes places Thomson squarely within the nineteenth-century tradition of the ether and indeed he could be found using ether theories as late as 1923.[15] Since Faraday tubes are directional, rising out of positive charge and ending in negative charge, Thomson's proposal meant there must be polarities within the molecule, in other words, in organic molecules as well as in inorganic molecules. Thus, two molecules of ethylene, for all the fact that they are *structurally* identical, are *electrically* different if a Faraday tube of force running between two central carbon atoms has a different direction in one ethylene molecule than in another ethylene molecule.

Harry S. Fry, a chemist at the University of Cincinnati, became one of several American chemists to use directional or arrow formulas for what he called "electromers," that is, organic molecules that are electrically different isomers, isomers beings chemical molecules with the same atomic composition but with differing physical or chemical properties.[16]

G. N. Lewis was another American chemist who became interested in Thomson's applications of electrons to chemistry. Lewis began thinking about these matters while studying at MIT in the early 1910s.[17] In 1916 he proposed an atom-model in which eight electrons are arranged at the corners of a cube rather than in spheres or circles. He further introduced the idea that a pair of electrons is the most fundamental of all possible electron groupings so that an electron octet consists of four "duplets."

Lewis suggested that two atoms may share an electron pair between them, constituting the chemical bond, but that these two electrons may distribute themselves along a range of distances between themselves and between the atoms. Thus, there is only one kind of bond, ranging from strongly polar to nonpolar in character, for both organic and inorganic molecules.[18] In contrast to Lewis, in 1919 Irving Langmuir gave the names "covalent" and "electrovalent" to chemical bonds, making a distinction between nonpolar and strongly polar bonds. Langmuir also reintroduced the notion of electrons layered in "sheaths" surrounding the nucleus rather than using Lewis's notion of the cube.[19]

Two years later, in 1921, Langmuir lectured before a joint session of the chemistry and physics sections of the British Association for the Advancement of Science (BAAS). Langmuir's BAAS lecture in Edinburgh brought the electron-pair theory to the attention of many British chemists who had hardly noticed it during the course of World War I and its aftermath.[20] Langmuir, like Lewis, was concerned about the need to explain why paired electrons would not repel each other, and Langmuir hoped that an answer might lie in Niels Bohr's new quantum mechanics of the electron.

Langmuir was one of few chemists who showed an early interest in Bohr's 1913 series of papers "On the Constitution of Atoms and Molecules." What initially intrigued Langmuir was Bohr's calculation of the heat formation of the hydrogen molecule, a problem on which Langmuir was working at the General Electric Research Laboratory in Schenectady, New York. But Langmuir's proposal of a "quantum force" that counterbalances Coulombic force got nowhere.[21]

Nor was there much interest at this time among the majority of chemists in Bohr's work, even though Bohr sought to improve his interpretation of the relationship between the periodic table and arrangements of electrons in the early 1920s by substituting the notion of "shells" for the earlier idea of "orbits" and, in the spirit of Thomson and Abegg, relating the filling of these shells to properties of groups within the periodic table.[22]

Many chemists claimed in the early 1920s, as has the philosopher Eric Scerri more recently, that Bohr's shell configurations were arrived at by intuition and by reference to well-known chemical and spectroscopic evidence, thus offering nothing new or important to chemists. Bohr's use of what he called the "Aufbauprinzip" met considerable criticism, as well, from physicists like Werner Heisenberg and Wolfgang Pauli for what they thought was its chemical and ad hoc basis.[23]

In addition to chemists' contempt, because Bohr's approach was based on a planar atomic model, it could not account for the classical tetrahedron valence-structure of the carbon atom and for the stereochemistry of organic molecules. Nor could Bohr account for covalent bonds, because the radii of the Bohr orbits were calculated on the basis of a Coulombic force model. Although Bohr discussed the molecules H_2, HCl, H_2O, and CH_4 in his 1913 papers, physicists and physical chemists who were extending Bohr's work in the 1920s concentrated on the hydrogen molecule and the hydrogen molecular ion H2+ which were of no particular interest to chemists.[24]

Many chemists, including G. N. Lewis, saw nothing helpful to chemistry in Bohr's dynamical model of electrons rotating in a circle at a right angle to the axis connecting two atoms. Lewis wrote a paper critical of Bohr's dynamic atom in 1917 as did J. J. Thomson in 1919. After reading Thomson's paper, Lewis wrote Thomson to express his delight at finding "what an extraordinary similarity there is between the ideas which were forced upon me chiefly from chemical considerations and those to which you have been led chiefly from physical considerations."[25]

By 1923, largely because of his own research interests, the English physical chemist Thomas M. Lowry thought it was time to organize a Faraday Society conference at Cambridge University on the "The Electronic Theory of Valence." The temperature was 86 degrees in the shade and the lecture theater was even hotter, accounting for some of the short tempers that flared at the meeting. G. N. Lewis delivered the opening address.[26]

In the previous year, Bohr had taken the opportunity of his Nobel Prize lecture in Stockholm to criticize attempts by Lewis, Thomson, Langmuir, and others to retain a static atom, arguing that a static distribution of charges must be unstable if Coulomb's law applies at all. Reports of Bohr's lecture were fresh in the minds of Lewis and other chemists at the Cambridge meeting.

In his opening address, Lewis expressed some openness toward reconsideration of a dynamically modeled atom, as did the Oxford chemist Nevil Sidgwick in a considerably more positive tone. Sidgwick suggested a reconciliation of the Lewis electron-pair and the Bohr dynamic atom in a molecule that contains pairs of electrons holding atomic nuclei together, while the electrons move on orbits of the Bohr-Sommerfeld type.[27] Within a few years Sidgwick published his widely read book *The Electronic Theory of Valency,* applying electronic theories to a broad range of chemical compounds.[28]

Sidgwick was unusual among chemists, however, in his considerable sympathy for early developments in quantum theory and his attempt to understand their possible implications for chemistry. In contrast, Lewis's initial reservations about quantum mechanics gained considerable currency in the early 1920s through his book *Valence and the Structure of Atoms and Molecules* which appeared in 1923. Toward the conclusion of the book, Lewis characterized the quantum theory as "the entering wedge of scientific bolshevism" into chemistry.[29]

REACTION MECHANISMS IN ORGANIC CHEMISTRY

The most heated arguments at Cambridge in the summer of 1923 had to do not with Bohr and quantum mechanics, however, but with the use of notions of polarity in explaining the chemistry of organic molecules. Part of the argument was over the very existence of the electron-pair bond. Bernard Flürscheim, an independent chemist with a laboratory in Hampshire, was still committed in the 1920s to the chemical theories and the affinity language of Johannes Thiele and Alfred Werner.

Flürscheim offered the view that any talk of an electron bond was foolishness and that "atoms are not linked by the interposition of electrons." Instead, Flurscheim employed the model of the spatial distribution of chemical affinity over the surface of an atom sphere, with the inference that this affinity is indefinitely divisible. This notion of affinity allowed for free, bound, and partial affinities among atoms.[30]

At the Cambridge meeting, Flürscheim's views were seconded by Jocelyn Thorpe, director of the chemical laboratories at Imperial College London. Christopher Ingold, a lecturer at Imperial College, also was identified with the affinity group, although he was not present at the Cambridge meeting because he was at that moment on his honeymoon in Wales with an Imperial College postgraduate student, Edith Hilda Usherwood.[31]

Thomas Lowry took the view at Cambridge in 1923 that "the electron has come and has come to stay" and that chemists and physicists now should work together to investigate the electronic structure of molecules. But, in contrast to Lewis, on the one hand, and Flürscheim, on the other, Lowry favored the hypothesis that a double bond between atoms may be composed of electrons constituting one covalent and one electrovalent bond. In organic chemistry, as in inorganic chemistry, chemical reaction may be due to forces between molecular ions that develop due to conditions at the moment of chemical activation. Lowry proposed a dynamic process that makes possible hydrogen transfers and electron transfers within a molecule, creating ionic charges and directing the course of chemical reactions.[32] His views were to have considerable influence in France, where they played a role in the development of a theory of reaction mechanisms by the Parisian chemists Albert Kirrmann and Charles Prévost in the late 1920s, but they were not immediately persuasive to Lewis or Flürscheim in 1923.[33]

Finally, another point of view was that of Arthur Lapworth and Robert Robinson. From as early as 1901 Lapworth had played with the idea of latent polarities or ionizations within the organic molecule, and by 1923 he had long been using a system of + and − labeling signs to express the polar characteristics of atoms within a molecule at the instant of chemical transformation. Lapworth especially emphasized what he called the enhanced positive polar character of hydrogen atoms in relation to a "key" carbonyl group.[34]

From 1916, when Robinson was teaching in Liverpool, through the 1920s, when Robinson joined Lapworth as a colleague in the chemistry department at Manchester, the two men worked both together and in parallel on theories of organic reaction mechanism, with Robinson coming to the conclusion in 1920 that the activation of molecules through a rearrangement of internal valences was "most probably synonymous with changes in position of the electrons."[35]

This was not yet a theory involving electron pairs, since Robinson represented a valence bond as two or three dotted lines still in the Thiele tradition. But by the end of 1921 Lapworth was dividing a valence bond into two, not three, partial valencies, each equal to one shared electron,[36] thereby making a transition from affinities and tubes of force to electrons. In 1922 Robinson coauthored a paper with W. O. Kermack interpreting partial valencies on an electronic basis, although they still allowed the possibility for three-electron bonds in conjugated or aromatic compounds like benzene.[37] In their correspondence and in the chemistry department's staff common room, Lapworth and Robinson could be found covering letter paper and backs of old envelopes not only with + and − signs but also with arrows indicating the dynamic ideas of the *drift* and *movement* of electrons from one part of a molecule to another.[38]

These signs, the "curly arrows" representing electron displacements along the carbon chain, first appeared in print in 1925.[39] Lapworth and Robinson used them to help predict substitution patterns in aromatic molecules in what had become a confrontation between the classical theory of affinities and the rival theory of electrons. This competition took place between Lapworth and Robinson, on the one hand, and Thorpe, Ingold, and Flürscheim, on the other.

It was at the Cambridge meeting in 1923 that the first challenge was made. Flürscheim claimed on that occasion that he had a list of over 100 facts that disproved the electronic theories of valence. For the next three years, laboratory results were published back and forth in a series of claims and counterdemonstrations that have been described elsewhere in detail, most recently in an engaging new biography of Christopher Ingold by the Canadian chemist Kenneth Leffek.[40]

By 1926 Ingold recognized that his laboratory group at Leeds had made several mistakes in analyzing the identity of reaction products (particularly, substitution in benzylamines). He also had learned a great deal from the arguments of his antagonists, as Robinson and Lapworth refined the predictions that they could make about aromatic substitution on the basis of electron displacements. In 1926 Ingold came around to his opponents' viewpoint in a surprising paper he coauthored with Hilda Usherwood Ingold. They dealt in the paper with the nitration of ortho aminophenol derivatives and rationalized the meaning of Flürscheim's classical use of free and bound affinity by the new electronic interpretation. Ingold and Ingold not only employed Robinson's notation of curly arrows but they also introduced a new delta notation for polarity within the molecule as well as other new terms. The Manchester group, somewhat stunned, became convinced that Christopher Ingold had stolen their show, referring to this paper as "Ingold's conversion."[41] (See figure 11.2.)

In a series of papers from 1926 to 1934 Ingold developed a system of explanation of reaction mechanisms in organic chemistry that incorporated nearly everyone's ideas and related electronic effects conceptually to the Werner-Thiele-Flürscheim affinity tradition. In doing this, Ingold distinguished two types of electronic displacement. First, the "inductive" effect, as discussed by Lewis and Langmuir, is due to unequal sharing of electrons in an electron pair; it is an effect analogous to electrostatic induction and it is designated by a straight arrow. A second effect, which Ingold called "tautomeric" in the late 1920s, is an electronic interpretation of the Flürscheim notion of residual affinity; this effect is an activation mechanism represented by the curved arrow of Robinson. The inductive effect may be permanent ("polarization") or temporary ("polarizability") as may be the tautomeric effect. In a paper on free radicals with H. Burton in 1929 Ingold reiterated that there can be a permanent tautomeric state which, as discussed below,

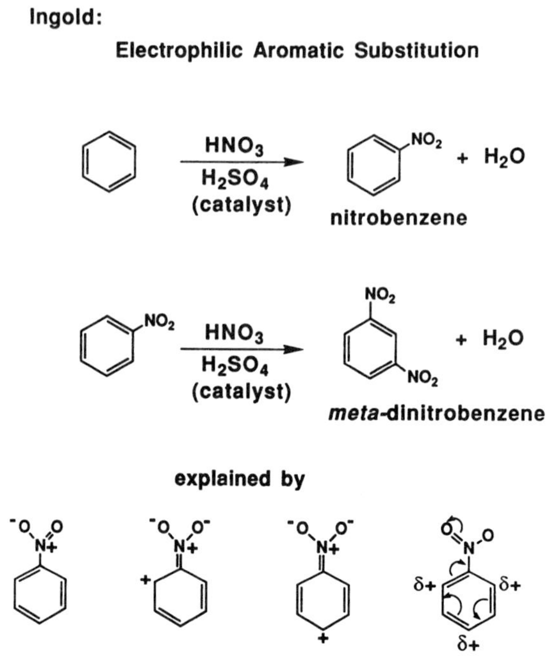

Ingold:

Electrophilic Aromatic Substitution

[Reaction 1: benzene + HNO₃ / H₂SO₄ (catalyst) → nitrobenzene + H₂O]

[Reaction 2: nitrobenzene + HNO₃ / H₂SO₄ (catalyst) → *meta*-dinitrobenzene + H₂O]

explained by

[Resonance structures of nitrobenzene showing delocalized charge distribution with $\delta+$ partial charges]

Figure 11.2
The Ingold notation.

corresponds to a nonlocalized distribution of electric charge, to which he gave the name "mesomerism" in 1933.[42]

Moving from Leeds to University College London in 1930, Ingold began a long and fruitful collaboration with Edward D. Hughes, who joined him from Wales. It was in 1934 that Ingold's summary of his work at Leeds and London appears in *Chemical Reviews,* introducing to a broad array of chemical readers what is the now-familiar terminology of "nucleophilic" and "electrophilic" reagents, eventually displacing forever Robinson and Lapworth's rival terminology of "anionoid" and "cationoid" reagents. Ingold created a compelling picture in chemistry of "nucleus-seeking" and "electron-seeking" reagents that worked by means of "substitution" (SN1, SN2) and "elimination" (E1, E2) of molecular fragments. His book *Structure and Mechanism in Organic Chemistry,* which appeared in 1953, summed up the electron theory of chemistry, not just organic chemistry, but all of chemistry: "chemical change is an electrical transaction and . . . reagents act by virtue of a constitutional affinity, either for electrons or for atomic nuclei. . . . Sharing economizes electrons."[43]

Benzene and Quantum Mechanics

Robert Kohler wrote of Ingold that he seemed a "traditional Organiker" at the end of 1924.[44] Recalling Kohler's remark, in combination with Ingold's own statement from 1941 that his goal was "to establish Organic Chemistry as a physical science by elucidating its processes in physical terms,"[45] Kent Schofield asks in a recent article, what changed Ingold into a *physical* organic chemist?

One clue comes from a lecture that Ingold gave at Imperial College in the fall of 1922 on "Some Aspects of the Quantum Theory." Here he described how Bohr's theory of 1913 gives a quantitative interpretation of the spectrum of hydrogen and how it may be applied to the specific heats of gases. As we have noted earlier, this problem of specific heats was the one that had interested Langmuir in Bohr's theory. Perhaps not coincidentally, the investigation of specific heat of gases was also the subject of research at this time of Edith Hilda Usherwood.[46]

A consistent theme in Ingold's interests from the early 1920s was the problem of the structure of benzene or rather the interpretation of its valencies. In his first published paper on the structure of benzene in 1922, Ingold stated his aim of unifying aliphatic and aromatic chemistry by studies of ring formation in unsaturated systems. At this time, he was using the classical notions of partial (Thiele) and oscillating (Kekulé) valency, in contrast to current work by Lowry, Howard Lucas, and Fritz Arndt who were speculating that there might be inner ionic charges in the benzene ring.[47]

These chemists were postulating that a molecule like benzene might have a structure in its "real" normal state that is different from any of the familiar valence-bond structures commonly used by chemists, particularly the Kekulé oscillating structure, the Dewar bridged structure, or the Armstong centric structure. Similarly, by 1926, Christopher and Hilda Ingold proposed in their joint paper that the use of noncharged benzene formulas and internally charged benzene formulas may express the "direction of imaginary gross changes which actually do not at any time proceed to more than a limited (in some cases an exceedingly small) extent."[48]

In the 1929 paper with Burton on free radicals, Christopher Ingold suggested that the forces responsible for the peculiarities of benzene valency might arise from a delocalization of electrons, the kind of delocalization that was permitted in Walter Heitler's and Fritz London's recent demonstration (1927) that the source of strength of the covalent bond lies in quantum resonance or the tendency of electrons to maximize their freedom from excessive localization.[49] Burton and Ingold wrote: "[W]e can ... refer to the microphysical equivalent (the exclusion principle) of the macrophysical law, which militates against the continued existence of intense, highly localised charge. In our view, this is also the ultimate cause of tautomeric change."[50]

In the triphenylmethy radical, "owing to the large number of possible positions for the free valency, the energy of degeneracy becomes comparable with the energy of the homopolar linking."[51]

In 1933 Ingold introduced the term "mesomerism," meaning "between the parts" for a time-independent tautomeric effect. In 1934 in an article in *Nature,* Ingold went on to clearly define the new *chemical* term "mesomerism," distinct from the classical term "tautomerism," saying that mesomerism refers to stable intermediate states explained *physically* by quantum resonance, referring back to the discussion of free radicals in 1929.

Ingold took considerable pains to stress preference for the word "mesomerism" over the terms "tautomerism" or "resonance" on the grounds that "mesomerism" ("between the forms") does not lead to the false impressions that alternate structural formulas are passing into each other rapidly like tautomers, for example, of enol and keto forms. Ingold also offered chemical evidence in support of the negative proposition of the nonexistence as separate chemical individuals of the isomeric or tautomeric forms often used to represent benzene.[52]

> We are here picturing the producing of real states [mesomeric forms] from unreal states [structural formulas] . . . . There can be no physical separation . . . between resonance vibrations and other electronic vibrations; it follows that the unperturbed structures . . . are only of the nature of intellectual scaffolding, and that the actual state is the mesomeric state. Chemical evidence in support of these ideas is extensive.[53]

By now, in 1934, Linus Pauling's work was well-known in which he had extended the notion of mechanical resonance to the molecule of benzene as well as to carbon dioxide and carbonate and nitrate ions. This was the work of his series of papers on the chemical bond from 1931 to 1933, in which, among other things, he explained the valences of carbon by what came to be called the "hybridization" of the s (spherical) and p (elliptical) wave functions for electrons.[54]

Quantum mechanics allowed the calculation of the relative contribution of each electron valence bond in the benzene ring to the benzene molecule, using alternative electronic structures of both the Kekulé and Dewar types. Each of the energy values for the alternate electronic structures is higher than the energy value for the real molecule, and thus the actual "resonance hybrid" is the stable form because it has the lowest energy value.[55]

Resonance explains the unusual stability of aromatic molecules in comparison to aliphatic unsaturated molecules.

Erich Hückel also had been working on the benzene problem in Leipzig, but he was using a different method than the valence-bond approach that had been developed by Heitler and London and initially preferred by Slater as well as by Pauling.[56] In contrast, Hückel employed the molecular-orbital approach of Friedrich Hund and Robert Mulliken who also were in Leipzig in 1930.[57]

Hückel and others argued that the molecular approach led to fewer misunderstandings about the chemical meaning of the quantum mechanical explanation of the conjugated bond system. Fritz London, who was in Oxford during 1933–1935, wrote a note to Sidgwick at Lincoln College in the spring of 1934 complaining about the frequent misreading of Pauling's paper on resonance. Too many people thought that Pauling's resonating mechanism was a restatement of Kekulé's hypothesis. Pauling's idea has nothing to do with Kekulé's oscillation hypothesis, wrote London. Rather, resonance merely means that the stationary state of a configuration of electrons with fixed nuclei cannot be represented by just one eigenfunction corresponding to a static electronic structure of the Lewis type.[58]

During 1932–1933 Ingold took an academic leave at Stanford and while there he spent a few days in Berkeley at Lewis's invitation, giving a lecture on organic radicals and talking extensively with Lewis and Gerald Branch.[59] Pauling was lecturing regularly at Berkeley during this period although there is no indication that the two met at this time.[60] As is well known, Pauling later emphasized the origins of the quantum mechanical theory of resonance in the chemical theory of mesomerism, an argument equally made by Ingold.[61]

> Classical structure theory was developed purely from chemical facts, without any help from physics [said Pauling in 1956]. The theory of resonance was well on its way toward formulation before quantum mechanics was discovered . . . the idea of resonance energy was then provided by quantum mechanics . . . but the theory or resonance in chemistry has gone far beyond the region of application in which any precise quantum mechanical calculations have been made . . . like the classical structure theory, [it] depends for its successful application largely upon a chemical feeling that is developed through practice.[62]

That the chemical feeling should be reinforced by physical methods is demonstrated in Ingold's own ongoing work on benzene in the 1930s. There was considerable evidence for the symmetry of the benzene molecule, rooted in chemical experience, x-ray crystallographic work, and Pauling's quantum-mechanical resonance theory. While spectrosopic theory demanded that there should be no coincidence lines, however, in Raman and infrared spectra for any molecule with a center of symmetry, experimental data indicated as many as twelve coincident frequencies for benzene.

Collaborating with coworkers in his laboratory at University College London, Ingold solved the anomaly by measuring spectra for benzene vapor, rather than benzene liquid, and the coincidences disappeared. Further investigations showed, too, that the spectra of benzene that is substituted with three deuterium atoms, destroying a center of symmetry, does show coincidences of spectra.[63]

While doing this work, Ingold corresponded with Fritz London and was assisted by Edward Teller. Teller stayed briefly with the Ingolds in their home in Edgeware, giving rise to a memorable story, although in 1993 Teller could not verify it. It was said that, when hit by a taxi in Canons Drive, the Ingolds' street, Teller responded to the taxi driver with the remark, "I hope that I have not damaged your taxi."[64]

Conclusion: Chemical Electrons and Chemical Revolutions

During the summer of 1941, after Teller had taken up a post at Georgetown University in Washington, D.C., he coauthored a book on the electronic theory of matter with the chemist Francis Rice. The book, *The Structure of Matter,* was published after World War II in 1949.[65] This book, like many others dealing with the applications of quantum mechanics, is strongly reductionist in tone. The introductory words are: "At present, atomic *physical theory* in principle enables us to calculate *all of the chemical* and most of the physical properties of matter and thus makes the science of experimental chemistry superfluous."[66] There follows a demurral by the authors that this calculation is not possible in practice.

It is not surprising that chemists resisted this kind of reductionist language, and there are many earlier instances of theoretical physicists' *hubris,* including well-known statements in the 1920s by Max Born and Paul Dirac.[67] The chemist R. P. Linstead, in a obituary essay for Jocelyn Thorpe in 1941, recalled with some (misplaced) nostalgia the era around 1900:

> It was a time when the conception of valency lacked the comparative precision which it has acquired today and when men of bold minds were able to express themsleves [sic] with freedom. The shadow of the physicist had

yet to fall upon the landscape, and organic chemists were still living in a happy Arcadian simplicity.[68]

Putting aside chemists' resentment at the occasional arrogance of physicists, however, most present-day chemists have a rather pragmatic view on the question of the electron theory as the foundation of chemistry. Brian Sutcliffe summarizes Pauling's incorporation of electrons and quantum mechanics into chemistry by saying,

> once the . . . ideas had been assimilated, chemists could carry on in much the same way as had been the case formerly but with resonance and hybrids added to their armory. And they did so with the comforting feeling that the most sophisticated theory in modern mathematical physics supported their actions.[69]

Thus, in this view, electron and quantum mechanics enriched chemical understanding without destroying it. There was no second chemical revolution and no triumph of physics in chemistry because nothing important in the subject matter of chemistry had to be discarded.

Charles Coulson was to become one of the leading theoretical chemists of the generation after Pauling and the most important popularizer of the molecular-orbital approach, which came to dominate theoretical and quantum chemistry after the 1950s.[70] But like Edward Frankland 100 years earlier, Coulson struck a cautionary tone in an after-dinner speech in Boulder, Colorado in 1960: "The concepts of classical chemistry were never completely precise. . . . Thus when we carry these concepts over into quantum chemistry we must be prepared to discover just the same mathematical unsatisfactoryness [sic]."[71]

As British chemists have noted in a recent book on modeling in chemistry, the adoption of semiempirical methods has been necessary in theoretical chemistry: a "threefold compromise between rigor, experiment, and intuition."[72]

The qualitative nature of chemical knowledge has been emphasized by Pauling, Wheland, Coulson, Ingold, and other twentieth-century chemists no less than it was emphasized by their predecessors in the eighteenth and nineteenth centuries. Jack Roberts, like many other modern-day chemists, is impressed with the mathematical tools of modern chemistry and the "fantastic use of number crunching." But Roberts, too, has his conservative side: "In my view, the results often seem sterile because, while lots of numbers are obtained, little or no qualitative understanding is provided of what those numbers mean."[73]

For chemists, for all the fact that the history of chemistry no longer figures explicitly in the chemistry curriculum, the history of chemistry is taken to be one of the gradual and systematic construction of chemical knowledge. Older chemistry journals still are consulted, because they are full of laws, facts, relationships, and procedures that remain valid.

The electron, when it came to chemistry in the early 1900s, provided an organizing principle for twentieth-century chemistry but not the only organizing principle. Quantum mechanics provided a mathematical framework but not the only mathematical framework. Physics became a part of chemistry, and many of the physical instruments and physical theories used by chemists have come to rely more and more upon the electron.

This development, however, is taken to have occurred gradually, fulfilling the presentiments of the great chemical heroes: Lavoisier and Dalton, who brought physical methods and principles into the practice of chemistry; Faraday and Berzelius, who concerned themselves with ions and electricity; Mendeleev, who had his own periodic *aufbauprinzip* for the building up of a table of elements; and Kekulé, who recognized the dynamical structure of benzene. Lewis, Langmuir, Robinson, Ingold, Pauling, Mulliken, and others brought the electron into chemistry but they made no revolution. Instead, they are seen to have fulfilled the expectations of a long chemical tradition.

NOTES

1. For fuller development of this theme and an introduction to the literature on the subject, see Mary Jo Nye, *From Chemical Philosophy to Theoretical Chemistry: Dynamics of Matter and Dynamics of Disciplines, 1800–1950* (Berkeley: University of California Press, 1993) and *Before Big Science: The Pursuit of Modern Chemistry and Physics, 1800–1940* (New York: Twayne, and London: Prentice-Hall, 1996), esp. chapter 5.

2. Brian T. Sutcliffe, "The Development of the Idea of a Chemical Bond," *International Journal of Quantum Chemistry*, 1996, 58: 645–655, on 645. For Franklin's remarks, see Edward Frankland, *Lecture Notes for Chemical Students: Embracing Mineral and Organic Chemistry* (London: John Van Voorst, 1866), 25.

3. Sutcliffe (1996), 646–647.

4. Alan J. Rocke, *The Quiet Revolution: Hermann Kolbe and the Science of Organic Chemistry* (Berkeley: University of California Press, 1993).

5. For the literature of "revolutions," see I. Bernard Cohen, *Revolution in Science* (Cambridge: Harvard University Press, 1985). On the "second scientific revolution," see Stephen G. Brush, *The History of Modern Science: A Guide to the Second Scientific Revolution, 1800–1950* (Ames: Iowa State University Press, 1988). Brush identifies the "Scientific Revolution" with the "fundamental change in all the sciences, such as occurred in the pe-

riod 1500–1800 as a result of the work of Nicolaus Copernicus, Galileo Galilei, Descartes, and Newton." The "second scientific revolution" is a phrase that Brush attributes to Thomas S. Kuhn for the "successful quantification of the Baconian sciences" primarily in the nineteenth century. Brush employs the term for the "equally revolutionary developments in physics in the period 1895–1925, in biology after 1859, in psychology in the period 1895–1920, and in the explosive [!] outcome of nuclear physics in 1945" (4–5). Frank Sulloway's *Born to Rebel* includes data from twenty-eight scientific debates, including only one controversy in chemistry, the "chemical revolution" associated with Lavoisier. In Frank Sulloway, *Born to Rebel: Birth Order, Family Dynamics, and Creative Lives* (New York: Pantheon, 1996), 384–385.

6. Howard Lucas, *Organic Chemistry* (New York: American Book Co., 1935). See John D. Roberts, "The Beginnings of Physical Organic Chemistry in the United States," *Bulletin for the History of Chemistry*, no. 19 (1996): *C. K. Ingold: Master and Mandarin of Physical Organic Chemistry*: 48–56, on 50. Roberts recalls that when he was an undergraduate at UCLA in 1936, the chairman of the chemistry department, William Conger Morgan, had been persuaded by a colleague to use Lucas's new textbook, but he told his students to skip the theoretical part in the first chapter because he himself could not understand it. In Jack D. Roberts, *The Right Place at the Right Time* (Washington, D.C.: American Chemical Society, 1990), 11. On the electron in organic chemistry, see G. V. Bykov, "Historical Sketch of the Electron Theories of Organic Chemistry," *Chymia*, 1965, 10: 199–253; and Jennifer Seddon, "The Development of Electronic Theory in Organic Chemistry," St. Hugh's College, Oxford University Honors Thesis, 1972.

7. G. E. K. Branch and Melvin Calvin, *The Theory of Organic Chemistry* (New York: Prentice-Hall, 1941). Quoted from Melvin Calvin, *Following the Trail of Light: A Scientific Odyssey* (Washington, D.C.: American Chemical Society, 1992), in Roberts (1996), on 49.

8. George Willard Wheland, *The Theory of Resonance* (New York: Wiley, 1941), followed by *Resonance in Organic Chemistry* (New York: Wiley, 1955), and *Advanced Organic Chemistry* (New York: Wiley, 1957). See Theodor Benfey, "Teaching Chemistry Embedded in History: Reflections on C. K. Ingold's Influence as Historian and Educator," *Bull. Hist. Chem.*, 1996, 19: 19–24, on 20. Also see the considerable discussion of Wheland and his later textbooks on resonance in organic chemistry in Buhm Soon Park, "Teaching the Resonance Theory: Pauling, Wheland, and the Interplay of Quantum Mechanics and Chemistry," August 1995, ms., 33 pages Park is completing a Ph.D. dissertation at Johns Hopkins University.

9. Edward Remick's *Electronic Interpretations of Organic Chemistry* (New York: Wiley, 1943) was the first book in North America to use the Ingold-Hughes interpretation of organic reaction mechanisms, which is discussed later in this chapter. See Kenneth T. Leffek, *Sir Christopher Ingold: A Major Prophet of Organic Chemistry* (Victoria, B.C.: Nova Lion Press, 1996), 219. More important than Branch and Calvin's book in establishing a well-defined field of physical organic chemistry was L. P. Hammett, *Physical Organic Chemistry* (New York: McGraw-Hill, 1940). See Martin D. Saltzmann, "The Development of Physical Organic Chemistry in the United States and the United Kingdom: 19191–1939, Parallels and Contrasts," *Journal of Chemical Education*, 1986, 63: 588–593, on 590; and Leon

Gortler, "The Physical Organic Community in the United States, 1925–1950," *Journal of Chemical Education,* 1985, 62: 753–757, on 755, 757.

One of the few early non-English-language texts incorporating electron theory into organic chemistry was a revised edition of Walter Hückel's *Theoretische Grundlagen der organischen Chemie,* published in 1948. See Walter Hückel, *Theoretische Grundlagen der organischen der Chemie,* 2 vols. (Leipzig: Akademische Verlagsgesellschaft, 1931), revised as Walter Hückel and F. Seel, eds. *Theoretical Organic Chemistry,* 2 vols. (Office of Military Government for Germany, Field Information Agencies Technical, British, French, U.S. ca. 1948), with French preface and German texts. Walter Hückel's brother was the theoretical physicist Erich Hückel who is discussed later in this chapter. See Ralph Oesper, "Walter Hückel," *Journal of Chemical Education,* 1950, 27: 625.

10. Quoted from a typewritten document, written by C. K. Ingold during 1965–1968, in Leffek (1996), 235.

11. This was during Pauling's junior year in college, 1920–1921. See Thomas Hager, *Force of Nature: The Life of Linus Pauling* (New York: Simon and Schuster, 1995); Ted and Ben Goertzel, *Linus Pauling: A Life in Science and Politics* (New York: Basic Books, 1995); and Barbara Marinacci, ed. *Linus Pauling in His Own Words. Selected Writings, Speeches and Interviews* (New York: Simon and Schuster, 1995).

12. Linus Pauling, *General Chemistry* (Pasadena: California Institute of Technology, 1941), in Ava Helen and Linus Pauling Papers, Oregon State University Special Collections, Valley Library: 449:1. This bound, lithographed volume became the basis for the *General Chemistry: An Introduction to Descriptive and Modern Chemical Theory* (San Francisco: William H. Freeman, 1947). I have written about the textbook in "From Student to Teacher: Linus Pauling and the Reformulation of the Principles of Chemistry in the 1930s," in Bernadette Bensaude-Vincent and Anders Lundgren, eds. *Communicating Chemistry: Textbooks and Their Audiences,* for a forthcoming issue of *Ambix.*

13. Linus Pauling and E. Bright Wilson, Jr., *Introduction to Quantum Mechanics with Applications to Chemistry* (New York: McGraw-Hill, 1935; Dover edition, 1985), 1.

14. See Richard Abegg, "Die Valenz und das periodische System. Versuch einer Theories der Molekular-Verbindungen," *Zeitschrift für anorganische Chemie,* 1904, 39: 330–380. In general, see Colin Russell, *The History of Valency* (New York: Humanities, Press, 1971) and Anthony N. Stranges, *Electrons and Valence: Development of the Theory, 1900–1925* (College Station: Texas A & M University Press, 1982).

15. J. J. Thomson, *The Electron in Chemistry* (London: Chapman and Hall, 1923).

16. J. J. Thomson, *Electricity and Matter* (Westminster: Constable, 1904), 117–120, 133–135. Harry Shipley Fry, *The Electronic Conception of Valence and the Constitution of Benzene* (London: Longmans, Green, and Co., 1921), 4–5, 48, 272–273, citing earlier papers from 1908–1911.

17. See John Servos, *Physical Chemistry from Ostwald to Pauling: The Making of a Science in America* (Princeton: Princeton University Press, 1990), 131–133; and Robert E. Kohler, Jr., "The Origin of G. N. Lewis's Theory of the Shared Pair Bond," *Historical Studies in the Physical Sciences,* 1971, 3: 343–376.

18. G. N. Lewis, "The Atom and the Molecule," *Journal of the American Chemical Society,* 1916, 38: 762–785. Lewis had lectured on the cubic octet as early as 1902. See Stranges, 173, 219.

19. Irving Langmuir, "The Arrangement of Electrons in Atoms and Molecules," *Journal of the American Chemical Society,* 1919, 41: 868–934; and "Isomorphism, Isoterism and Covalence," ibid.: 1543–1559. When criticized by Lewis for allowing chemists to refer to the electron-pair valence-bond theory as the "Lewis-Langmuir theory," Langmuir reminded Lewis that Johannes Stark had proposed in 1908 that a pair of electrons held in common between adjacent atoms constitutes the valence bond. In a letter from Irving Langmuir to G. N. Lewis, 3 April 1920, G. N. Lewis Papers, Bancroft Library, University of California at Berkeley.

20. Irving Langmuir, "The Structure of Molecules," *BAAS Reports. Edinburgh. 1921* (1922): 468–469.

21. On this, see Helge Kragh, "Bohr's Atomic Theory and the Chemists, 1913–1925," *Rivista di storia della scienza,* 1985, 2: 463–485, on 470, 477–481.

22. Niels Bohr, "On the Constitution of Atoms and Molecules, Pt. I, Binding of Electrons by Positive Nuclei," *Philosophical Magazine,* 1913, 26: 1–25; "Pt. II, Systems containing Only a Single Nucleus," ibid.: 476–502; "Pt. III, Systems Containing Several Nuclei," ibid.: 857–875. See John Heilbron and Thomas S. Kuhn, "The Genesis of the Bohr Atom," *Historical Studies in the Physical Sciences,* 1969, 1: 211–290.

23. See Eric Scerri, "Correspondence and Reduction in Chemistry," in S. French and H. Kamminga, eds. *Correspondence, Invariance and Heuristics* (Dordrecht: Kluwer, 1993): 45–64, on 48–51. Scerri calls attention to Bohr's arrangement of electrons for nitrogen, oxygen, and fluorine, for example.

24. See Kragh (1985).

25. Letter from G. N. Lewis to J. J. Thomson, 13 August 1919, G. N. Lewis Papers, Bancroft Library, University of California at Berkeley. Quoted in Nye (1993), 135. See G. N. Lewis, "The Static Atom," *Science,* 1917, 46: 297–302; and J. J. Thomson, "On the Origin of Spectra and Planck's Law," *Philosophical Magazine,* 1919, 37: 419–446.

26. G. N. Lewis, "The Electronic Theory of Valency," in *Transactions of the Faraday Society,* 1923, 19: 450–543. For an account of the conference, see G. Norman Burkhardt, "Arthur Lapworth and Others: Structure, Properties, and Mechanism of Reactions of Carbon Compounds: Some Developments 1898 to 1939, with Particular Reference to Arthur Lapworth and His Work," typescript, 150 pages, with addenda, written 1973–1978, bound and deposited at the Royal Society, London, 1980: on 89.

27. See Sutcliffe (1996), 649, as well as papers in volume 19 of *Transactions of the Faraday Society* (1923).

28. Nevil V. Sidgwick, *The Electronic Theory of Valence* (London: Oxford University Press, 1927). See Keith J. Laidler, "Contrasts in Chemical Style: Sidgwick and Eyring," Dexter Award Address, American Chemical Society, 15 April 1997, 11 page ms.

29. Quoted in Sutcliffe (1996), 649.

30. See K. Schofield, "The Development of Ingold's System of Organic Chemistry," *Ambix*, 1994, 41: 87–107, on 91, quoting from Bernard Flürscheim, "The Relation between the Strength of Acids and Bases and the Quantitative Distribution of Affinity in the Molecule," *Journal of the Chemical Society, Transactions*, 1909, 95: 178–734, on 718, a view reitered in *Trans. Faraday Soc.*, 1923, 19: 531–535.

31. For valuable information about the life and work of Edith Hilda Usherwood Ingold, see Leffek (1996), esp. 58–73.

32. Thomas Lowry, "Applications in Organic Chemistry of the Electronic Theory of Valence" and "Intramolecular Ionisation in Organic Compounds," *Trans. Faraday Soc.*, 1923, 19: 485–487 and 487–496. Also see Lowry, "Preuves expérimentales de l'existence des doubles liaisons semi-polaires," *Bulletin de la Société Chimique de France*, 1926, 39: 203–206.

33. See Charles Prévost and Albert Kirrmann, "Essai d'une théorie ionique des réactions organiques: Premier mémoire," *Bulletin de la Société Chimique de France*, 1931, 49: 194–243; "Essai . . . Deuxième mémoire," *Bull. Soc. Chim. France*, 1931, 49: 1309–1368; and "La tautomérie anneau-chaine, et la notion de synionie," *Bull. Soc. Chim. France*, 1933, 53: 253–260; discussed in Nye (1993), 151–157.

34. Arthur Lapworth, "The Form of Change in Organic Compounds, and the function of the alpha-meta-Orientating Groups," *Transactions of the Chemical Society*, 1901, 79, Pt. II: 1265–1284, esp. 1265–1267, 1276–1277; and Arthur Lapworth, "Latent Polarities of Atoms and Mechanism of Reaction, with Special Reference to Carbonyl Compounds," *Memoirs of the Manchester Literary and Philosophical Society*, (1919–1920), 64, no. 3: 1–16. All this is discussed in Nye (1993) and in Nye, "Chemical Explanation and Physical Dynamics: Two Research Schools at the First Solvay Chemistry Conferences, 1922–1928," *Annals of Science*, 1989, 46: 461–480.

35. Robert Robinson, "The Conjugation of Partial Valences," *Mem. Manchester LPS*, 1920, 64, no. 4: 1–14.

36. Letter from Lapworth to Robinson, 30 December 1921, Robert Robinson Papers, Royal Society, London: D.40; also see Burkhardt (1980), 88–89. Lapworth, "A Theoretical Derivation of the Principle of Induced Alternate Polarities," *Journal of the Chemical Society*, 1922, 121: 416–427.

37. W. O. Kermack and Robert Robinson, "An Explanation of the Property of Induced Polarity of Atoms and an Interpretation of the Theory of Partial Valencies on an Electronic Basis," *Journal of the Chemical Society*, 1922, 121: 427.

38. Burkhardt (1980), 31.

39. Robert Robinson, "Polarisation of Nitrosobenzene," *Journal of the Society of Chemistry and Industry* [later, *Chemisry and Industry*], 1925, 44: 456–458.

40. Leffek (1996), 74–95. Also see J. Shorter, "Electronic Theories of Organic Chemistry: Robinson and Ingold," *Natural Product Reports: Royal Society of Chemsitry*, 4 (1987), 61–66. Flürscheim's challenge is in *Trans. Faraday Soc.*, 1923, 19: 531–533.

41. E. H. Ingold and C. K. Ingold, "The Nature of the Alternating Effect in Carbon Chains. Part V. A Discussion of Aromatic Substitution with Special Reference to the Respective Roles of Polar and Non-Polar Dissociation; and a Further Study of the Relative Directive Efficiencies of Oxygen and Nitrogen," *Journal of the Chemical Society*, 1926, 129: 1310–1328. See Burckhardt (1980), 140. In a letter meant to reassure Robinson about Ingold, Thorpe wrote Robinson: "When I read the *electronic* part I regarded it as based on your views and, in fact, to offer support to them on the experimental side. It seemed to me that your views were so well known that no one would fail to see the source of inspiration." Letters from Jocelyn Thorpe to Robert Robinson, 4 July 1926, and 11 July 1926, Royal Society, London, Robinson Papers: D.33, quoted in Nye (1993), 192.

42. H. Burton and C. K. Ingold, " The Existence and Stability of Free Radicals," *Proceedings of the Leeds Philosophical Society*, 1929, 1: 421–431, and C. K. Ingold, "Significance of Tautomerism and of the Reaction of Aromatic Compounds in the Electronic Theory of Organic Reactions," *Journal of the Chemical Society*, 1933: 1120–1127, discussed in Saltzmann (1996), 29–30. Also, see Nye (1993), 207–211.

43. Ingold, *Structure and Mechanism in Organic Chemistry* (London: Bell, 1953), 3–4.

44. Robert E. Kohler, "The Lewis-Langmuir Theory of Valence and the Chemical Community, 1920–1928," *Historical Studies in the Physical Sciences*, 1975, v. 6, cited in Schofield (1994), 93.

45. Ingold, *Trans. Faraday Soc.*, 1941, 37: 602.

46. Leffek (1996), 54–55.

47. See E. Campaigne, "The Contributions of Fritz Arndt to Resonance Theory," *Journal of Chemical Education*, 1959, 36: 336–339. James B. Conant also speculated on uses of the Lewis electron theory in organic chemistry about this time. See Martin Saltzmann, "C. K. Ingold's Development of the Concept of Mesomerism," *Bull. Hist. Chem.* (1996): 25–32, on 25.

48. Ingold and Ingold (1926), 1312.

49. C. K. Ingold and H. Burton, "The Existence and Stability of Free Radicals," *Proceedings of the Leeds Philosophical Society*, 1929, 1: 421, cited by Ingold in *Structure and Mechanism*, 83–84 and by C. W. Shoppee, "Christopher Kelk Ingold," *Biographical Memoirs of Fellows of the Royal Society*, 1972, 18: 349–410, 355. On the development of the theory of exchange forces, see Cathryn Carson, "The Peculiar Notion of Exchange Forces. I: Origins in Quantum Mechanics, 1926–1928" and "II: From Nuclear Forces to QED, 1929–1950," *Studies in the History and Philosophy of Modern Physics*, 1996, 27: 23–45 and 99–131.

50. Quoted in Saltzmann (1996), 29.

51. Ibid., 30.

52. Ingold, *Chemical Reviews*, 1934, 252–253.

53. Ingold, *Chemical Reviews* (1934), quotation from 250–252; also see 256.

54. Linus Pauling, "The Nature of the Chemical Bond: Applicatoins of Results Obtained from Quantum Mechanics and from a Theory of Paramagnetic Susceptibility to the Structure of Molecules," *Journal of the American Chemical Society,* 1931, 53: 1367–1400.

55. Linus Pauling and George W. Wheland, "The Nature of the Chemical Bond. V. The Quantum-Mechanical Calculation of the Resonance Energy of Benzene and Naphthalene and the Hydrocarbon Free Radicals," *Journal of Chemical Physics,* 1933, 1 : 362–374 and other papers.

56. See Nye (1993), 243–246.

57. On the different methods, see Kostas Gavroglu and Ana Simoes, "The Americans, the Germans, and the Beginnings of Quantum Chemistry: The Confluence of Diverging Traditions," *Historical Studies in the Physical and Biological Sciences,* 1994, 25: 47–110; along with Ana Simoes, "Converging Trajectories, Diverging Traditions: Chemical Bond, Valence, Quantum Mechanics, and Chemistry, 1927–1937," University of Maryland Ph.D. thesis, 1993; and Kostas Gavroglu, *Fritz London: A Scientific Biography* (Cambridge: Cambridge University Press, 1995). Stephen G. Brush recently completed a manuscript on the predictive power of theories of benzene from Kekulé through Pauling and Mulliken to the present time: "Dynamics of Theory Change in Chemistry: The Benzene Problem," 115 page ms.

58. Letter from Fritz London to Nevil Sidgwick, 10 May 1934, Sidgwick Papers, Lincoln College, Oxford.

59. Leffek (1996), 137.

60. Beginning in 1929, Pauling lectured in Berkeley every year for five years. See Gavroglu and Simoes, 81; and Hager, *Force of Nature*.

61. Pauling's statements date from 1956, at a time when Pauling's benzene resonance theory was under attack in the Soviet Union for its alleged "idealism." See Loren Graham, "A Soviet Marxist View of Structural Chemistry: The Theory of Resonance Controversy," *Isis,* 1964, 55: 20–31. The triumph of the molecular orbital theory over Pauling's valence bond/atomic orbital theory may be interpeted as due, in part, to the triumph of the argument against "idealism" in favor of "realism."

62. Pauling, in "The Nature of Resonance," in [Sir] Alexander Todd, ed. *Perspectives in Organic Chemistry* (New York: Interscience Publishers, 1956): 1–8, on 7. Also see Ingold, "Mesomerism and Tautomerism," *Nature,* 1934, 133: 946–947, on 947.

63. For the theory of spectra, see Leffek (1996), 141–143. Famously among British chemists, Ingold filled an entire issue of the *Journal of the Chemical Society* with this work: eight consecutive papers in the *Journal of the Chemical Society,* 1936, 912–986. See Derek H. R. Barton, "Ingold, Robinson, Winstein, Woodward, and I," *Bull. Hist. Chem.* (1996): 43–47, on 45.

64. Teller thought this was a great story. It originated with Keith Ingold. See Leffek (1996), 151 and 248, n. 56. While Christopher Ingold was at Leeds, Kathleen Lonsdale, who was in the Physics Department, carried out her first crystal structure determination, concluding that the benzene molecule is flat and symmetrical (Leffek (1996), 104.

65. Francis Owen Rice and Edward Teller, *The Structure of Matter* (New York: Wiley, 1949).

66. Ibid., 1.

67. Max Born, *The Constitution of Matter: Modern Atomic and Electron Theories,* trans. from 2nd. German ed., E. W. Blair and T. S. Wheeler (London: Methuen, 1923), 78; and Paul Dirac, "Quantum Mechanics of Many-Electron Systems," *Proceedings of the Royal Society of London,* 1929, A123: 714–733, on 714. Quoted in Nye (1993), 229 and 238.

68. Quoted in Schofield (1994), 88 from G. A. R. Kon and R. P. Linstead, "Sir Jocelyn Field Thorpe 1872–1940," *Journal of the Chemical Society,* 1941: 444–464, on 458.

69. Sutcliffe (1996), 649.

70. C. A. Coulson's classic text is titled, simply, *Valence* (Oxford: Clarendon Press, 1952).

71. Charles Coulson, "The Present State of Molecular Structure Calculations," *Reviews of Modern Physics,* 1960, 32: 170, quoted in Sutcliffe (1996), 651.

72. Colin J. Suckling, Keith E. Suckling, and Charles W. Suckling, *Chemistry through Models: Concepts and Applications of Modelling in Chemical Science, Technology, and Industry* (Cambridge: Cambridge University Press, 1978), 65, 135.

73. John D. Roberts, (1996), 53.

12

The Physicists' Electron and Its Appropriation by the Chemists
Kostas Gavroglu

In this chapter I argue that the historian's main themes in discussing the electron in chemistry are predominantly philosophical. The process of appropriation of the electron by chemistry is a process related to the praxis of its practitioners with respect to their views on theory building, the use of mathematics, and their relations to physics. In trying to understand the issues involved in the appropriation of the electron by chemistry, one is obliged to discuss a number of questions that eventually help to articulate the theoretical particularity of (quantum) chemistry. Discussing the role of a theoretical entity in a discipline becomes, in a way, synonymous with tracing the theoretical particularity of that discipline.

Notwithstanding the less or more violent invasions of mathematics into chemistry during the last two centuries and the remarkable successes of such invasions, the chemists' culture has been the culture of the laboratory; their theoretical constructs were always very sensitive to the exigencies of the laboratory, and theory building strongly depended on using as inputs the experimentally measured values of various parameters. The use of semi-empirical methods in constructing theoretical schemata has always had a far stronger legitimacy in chemistry than in physics. Nevertheless, a rather misplaced emphasis exclusively on the context of the laboratory, which conditioned the culture of chemistry, contributed to the concealment of the *theoretical* particularity of chemistry.

I shall not be concerned with the situation of the period following the discovery of the electron and its initial use for providing an interpretative framework for chemical as well as physical phenomena. Nor will I be discussing the attempts of Niels Bohr and G. N. Lewis to endow the electron with those quasi-classical properties which, eventually, led to the disappearance of the antithesis between the "physicists' atom" and the "chemists' atom." It can, in fact, be claimed that up to the formulation of quantum mechanics, there appears to be a consensus between chemists and physicists about the electron's benevolent role in both disciplines and in erasing the traces of existing or projected problems between the two disciplines.

What is intriguing to discuss, though, is the situation after the Schrödinger equation when physics as a whole (re)asserted—if not (re)gained—its status of being the paradigmatic science. It was the period when the triumph of the new physics turned the electron into an entity "belonging" to physics and being under the physicists' prerogative. And at the same time it was a period where chemistry started living under the specter of being reduced into physics. The reason that this period is particularly interesting here is that one witnesses chemists appropriating an entity that had become a necessary part of the physicists' theoretical framework, and it was not the case—as in the earlier periods—when both physicists and chemists had the same jurisdiction over the particular theoretical entity.

In this respect I shall, almost exclusively, deal with quantum chemistry and discuss a number of issues that come to the fore through the appropriation of the electron by the chemists and which also underline the theoretical particularity of chemistry. To discuss the appropriation of electrons by chemists after quantum mechanics means to discuss analytically the role of electrons in the various theoretical schemata about valence. And the discussion of these schemata involves issues such as *the role of theory in chemistry, the strategies of theory building, and the legitimization of the dominant discourse of quantum chemistry.*

These issues are not wholly independent from the implications of reductionism and realism for chemistry. Nevertheless, in examining the ways chemists attempted to deal theoretically with the classic problems of chemistry, the historian is invariably confronted with the chemists' particular attitude on how to construct a theory in chemistry, on what to "borrow" from physics and what is the methodological status of empirical observations for theory building. The choices made by the chemists and the schemata they proposed brought into being new research traditions, articulated new strategies of experimental manipulation, implied a different role for mathematics in each tradition, and gave rise to different styles of research within these traditions. It is the confluence of all these processes that eventually became decisive in forming the theoretical particularity of chemistry.

These characteristics, instead of being considered as processes contributing to the theoretical particularity of chemistry, are often viewed as the shortcomings of chemistry with respect to physics. Chemistry, it has been claimed, has all these characteristics because it cannot by itself and without becoming physics develop a one "true" theory at any period nor can it fully assimilate mathematics into its practice. If, however, one accepts such an argument and considers these characteristics as shortcomings rather than expressions of the theoretical particularity of chemistry, then one should claim that the history of chemistry is to a large extent a long history of devising empirical rules and innovating approximate and phenomenological theoretical schemata, and that to reach a "serious" theoretical level, chemistry always depends on physics and follows developments in physics. Such an unquestioned reign of reductionism implies a rather warped history of chemistry: the history of chemistry should be nothing but a long wait, liberating chemistry from itself by having physics provide the hoped for "true" basis for chemistry. But such considerations have no *rapport* with the history of chemistry. If they did, then the chemists' self-consciousness would be nothing more than a shared collective false consciousness. It is rather presumptuous to believe that a community and its culture could be formed by sharing the belief that what they think they do, is a job others—the physicists—could do "more" properly. Those who have physicists dreaming of a final and finished theory, by the very same token, have chemists traverse their history by living the nightmare of temporary and unfinished theories.

THE SPECTRE OF REDUCTIONISM

Though it is acknowledged that a reductionist view of biology and the social sciences is rather problematic, there is a peculiar uneasiness about chemistry. This is more so if we realize that in the case of chemistry—more than in any other discipline—the issues related to reductionism, scientific realism, and theory building are strongly interrelated and none can be dealt with separately from the rest.

Dirac's pronouncement in 1929 that since the underlying laws governing the behavior of electrons became known, to do chemistry meant to deal with equations that were in principle soluble even though in practice they may only produce approximate solutions, has often been the starting point of modern discussions about reductionism in chemistry. Dirac's comment appeared to have given the misleading impression that the question of reduction has arisen only after the advent of quantum mechanics. Furthermore, the absoluteness with which Dirac expressed his view, and the way it has been raised by some to the status of a dogma, set the tone for the ensuing discussions: the reactions, on the whole, to such a claim that all chemistry is physics, has been to strive for an argument about the absolute and full autonomy of chemistry. And as a result what had been neglected was the realization that the methodological significance of raising the discussion of reductionism in chemistry lies in the implicit understanding that what is at stake is neither the absolute reduction of chemistry to physics nor the absolute autonomy of chemistry. The two extreme cases (full reduction of chemistry to physics and no reduction at all) are both historically wrong, philosophically naïve, and lead to a methodological deadlock. The challenge lies in the ability to articulate the necessarily intermediate position of the relative autonomy of chemistry with respect to physics.

In the discussions about scientific realism, there is an implicitly shared set of values that undermine the possible contributions chemistry can provide to these philosophical issues. The special role of mathematics in physics renders the philosophical problems related to scientific realism to be, at least, unambiguously expressed. It is claimed that physics deals with the fundamental entities of the world and there are no intrinsic limitations as to how deep it could probe. Whether it studies the planets, billiard balls, atoms, nuclei, electrons, quarks, or superstrings, it is still physics, and the change of scale—though crucial to the problem of realism—does not oblige the change of the discipline itself as would be the case in biology and chemistry. But the view which confines the study of realism predominantly to the problems of physics is more a matter of convenience than something that has a serious theoretical justification, especially since such an attitude neglects the theoretical particularity of chemistry—in fact, it supposes an absolute reductionism of chemistry to physics. When talking about quarks, superstrings or the big bang, physicists do not appear to be particularly bothered by pushing ontological considerations to the background. Chemists—or practitioners of any discipline who strive to assert its autonomy and nonreducibility—do not and cannot share such an attitude with physicists for it comes into strong conflict with the constitutive principles of their practice. At times, and quantum chemistry is such a case, ontological commitments become a prerequisite for drawing disciplinary boundaries.

Ever since the end of the nineteenth century, chemists have been debating whether their science may not be the "science of bodies that do not exist." Or whether the unsettling discovery of radium implied that "in relation to the ponderable, we seem to be creating a chemistry of phantoms." The history of chemistry is also a history of the attempts of the chemists to establish its *relative* autonomy with respect to physics. Hence, and unlike the physicists, the chemists are obliged to proceed to ontological commitments that are unambiguous and clearly articulated, and they have little or no tolerance to an attitude that stipulates these may be temporary commitments. Otherwise the chemist would be at a loss about the underlying ontology and would never be sure whether chemistry should be doing the describing and physics the explaining. The chemists have passionately debated these issues, and the myth of the reflective physicist and the more pragmatic chemist is, if anything, historically untenable.

Van Fraassen has asserted that descriptive excellence at the observational level is the only genuine measure of any theory's success and that one's acceptance of a theory should create no ontological commitment whatsoever beyond the observational level. This argument has been counteracted by the claim that observational excellence or empirical adequacy is only one of the epistemic virtues among others of equal or comparable importance. And it was also argued that the ontological commitments of any theory are totally blind to the distinction of what is and what is not humanly observable and so should be our own ontological commitments. How can this *problematique* be recast to accommodate chemistry whose theoretical particularity is such as to admit the simultaneous existence of more than one theoretical schemata with "equivalent" empirical adequacy? Claiming that these considerations refer to mature and closed theories is a way of avoiding having to deal with the problem by talking, in effect, about physics rather than chemistry.

The Mysterious Bond

The use of the electron by the chemists offers an almost ideal ground for tracing these considerations that define the context of appropriation. The resolution of the paradox concerning the conflicting ontologies implied by the chemists' static atom and the physicists' dynamic atom was mediated by the electron. It happened through a two-step process: Bohr's early papers provided a theoretical framework for decoding the spectroscopic data, and Lewis's proposal concerning electron pairs provided a quasi-theoretical justification for explicating the covalent bond. The implied antithesis between the chemists' atom and the physicists' atom was dealt with only after a truly weird property of the electron (discrete orbits) was supplemented by another weird property of the electron (pairing). It should be emphasized that the atom regained its status as an entity independent of the phenomena it was used to describe only after the electron was first used by Bohr to account for spectroscopy and, then, supplemented by Lewis's proposal of electron pairs as a result of systematizing a large amount of empirical data (even number of electrons in molecules).

What was the situation during the later phase after the discovery of spin and the advent of quantum mechanics? Though spin had been used for the further decoding of the spectroscopic data, it was the realization that spin *together* with the Pauli exclusion principle can explain a whole class of phenomena, namely the covalent bonding in chemistry, which constituted the great triumph of this particular quantum number. This was shown for the first time in the joint paper of Heitler and London in 1927, but their approach appeared impotent since it was impossible to have analytical solutions for any molecule with more than two electrons. But chemistry has never been in need of explanatory schemata *at the expense* of specific solutions to real problems. Chemists have never been content with assurances that equations are *in principle soluble*. It is the demand for specifics, the demand for everything to be directly translatable into the language of the laboratory which brought about the demise of a successful schema which was the Heitler-London theory. The schema proposed by Pauling to overcome this difficulty necessitated a new concept, that of resonance. The schema, though very dear to the chemists, eventually started to wane. And what dominated the scene was Mulliken's molecular orbitals, formulated concurrently with Pauling's resonance, being more or less a schema full of rules and empirically found numbers.

Let me be more analytic about the situation after the advent of quantum mechanics. Among the first successful applications of quantum mechanics were the calculation of the energy levels of the hydrogen atom and of the hydrogen molecular ion. By the beginning of 1927 many physicists and chemists talked about the possibilities of quantum mechanics to deal with chemical problems. The simplest but deeply intriguing chemical problem was the formation of the hydrogen molecule. The "mechanism" responsible for a such a formation—the homopolar bond—was quite puzzling, since it joined two electrically neutral atoms to form a molecule. Earlier on, there had been quite a few suggestions to explain the homopolar bond, but it was only since 1916 and within the framework of the old quantum theory, that some remarkably insightful proposals by Gilbert Newton Lewis provided a rather simple rule to deal with such a puzzling bond. Lewis had proposed that chemical bonding—both the ionic type as well as the homopolar type—could be explained in terms of shared electron pairs. It was a semiempirical "theory." He had started from the observation that almost all the molecules had an even number of electrons. By requiring to have eight electrons in the outer orbits of each atom in a molecule, he argued two neutral atoms could only be joined together by the sharing of pairs of electrons between them.

Toward the end of 1926, it appeared that many chemists were slowly becoming aware of the amazing explanatory power of quantum mechanics, yet it was difficult for them to see how this newly developing explanatory framework would be assimilated into the culture of chemistry. Quite a few of them became rather apprehensive that such an assimilation may bring lasting and not altogether welcome changes to their culture; but for some it was a risk worth taking. N. V. Sidgwick in his influential book *The Electronic Theory of Valency* would have no inhibitions about letting the new quantum mechanics invade the realm of chemistry. He expressed an unreserved enthusiasm about the new quantum mechanics and adopted Lewis's theory for the nonpolar bond. Faced with the full development of the new mechanics by Heisenberg and Schrödinger, but not with an application of the theory to a chemical problem, Sidgwick in the first lines of the preface to his book attempted to clarify the methodologial stumbling block that he sensed would be in the way of his fellow chemists. He talked of the courses open to the chemist in developing a theory of valency. One alternative, he thought, would be to use symbols with no definite physical connotation to express the reactivity of the atoms in a molecule and "leave it to the subsequent progress of science to discover what realities these symbols represent." But the chemist could also adopt the concepts of atomic physics and try to explain the chemical facts in terms of these. Sidgwick chose the latter path; he was obliged to accept the physical conclusions in full, and he was convinced that he "must not assign to these entities properties which the physicists have found them not to possess." As he emphasized, the chemist must not use the terminology of physics unless "he is prepared to recognize its laws." This was in 1927, just prior to the work of Walter Heitler and Fritz London that would remove any ambivalence in the feelings of the chemists about the use of the new quantum mechanics in chemistry.

The First Applications of Quantum Mechanics to Chemistry

In 1926 London applied for a "fellowship in science" from the International Education Board, which later became part of the Rockefeller Foundation. London was born in Breslau (now Wroclaw in Poland) in 1900 into a Jewish family. His father was professor of mathematics at the universities of Breslau and, then, Bonn. London matriculated from the University of Munich in 1921 and wrote his thesis in philosophy with one of the most well-known phenomenologists of the period, Alexander Pfander. His thesis dealt with deductive systems. After a short stint in teaching at the gymnasium, London in 1923 started doing physics, first with Born at Göttingen and then with Sommerfeld at Munich, first publishing a paper in spectroscopy and then working on the theory of transformations in quantum theory. In December 1926 he was granted a fellowship and he wrote to the board that he intended to go to Zurich and work with Schrödinger. Fritz London's decision to go to Zurich hoping to work with Schrödinger would bring a lasting change to his scientific agenda. He would no longer work on the problems related with quantum theory and he would start work on problems requiring the application of quantum mechanics—first to chemistry and then to low temperature physics. He arrived in Zurich in April 1927. In Zurich and in Berlin, London did not work with Schrödinger, since Schrödinger hardly ever collaborated with anyone else. But during his short stay in Zurich, London together with Walter Heitler managed to solve one of the outstanding problems of chemistry by using the wave mechanical methods of Schrödinger. They showed that the mysterious chemical binding of two *neutral* hydrogen atoms to form a hydrogen molecule could *only* be undertood in terms of the principles of the new quantum mechanics.[1]

Undoubtedly the simultaneous presence of both Heitler and London in Zurich was one of those unplanned happy coincidences. Walter Heitler was born in Karlsruhe in 1904 to a Jewish family and his father was a professor of engineering. His interest in physical chemistry grew while he attended lectures on the subject at the Technische Hochschule and through these lectures in physical chemistry he came into contact with quantum theory. He had also acquired a strong background in mathematics. Wishing to work in theoretical physics, he first went to Berlin but found the atmosphere not too hospitable especially since a student was left to himself to choose a problem and write a thesis. Only after its completion would the "great men" examine it. After a year in Berlin he went to Munich and completed his doctoral thesis with Karl Herzberg on concentrated solutions. The writing of his thesis coincided with the development of the new quantum mechanics, but because of the kind of problems he was working on he never had the opportunity to study the new developments in any systematic manner. After completing his thesis, Sommerfeld helped him secure funding from the International Education Board, and he went to Copenhagen to work with J. Bjerrum on a problem about ions in solutions. He was not particularly happy in Copenhagen. Determined to work in quantum mechanics, he convinced Bjerrum, the Education Board and Schrödinger to allow him to spend the second half of the period for which he received funding in Zurich.

When they met in Zurich, Heitler and London decided to calculate the "van der Waals" forces between two hydrogen atoms. Nothing indicates that London and Heitler were either given the problem of the hydrogen molecule by Schrödinger or that they had detailed discussions with the latter while they were proceeding with their calculations. Linus Pauling, who was also in Zurich during the same time as Heitler and London, notes that neither he nor Heitler and London discussed their work with Schrödinger, who nonetheless did know what they were all working on as witnessed by Robert Mulliken's visit to Zurich in 1927. Schrödinger had told Mulliken that there were two persons working at his institute and who had some results "which he thought would interest me very much; he then introduced me to Heitler and London whose paper on the chemical bond in hydrogen was published not long after."[2]

Binding Forces

Heitler and London's initial aim was to calculate the interaction of the charges of two atoms. They were not particularly encouraged by their first results, since the attraction due to the "Coulomb integral" was too small to account for the homopolar bond between two hydrogen atoms. But they were puzzled by the presence of the "exchange integral" whose physical significance was not evident at all. Heisenberg's work on the quantum mechanical resonance phenomenon which had already been published was not of particular help to Heitler and London, since the exchange was part of the resonance of two electrons one of which was in the ground state and the other was excited and both were in the same atom. Heitler remembered that they were stuck and "we did not know what it meant and did not know what to do with it."

> Then one day was a very disagreeable day in Zurich; [there was the] Fohn. It's a very hot south wind, and it takes people different ways. Some are very cross . . . and some people just fall asleep. . . . I had slept till very late in the morning, found I couldn't do any work at all . . . went to sleep again in the afternoon. When I woke up at five o'clock I had clearly—I still remember it as if it were yesterday—the picture before me of the two wave functions of two hydrogen molecules joined together with a plus and minus and with the exchange in it. So I was very excited, and I got up and thought it out. As soon as I was clear that the exchange did play a role, I called London up; and he came as quickly as possible. Meanwhile I had already started developing a sort of perturbation theory. We worked together until rather late at night, and then by that time most of the paper was clear. . . . Well, I am not quite sure if we knew it in the same evening, but at least it was not later than the following day that we knew we had the formation of the hydrogen molecule in our hands. And we also knew that there was a second mode of interaction which meant repulsion between two hydrogen atoms—also new at the time—new to the chemists too. Well the rest was then rather quick work and very easy, except, of course, that we had to struggle with the proper formulation of the Pauli principle, which was not at that time available, and also the connection with spin. . . . There was a great deal of discussion about the Pauli principle and how it could be interpreted.[3]

The "mechanism" responsible for the bonding between the two neutral hydrogen atoms was the pairing of the electrons which became possible only when the relative orientations of the spins of the electrons were antiparallel. To form an electron pair it did not suffice to have only energetically available electrons, but the electrons had to have the right spin orientations. The homopolar bonding turned out to be a pure quantum effect, since its explanation depended wholly on the electron spin which had no classical analogue. As Heitler and London noted in their paper, such a result could only be described very artificially in classical terms. They found the bond energy to be 72.3 kcals and the internuclear distance 0.86 Angstroms to be compared with the experimental values of 109.4 and 0.74 respectively. The numerical results for the interatomic distance and the binding potential derived by the original calculation of Heitler and London did allow for further more exact approximations. Their goal was not to calculate the most exact numerical values possible, but "to gain insights into the physical conditions of the homopolar bond."[4]

Heitler and London soon realised that the proposed exchange mechanism obliged them to be confronted with a fundamentally new phenomenon. As Heitler remembered, they had to answer questions posed by experimental physicists and chemists, like "What is really exchanged? Are the two electrons really exchanged? Is there any sense in asking what the frequency of exchange is?"

> It became gradually clear to me that it has to be taken as a fundamentally new phenomenon that has no proper analogy in older physics. But I think the only honest answer today is that the exchange is something typical for quantum mechanics, and should not be interpreted—or should not try to interpret it—in terms of classical physics.[5]

Both London and Heitler in all their early writings repeatedly stressed this "nonvisualizability" of the exchange mechanism. It is one aspect of their work that in the name of didactic expediency has been consistently misrepresented.

The Paulie Principle

Though it appeared that the treatment of the homopolar bond of the hydrogen molecule was an "extension" of the methods successfully used for the hydrogen molecular ion, there was a difference between the two cases that lead to quite radical implications. It was the role of the elusive Pauli principle. In the case of the hydrogen molecule ion, its solution was a successful application of the Schrödinger equation where the only forces determining the potential are electromagnetic. A similar approach to the problem of the hydrogen molecule leads to a mathematically well-defined, but physically meaningless solution where the attractive forces could not be accounted for. There was a need for an additional constraint, so that the solution would become physically meaningful. At least part of the theoretical significance of the original work of Heitler and London was that this additional constraint was not in the form of any further assumptions about the forces involved. Invoking the Pauli exclusion principle as a further constraint led to a quite amazing metamorphosis of the physical content of the mathematical solutions. These solutions became physically meaningful and their interpretation in terms of the Pauli principle brought about the new possibilities provided by the electromagnetic interaction.

London, in his subsequent publications, proceeded to a formulation of the Pauli principle for cases with more than two electrons and which was to become more convenient for his later work in group theory: the wave function can at most contain arguments symmetric in pairs; those electron pairs on which the wave function depends symmetrically have antiparallel spin. He considered spin to be the constitutive characteristic of quantum chemistry. And since two electrons with antiparallel spin are not identical, the Pauli principle did not apply to them and one could, thus, legitimately choose the symmetric solution. With the Pauli principle it became possible to comprehend "valence" saturation and as it will be argued in the future work of both Heitler and London, spin would become one of the most significant indicators of valence behavior and, in the words of Van Vleck, would forever be "at the heart of chemistry."[6]

REACTIONS TO THE HEITLER-LONDON PAPER

Right after the publication of the Heitler-London paper, it became quite obvious that it was opening a new era in the study of chemical problems. But it also signified the formation of a subdiscipline—that of quantum chemistry. W. M. Fairbank, who was London's colleague at Duke University in 1953 and the coauthor with C. W. F. Everitt of the entry on Fritz London in the *Dictionary of Scientific Biography,* recalled London telling him that Schrödinger did not expect that his equation would be able to describe the whole of chemistry as well. The authors themselves were not expecting to find any such force as London told A. B. Pippard, since they had started working on the problem as a problem in van der Waals forces.[7] Born and Franck were enthusiastic about the paper. Sommerfeld had a rather cool reaction, but he also became enthusiastic once Heitler met with him and explained certain points. The application of quantum mechanics led not only to the conclusion that two hydrogen atoms formed a molecule, but also that such was not the case with two helium atoms. Such a "distinction is characteristically chemical and its clarification marks the genesis of the science of subatomic theoretical chemistry" remarked Pauling.[8] A similar view with a slightly different emphasis, was put forward by van Vleck

> Is it too optimistic to hazard the opinion that this is perhaps the beginnings of a science of "mathematical chemistry" in which chemical heats of reaction are calculated by quantum mechanics just as are the spectroscopic frequencies of the physicist? . . . The theoretical computer of molecular energy levels must have a technique comparable with that of a mathematical astronomer.[9]

Louis de Broglie and Max von Laue both regarded the paper as a classic work. Eugene Wigner was "impressed with the skill" of London's papers.[10] In their book on quantum mechanics for chemists, Linus Pauling and E. Bright Wilson hailed the paper as the "greatest single contribution to the clarification of the chemists' conception of valence"[11] which had been made since Lewis's suggestion of the electron pair. Heisenberg in an address to the Chemical Section of the British Association for the Advancement of Science in 1931 considered the theory of valence of Heitler and London to "have the great advantage of leading exactly to the concept of valency which is used by the chemist." Buckingham quoted McCrea who recalled his own attempts to solve the problem of the hydrogen molecule bond.

> This was the most important problem I considered these days, but I got nowhere. Then one day in 1927, I was able to tell Fowler that a paper by Walter Heitler and Fritz London apparently solved the problem in terms of a new concept: a quantum mechanical exchange force. He grasped the idea at once, and bade me to expound it in the next colloquium—which is how quantum chemistry came to Britain.[12]

In April 1926, Linus Pauling supported by a Guggenheim Fellowship, arrived in Munich where he planned to work at the Institute of Theoretical Physics. He was twenty-five years old and had already received his doctorate from the California Institute of Technology working with Roscoe Dickinson on the structure of molybdenite. When he met Sommerfeld, the latter suggested that Pauling work on the electron spin. Pauling did not follow this advice, since his main interest was in chemistry. When he was in Munich he discussed the problem of the chemical bond with various people and he himself thought it could be solved by using a Burrau-like approach and the Pauli principle. He was impressed by Condon's treatment of the hydrogen molecule, whose clever numerology and "empirical method" were not at all adverse to Pauling especially since he "got results as good as Heitler and London got later."[13]

After Munich and before going to Copenhagen in late 1927, he spent three months in Zurich where he also met Heitler and London. Right after the appearance of the Heitler-London paper, Pauling published a short note to bring attention to an unforgivable omission: Lewis was nowhere mentioned in the paper and Pauling wanted to emphasize that the Heitler-London approach "is in simple cases entirely equivalent to G. N. Lewis's successful theory of the shared electron pair, advanced in 1916 on the basis of purely chemical evidence,"[14] acknowledging at the same time that the quantum mechanical explanation of valence was more powerful than the old picture. In this paper Pauling mentioned for the first time that the changes in quantization may play a dominant role in the production of stable bonds in the chemical compounds. That was the first hint about the hybridization of orbitals. Perturbations to the quantized electronic levels may produce directed atomic orbitals whose overlapping would be better suited for the study of chemical bonds. Pauling, in that note, suggested the direction along which he would move to derive some new results and he explicitly stated his methodological commitments. When Pauling informed Lewis about his short note,[15] Lewis's response contains what is, perhaps, his most insightful statement about valence.

> I was very much interested in your paper as I had been in London's, and there is much in both papers with which I can agree. . . . I am sorry that in one regard my idea of valence has never been fully accepted. It was an essential part of my original theory that the two electrons in a bond completely lose their identity and can not be traced back to the particular atom or atoms from which they have come; furthermore that this pair of electrons is the only thing which we are justified in calling a bond. Failure to recognize this principle is responsible for much of the confusion prevailing in England on this subject.[16]

In 1928 there appeared two review articles that exerted a strong influence among the chemists. Both were published in the *Chemical Reviews,* they were written by Americans, Linus Pauling and John Van Vleck, and both had the explicit aim of "educating" the chemists in the ways of the new mechanics. In Pauling's article the Heitler-London treatment of the structure of the hydrogen molecule was considered as "most satisfactory" and it was repeatedly stated that spin and resonance will provide a satisfactory explanation of chemical valence. Pauling felt that the agreement with the qualitative conclusions of Lewis was one of the advantages of the work by Heitler and London.

Van Vleck's (1928) review of quantum mechanics concentrated on explaining the principles and the internal logic of the new theory. He was quite sympathetic to matrix mechanics. In this paper Van Vleck gave full credit to the work of Heitler and London whose results were "already yielding one of the best and most promising theories of valency."[17] The achievements of quantum mechanics in physics were summarized in ten points and the section about chemistry was appropriately titled "What the quantum mechanics promises to do for the chemist." Great emphasis was placed on the importance of spin for chemistry and it was shown that the Pauli exclusion principle could provide a remarkably coherent explanation of the periodic table. Some years later, its importance was stressed even more dramatically: "The Pauli exclusion principle is the cornerstone of the entire science of chemistry."[18] Nevertheless, if quantum mechanics was to be of any use in chemistry, one should go further than the periodic table and understand which atoms can combine and which cannot. Van Vleck propounded his reductionist attitude and thought the dynamics that were so successful in explaining atomic energy levels for the physicist should also be successful in calculating molecular energy levels for the chemist.

POLYELECTRONIC MOLECULES AND THE APPLICATION OF GROUP THEORY TO PROBLEMS OF CHEMICAL VALENCE

The first indications that the work they started in their joint paper could be continued by using group theory are found in a letter to London by Heitler in late 1927. In September 1927 Heitler had become Born's assistant at Göttingen and wrote to London saying how excited he was about the physics at Göttingen and especially about Born's course in quantum mechanics where everything was presented in the matrix formulation and then one derived "God knows how, Schrödinger's equation."[19] Heitler felt that the only way the many-body problem could be dealt with was with group theory and outlined his program to London in two long letters.

His first aim was to clarify the meaning of the line chemists drew between two atoms. His basic assumption was that every bond line meant exchange of two electrons of opposite spin between two atoms. He examined the case with the nitrogen molecule and, in analogy with the hydrogen case, among all the possibilities, the term containing the outermost three electrons of each atom with spins in the same direction (i.e., ↑↑↑ and ↓↓↓) was picked out as signifying attraction. He felt that the general proof for something like this cannot be given, except by group theory. His hunch was that it involved the theory of reducible representations, which as representations of the totality of permutations cannot be further reduced. "Let us assume for the moment that the two atomic systems ↑↑↑↑↑ ... and ↓↓↓↓↓ ... are always attracted in a homopolar manner. We can, then, eat Chemistry with a spoon."[20]

This overarching program to explain all of chemistry got Heitler into trouble more than once. Wigner used to tease him, since he was sceptical that the whole of chemistry could be "derived" in such a way. Wigner would ask Heitler to tell him what chemical compounds between nitrogen and hydrogen could his theory predict, and "since he did not know any chemistry he couldn't tell me."[21] Heitler confessed as much in his interview: "The problem was to understand chemistry. This is perhaps a bit too much to ask, but it was to understand what the chemists mean when they say an atom has a valence of two or three or four." Both London and Heitler believed that all this must be now within the reach of quantum mechanics.

London agreed with Heitler that group theory may provide many clues for the generalization of the results derived by perturbation methods. The aim of such a program was to prove that from all the possible combinations of spins between atoms, only one term provides the necessary attraction for molecule formation. It took London a while to get familiarized with the new possibilities provided by group theory and he was not carried away by the spell of the new techniques—as Heitler was in the company of Wigner and Weyl at Göttingen.[22] "He thought it was too complicated and wanted to get on in his own more intuitive way."[23]

In Göttingen Heitler started to study group theory intensively. Wigner's papers had already appeared and there was a realization that group theory could be used for classifying the energy values in a many-body problem.

In a significant paper, Heitler and Rumer (1931), were able to study the valence structures of polyatomic molecules and find the closest possible analogue in quantum mechanics to the chemical formula that represented the molecule by fixed bonds between two adjoining atoms. They found that the emerging quantum mechanical picture was more general and that the bonds were not strictly localized. Nevertheless, the dominant structure was, in general, the one corresponding to the chemical formula. But there were other structures that were also significant and these structures were quite useful in understanding chemical reactions. Heitler thought that it was in this paper that he could understand what the chemists meant by a chemical formula. There were a few new things for the chemists; one was that to each chemical formula there corresponded a wave function, but the wave function was not such that it corresponded to one chemical formula alone, but was in general a combination of several. London was the first a long time before the Heitler-Rumer paper who showed that the activation energies in the treatment of the three hydrogen atoms could only be understood in terms of quantum mechanics. "Later Pauling called this a resonance between several structures, which is a name perhaps not quite in agreement with the use of the word resonance by physicists . . . a further point which was violently objected to by the chemists was that both London and I stated that the carbon atom with its 4 valences must be in an excited state . . . all this was later accepted by the chemists, but at that time I don't think the chemists did find this of much use for them."[24]

The theory of the irreducible representations of the permutation group provided the possibility of dealing *mathematically* with the problems of chemical valence in view of the difficulties with the many-body problems. And though this unavailability of reliable methods for tackling many-body problems haunted London all his life, many years later this difficulty was strangely liberating and helped him to articulate the concepts related to the macroscopic quantum phenomena in order to supercede such difficulties.

Convinced that it was impossible to continue his work in chemical valence by more analytic methods, London eventually turned to group theory. But not everyone—even among those who thought of London's work very highly—was enthusiastic about the group theoretical techniques. Douglas Hartree was one such person.

> I am afraid that having studied physics, not mathematics, I find group theory very unfamiliar, and do not feel I understand properly what people are doing when they use it. [In England, "Physics" usually means "Experimental Physics"; until the last few years "Theoretical Physics" has hardly been recognized like it is here [at the time Hartree was in Copenhagen], and, I understand, in your country. In Cambridge particularly the bias has been very much towards the experimental side, and most people now doing research in theoretical physics studied mathematics, not physics]. I have been waiting to see if the applications of group theory are going to remain of importance, or whether they will be superseded, before trying to learn some of the theory, as I do not want to find it is going to be of no value as soon as I begin to understand something about it! Is it really going to be necessary for the physicist and chemist of the future to know group theory? I am beginning to think it may be.[25]

Born had also responded in surprisingly negative manner to the use of group theory. His objection was not because group theory was not easy to use, but because, "in reality it is not in accord with the way things are."[26]

London's first major paper on the application of the group theoretical methods to problems of chemical bonding appeared in 1928. London's group theoretical approach to chemical valence was formed around three axes. First, anything that may give a rather strong correlation between qualitative assessments of a theoretical calculation and the "known chemical facts" provided a rather strong backing for the methodological correctness of the approach chosen by expressing the observed regularities as rules. Second, since analytic calculations were hopelessly complicated and in most cases impossible, the use of group theoretical methods was especially convenient when one was dealing with the valence numbers of polyelectronic atoms, since the outcome was expressed either as 0 or in natural numbers. Third, the overall result was that the interpretation of the chemical facts was compatible with the conceptual framework of quantum mechanics.

Chemists as Physicists?

Among the meetings where questions related to chemical bonding and valence were exhaustively discussed, two in particular were quite suggestive of the changes occurring among chemists. The first was the "symposium on atomic structure and valence" organized by the American Chemical Society in 1928 at St. Louis. The second was organized by the Faraday Society in 1929 at Bristol and its theme was "molecular spectra and molecular structure."

G. L. Clark's opening remarks in the 1928 meeting of the American Chemical Society noted some of the difficulties associated with atomic physics, but ascribed them to the failure of the chemists to test "their well-founded conceptions with the facts of physical experimentation, and that far too few physicists inquired critically into the facts of chemical combination." He thought that physicists and chemists were "firmly entrenched, each in his own domain, a certain long-range firing of static cubical atoms against infinitesimal solar atoms has ensued, with few casualties and few peace conferences."[27]

Clark was not alone in attempting to specify the newly acquired consciousness about this strange relationship between the physicists and the chemists. Worth Rodebush, one of the first to receive a doctorate in 1917 from the newly established Department of Chemistry at Berkeley under the chairmanship of Lewis, went a step further than Clark. He asserted that the divergent paths of physicists and chemists were being drawn together after the advent of quantum theory and especially after Bohr's original papers. But in this process the physicist appeared to have yielded more ground than the chemist and the former appeared to have learned more from the latter than was the case with the chemists. Rodebush gracefully remarked that it was to the credit of the physicist that he could now calculate the energy of formation of the hydrogen molecule by using the Schrödinger equation. But the outstanding tasks for a theory of valence was to predict the existence and absence of various compounds, and the nature of valence which can be expressed by a series of small whole numbers leading to the law of multiple proportions. The "brilliant theories" of Lewis accounted for the features of valence "in a remarkably satisfactory manner, at least from the chemist's point of view."[28] London's group theoretical treatment of valence was considered as an important piece of work even though it did not provide answers to all the queries of the chemist as, for example, the differences in degree of stability between chemical compounds.

Perhaps the most cogent manifestation of what would become the characteristic approach of the American chemists was Harry Fry's contribution to this symposium. He attempted to articulate what he called the pragmatic outlook. He started by posing a single question. What would be the kind of modifications to the structural formulas so as to conform with the current concepts of electronic valency? Such a question, he suggested, should by no means lead to a confusion of the fundamental purpose of a structural formula which is to present the number, the kind, and the arrangement of atoms in a molecule as well as to correlate the manifold chemical reactions displayed by the molecule. "The opinion is now growing that the structural formula of the organic chemist is not the canvas on which the cubist artist should impose his drawings which he alone can interpret.... On the grounds that practical results are the sole test of truth, such simple system of electronic valence notation may be termed 'pragmatic.'"[29]

"Chemical pragmatism" resisted the attempts to embody in the structural formulas what Fry considered to be metaphysical hypotheses: questions related to the constitution of the atom and the disposition of its valence electrons. It was the chemical behavior of molecules that was the primary concern of the pragmatic chemist, rather than the imposition of an electronic system of notation on these formulas which was further complicated by the metaphysical speculations involving the unsolved problems about the constitution of the atom. Fry had to admit the obvious fact that as the chemists would know more about the constitution of the atom, they would be able to explain more fully the chemical properties. He warned, though, that premises lying outside the territory of sensation experience are bound to lead to contradictory conclusions quoting Kant and becoming, surely, the only chemist to use Kant's ideas to convince other chemists at a Conference about an issue in chemistry!

The Heitler-London approach to the homopolar bond was neither the only approach nor the most practical one for molecules with many electrons.

Their group theoretical proposals were mathematically quite involved and not conducive to produce quantitative results. The methods that could produce such quantitative results were developed by two Americans—Linus Pauling and Robert Mulliken. As these methods were being developed, Heitler and London perceived them as antagonistic to their own and progressively realized that the chemical community was less and less willing to adopt its own approach to the chemical bond. Drawing up a program to examine the nature of the chemical bond presupposed a particular attitude on how to construct a theory in chemistry, on how much one "borrowed" from physics, and what the methodological status of empirical observations for theory building was. Concerning the work of Heitler, London, Pauling, and Mulliken, two different research traditions were formed. Heitler and London insisted on an approach that, even though it was not as reductionist as Dirac's pronouncement of 1929, followed this path of orthodoxy. Pauling and Mulliken had a strong inkling to semiempirical methods whose only criterion for acceptability was their practical success. To suppose that the question of a stronger command over the mathematical details was the sole differentiating criterion between the two traditions is quite misleading. The difference could be understood by examining the two different styles for doing quantum chemistry. It was a matter of explicating the internal, theoretical, and methodological coherence of each proposed schema, and realizing that they constitute two diverging programs. In 1935, London and Heitler reestablished contact and started writing frantically to each other, trying to see what could be saved of their common approach from the onslaught of the new methods. Before discussing their correspondence, let us discuss Pauling's and Mulliken's approaches.[30]

LINUS PAULING'S RESONANCE STRUCTURES

Almost everything in the series of Pauling's papers starting in 1931 and titled *The Nature of the Chemical Bond* are included in his book of the same title. Pauling proposed six rules for the electron-pair bond. Not all of these rules were derived from first principles, but they were mostly inferred from rigorous treatments of the hydrogen molecule, the helium atom, and the lithium atom. Pauling exploited maximally the quantum mechanical phenomenon of resonance and was eventually in a position to formulate a comprehensive theory of chemical bonding. The success of the theory of resonance in structural chemistry consisted in finding the actual structures of various molecules as a result of resonance among other "more basic" structures. In the same manner that the Heitler-London approach provided a quantum mechanical explanation of the Lewis electron pair mechanism, the quantum mechanical theory of resonance provided a more sound theoretical basis for the ideas of tautomerism, mesomerism, and the theory of intermediate state.

Differences in the assessment of the methodological and ontological status of resonance were the object of a dispute between Pauling and George Wheland, who more than anyone else worked toward the extension of resonance theory to organic molecules. Wheland, in his book *The Theory of Resonance and Its Applications to Organic Molecules* dedicated to Pauling, argued that the resonance concept was a "man-made concept"[31] in a more fundamental way than in most other physical theories. This was his way to counter the widespread view that resonance was "a real phenomenon with real physical significance," which he classified as one example of the nonsense that organic chemists were prone to.[32]

> What I had in mind was, rather, that resonance is not an intrinsic property of a molecule that is described as a resonance hybrid, but is instead something deliberately added by the chemist or the physicist who is talking about the molecule. In anthropomorphic terms, I might say that the molecule does not know about resonance in the same sense in which it knows about its weight, energy, size, shape, and other properties that have what I call real physical significance. Similarly . . . a hybrid molecule does not know how its total energy is divided between bond energy and resonance energy. Even the double bond in ethylene seems to me less "man-made" than the resonance in benzene. The statement that the ethylene contains a double bond can be regarded as an indirect and approximate description of such real properties as interatomic distance, force constant, charge distribution, chemical reactivity, and the like; on the other hand, the statement that benzene is a hybrid of the two Kekulé structures does not describe the properties of the molecule so much as the mental processes of the person who makes the statement. Consequently, an ethylene molecule could be said to know about its double bond, whereas a benzene molecule cannot be said, with the same justification, to know about its resonance. . . . Resonance is not something that the hybrid *does,* or that could be "seen" with sufficiently sensitive apparatus, but is instead a description of the way that the physicist or chemist has arbitrarily chosen for the approximate specification of the true state of affairs.

Pauling could not disagree more. For him, the double bond in ethylene was as "man-made" as resonance in benzene. Pauling summarized their divergent viewpoints by saying that Wheland appeared to believe that there was a "quantitative difference" in the man-made character of resonance theory when compared to ordinary structure theory, but he could not find such a difference. He asserted that Wheland made a disservice to resonance theory by overemphasizing its "man-made character."[33] Wheland conceded that resonance theory and classical structural theory were qualitatively alike, but he still defended, contrary to Pauling, that there was a "quantitative difference" between the two. He viewed his disagreement with Pauling as a result of different value judgments on what he classified as philosophical rather than scientific matters.

Nevertheless, acknowledging or denying the existence of differences between resonance theory and classical structural theory depended on their different assessments of the role of alternative methods of studying molecular structure. Wheland equated resonance theory to the valence-bond method and viewed them as alternatives to the molecular-orbital method. Pauling conceded that the valence-bond method could be compared with the molecular-orbital method but not with the resonance theory which was largely independent of the valence-bond method. For Pauling the theory of resonance was not merely a computational scheme. It was an extension of the classical structure theory, and as such it shared with its predecessor the same conceptual framework. If one accepted the concepts and ideas of classical structure theory one had to accept the theory of resonance. How could one reject their common conceptual base if they had been largely induced from experiment?[34]

> I think that the theory of resonance is independent of the valence-bond method of approximate solution of the Schrödinger wave equation for molecules. I think that it was an accident in the development of the sciences of physics and chemistry that resonance theory was not completely formulated before quantum mechanics. It was, of course, partially formulated before quantum mechanics was discovered; and the aspects of resonance theory that were introduced after quantum mechanics, and as a result of quantum mechanical argument, might well have been induced from chemical facts a number of years earlier.

This discussion with Wheland prompted Pauling to make his position about these issues public. More than the question of the artificiality of the resonance concept, to which he alluded briefly in his Nobel lecture[35], he wanted to state as clearly as possible his views on theory building. A revised version of the arguments brought about in the discussion with Wheland appeared in *Perspectives in Organic Chemistry*[36] and later on, in the third edition of *The Nature of the Chemical Bond*.[37] In the preface, Pauling pointed out that the theory of resonance involves the "same amounts of idealization and arbitrariness as the classical valence-bond theory." Pauling added a whole section in the new edition to discuss this question. His manifesto was called "The Nature of the Theory of Resonance." There, he argued that the objection concerning the artificiality of concepts applied equally to resonance theory as to classical structure theory. To abandon the resonance theory was tantamount to abandoning the classical structure theory of organic chemistry. Were chemists willing to do that? According to Pauling, chemists should keep both theories because they were chemical theories and as such possessed "an essentially empirical (inductive) basis."[38]

> I feel that the greatest advantage of the theory of resonance, as compared with other ways (such as the molecular-orbital method) of discussing the structure of molecules for which a single valence-bond structure is not enough, is that it makes use of structural elements with which the chemist is familiar. The theory should not be assessed as inadequate because of its occasional unskillful application. It becomes more and more powerful, just as does classical structure theory, as the chemist develops a better and better chemical intuition about it. . . . The theory of resonance in chemistry is an essentially qualitative theory, which, like the classical structure theory, depends for its successful application largely upon a chemical feeling that is developed through practice.

In 1947 Charles Coulson, a mathematician turned chemist who taught for many years at Oxford and whose book *Valence* became a standard textbook, wrote an article in a semipopular magazine on what he thought was resonance.[39]

> Is resonance a *real* phenomenon? The answer is quite definitely no. We cannot say that the molecule has either one or the other structure or even that it oscillates between them. . . . Putting it in mathematical terms, there is just one full, complete and proper solution of the Schrödinger wave equation which describes the motion of the electrons. Resonance is merely a way of dissecting this solution: or, indeed, since the full solution is too complicated to work out in detail, resonance is one way—and then not the only way—of describing the approximate solution. It is a "calculus," if by calculus we mean a method of calculation; but it has no physical reality. It has grown up because chemists have become used to the idea of localized electron pair bonds that they are loath to abandon it, and prefer to speak of a superposition of definite structures, each of which contains familiar single or double bonds and can be easily visualizable.

The question as to the ontological status of resonance was not an issue that was confined to this exchange between Pauling and Wheland. Pauling's theory of resonance was viciously attacked in 1951 by a group of chemists in the Soviet Union in their report of the Commission of the Institute of Organic Chemistry of the Academy of Sciences.[40] As they themselves stressed, their main objection was methodological. They could not accept that by starting from conditions and structures that did not correspond to reality one could be led to meaningful results. Of course, they discussed analytically the work of A. M. Butlerov who in 1861 had proposed a materialist conception of chemical structure: this was the distribution of the action of the chemical force, known as affinity, by which atoms are united into molecules. He insisted that any derived formula should express a real substance, a real situation. According to the report, Pauling was moving along different directions. For him a chemical bond between atoms existed if the forces acting between them were such as to lead to the formation of an aggregate with sufficient stability to make it convenient for the chemist to consider it as an independent molecular species. To these chemists Pauling's operational definition was totally unacceptable.

> In this treatment the objective criterion of reality of the molecule and of the chemical bond vanishes. Since the definition of the molecule and the chemical bond given by Pauling is methodologically incorrect, it naturally leads, when logically developed, to absurd results.[41]

It is interesting to note the initiative of the New York chapter of the National Council of Arts, Sciences, and Professions to organize a meeting on the subject. It was proposed that the meeting would have the form of a debate where N. D. Sokolov from Moscow, Charles Coulson, and Linus Pauling would each contribute a paper and there would follow a discussion of the points raised in the communications. Coulson felt that the best way would be for Sokolov and Pauling to present their viewpoints and that himself would make a series of comments. Each party would be asked to provide answers to the following questions: What is the resonance theory? What is the evidence in proof or disproof of the resonance theory? Is the convenience of the theory a proof or a corroboration of the theory? Is the resonance theory essentially a theory with physical meaning, or a mathematical technique or both? Has the resonance theory a basis in related sciences, such as physics? Is the resonance theory applicable in all aspects of chemical valence or is it in conflict?[42] The meeting did not take place because of the unwillingness of the Soviets, but the points that each party would have had to address were indicative of the uncertainties involved as to the methodological significance and the ontological status of resonance in quantum chemistry.

What Pauling greatly emphasized was not the arbitrariness of the concept of resonance, but its immense usefulness and convenience which "make the disadvantage of the element of arbitrariness of little significance."[43] This according to Pauling became the constitutive criterion for theory building in chemistry. It was the way, as he had noted, to particularize Bridgman's operationalism in chemistry. Pauling, as he noted in his interview, felt more at ease with the Schrödinger approach than with matrix mechanics and did not worry about questions of interpretation of quantum mechanics. "I tend not to be interested in the more abstruse aspects of quantum mechanics. I take a sort of Bridgmanian attitude toward them."

In his analysis of resonance, Pauling expressed in the most explicit manner his views about theory building in chemistry. He asserted that the theory of resonance was a chemical theory, and, in this respect, it had little in common with the valence-bond method of making approximate quantum mechanical calculations of molecular wave functions and properties. Such a theory was "obtained largely by induction from the results of chemical experiments."[44] The development of the theory of molecular structure and the nature of the chemical bond, Pauling asserted in his Nobel speech in 1954, "is in considerable part empirical—based upon the facts of chemistry—but with the interpretation of these facts greatly influenced by quantum mechanical principles and concepts"[45]

Both the discussions with Wheland and the attack against his theory by chemists in the Soviet Union prompted Pauling to include a discussion of the character of theory in chemistry in the third edition of his book in 1960. The theory of resonance was not simply a theory embodying exact quantum mechanical calculations. Its great extension has been "almost entirely empirical, with only the valuable and effective guidance of fundamental quantum mechanical principles." Pauling emphasized that the theory of resonance in chemistry was an essentially qualitative theory which, "like the classical structure theory, depends for its successful application largely upon a chemical feeling that is developed through practice."[46] Pauling himself has repeatedly stressed the rather empirical character of his theory of resonance.

> My work on the nature of the chemical bond and its application to the structure of molecules and crystals has been largely empirical, but for the most part guided by quantum mechanical principles. I might even contend that there are four ways of discussing the nature of the chemical bond: the Hund-Milliken way, the Heitler-London way, the Slater-Pauling way, and the Pauling semiempirical way.[47]

Robert Mulliken's Molecular Orbitals

Though the method of molecular orbitals was first introduced by Hund, it was Mulliken who provided both the most thorough treatment of the different kinds of molecules as well as the theoretical and methodological justifications for legitimizing the molecular-orbital approach. Mulliken was born in 1897 and received his doctorate from the University of Chicago in 1921 working with D. W. Harkins on isotope separation especially of mercury. He had worked at the University of Chicago and at Harvard as a National Research Fellow and by 1926 he was an assistant professor at New York University. When he did his foundational work on the method of molecular orbitals, he had moved to the University of Chicago and had spent some months travelling in Europe before his extended stay there in 1930 as a Guggenheim Fellow.

After his work on band spectra and the assignment of quantum numbers to electrons in molecules, Mulliken was getting ready for "an attack on the Heitler and London theory of valence" as he wrote to Birge in 1931, since he was becoming more and more convinced that "one can understand chemical binding decidedly better and more intimately, by a consideration of molecular electron configurations than by Heitler and London's method."[48] Mulliken proceeded to formulate his approach to the problem of valence in a series of papers in 1932. The theory was, in a way, the outcome of a program whose aim was to describe and understand molecules in terms of (one-electron) orbital wave functions of "distinctly molecular character." The attempt was to articulate the autonomous character of molecules through a process that depended on the extensive data on band spectra and on analogies with atoms. His theory became an alternative mode to the treatment of the problem of valence by Heitler, London, Pauling, and Slater. Holding the view that the concept of valence itself is one that should not be held too sacred, Mulliken assumed that there were not only bonding and nonbonding electrons, but also "antibonding electrons, that is, electrons which actively oppose the union of the atoms."[49] For his was the "molecular" point of view where the emphasis was on the existence of the molecule as a distinct and autonomous entity and not as a union of atoms held together by valence bonds. Therefore, from such a "molecular " point of view the understanding of the mechanism of uniting atoms became of secondary importance.

Unshared electrons were described in terms of atomic orbitals and the notion of molecular orbitals was introduced to describe shared electrons. Electrons were divided into three categories according to their roles in the binding process: shared electrons (at least for diatomic molecules) were either bonding or antibonding electrons, unshared electrons were the nonbonding electrons. The latter occured in diatomic molecules only when accompanied by a larger number of bonding electrons. In Mulliken's method molecular orbitals were conceived as "entities quite independently of atomic orbital."[50]

Mulliken urged the distinction between Heitler and London's valence theory and their "valuable perturbation-method for calculating energies" of molecule formation. He thought that the theories of Heitler and London, Pauling and Slater (HLPS) might be called electron-pairing theories whereas Lewis's theory was an electron-pair theory.[51] Mulliken time and again emphasized that the concept of the bonding molecular orbitals was more general, more flexible, and certainly more "natural" than the Heitler-London electron-pair bonding—even though the latter may turn out to be more convenient for quantitative results for a number of problems.

The assignment of the various quantum numbers to the molecular orbitals led to an alternative explanation of homopolar valence that did not depend on resonance but rather on the redefinition of the notion of premotion to be used for the one-nucleus viewpoint of the nuclei in the molecule. Then, bonding-electrons became, in effect, unpremoted electrons whereas antibonding electrons were strongly premoted electrons. Therefore, chemical combination of the homopolar type was the result of the shrinkage and consequent energy-decrease of atomic orbitals in the fields of the neighboring nuclei when such orbitals were shared with little or no premotion. It was shown that the role of the exchange integrals of Heitler and London corresponded to the electron density of the molecular orbitals: bonding orbitals had a higher electron density, antibonding orbitals had a lower density in the regions between the nuclei than the densities that would have resulted by the overlapping of the electron densities of the orbitals of isolated atoms. Mulliken placed himself in the antipodes of Heitler and London. He insisted that the occurrence of the electron pairs in the molecules had "no fundamental connection with the existence of chemical binding." The Pauli principle could adequately explain the fact that each type of molecular orbit can be occupied by just two electrons.

These remarks by Mulliken about the Heitler-London approach were less than graciously received by London who urged Heitler to look up the assessment of their work by Mulliken and implored him to "judge for yourself whether we are neglecting something or not, when we leave unanswered these kinds of distortions. And they are not at all isolated cases."[52] Mulliken had recognized that the Heitler-London approach produced results in agreement with the experimentally measured values for the ground states of simple molecules, but he warned that the "experimental data show that it is unsafe to generalize too far from calculations made for a limited number of cases." Mulliken's main objection was methodological. The approach of Heitler and London required long calculations to make the quantitative predictions, but "qualitative predictions can usually be made much more easily by a consideration of electron configurations of atoms and molecules."[53] It was these kinds of pronouncements that deeply angered London. He did not mind there being a theory "superior" than his own approach, but one had to play the game according to the rules and not devise new rules along the way. So much the worse when these rules were nothing but rationalizations of experimental data!

Earlier, in 1928, Mulliken had made some attempts to give due credit to the work of Heitler and London. He considered their joint work and the subsequent papers using group theory together with Hund's papers as promising "at last a suitable theoretical foundation for an understanding of the problems of valence and of the structure and stability of molecules."[54] A year later in a presentation of London's group theoretical approach, Mulliken (1929), thought that London's theory was a translation of Lewis's theory into quantum mechanical language, but such credit slowly waned. In 1933 he did not refer at all to the Heitler-London paper but rather to the theory of Slater and Pauling which, together with the molecular-orbital approach, he considered to illuminate the Lewis theory from more or less complementary directions. Mulliken became progressively more reserved about the usefulness of the Heitler-London paper. In his Nobel Prize speech of 1966 he referred to the paper as merely initiating an alternative approach to the molecular orbital method. He did not even recognize that it provided the quantum mechanical explanation of the Lewis schema, since the "electrons in the chemical molecular orbitals represent the closest possible quantum mechanical counterpart to Lewis's beautiful prequantum valence theory."[55] London's work in group theory, however, was nowhere mentioned in Mulliken's 1932 article in *Chemical Reviews* where Mulliken expressed in a detailed manner his objections to the Heitler-London method and theory.

In a paper titled "On the method of molecular orbitals" published in 1935, Mulliken expresseed his views on what he considered to be the most characteristic and differentiating aspects of his theory. The Heitler-London method "follows the ideology of chemistry and treats every molecule, so far as possible, as composed of definite atoms. . . . It has had the notable success as a qualitative conceptual scheme for interpreting and explaining empirical rules of valence and in semiquantitative, mostly semiempirical calculations of energies of formation."[56] His method of molecular orbitals departed from "chemical ideology . . . and treats each molecule, so far as possible, as a unit." This seemingly terminological difference highlighted the more theoretical issues involved in the study of molecular physics. "It is the writer's belief that, of the various possible methods, the present one may be the best adapted to the construction of an exploratory conceptual scheme within whose framework may be fitted both chemical data and data on electron levels from electron spectra."[57]

Mulliken had realized that one of the reasons for the poor quantitative agreement using the molecular orbital approach was because of the inability of this theory to include the details of the interactions between the electrons. But even though their inclusion would make a theoretical calculation from first principles an impossible job, "their qualitative inclusion has always formed a vital part of the method of molecular orbitals used as a conceptual scheme for the interpretation of empirical data on electronic states of molecules."[58] Such considerations, in fact, led to the qualitative explanation of the paramagnetism of oxygen—one of the main weaknesses of the valence bond approach.

Differences in Opinion or Different Choices?

Apart from the letters they exchanged in late 1927 about the possibilities offered by group theory to the problems related to the chemical bond, there was no correspondence between Heitler and London until 1935. The contacts between them were prompted by the developments of Pauling's resonance theory and Mulliken's molecular-orbital approach. The correspondence between them is quite revealing. It shows the attitude of each about the possible development of the approach laid down in their common paper, and the tension between them as well as the search for the means to consolidate their theory at a time when the Americans appeared to be taking over the field of quantum chemistry. Their correspondence also reflects the different styles of the intellectual heritage of each. Faithful to the Göttingen spirit, Heitler was "more mathematical," while London continued in the Berlin tradition of theoretical physics and his inclination was to examine intuitive proposals. Both had left Germany after the rise of the Nazis and had settled in England. Heitler was in Bristol, and London in Oxford. Both were working on different problems far removed from the problem of chemical bonding. Heitler was working on the theory of radiation, and London had just published the paper with Heinz on superconductivity. Recent publications, however, showed a marked preference in the methods proposed by Pauling and Mulliken over their own. In 1935 Heitler and London started writing to each other trying to "save" their own schema.

In their initial letters, Heitler and London discussed the possibility of writing an article in *Nature* to present their old results and include some new aspects that had not been emphasized properly in their earlier papers. These were the activation of spin valence and the possibility of a bond that would not depend on spin saturation.[59] Heitler's attitude concerning the approach by Slater and Pauling was that they were correct about the principles they adopted and he was quite sympathetic about the direction of their researches, even though a series of results did not follow exclusively from their theory. He thought a polemic against them was quite unjustified. "I simply find that the importance of this theory has been monstrously overrated in America."[60] Heitler and London felt that among the missed opportunities was their lack of insistence about the oxygen molecule: "It is only due to our negligence that now comes v. Vleck (after the publication of the matter!) and writes that O_2 is a triumph of Mulliken-Hund, because in our theory is 'less elementary.'"[61] London insisted that the "essence of a discovery is to know what one is doing."[62]

Doubts were expressed for the first time about the character of the attractive forces. It was conceivable that these forces may not be only due to spin. There were other attractive forces with the same order of magnitude and those did not follow from their original theory of spin valence. It was wrong to believe that these forces could originate only from the directional degeneracy of the ground state. It may be the case that these forces resulted in the formation of a molecule only if there were also spin valences. At this point nothing much could be said about the claim that these forces did not have the characteristics of valence. They admitted that they did not know much about what happened when more than two atoms were near each other since the mechanism of these forces differed from the mechanism of spin valences. The chemists, they thought, were not so fussy about it and named as valence forces whatever forces formed molecules. "This is exactly our job. To say that there are also other forces of molecule formation except our old ones and which phenomena of chemical valence depend on those, and, especially, that our old scheme can be extended." Heitler's feeling was that there had been no attack against them by the Americans except for the case of the oxygen molecule whose diamagnetism they could not explain. The nucleus of their theory was the spin valence and he insisted that their theory was the only one that explained the mechanism of repulsion in a qualitatively exact manner. "You could perhaps include the above discussion under the title: Delineation of completeness. . . . In any case, we should stress that the extension could be realized on the basis of our theory and, substantially, it includes whatever one could wish (this last thing only as a footnote for us)."[63]

London's answer was not exactly an eulogy to the chemical profession.

> The word "valence" means for the chemist something more than simply forces of molecular formation. For him it means a substitute for these forces whose aim is to free him from the necessity to proceed, in complicated cases, by calculations deep into the model. It is clear that this remains

wishful thinking. Also the fact that it has certain heuristic successes. We can, also, show the quantum mechanical framework of this success . . . the chemist is made out of hard wood and he needs to have rules even if they are incomprehensible.[64]

They progressively realized that part of the problem was their isolation and this realization bred even more frustration. The fact that they had not even been attacked was not an indication of the acceptance of their theory. Their feeling was that their theory may have been forgotten or that it "can be combatted much more effectively by the conscious failure to appreciate and avoid mentioning it."[65]

Heitler did not agree with London that their theory "is fought by the most unfair and secretive means."[66] "It may be true for some people in America. Not all people, however, are rascals (e.g., I would not believe it for van Vleck), but only silly and lazy. And we should accept that our theory was quite complicated. I would gladly like to look at the books of Sidgwick and Pauling. I cannot get them here."[67]

Heitler visited Oxford at the beginning of December 1935. Both were now fully aware that the Americans were starting to dominate the field. As soon as Heitler was back in Bristol, he read a paper by Wheland. In the paper it was noted that the Heitler-London-Slater-Pauling method was developed originally by Slater as a generalization of Heitler and London's treatment of H_2 and has since been simplified by Pauling. These thoughts exhausted the last vestiges of tolerance displayed by the more "objective" of the two. Heitler was vitriolic in his response to London:

> I propose in the future to talk only about the theory of Slater-Pauling of the chemical bond, since, in the last analysis [it explained only] the H_2—well now what can this be compared with the feats of the Americans. . . . I am afraid that the reading of the papers that we have voluntarily undertaken shall be the purgatorium of our souls. If you cannot restrain me, I think, I will write a very clear letter to this Pauling (he should give a better upbringing to his students). . . . It would be really good to write something which will mostly have those things they are stealing in America. Do not think I am exasperated because (in the case of Pauling) it involves my paper with Rumer, but because of our common cause. Your achievements disappear equally in the lies. . . . For van Vleck I notice that his papers are more dignified than his report and does not thank Slater and Pauling for free.[68]

They decided to find an excuse to write to Pauling admitting that the work of Pauling and Slater did, in fact, go beyond the version of their original theory. London took it upon himself to read their papers carefully. "We should find many points where it will be evident that the passages were written in bad faith. . . . The best thing would be to have as an excuse a substantial question or a criticism to Pauling's papers." Slater's "shameless behaviour starts from 1931." Slater had claimed that the theory of Heitler and Rumer was valid only when the bond energy was small with respect to multiplet dissociation and, therefore, it had no physical meaning. This, according to Heitler, was not correct and it was because Slater confused multiplet dissociation with the separation of terms via the Coulomb interaction.

They planned to meet in London in mid-February. They were both back to study chemistry. Heitler felt that Slater and Pauling were "so proud about something which is not so bad, but which, under no circumstances, is so distinguished. It gives a general formula for the bond, that corresponds to the pair bonding and the repulsion of the valence lines."[69] Their approximation was as rough as Heitler's semiclassical theory and it is superior only because it included the directional properties. One, however, totally lost the activation energy and the nonadditivity of the bond energy.

> It is needless to say that it is fully based on our ideas. . . . We should not, though, fall into the error and regard this work bad or insignificant (as these people do ours). It is a branching from our work, from about the point where we strictly suppose that the atoms are in only one state. . . . Generally, I believe that we did the mistake not to give more concrete applications of the theory, it was a mistake to leave it to the chemists (who are nearer to this kind of work). . . . I do not find, though, that our direction is not being given any attention (apart from the details) in Europe.[70]

There are not many places where we can read the opinions of either of them concerning the molecular orbital approach. Heitler thought that their basic objection with "Hund's people"—who both agree are not the most unpleasant of their enemies—was not related so much with the actual results derived by this method. Sufficient patience with the calculations and a lot of semiempirical considerations gave, in fact, correct results. "Nevertheless, no one could name this a general theory—much less a valence theory—since all the general and substantive points are forever lost."[71]

An article by Lennard-Jones gave the opportunity for some further clarifications of London's position. Lennard-Jones in all his publications preferred the "one-electron-orbital-bonds" and presented the version by Heitler and London "as not so beautiful and as inadequate." London asked Born's advice on how to proceed and get out of the deadlock they found themselves in. London felt that the big publicity of the molecular orbital theory due to its simplicity harmed the reliability of their valence theory. "Maybe it was a mistake that we never expressed the objections we had from the beginning on questions of principle concerning the approach of Lennard-Jones-Mulliken. Both of us thought it as superfluous, because we had both 'transcended' this same phase of Lennard-Jones-Mullikan in the beginning of our observations in 1927, and we were very proud when we realized that we get the exchange degeneracy because of the similarity of the electrons." They thought that the molecular orbital approach was inherently contradictory "and maybe, for this reason we did not take it seriously. Recently, I talk very often with Heitler for this lost ground and repeatedly we tried to find a way to make up for it. We continuously fail. . . . We have, undoubtedly, made a mistake by not taking seriously our competitors. . . . The situation had become clear since 1932–33, when we should have thought to find new issues and not make enemies with our polemics."[72]

London's move to Paris and the incomparably more pleasant conditions there in comparison to Oxford; Heitler's success with his book and his work in quantum electrodynamics; London's success with the theory of superconductivity—somehow, one cannot help but feel that both of them could now afford to be gracious. Suddenly, as if by magic, there was no more talk about these issues—maybe all the reading and the discussions became the "purgatorium of our souls" as Heitler had suggested in one of his first letters.

Quantum chemistry developed an autonomous language with respect to physics and, above all, its eventual acceptance by the chemists involved a series of issues concerning the way chemists (should) practice the new subdiscipline. Coulson has talked of the period through World War II as a period when the chemists were concerned "in escaping from the thought forms of the physicist."[73] Usually, the two different approaches to the basic problem of quantum chemistry—that of valence—are compared with each other and then the relative merits and disadvantages of each method are assessed. What appeared to be disputes over methods were, in fact, discussions concerning the collective decision of the chemical community about methodological priorities and ontological commitments. In many instances the scientific papers had a strong rhetoric "propagandizing" various changes in the chemists' culture. During the 1930s the discussions and disputes among chemists were to a large extend about the new legitimizing procedures and consensual activities to be incorporated into the chemists' culture.

These differences were rather eloquently expressed by Van Vleck in a review article he wrote in 1935 with Sherman. Anyone who was looking for straightforward calculations from the basic postulates of quantum mechanics was bound to be disappointed. How, then, van Vleck asked can it be said that we have a quantum theory of valence? To give a satisfactory answer one "must adopt the mental attitude and procedure of the optimist rather than the pessimist." The latter demands rigorous calculations from first principles and does not allow questionable approximations or appeals to empirically known facts. The optimist is content with approximate solutions of the wave equation and "he appeals freely to experiment to determine constants, the direct calculation of which would be too difficult." The optimist's attitude is that the approximate calculations "give one an excellent 'steer'" and a very good idea of "how things go," permitting the systematization and understanding of what would otherwise be a maze of experimental data "codified by purely empirical valence rules."[74] It is not clear whether Van Vleck and Sherman used the words "optimist" and "pessimist" for pedagogic purposes and to imply their own preferences despite the fact that they promised to "adopt a middle ground between the two extreme points of view." Whether the "optimists" referred to the enthusiastic Americans and the "pessimists" to the reserved Germans, remains a guess.

What I have attempted to show here is the many strategies of appropriating the electron in chemistry, and that part of the theoretical particularity of chemistry is that it may become possible to accommodate—at least during the initial phases of the formation of quantum chemistry—in chemistry more than one theoretical schemata with equivalent empirical adequacy. The Heitler-London valence bond theory, the group theoretical extensions of their theory, Pauling's resonance theory, and Mulliken's molecular orbital theory are some of the more successful theoretical schemata. The copresence of all these schemata was not free of disagreements. The genesis and development of quantum chemistry as an autonomous discipline owed much to those chemists who successfully managed to escape from the "thought forms of the physicists" by explicitly addressing issues such as the role of theory in chemistry and the methodological status of empirical observations. My impression is that the story of the electron in chemistry after the advent of quantum mechanics is the story of the chemists trying to explain to the rest of the world what theory is in chemistry.

NOTES

1. For details see Gavroglu (1995).

2. Mulliken (1965), S7.

3. Interview of Heitler by John Heilbron.

4. Heitler, London (1927).

5. Interview of Walter Heitler at *A.H.Q.P.*

6. Van Vleck (1970), 240.

7. Private communication by Professor Sir Brian Pippard.

8. Pauling (1928a), 174.

9. Van Vleck (1928), 506.

10. Interview with E. Wigner by T. S. Kuhn, 4 December 1963. *A.H.Q.P.*

11. Pauling, Wilson (1935), 340. Banz (1980) has discussed the Heitler-London paper as a case study to examine a series of claims in the philosophy of science.

12. Buckingham (1987), 113.

13. Interview of Linus Pauling with John Heilbron, 27 March 1964. *A.H.Q.P.* An alternative approach to the valence bond method was first attempted by Edward Condon (1927) a few months before the appearance of the paper by Heitler and London.

14. Pauling (1928b), 359.

15. Linus Pauling to G. N. Lewis, 7 March 1928 . G. N. Lewis papers at the College of Chemistry, University of California at Berkeley. "It pleases me very much that in the new atomic model the salient features of the Lewis atom have been reproduced as much as those of the Bohr atom."

16. G. N. Lewis to Linus Pauling, 1 May 1928. Pauling papers, Oregon State University.

17. Van Vleck (1928), 500.

18. Van Vleck and Sherman (1935), 173.

19. Heitler to London, September(?) 1927. Fritz London Archives, University Library, Duke University, North Carolina, U.S.A.

20. Heitler to London, September (?) 1927.

21. Interview with E.Wigner by T. S. Kuhn, 4 December 1963, 14.

22. Heitler to London, 7 December 1927.

23. Interview of Walter Heitler by John Heilbron, 18 March 1963. *A.Q.H.P.*

24. From the interview with Walter Heitler. *A.Q.H.P.*

25. Hartree to London, 16 September 1928.

26. Born to London, 8 October 1930.

27. Clark (1928), 362.

28. Rodebush (1928), 513.

29. Fry (1928), 558–559.

30. For a detailed discussion of Linus Pauling and Robert Mulliken, see Ana Simoes (1993) and Gavroglu, Simoes (1994).

31. Wheland (1944).

32. Wheland to Pauling, 20 January 1956. Pauling papers.

33. Pauling to Wheland, 26 January and 8 February 1956. Wheland papers, University of Chicago.

34. Ibid.

35. Pauling (1955).

36. Pauling (1956).

37. Pauling (1960).

38. Ibid., 219–220.

39. Coulson (1947), 47.

40. Kursanov et al. (1952); Tatevskii, Shakhparanov (1952); Moyer Hunsberger (1954).

41. Kursanov et al. (1952).

42. M. V. King to Pauling, 23 January 1953; letter Coulson to Pauling, 9 October 1953; letter Coulson to King, 18 January 1954; letter King to Pauling, 9 February 1954. Pauling Papers.

43. Pauling (1960), 95.

44. Pauling (1939), 219.

45. Pauling (1955), 92.

46. Pauling (1960), 220.

47. Pauling, private communication.

48. Mulliken to Birge, 26 March 1931. Birge correspondence, The Bancroft Library, University of California Berkeley.

49. Mulliken (1931), 349.

50. Mulliken (1932b), 51.

51. Ibid., 54–55.

52. F. London to W. Heitler, end(?) of November 1935.

53. Mulliken (1932a), 30.

54. Mulliken (1928), 189.

55. Mulliken (1967), 157.

56. Mulliken (1935), 376.

57. Ibid., 377.

58. Ibid., 378.

59. Heitler to London, 4 November 1935.

60. Heitler to London, 12 November 1935.

61. Heitler to London, 7 November 1935.

62. London to Heitler, 6 November 1935.

63. Heitler to London, 12 November 1935.

64. London to Heitler, October or November 1935.

65. London to Heitler, October or November 1935.

66. London to Heitler, 17 November 1935.

67. Heitler to London, 22 November 1935.

68. Heitler to London, beginning of December 1935.

69. Heitler to London, 6 February 1936.

70. Heitler to London, 6 February 1936.

71. Heitler to London, 7 October 1936.

72. London to Born, 1 October 1936.

73. Coulson (1970), 259 and 287. Coulson was a mathematician by training and the author of one of the standard books on valence theory. He was the first student of Lennard-Jones at Cambridge when the latter moved from Bristol to Cambridge in 1928 to the first (and for many years only) chair of theoretical chemistry in Britain.

74. Van Vleck, Sherman (1935), 168–169.

References

A.H.Q.P. Archives for the History of Quantum Physics, American Institute of Physics.

Banz, D. A., (1980). "The Structure of Discovery: Evolution of Structural Accounts of Chemical Bonding." In T. Nickles (ed.), *Scientific Discovery: Case Studies* (Reidel Publishers, Dordrecht, 1980), 291–329.

Buckingham, A. D., 1987. "Quantum Chemistry," in Kilminster (1987): 112–117.

Clark, G. L., 1928. "Introductory Remarks in the Symposium on Atomic Structure and Valence," *Chemical Review* 5: 361–364.

Condon, E. U., 1927. "Wave Mechanics and the Normal State of the Hydrogen Molecule," *Proceedings of the National Academy of Sciences* 13: 466–470.

Coulson, C. A., 1947. "The Meaning of Resonance in Quantum Chemistry," *Endeavour* 6: 42–47.

Coulson, C. A., 1970. "Recent Developments in Valence Theory–Symposium Fifty Years of Valence," *Pure and Applied Chemistry* 24: 257–287.

Fry, H. S., 1928. "A Pragmatic System of Notation for Electronic Valence Conceptions in Chemical Formulas," *Chemical Reviews* 5: 557–568.

Gavroglu, K., and A. Simoes, 1994. "The Americans, the Germans and the Beginning of Quantum Chemistry: The Confluence of Diverging Traditions," *Historical Studies in the Physical Sciences* 25: 47–110.

Gavroglu, K., 1995. *Fritz London, A Scientific Biography,* Cambridge: Cambridge University Press.

Heitler, W., and F. London, 1927. "Wechselwirkung neutraler Atome und homopolare Bindung nach der Quantenmechanik," *Zeitschrift für Physik* 46: 455–477.

Kilminster, C. W., ed., 1987. *Schrödinger—Centenary Celebration of a Polymath,* Cambridge: Cambridge University Press.

Kursanov, D. N., M. G. Gonikberg B., Dubinin, M. I. Kabachnik, E. D. Kaveraneva, E. N. Prilezhaeva, N. D. Sokolov, and R. Kh. Freidlina, 1952. "The Present State of the Chemical Structural Theory." Translation by I. S. Bengelsdorf in *Journal of Chemical Education* 78: 2–13.

Moyer Hunsberger, I., 1954. "Theoretical Chemistry in Russia," *Journal of Chemical Education* 80: 504–514.

Mulliken, R. S., 1928. "The Assignment of Quantum Numbers for Electrons in Molecules. Part I." *Physical Review* 32: 186–222.

Mulliken, R. S., 1931. "Bonding Power of Electrons and Theory of Valence," *Chemical Reviews* 9: 347–388.

Mulliken, R. S., 1932a. "The Interpretations of Band Spectra," *Reviews of Modern Physics* 4: 1–86.

Mulliken, R. S., 1932b. "Electronic Structures of Polyatomic Molecules and Valence. II. General Considerations." *Physical Review* 41: 49–71.

Mulliken, R. S., 1935. "Electronic Structures of Polyatomic Molecules and Valence.VI. On the Method of Molecular Orbitals," *Journal of Chemical Physics* 3: 375–378.

Mulliken, R. S., 1965. "Molecular Scientists and Molecular Science: Some Reminiscences," *Journal of Chemical Physics* 43: S7.

Mulliken, R. S., 1967. "Spectroscopy, Molecular Orbitals and Chemical Bonding," *Science* 157, 7 July: 17.

Pauling, L., 1928a. "The Application of the Quantum Mechanics to the Structure of the Hydrogen Molecule," *Chemical Reviews* 5: 173–213.

Pauling, L., 1928b. "The Shared-Electron Chemical Bond," *Proceedings of the National Academy of Sciences* 14: 359.

Pauling, L., 1939. *The Nature of the Chemical Bond,* Ithaca: Cornell University Press.

Pauling, L., 1955. "Modern Structural Chemistry," in *Les Prix Nobel,* Stockholm.

Pauling, L., 1956. "The Nature of the Theory of Resonance," in Todd (1956).

Pauling, L., 1960. *The Nature of the Chemical Bond,* 3rd. Ed., Ithaca: Cornell University Press.

Pauling, L, and E. B. Wilson, 1935. *Introduction to Quantum Mechanics with Applications to Chemistry,* New York: McGraw-Hill.

Rodebush, W. H., 1928. "The Electron Theory of Valence," *Chemical Reviews* 5: 509–531.

Simoes, A., 1993. "Converging Trajectories and Diverging Traditions: Chemical Bond, Valence, Quantum Mechanics and Chemistry 1927–1937," Ph.D. Thesis, University of Maryland.

Tatevskii, V. M., and M. I. Shakhparanov, 1952. "About a Machistic Theory in Chemistry and its Propagandists," *Journal of Chemical Education* 78: 13–14.

Todd, A., ed., 1956. *Perspectives in Organic Chemistry, Dedicated to Sir Robert Robinson,* New York: Interscience Publishers.

Van Vleck, J., 1928. "The New Quantum Mechanics," *Chemical Reviews* 5: 467–507.

Van Vleck, J., 1970. "Spin, the Great Indicator of Valence Behaviour," *Pure and Applied Chemistry* 24: 235–255.

Van Vleck, J., and A. Sherman, 1935. "The Quantum Theory of Valence," *Reviews of Modern Physics* 7: 167–228.

Wheland, G. W., 1944. *The Theory of Resonance and Its Applications to Organic Molecules* New York: John Wiley and Sons.

IV

Philosophical Electrons

13

Who Really Discovered the Electron?
Peter Achinstein

Two Problems with Identifying J. J. Thomson as the Discoverer

Heroes are falling in this age of revisionist history. Thomas Jefferson, according to one recent authority, was a fanatic who defended the excesses of the French Revolution. Einstein was not the saintly physicist we were led to believe but was mean as hell to his first wife. And, more to the present purpose, J. J. Thomson really didn't discover the electron. So claim two recent authors, one a contributor to this volume, Theodore Arabatzis, in a 1996 article on the discovery of the electron,[1] and the other my very talented colleague, Robert Rynasiewicz, at a February 1997 A.A.A.S. symposium in honor of the 100th anniversary of the discovery.

I would like my heroes to retain their heroic status, however, my aim in this chapter is not to defend Thomson's reputation but to raise the more general question of what constitutes a discovery. My strategy will be this. First, I want to discuss why anyone would even begin to doubt that Thomson discovered the electron. Second, I want to suggest a general view about discovery. Third, I will contrast this with several opposing positions, some of which allow Thomson to retain his status, and others of which entail that Thomson did not discover the electron; I find all of these opposing views wanting. So who, if anyone, discovered the electron? In the final part of this chapter I will say how the view I develop applies to Thomson and also ask why we should care about who discovered the electron, or anything else.

Let me begin, then, with two problems with identifying Thomson as the discoverer of the electron. The first is that before Thomson's experiments in 1897 several other physicists reached conclusions from experiments with cathode rays that were quite similar to his. One was William Crookes. In 1879, in a lecture before the British Association at Sheffield, Crookes advanced the theory that cathode rays do not consist of atoms, "but that they consist of something much smaller than the atom—fragments of matter, ultra-atomic corpuscles, minute things, very much smaller, very much lighter than atoms—things which appear to be the foundation stones of which atoms are composed."[2]

So eighteen years before Thomson's experiments, Crookes proposed two revolutionary ideas essential to Thomson's work in 1897: that cathode rays consist of corpuscles smaller than atoms, and that atoms are composed of such corpuscles. Shouldn't Crookes be accorded the title "discoverer of the electron"?

Another physicist with earlier views about the electron was Arthur Schuster. In 1884, following his own cathode ray experiments, Schuster claimed that cathode rays are particulate in nature and that the particles all carry the same quantity of electricity.[3] He also performed experiments on the magnetic deflection of the rays, which by 1890 allowed him to compute upper and lower bounds for the ratio of charge to mass of the particles comprising the rays. Unlike Thomson (and Crookes in 1879), however, Schuster claimed that the particles were negatively charged gas molecules.

Philipp Lenard is still another physicist with a considerable claim to be the discoverer of the electron. In 1892 he constructed a cathode tube with a special window capable of directing cathode rays outside the tube. He showed that the cathode rays could penetrate thin layers of metal and travel about half a centimeter outside the tube before the phosphorescence produced is reduced to half its original value. The cathode rays, therefore, could not be charged molecules or atoms, since the metal foils used were much too thick to allow molecules or atoms to pass through.

Other physicists as well, such as Hertz, Perrin, and Wiechert, made important contributions to the discovery. Why elevate Thomson and say that he discovered the electron? Why not say that the discovery was an effort on the part of many?

The second question is: Even assuming that Thomson discovered something, was it really the *electron*? How could it be, since Thomson got so many things wrong about the electron? The most obvious is that he believed that electrons are particles or corpuscles (as he called them), *and not waves*. In a marvelous twist of history, Thomson's son, G. P. Thomson, received the Nobel Prize for experiments in the 1920s demonstrating the wave nature of electrons. Another mistaken belief was that electrons are the only constituents of atoms. Still others were that the charge carried by electrons is not the smallest charge carried by charged particles, and that the mass of the electron, classically viewed, is entirely electromagnetic, a view Thomson came to hold later. Why not deny that Thomson discovered anything at all, since nothing exists that satisfies his electron theory? To deal with these issues something quite general needs to be said about discovery.

What Is Discovery?

The type of discovery with which I am concerned is discovering some thing or type of thing (for example, the electron, the Pacific Ocean), rather than discovering that something is the case (for example, that the electron is negatively charged). Later I will consider a sense of discovering some thing X that requires a knowledge that it is X, as well as a sense that does not.

My view has three components, the first of which is ontological. Discovering something requires the existence of what is discovered. You cannot discover what doesn't exist—the ether, the Loch Ness monster, or the fountain of youth—even if you think you have. You may discover the *idea* or the *concept* of these things. Everyone may think you have discovered the things corresponding to these ideas or concepts. They may honor you and give you a Nobel Prize, but if these things don't exist, you haven't discovered them.

The second component of discovery is *epistemic*. A certain state of knowledge is required. If you are to be counted as the discoverer of something, not only must that thing exist but also you must know that it does. Crookes in 1879 did not discover electrons because he lacked such knowledge; his theoretical claim that cathode rays consist of subatomic particles, although correct, was not sufficiently established to produce the knowledge that such particles exist. Not just any way of generating knowledge, however, will do for discovery. I may know that something exists because I have read that it does in an authoritative book. Discovery, in the sense we are after, requires that the knowledge be firsthand.

What counts as "firsthand" can vary with the type of object in question. With physical objects such as electrons one might offer this rough characterization: knowledge that the objects exist is generated, at least in part, by observing those objects or their direct effects. This knowledge may require rather strenuous inferences and calculations from the observations. (Scientific discovery is usually not like discovering a cockroach in the kitchen or a nail in your shoe.) As noted, discovery involves not just any observations that will produce knowledge of the object's existence, but observations of the object itself or its direct effects. I may come to know of the existence of a certain library book by observing a computer screen in my office which claims that the library owns it. I may discover that the book exists by doing this. But I may never discover the book itself if I can't find it on the shelf. In discovering the book at least among the things that make me know that it exists is my seeing it. Finally, for discovery, the knowledge in question involves having as one's reason, or at least part of one's reason, for believing that X exists the belief that it is X or its direct effects that have been observed. My knowledge that electrons exist may come about as a result of my reading the sentence "Authorities say that electrons exist" on my computer screen. What is on my screen is a direct effect of electrons. But in such a case, I am supposing, my reason for believing that electrons exist does not include the belief that I have observed electrons or their direct effects on the screen.[4]

Putting together these features of the second (epistemic) component of discovery, we can say that someone is in an *epistemic state necessary for discovering X* if that person knows that X exists, observations of X or its direct effects caused, or are among the things that caused, that person to believe that X exists, and among that person's reasons for believing that X exists is that X or its direct effects have been observed. More briefly, I will say that such a person knows that X exists from observations of X or its direct effects.

The third component of discovery is priority. If I am the discoverer of something, then the epistemic state I have just described must be a "first." I put it this way because it is possible to relativize discovery claims to a group or even to a single individual. I might say that I discovered that book in the library last Tuesday, meaning that last Tuesday is the first time for me. It is the first time I knew the book existed by observing it, even though others knew this before I did. I might also make a claim such as this: I was the first member of my department to discover the book, thereby claiming my priority over others in a certain group. Perhaps it is in this sense that we say that Columbus discovered America, meaning that he was the first European to do so. And, of course, the relevant group may be the entire human race. Those who claim that Thomson discovered the electron mean, I think, that Thomson was the first human to do so.

There is a rather simple way to combine these three components of discovery, if we recognize that knowing that something exists entails that it does, if we confine our attention to discovering physical objects (rather than such things as facts, laws, or proofs), and if we employ the previously introduced concept of an epistemic state necessary for discovery. The simple way is this: P discovered X if and only if P was the first person (in some group) to be in an epistemic state necessary for discovering X. That is, P was the first person (in some group) to know that X exists, to be caused to believe that X exists from observations of X or its direct effects, and to have as a reason for believing that X exists that X or its direct effects have been observed.

Before contrasting this with opposing views, and applying this to Thomson and the electron, some points need clarification.

First, on this account, to discover X, you don't need to observe X directly. It suffices to observe certain causal effects of X that can yield knowledge of X's existence. If I see a cloud of dust moving down the dirt road that is obviously being produced by a car approaching, then I can discover a car that is approaching even though I can't see the car itself, but only the cloud of dust it is producing as it moves. It is not sufficient, however, to come to know of X's existence via observations of just any sort. If I read a letter from you saying that you will be driving up the dirt road to my house at noon today, and I know you to be someone who always keeps his word, that, by itself, does not suffice for me to say that I discovered a car that is approaching at noon, even if I know that the car is approaching. Discovering the car requires observations of the car or its direct effects.

Second, this will prompt the question "What counts as observing direct effects?" Some physicists want to say that the tracks produced by electrons in cloud chambers are direct effects, because electrons, being charged, ionize gas molecules around which drops of water condense forming the tracks. By contrast, neutrons, being neutral, cannot ionize gas molecules and hence do not leave tracks. They are detected by bombarding charged particles that do leave tracks. More recently detected particles, such as the top quark, involve many different effects that are less direct than these.[5] This is a complex issue that cannot be quickly settled.[6] What appears to be involved is not some absolute idea of directness, but a relative one. Given the nature of the item whose effects it is (for example, if it is a neutron it cannot produce a track but must interact with charged particles that do produce tracks), this degree of directness in detecting its effects not only yields knowledge that the item exists but also furnishes the best, or one of the best, means at the moment available for obtaining that knowledge.

Third, on this account, the observations of X or its effects need not be made by the discoverer, but by others. What is required is only that the discoverer be the first to know that X exists from such observations. The planet Neptune was discovered independently by Adams and Leverrier from observations of the perturbations of Uranus caused by Neptune. These observations were made by others, but complex calculations enabled these astronomers to infer where the new planet could be observed in the sky. The first actual telescopic observation of Neptune was made not by either of these astronomers but by Johann Galle at the Berlin Observatory. Although Galle may have been the first to see Neptune, he is not its discoverer, because he was not the first to come to know of its existence from observations of Neptune or its effects.

Fourth, to discover X it is not sufficient simply to postulate, or speculate, or theorize that X exists. In 1920 Rutherford theorized that neutrons exist, but Chadwick in 1932, not Rutherford in 1920, is the discoverer of these particles. There were before 1932 experimental results that allowed the existence of this particle to be known.

In connection with the electron, there are two physicists whose names I have not mentioned so far: Larmor and Lorentz. Both had theories about what they called electrons. Setting aside questions about whether they were referring to what we call electrons, one reason these physicists are not the discoverers of electrons is, I think, epistemic. Although their theories explained experimental results, such results were not sufficiently strong to justify a knowledge-claim about the electron's existence. Their claims about electrons were primarily theory driven.

Fifth, this view allows there to be multiple discoverers, as were Adams and Leverrier. They came to be in the appropriate epistemic states at approximately the same time. It allows a cooperative group of scientists, rather than the scientists individually in that group, to be the discoverers—as in the recent case of the top quark. And it allows scientists to make contributions to the discovery of X without themselves being discoverers or part of a group that discovered X. Plucker did not discover the electron, though in 1859 he made a crucial contribution to that discovery—the discovery of cathode rays.

Sixth, we need to distinguish two ways of understanding the phrase "knowing that X exists" in my definition of discovery and hence two senses of discovery. Suppose that while hiking in the Rockies, I pick up some shiny stones. You inform me that I have discovered gold. This could be true, even if I don't know that it is gold. In this case by observing the stones I have come to know of the existence of something that, unknown to me, is gold. That is one sense in which I could have discovered gold. Of course, I might also have come to know that these objects are gold. That is another sense in which I could have discovered gold.[7]

The same applies to discovering the electron. To say that Thomson discovered the electron might mean only that by suitable observations he came to know of the existence of something that happens to be the electron, even if he didn't realize this. Or it might mean something stronger to the effect that he came to know that the thing in question has the electron properties (whatever those are). I speak here of the latter as the "stronger" sense of discovery and the former as the "weaker."

CONTRASTING VIEWS OF DISCOVERY

The present view of discovery will now be contrasted with several others, including ones suggested by two historians of science who have discussed the history of the discovery of the electron. Although the primary focus of these authors is historical and not philosophical, what they claim about Thomson suggests more general views about what counts as a discovery. These more general views provide sufficient conditions for discovery, or necessary ones, or both. I want to indicate how these views conflict with mine, and why I reject them both as generalizations about discovery and as particular views about what made, or failed to make, Thomson the discoverer of the electron.

Manipulation-and-Measurement View
At the end of her important 1987 paper on Thomson, Isobel Falconer writes:

> In the light of this reinterpretation of Thomson's work is there any sense remaining in which he can be said to have "discovered the electron"? Arriving at the theoretical concept of the electron was not much of a problem in 1897. Numerous such ideas were "in the air." What Thomson achieved was to demonstrate their validity experimentally. Regardless of his own commitments and intentions, it was Thomson who began to make the electron "real" in Hacking's sense of the word. He pinpointed an experimental phenomenon in which electrons could be identified and methods by which they could be isolated, measured, and manipulated.[8]

Several things are suggested here, but one is that Thomson discovered the electron because he was the first to design and carry out experiments in which electrons were manipulated and measured. We might recall that, on Hacking's view, to which Falconer alludes, "if you can spray them they are real."[9] On the more sophisticated version suggested by Falconer in this passage, if you can manipulate them in such a way as to produce some measurements they are real; and if you are the first to do so, you are the discoverer. Such a view needs expanding to say what counts as "manipulating" and "measuring." I will not try to do so here, but will simply take these ideas as reasonably clear. It appears obvious that Thomson manipulated electrons by means of magnetic and electric fields and that he measured their mass-to-charge ratio.

Important Classification View
This view is suggested by an earlier passage in Falconer's paper. Discussing the experimental work of Wiechert, she writes:

> Wiechert, while realizing that cathode ray particles were extremely small and universal, lacked Thomson's tendency to speculation. He could not make the bold, unsubstantiated leap, to the idea that particles were constitutents of atoms. Thus, while his work might have resolved the cathode ray controversy, he did not "discover the electron."[10]

This suggests that, despite the facts that both Wiechert and Thomson manipulated the electron in such a way as to obtain a mass-to-charge ratio and that both physicists claimed that cathode particles were "extremely small and universal," Thomson, and not Wiechert, is the discoverer of electrons because Thomson but not Wiechert got the idea that cathode particles are constituents of atoms. Although Falconer does not say so explicitly, perhaps what she has in mind is that Thomson's identification of cathode particles as universal constituents of atoms is what is important about electrons. Generalizing from this, you are the discoverer of X when you are the first to arrive at an important (and correct) classification of X. The question remains as to what counts as an "important" classification—a major lacuna, as I will illustrate in a moment. However this is understood, it should include Thomson's classification of electrons as constituents of atoms.

Social Constructivist View

Social constructivism is a broad viewpoint pertaining to many things, including the reality of scientific objects themselves such as electrons (they are "socially constructed" and have no reality independently of this). There is, however, a much narrower social constructivist view that is meant to apply only to scientific discovery. On this view, whether some scientist(s) discovered X depends on what the relevant scientific community believes. This view is adopted by Arabatzis prior to his historical discussion of the work of Thomson and others on the electron. He writes:

> A final approach [to discovery]—and the one I favour—takes as central to the account the perspectives of the relevant historical actors and tries to remain as agnostic as possible *vis-à-vis* the realism debate. The criterion that this approach recommends is the following: since it is the scientific community (or its most eminent representatives) which adjudicates discovery claims, an entity has been discovered only when consensus has been reached with respect to its reality. The main advantage of this criterion is that it enables the reconstruction of past scientific episodes without presupposing the resolution of pressing philosophical issues. For historical purposes, one does not have to decide whether the consensus reached by the scientific community is justifiable from a philosophical point of view. Furthermore, one need not worry whether the entity that was discovered (in the above weak sense) can be identified with its present counterpart.[11]

Although in this passage Arabatzis claims that there is a discovery only when the community believes there is, he also says that the main advantage of his criterion is that it avoids the issue of whether the concensus reached is justified, and the issue of whether the entity that was discovered is the same as the one scientists now refer to. Accordingly, the view suggested is a rather strong one, to the effect that consensus is both necessary and sufficient for discovery. (At least, that is the social constructivist view about discovery that I will consider here.) Thomson discovered the electron if he is generally regarded by physicists as having done so. The physicists who so regard him may have different reasons for doing so, but these reasons do not make him the discoverer: simply their regarding him as such does. Even if the reasons are false (in some "absolute," non-consensual sense), he is still the discoverer, unless the physics community reaches a different concensus.

Different Contributions View
According to this idea, there are discoveries in science, including that of the electron, that are not made by one person, or by several, or by any group, but involve various contributions by different people. We need to replace the question "who discovered the electron?" with more specific questions about who made what contributions to the discovery. We might note that in 1855 Geissler contributed by inventing a pump that allowed much lower gas pressures to be produced in electrical discharge tubes; that in 1859 using this pump, Plucker found by experiment that when the pressure is reduced to .001 mm of mercury, the glass near the cathode glows with a greenish phosphorescence and the position of the glow changes when a magnetic field is introduced; that in 1869 Plucker's student Hittorf found that if a solid body is placed between the cathode and the walls of the tube it casts a shadow, from which he concluded that rays are emitted from the cathode that travel in straight lines. This story could be continued with experimental and theoretical contributions by Crookes, Larmor, Lorentz, Hertz, Goldstein, Schuster, and so forth, culminating with the experiments of Thomson—or well beyond if you like.

Now, it is not that all the people mentioned, or even several of them, or a group working together, discovered the electron. Plucker didn't discover the electron, nor was he one of several people or a group that did. Still the electron was discovered, but it was not the sort of discovery made by one individual, or several, or a group. Rather it was the sort of discovery that involved different contributions by different persons at different times. Thus, Arabatzis writes:

> Several historical actors provided the theoretical reasons and the experimental evidence which persuaded the physics community about its [the electron's] reality. However, none of those people discovered the electron. The most that we can say is that one of those, say Thomson, contributed significantly to the acceptance of the belief that "electrons" denote real entities.[12]

True Belief View
According to this view, you have discovered something only if what you believe about it is true or substantially true. Despite Lord Kelvin's claim to know

various facts about the luminiferous ether,[13] that entity was not discovered by nineteenth century wave theorists (or by anyone else), since what was believed about it, including that it exists and that it is the medium through which light is transmitted, is false. Similarly, in this view, Thomson did not discover the electron since quite a few of his core beliefs about electrons (or what for many years he called corpuscles) were false. His corpuscles, he later thought, were entirely electrical, having no inertial mass; they were arranged in stationary positions throughout the atom; they were the only constituents of atoms; they were not waves of any sort; and they were not carriers of the smallest electric charge. So, if he discovered anything at all, it was not the electron.

Now, I reject each of these five views about discovery, both in the generalized forms I have given them and as ones applicable to the case of Thomson and the electron. Although manipulation and measurement are frequently involved in a discovery, they are neither necessary nor sufficient. Galileo discovered mountains and craters on the moon without manipulating or measuring them (in any reasonable sense of these terms). Moreover, the manipulation and measurement view would too easily dethrone Thomson. Many physicists before Thomson in 1897 manipulated electrons in the sense that Thomson did; that is, they manipulated cathode rays, and did so in such a way as to produce measurements. As noted, in 1890 Schuster conducted experiments involving magnetic deflection of electrons in which he arrived at upper and lower bounds for their ratio of mass to charge. Lenard's experiments manipulated cathode rays out of the tube and measured the distance they traveled. Perhaps one can say that Thomson's manipulations yielded better and more extensive measurements. But why should that fact accord him the title "discoverer?" Manipulations and measurements after Thomson by Seitz in 1901 and by Rieger in 1905 gave even more accurate measurements of the mass-to-charge ratio. Yet none of these physicists is regarded as having discovered the electron.

The second view—"important classification"—fails to provide a sufficient condition for discovery since you can arrive at an important classification of X's without discovering them. You can postulate their existence on largely theoretical grounds, and describe important facts about them, without "confronting" them sufficiently directly to count as having discovered them. In the early 1930s Pauli hypothesized the existence of a neutral particle, the neutrino, in order to account for the continuous distribution of energy in beta decay. But the neutrino was not discovered until there was a series of experiments, beginning in 1938, that established its existence more directly.

Whether the important-classification view fails to provide even a necessary condition for discovery is more difficult to say because of the vagueness in the notion of important classification. Roentgen discovered x-rays in 1895 without knowing that they are transverse electromagnetic rays. Although he speculated that they were longitudinal vibrations in the ether, he did not claim to know this (nor could he know this) and for this reason, and to distinguish them from other rays, he called them x-rays. Did he fail to arrive at a sufficiently important classification? Or shall we say that the fact that he discovered that x-rays are rays that travel in straight lines, that have substantial penetrating power, that cannot be deflected by an electric or magnetic field, and so forth, is sufficient to say that he arrived at an important classification?

Similarly, in the case of the electron, isn't the fact that the constituents of cathode rays are *charged particles smaller than ordinary ions* an important classification? If so, then Crookes in 1879 deserves the title of discoverer. Is it that the classification "constituent of all atoms" is more important than "being charged particles smaller than ordinary ions," and so Thomson rather than Crookes deserves the honor? Crookes, indeed, claimed that he, not Thomson, first arrived at the classification "constituent of all atoms." Moreover, why choose this classification rather than something more specific about how these constituents are arranged in atoms? If so, then Rutherford or Bohr should be selected, not Thomson, whose plum-pudding model got this dead wrong.

The crucial question concerning the present view is whether you could know that X exists from observations of X without knowing an important classification of X. In the weaker sense of discovery I distinguished earlier, one could discover X without knowing very much about X, including that it is X. (Recall my discovering gold.) The stronger sense involves knowing that it is X. But what important classification one needs to know to know that something is X I'll leave to important classification theorists.

The view I propose also contradicts the social constructivist account of discovery, since in my view there is, or at least can be, a fact of the matter about who discovered what that is independent of who the scientific community regards as the discoverer. This is because there is, or can be, a fact of the matter about who was the first to be in an epistemic state necessary for discovery. Being regarded by the scientific community as the discoverer of X is neither necessary nor sufficient for being the discoverer of X. No doubt scientific discoverers wish to be recognized by the scientific community for their discovery. Perhaps for some a discovery without recognition is worthless. But this does not negate the fact of discovery itself. Nor is this to deny that a discovery that is and remains unknown except to the discoverer will have little chance of advancing science, which depends on public communication. That is one reason scientists make their discoveries public. Although publicity helps to promote the discovery and the recognition for it, neither publicity nor recognition creates the discovery. Finally, one can relativize discovery claims to a group. I can be the first in my department to discover a certain book in the library; Columbus the first European to discover America; and so forth. This is not social constructivism, however, since there is a fact of the matter about discovery within a group that is independent of the beliefs of the members of the group. Either I was or I wasn't the first in my department to discover that book, no matter what views my colleagues have about my discovery.

Two of the views of discovery that contrast with mine deny the claim that Thomson discovered the electron: the "different-contributions" view and the "correct-belief" view. Briefly, my response to these views is this. The fact that various people made contributions to the discovery of the electron does not, on my account, necessarily preclude the fact that Thomson discovered the electron. All this means is that various people helped make it possible for Thomson to be the first to achieve an epistemic state necessary for discovery. Nor, finally, does getting into that epistemic state about some X require that all or most of your beliefs about X be true. Suppose that while walking along a road I discover a person lying in the ditch beside the road. Suppose that, after observing the person, I come to believe that the person is a woman, quite tall, at least fifty years old, with blond hair, and wearing a gray jacket. Suppose, finally, that I am quite wrong about these beliefs. The person in the ditch is actually a man, five feet tall, thirty years of age, with dark hair, and wearing no jacket at all. I can still be said to have discovered the person in the ditch, despite the fact that what I believe about the person in the ditch is substantially false. So I reject the general rule that you have discovered X only if what you believe about X is true or substantially true.

Did Thomson Discover the Electron?

Having proposed an account of discovery and disposed of some others, we are now in a position to take up this question. To begin with, I think my account helps us to see why we refrain from attributing this discovery to some of the other physicists mentioned. For example, claims made about the electron by Crookes, Larmor, and Lorentz, even if many were correct, were primarily theory-driven, not experimentally determined. This is not to say that Thomson had no theoretical beliefs about electrons. Falconer and Feffer[14] claim that he probably believed that they are not discrete particles with empty spaces between them, but certain configurations in an all-pervading ether. But that is not enough to put him in the same category as some of the more theoretically driven physicists. The question is whether Thomson was the first to know that electrons exist from observations of them or their direct effects.

Let me divide this question into three parts. First, in 1897 did Thomson know that electrons exist? Second, if he did, did he know this from observations of electrons or of their direct effects? Third, was he the first to know this from such observations? If the answer to all three questions is "yes," then Thomson retains the honor usually accorded to him.

In 1897 did Thomson know that electrons exist? Well, what did he claim to know in 1897? Here is a well-known passage from his October 1897 paper:

> As the cathode rays carry a charge of negative electricity, are deflected by an electrostatic force as if they were negatively electrified, and are acted on by a magnetic force in just the way in which this force would act on a negatively electrified body moving along the path of these rays, I can see no escape from the conclusion that they are charges of negative electricity carried by particles of matter.[15]

Thomson continues: "The question next arises, What are these particles? are they atoms, or molecules, or matter in a still finer state of subdivision. To throw some light on this point, I have made a series of measurements of the ratio of the mass of these particles to the charge carried by it" (384).

Thomson then proceeds to describe in some detail two independent experimental methods he employed to determine the mass-to-charge ratio. At the end of this description he concludes: "From these determinations we see that the value of m/e is independent of the nature of the gas, and that its value 10^{-7} is very small compared with the value 10^{-4}, which is the smallest value of this quantity previously known, and which is the value for the hydrogen ion in electrolysis."

He continues:

> Thus, for the carriers of electricity in the cathode rays m/e is very small compared with its value in electrolysis. The smallness of m/e may be due to the smallness of m or the largeness of e, or to a combination of these two. That the carriers of the charges in the cathode rays are small compared with ordinary molecules is shown, I think, by Lenard's results as to the rate at which the brightness of the phosphorescence produced by these rays diminishes with the length of the path travelled by the ray. (392)

After a little more discussion of Lenard's experimental results, Thomson concludes: "The carriers, then, must be small compared with ordinary molecules."

In sum, in 1897 Thomson claimed to know these facts:

1. That cathode rays contain charged particles. (This he claimed to know from his experiments showing both the magnetic and the electrostatic deflection of the rays.)
2. That the ratio of mass to charge of the particles is approximately 10^{-7}, which is much smaller than that for a hydrogen atom. (The 10^{-7} ratio he claimed to know from experiments of two different types involving magnetic and electrical deflection.)
3. That the particles are much smaller than ordinary molecules. (This he claimed to know from his own experiments yielding a mass-to-charge ratio smaller than that for the hydrogen atom, together with Lenard's experiments on the distance cathode rays travel outside the tube, which is much greater than that for hydrogen ions.)

Did he know these facts? He certainly believed them to be true: he says so explicitly. Are they true? True enough, if we don't worry about how much to pack into the notion of a particle. (Clearly Thomson had some false beliefs about his particles, in particular that they lacked wave properties.) Was he justified in his beliefs? His experimental reasons for claims 1. and 2. are quite strong, that for the smallness of the particles is perhaps slightly less so (but I think better than Heilbron alleges in his article on Thomson in the *Dictionary of Scientific Biography,* 367). If justified true belief is normally sufficient for knowledge, then a reasonable case can be made that Thomson knew the facts in question in 1897.

To be sure, there are other claims Thomson made in 1897 concerning which one might not, or could not, attribute knowledge to him. Perhaps one of the former sort is the claim that the charged particles are constituents of all atoms. Indeed, Thomson's explicit argument here appears a bit more tentative and less conclusive than those for the three claims above. It is simply an explanatory one to the effect that if atoms are composed of the particles whose existence he has already inferred, then this would enable him to explain how they are projected from the cathode, how they could give a value for m/e that is independent of the nature of the gas, and how their mean free path would depend solely on the density of the medium through which they pass. In general, explanatory reasoning does not by itself establish the claims inferred with sufficient force to yield knowledge. And, finally, there are the claims that the particles are the only constituents of atoms, and are arranged in accordance with a model of floating magnets suggested by Mayer. Both claims, being false, are not claims that Thomson or anyone else could know to be true.

Like knowing that there is a person in the ditch, however, not every belief about that person needs to be true or known to be true. If in 1897

Thomson knew that cathode rays contain charged particles, whose ratio of mass to charge is 10^{-7} and that are much smaller than ordinary molecules, then I think it is reasonable to say that in 1897 he knew that electrons exist at least in the weaker of the two senses discussed earlier. He knew of the existence of things that happen to be electrons. Electrons are the charged particles in question. Knowing these particular facts about them entails knowing that they exist. Whether he knew that electrons exist in the stronger sense is a question I will postpone for a moment.

The second of my three questions is whether Thomson knew what he did from observations of electrons or their direct effects. I suggest the answer is clearly "yes." Those were electrons in his cathode tubes, and they did produce fluorescent effects and others that he observed in his experiments. Despite various theoretical assumptions, his conclusions about electrons are primarily experiment driven.

The final of the three questions concerns priority. Was Thomson the first to be in the appropriate epistemic state? Was he the first to know that electrons exist in the weaker sense of this expression? Was he the first to know of the existence of things that happen to be electrons? He was clearly not the first to know of the existence of cathode rays which happen to be, or to be composed of, electrons. But that is not the issue here. Was he the first to know, by experimental means, of the existence of the things that happen to be the constituents of cathode rays, that is, electrons? That would be a more important question, albeit a question of discovery in the weaker sense. How do you demonstrate the existence of the constituents of cathode rays? Not simply by showing that cathode rays exist. Thomson demonstrated their existence by showing that charged particles exist comprising the rays, and he did so by means of experiments involving the direct effects of those charged particles. Was he the first to do so?

The answer I would offer is a less than decisive "maybe." Other physicists, including Schuster, Perrin, Wiechert, and Lenard, had conducted experiments on cathode rays which yielded results that gave support to the claim that the constituents of cathode rays are charged particles. Moreover, these experiments involved observing the electron's direct effects. It might be argued that although these other physicists provided such experimental support, that support was not strong enough to produce knowledge. One might claim that Thomson's refinements of Perrin's experiment, and more importantly his achievement of producing electrostatic deflection of the rays, and his determination of m/e, showed conclusively, in a way not shown before, that cathode rays contain charged particles. (This is what Thomson himself claims in his October 1897 paper.) If this is right, then one can say that, in the weaker sense of discovery, Thomson discovered the electron. Although others before him had produced experimental evidence of its existence, he was the first to produce evidence sufficient for knowledge.

This, however, is a controversial priority claim. It was vehemently denied by Lenard who claimed that his own experiments prior to Thomson's conclusively proved the existence of electrons.[16] It was also denied, albeit less vehemently, by Zeeman who claimed that he determined the ratio of mass to charge before Thomson.[17] Finally, Emil Wiechert makes claims about the constituents of cathode rays that are fairly similar to Thomson's in a paper published in January 1897, before Thomson's papers of April and October of that year.[18] In this paper Wiechert explicitly asserts that cathode rays contain charged particles that are much smaller than ordinary molecules, and from experiments involving magnetic deflection of cathode rays he determines upper and lower bounds for the mass-to-charge ratio of the particles. Unlike Thomson, however, Wiechert did not produce electrostatic deflection, he did not obtain two independent means for arriving at his determination of mass to charge, and he did not produce precise values. The issue, as I have defined it, is simply this: even though others had provided some experimental evidence for the existence of charged particles as the constituents of cathode rays, were Thomson's experiments the first to *conclusively* demonstrate this? Were they the first on the basis of which knowledge of their existence could be correctly claimed? If so, he discovered the electron. If not, he didn't.

One might make another claim. Relativizing discovery to the individual, one might say that Thomson first discovered the electron (for himself) in 1897, whereas others had done so a bit earlier. One might then say that Thomson was among the first to discover the electron (for himself). Perhaps this is what Abraham Pais has in mind when, as reported in the *New York Times* (29 April 1997), he claims that Thomson was a, not the, discoverer of the electron. The others Pais mentions are Wiechert and Kaufmann.

Strong Discovery

What about the stronger sense of discovery, the sense in which if I discover gold, then I know it is gold? To those seeking to deny the title "discoverer of the electron" to Thomson, one can concede that he did not know that the constituents of cathode rays have all the properties that electrons do. If this is required, the electron has yet to be discovered, since presumably no one knows all the properties of electrons. Obviously, this is not required for knowledge in the stronger sense. I can know that I have discovered gold without knowing all the properties of gold. Indeed, I can know that I have discovered gold without knowing any of the properties of gold. If an expert, after examining it, tells me it is gold, then I think I know it is. Clearly, however, Thomson did not know *in this way* that the constituents of cathode rays are electrons. So what must one know to know that the items in question are electrons? That is a problem. (A similar problem was raised concerning the "important classification" view.)

There is a further problem here with the question of whether Thomson knew that the constituents of cathode rays are electrons. Putting the question that way presuppposes some established concept of electron. And the question appears to be whether what Thomson discovered (in the weaker sense) fits that concept and whether Thomson knew this. By analogy, to ask whether I discovered gold (in the stronger sense) is to presuppose that these objects satisfy some established concept of gold and whether I know that they do. Which concept of electron is meant in the question about Thomson? In 1897 there was no established concept. Stoney, who intoduced the term "electron," used it to refer to an elementary electric charge, but Thomson was not talking about this. Nor was his claim that the constituents of cathode rays are Lorentz's electrons, which in 1895 Lorentz claimed were ions of electrolysis. (In fact, Thomson never used the term "electron" until well into the twentieth century.) Nor did Thomson claim that the cathode ray constituents have the properties we currently attribute to electrons.

So the question "did Thomson know that the constituents of cathode rays are electrons" is, I think, ambiguous and misleading. Instead, I suggest, it is better simply to ask what facts, if any, about the constituents of cathode rays Thomson knew, when he knew them, and when others knew them.

Briefly, let's take four central claims that Thomson made about cathode ray constituents in his October 1897 paper: first, they are charged particles; second, their ratio of mass to charge is approximately 10^{-7}; third, they are much smaller than ordinary atoms and molecules; and, fourth, they are constituents of atoms. Earlier I said that it is reasonable to suppose that Thomson knew the first three of these facts in 1897, but not the fourth. He came to know them during that year as a result of his experiments with cathode rays. I also said that one might claim that Thomson was the first to demonstrate conclusively that the constituents of cathode rays are charged particles, though this is controversial. At least he was among the first to do so.

With regard to the second claim—that the ratio of mass to charge of these particles is approximately 10^{-7}—Wiechert had arrived at upper and lower bounds before Thomson. In defense of Thomson, one might say that his determinations were more precise and were based on two independent experimental methods.

With respect to the third claim—that the cathode particles are much smaller than atoms and molecules—perhaps Lenard is correct in claiming knowledge of this prior to Thomson. Indeed, Thomson made important use of Lenard's absorption results in his own arguments that cathode particles are smaller than atoms. And if Wiechert's arguments are sufficiently strong, he too has some claim to knowledge before Thomson.

Finally, the fourth claim—that cathode particles are constituents of atoms—is, it is probably fair to say, one that Thomson did not know the truth of in 1897, although he gave explanatory arguments in its favor.

What Is So Important about Who Discovered the Electron (or Anything Else)?

Was there a discoverer of the electron? Was it Thomson? Have I needlessly complicated the issue with recondite distinctions that permit no definitive answers, as philosophers are wont to do? I think the issue is complicated, much more so than when I first began to think about it. The complications arise for two reasons. The concept of discovery itself is complex, requiring philosophical attention. And the historical facts about who knew what and when are complex. So who discovered the electron is usefully addressed by joint efforts of philosophers and historians of science. Before turning to the question raised in the title of this concluding section, let me summarize and simplify what I have said so far.

First, the philosophical account of discovery that I propose involves three factors: ontological (the thing discovered must exist); epistemic (the discoverer must be in a certain knowledge-state with respect to it); and priority (this state must be a first). Second, contrary to the opposing views mentioned, discovery does not require either manipulation and measurement of the item discovered, or the idea of an important classification, or group recognition or consensus; nor is any of these sufficient. Furthermore, one can be the discoverer of some entity even if one's beliefs about it are substantially false and even if many persons contributed to making that discovery possible.

Third, my account distinguishes a weak and a strong sense of discovery. In the weak but not the strong sense one can discover X even if the discoverer does not know it is X that has been discovered.

Fourth, in virtue of Thomson's magnetic deflection experiments, which were better than Perrin's, his electrostatic deflection experiments, which had not been achieved before, and his two independent determinations of m/e (better and more precise than the results produced by Wiechert), some case might be made that Thomson discovered the existence of charged particles that are electrons. Relativizing discovery to the individual, we can at least say that, in the weak sense of discovery, he was among the first to discover them. As far as the strong sense is concerned, it may be better to replace the question "did Thomson know that the constituents of cathode rays are electrons?" with the question "What facts about the constituents of cathode rays did Thomson know and when?" A case can be made that Thomson was the first to demonstrate, from experimental results, in a way producing knowledge, that the constituents of cathode rays are charged particles and that their mass-to-charge ratio is 10^{-7}.

Now, why do, or should, we care about who discovered the electron, or any other entity? The question arises especially for my account of discovery. On that account, the fact that something has been discovered by someone does not by itself imply that what is discovered, or by whom, is important or interesting, even to the discoverer. (I may have discovered yet one more paperclip on the floor.) The importance of the discovery will depend on the item discovered and on the interests of the discoverer and of the group or individual to whom the discovery is communicated. Discovering a universal particle such as the electron, which is a constituent of all atoms, is obviously more important, especially to physicists, than discovering yet one more paperclip on the floor is to them or to me.

Not only can the object discovered be of importance, but so can the method(s) employed. In his discovery of the electron (at least for himself) Thomson discovered a way to produce electrostatic deflection of the cathode rays, which had not been achieved before. Using this he devised a new independent way to obtain a fundamental measurement of mass-to-charge.

There is another point worth emphasizing about discoveries of certain entities, particularly those that are too small, or too far away, or otherwise too inaccessible to be observed directly. Scientists may have theoretical reasons for believing that such entities exist. These theoretical reasons may be based on observations and experiments with other entities. Sometimes such reasoning is sufficiently strong to justify a claim to know that the entity exists. Yet there is still the desire to find it, to discover it, by observing it as directly as possible. (Although the case of the electron does not illustrate this, one that does fairly closely is that of the top quark, whose existence was inferred from the "standard model" before it was detected experimentally.[19]) This need not increase the degree of confidence in its existence significantly if at all over what it was before. So why do it?

One reason may simply derive from a primal desire or curiosity to "see" or detect something by confronting it more or less directly. Another more important reason is to discover new facts about it, which is usually facilitated by observing it or its effects, and which may allow the theory that entailed its existence to be extended. It will also provide additional support for that theory without necessarily increasing the degree of probability or confidence one attaches to that theory.[20]

Why should we care about who, if anyone, was the discoverer, that is, about who was the first to be in an appropriate epistemic state for discovery with respect to some entity? It depends on who the "we" is and on what is discovered. As noted, not all discoveries and discoverers are of interest to all groups; some may be of interest to none. If what is discovered is important to some community, and if there was a discoverer, whether a person or a group, then simply giving credit where credit is due is what is appropriate and what may act as a spur to future investigations. In this regard discovery is no different from other achievements. If accomplishing something (whether flying an airplane, or climbing Mt. Everest, or discovering the electron) is valuable to a certain community, and some person or group was the first to do it, or if several persons independently were the first, then such persons deserve to be credited and perhaps honored and rewarded by the community, especially to the extent that the accomplishment is important and difficult. Generally speaking, more credit should be given to such persons than to those who helped make the achievement possible but did not accomplish it themselves.

Whether Thomson deserves the credit he received for being the (or a) discoverer of the electron is, of course, of interest to him and to other contemporaries such as Lenard, Zeeman, and Crookes, who thought they deserved more credit. It should also be of interest to subsequent physicists, historians of physics, and authors of textbooks who write about the discovery. The answer to the question of who discovered the electron, and hence who deserves the credit, is, I have been suggesting, not so simple. Part of that answer depends upon establishing who knew what, when, and how, which in the electron case is fairly complex. The other part depends on establishing some reasonably clear concept of discovery. In this chapter I have attempted to contribute to each task, particularly the latter.

Finally, credit is deserved not only for discovering the existence of an important entity, but for other accomplishments with respect to it as well. Even if Lenard has some claim to priority for the discovery that cathode ray constituents are smaller than atoms, and even if in 1897 Thomson's arguments that his corpuscles are constituents of all atoms are not conclusive, we can admire and honor Thomson, among other reasons, for the experiments leading to the conclusions he drew, for the conclusions themselves, and for proposing and defending a bold idea that revolutionized physics: that the atom is not atomic.[21]

Notes

1. Theodore Arabatzis, "Rethinking the 'Discovery' of the Electron," *Studies in History and Philosophy of Science* 27B (December 1996); 405–435.

2. William Crookes, "Modern Views on Matter: The Realization of a Dream" (an address delivered before the Congress of Applied Chemistry at Berlin, 5 June 1903); 231. *Annual Report of the Board of Regents of the Smithsonian Institution* (Washington D.C., Government Printing Office, 1904). In this paper Crookes quotes the present passage from his 1879 lecture.

3. See Arthur Schuster, *The Progress of Physics* (Cambridge: Cambridge University Press, 1911), 61.

4. I am indebted to Kent Staley for this example and this point.

5. For an illuminating discussion, see Kent Staley, *Over the Top: Experiment and the Testing of Hypotheses in the Search for the Top Quark* (Ph.D. dissertation, Johns Hopkins University, 1997).

6. For more discussion, see my *Concepts of Science* (Baltimore: Johns Hopkins University Press, 1968), chapter 5.

7. In philosophical jargon this distinction corresponds to that between referential transparency and opacity in the expression "discovering X" (and in "knowing that X exists"). In the referentially transparent sense, but not the referentially opaque sense, if I have discovered X, and if X = Y, then it follows that I have discovered Y.

8. Isobel Falconer, "Corpuscles, Electrons and Cathode Rays: J. J. Thomson and the 'Discovery of the Electron,'" *British Journal for the History of Science* 20 (1987): 241–276; quotation on 276.

9. Ian Hacking, *Representing and Intervening* (Cambridge: Cambridge University Press, 1983), 23.

10. Falconer, "Corpuscles, Electrons and Cathode Rays," 251. I would take issue with Falconer here. In a paper of January 1897 Wiechert does indeed claim that cathode particles are constituents of atoms. Emil Wiechert, *Physikalisch-Ökonomischen Gesellschaft*, Königsberg, 7 January 1897, 3–16.

11. Arabatzis, "Rethinking the 'Discovery' of the Electron," 406.

12. Ibid., 432. Is Arabatzis's "different contributions" view about the electron compatible with what I take to be his more general social constructivist position about discovery? I believe so. The combined view would be that in general someone is the discoverer of something when and only when there is consensus about who discovered what; in the electron case, however, there is no such consensus about any one person, only (at most) about who made what contributions toward the discovery.

13. Kelvin wrote: "We know the luminiferous ether better than we know any other kind of matter in some particulars. We know it for its elasticity; we know it in respect to the constancy of the velocity of propagation of light for different periods." *Kelvin's Baltimore*

Lectures and Modern Theoretical Physics, ed. Robert Kargon and Peter Achinstein (Cambridge, MA: The MIT Press, 1987), 14.

14. Stuart M. Feffer, "Arthur Schuster, J. J. Thomson and the Discovery of the Electron," *Historical Studies in the Physical and Biological Sciences* 20 (1989): 33–51.

15. J. J. Thomson, "Cathode Rays," *Philosophical Magazine and Journal of Science* (October 1897), reprinted in Mary Jo Nye, ed., *The Question of the Atom* (Los Angeles: Tomash Publishers, 1986), 375–398; quotation on 384.

16. Philipp Lenard, *Wissenschaftliche Abhandlungen,* vol. 3 (Leipzig: S. Hirzel, 1944), 1.

17. See Arabatzis, "Rethinking the 'Discovery' of the Electron," p. 423.

18. Wiechert, *Physikalisch-Ökonomischen Gesellschaft.*

19. See Staley, *Over the Top.*

20. For arguments that evidence can provide support for a theory without necessarily increasing its probability, see my *The Nature of Explanation* (New York: Oxford University Press, 1983), chapter 10.

21. I am indebted to Wendy Harris for helping me express the views I want, to Robert Rynasiewicz and Kent Staley for stimulating discussions in which they tried their best to dissuade me from expressing those views, to Ed Manier who, when I presented an earlier version of this paper at Notre Dame, raised the question that forms the title of section six, and to the editors of this volume for helpful organizational suggestions and for convincing me to tone down my anti-social-constructivist sentiments.

14

HISTORY AND METAPHYSICS: ON THE REALITY OF SPIN
Margaret Morrison

Philosophical questions about the nature of reality, or the kinds of conditions required for something to count as "real," typically fall within the domain of metaphysics or ontology. Depending on one's views about the appropriateness of pure metaphysics for addressing problems in the philosophy of science, these kinds of metaphysical questions can take on an epistemological dimension. For example, how could we possibly affirm the existence of a particular entity independently of the ways in which we could find out about it. Thus, answers to questions about what is "real" may be determined, in part, by the kinds of things we are in a position to know or verify. This in turn depends not only on the sophistication of our cognitive structures but on the kinds of technological machinery we are able to utilize in scientific investigation. Using current technology, experimentalists are unable to verify the existence of the Higgs particle, yet it is presumably something that is verifiable in principle given the requisite conditions. In that sense the question of its reality is intimately linked to the technology necessary for its detection.

In addition to the role of technology and experiment there is another way in which epistemological issues inform, or even determine, metaphysical ones—namely, through theory. At first this may seem like a rather odd claim to make; in fact, much of the recent work done in history and philosophy of science has emphasized the importance of experiment and experimental practice in constituting a culture distinctly different from theory. While theoretical work supposedly dealt with abstract descriptions of physical systems, experiment provided the context where debates about the reality of entities and effects took on significance. Some of the more influential work on experiment (Buchwald 1994, Franklin 1986, Galison 1987, and Hacking 1983) tells us how laboratory practices can influence the interpretation of theories, how entities that were once thought to be intimately connected to theory can take on a life of their own and hence survive the Kuhnian problem of incommensurability, and how experiment can provide the foundation for a realist interpretation of scientific objects like elementary particles. Although there is much to admire in this philosophy of experiment, it has succeeded in forcing "theory" into a less prominent position in discussions about the material culture of the natural sciences. The philosophical moral I want to argue for here is a renewed appreciation for the role of theory in establishing the reality of scientific entities and effects. In other words, by focusing on "experimental" arguments we are telling only half the story. The ways in which theory influences questions about the reality of these entities is equally complex and important. Here I am not referring to a realism about a specific theory itself (the issue of whether it truly describes the physical world) but rather the ways in which general theoretical commitments can play a role in shaping our scientific ontology.

Traditionally the relation between theory and ontology has been a problematic one, especially in post-Kuhnian philosophy of science. There appears to be no basis for isolating entities as things distinct from their place within a theoretical framework. Indeed, the difficulties in separating theories and things in a coherent way was, for many, one of the motivating issues in developing a philosophy of experiment. The story I want to tell about the connection between theory and ontology, however, bears little, if any, similarity to Kuhnian holism. Instead, I want to show how answers to ontological/metaphysical questions about scientific entities can depend on the *theoretical* history of that entity. Theoretical histories are necessary for answering such questions because it is sometimes impossible to isolate an exact moment or specific procedure that transformed the entity or property from having a mere existence on paper to acquiring a more robust nature as something physically real. I want to make clear, however, that this is not a claim about how theory provides a meaning or interpretation for a particular entity or effect; rather, I want to maintain that the reality ascribed to entities is often the result of their evolution in a theoretical history. While attributing "reality" to objects or properties is usually seen as the result of some type of experimental verification, my goal here is simply to show that theory and theoretical histories can play a similar kind of legitimating role. Consequently issues of ontology and metaphysics become infused not only with epistemology but with history as well.[1]

The particular example I want to focus on is how spin came to be understood and accepted as something physically real. Spin is especially interesting for a number of reasons. First and perhaps most importantly, its physical nature is still not well understood, yet no one denies that it is a real property of the electron and other elementary particles. Second, there is a good deal of ambiguity surrounding its connection with relativity theory. Dirac claimed to have shown that spin was a consequence of relativity, something that appears unlikely given that other relativistic particles have spin zero. These theoretical issues figure importantly in the history of the development of spin and the story of its transformation into a real property. What I want to show is how that reality was established in stages, for different reasons by different actors. The connection of spin with relativity, quantum and classical mechanics all played a role in that transformation, yet its connection with those theories did not endow spin with an unambiguous physical interpretation. Hence its status as a real property emerged from a rather curious mixture of theoretical dependencies without the accompanying advantage of a full theoretical interpretation. Because the reality of spin cannot be separated from its historical evolution, it serves as a model for the kind of "historical metaphysics" I want to argue for.

Background

In 1976, fifty years after the discovery of what we now take to be an essential feature of the electron, Samuel Goudsmit (1976) wrote an article entitled "It Might as well be Spin." The article was aptly named not only from the point of view of the physics of the 1920s but also from a contemporary perspective. According to the received view the spin of the electron, or another elementary particle, is a mysterious internal angular momentum for which there is neither a classical analog nor a concrete physical picture. As Goudsmit's title suggests, the very idea of attributing spin to the electron was not something whose interpretation was obvious. Despite its puzzling features, spin was accepted as a real property of the electron almost immediately after it was proposed. Even Pauli who was initially skeptical of Goudsmit and Uhlenbeck's (1925, 1926) account, calling it a new Copenhagen heresy, was converted to the idea four months after the initial paper appeared. Yet, the question remains as to whether there is a satisfactory way of understanding spin in physical terms. An obvious answer is that we have a perfectly good understanding of spin as given by the mathematics of group theory. Consequently, we can understand the reality of spin in the same way as we understand gauge fields. Albeit technically correct, I nevertheless see this as an unsatisfactory physical explanation. Although my goal is to show the connection between the reality of spin and it's theoretical history, that task also involves investigating questions about its physical interpretation and how that influences its status as a real property. Many of the early problems surrounding spin arose from questions about its compatibility with relativity. Goudsmit and Uhlenbeck's (1925b) initial paper contained a fundamental conflict with relativity that was almost certainly one of the reasons Pauli opposed it. Interestingly enough the connection between spin and relativity is something that is still a source of debate. It is not uncommon to find contemporary texts referring to spin as a consequence of relativity theory. Dirac (1928) himself thought he had shown just that in his paper on the relativistic wave equation; yet the relation between spin and relativity continues to be a source of confusion.

I want to begin by going back to some of the initial work on spin done by Pauli (1925), Kronig, and Compton (1921) before the "discovery" by Goudsmit and Uhlenbeck. From the history we can see that one of the arguments for rejecting spin in the 1920s was based on the lack of a straightforward physical interpretation. Yet, the difficulties in understanding its physical properties, especially with respect to the nature of the electron and its relativistic features, were quickly overshadowed. Spin became a real property existing in nature, not just a mathematical tool for solving physical problems. Although this reality had been firmly established by 1926, questions of interpretation remained. In the latter portion of the chapter I discuss some of these interpretive problems, problems which curiously appear to have had little bearing on the ontological status of spin.

The Electron Magnet Gospel

The idea of an electron as a tiny spinning gyroscope can be traced back to the work of Abraham (1903) and later Compton (1921). Abraham's paper provided a detailed study of the dynamics of the electron, assuming it to be a spherical rigid object with an homogenous surface or volume charge. Although the electron, conceived in this way, would prove important for Goudsmit and Uhlenbeck in formulating their spin hypothesis, it was Compton who initially suggested that the electron, "spinning like a tiny gyroscope" was probably the ultimate magnetic particle. Compton's picture was influenced by the work of Parsons (1915) who suggested that many of the magnetic properties of matter can receive a satisfactory explanation by assuming that the electron is a continuous ring of negative electricity spinning rapidly about an axis perpendicular to its plane and therefore possessing a magnetic moment as well as an electric charge. Compton assumed instead that the electron had a more nearly isotropic form with a strong concentration of electric charge and mass near its center. On the basis of several experimental results (the Richardson-Barnett effect, experiments on the diffraction of x-rays by magnetic crystals, and the curvature of tracks of beta rays) Compton supported the idea that the electron, rather than any group of atoms, was the ultimate magnetic particle. Regardless of whether one adopted classical or quantum arguments the thermal motions of the electron would give it an appreciable magnetic moment. According to the quantum hypothesis, at absolute zero the average amount of energy retained by a particle ($1/2 \hbar v$) for each degree of freedom for motion would correspond to an angular momentum of $\hbar/2\pi$ for a rotating system. An electron spinning with this angular momentum will have exactly the magnitude of magnetic moment required to account for ferromagnetic properties.

The views of Abraham would figure prominently in Goudsmit and Uhlenbeck's spin hypothesis, however, it was the work of Pauli that provided the direct motivation for their account. At the time the problem that interested Pauli was understanding the anomalous Zeeman effect (the multiplet component splitting of spectral lines emitted by an atom in an external magnetic field); a phenomenon that did not follow the patterns predicted by Lorentz on the basis of classical theory. Pauli's (1925) famous paper contained an examination of the doublet structure of the alkali spectrum whose cause had been attributed to a finite angular momentum of the atomic core. This core model was initially developed by Heisenberg (1921), who suggested that in alkalis there is one valence electron and an inner complex of electrons which he called the "Atomrumpf," the atomic core, with both the valence and core electron having an angular momentum of 1/2 (in units of $\hbar/2\pi$). Heisenberg's model was a kind of physical interpretation for Landé's (1921) original idea of explaining the anomalous Zeeman effect by assuming that angular momentum quantum numbers can take on half-integer values. Landé then took up the idea of a core angular momentum quantum number that he called R which, for alkalis, equaled 1/2. This enabled him to deduce that the gyromagnetic ratio, and hence the Landé g factor, should have the value 2 for the core instead of the classical prediction of 1.

Pauli felt that, although this could account for the anomalous splitting, it produced no real theoretical understanding based on fundamental classical or quantum assumptions about the electron. He was convinced that there should be a connection between the theory of the multiplet structure and the problem of building up the periodic table of elements. In particular there appeared to be no answer to the question of why all the electrons for an atom in its ground state were not bound in the innermost shell. That is, no one could explain the closing of the electronic shells.

According to Heisenberg's "core model" the properties of the core were the result of a magnetomechanical anomaly of this K-shell, an anomaly grounded on the assumption that the quotient of the magnetic moment and the angular momentum of the shell was twice as large as the predicted classical value. What Pauli did was calculate the relativistic mass variation of the K-shell electrons and found that the magnetic moment/angular momentum ratio given by the core theory had to be multiplied by a correction factor γ, the time average value of $(1 - v^2/c^2)^{1/2}$ taken over a complete orbit of the revolving electron. If one assumed the correctness of the core theory then the value of $\rho = (e/2M_0 c)g$ (the gyromagnetic ratio) became a slowly decreasing function of the atomic number. The problem was that neither the normal nor anomalous Zeeman effect showed any influence of such a decreasing factor. Even if one assumed a doubling of the ratio of magnetic moment to mechanical momentum for electrons in the K-shell, and a compensation for the classically computed relativistic effect of the velocity, the electron would have to change its magnetic moment as soon as it left or entered the shell; an extremely improbable occurrence. So, although the Landé formula for g worked experimentally, Pauli concluded that the core theory was no longer tenable.

His solution was to assume that the closed shells have no angular momentum or magnetic moment which, for alkalis, meant that the core angular momentum R of the atom and its energy change in a magnetic field is due not to the core but only to the valence electron. As a result the anomalous Zeeman effect was caused by a "peculiar not classically describable two-valuedness of the quantum properties of the valence electron." This nonclassical two-valuedness (which would later be called spin) became the foundation for the exclusion principle and Pauli's hypothesis of a fourth quantum number for the electron. Because there was no qualitative picture that connected Pauli's new formalism with an atomic model, the question was how to interpret this new quantum number.

At this point the argument eventually put forward by Goudsmit and Uhlenbeck appears almost obvious. Since it was thought that every quantum number corresponded to a degree of freedom, why not consider the fourth quantum number as another degree of freedom—the rotation of the electron about its own axis. In essence, the electron had a spin 1/2; why 1/2?—because according to Landé angular momentum quantum numbers can take on 1/2 integer values.[2] But, because the point electron had only three degrees of freedom, a physical interpretation for this fourth degree of freedom required a different conception of the electron. Initially it appears odd that Pauli himself didn't simply interpret this new quantum number as the intrinsic angular momentum of the electron. Yet, his remark about the nonclassical two-valuedness of the electron speaks for itself. Spin is essentially a classical notion; how could one reconcile this two-valuedness with something like the classical idea of rotation?

Although Pauli would shun this obvious dynamical interpretation, it was the *desire* for a dynamic interpretation that led Goudsmit and Uhlenbeck to suppose that the electron spun about its own axis. That is to say, they wanted some way of making sense of the idea of a fourth quantum number. Goudsmit suggested that if the angular momentum of the electron was $\hbar/2$, one immediately had a picture of the alkali doublets as the two ways the electron could rotate with respect to its orbital motion. In fact, if one assumed that the gyromagnetic ratio for the rotation was twice the classical value, so that the magnetic moment was

$$2(e/2Mc)\hbar/2 = \text{One Bohr Magneton},$$

then the properties formerly attributed to the core were now properties of the electron. As a result, the simple "anschaulichen" features of the original Rumpf-Modell were thus reconciled with Pauli's ideas.

Several authors, including Van der Warden (1960), Jammer (1966), and Serwer (1977), have suggested that the elements necessary for electron spin were already implicit in Pauli's paper on the exclusion principle. Be that as it may, there is also a simple and rather obvious explanation as to why the notion of spin required much more than what was provided by the Pauli framework; that is, why spin required a different way of conceptualizing the problem. The Bohr-Sommerfeld theory of the electron with three quantum numbers was applicable only to the hydrogen atom. Pauli's account of the valence electron with four quantum numbers was formulated specifically for alkalis and more complicated atoms. Hence, one of the peculiarities of Pauli's two-valuedness was that it was attributable to only the alkalis. Part of the motivation behind Goudsmit's and Uhlenbeck's idea stemmed from the fact that there appeared to be no reason why the hydrogen atom should behave any differently than the alkalis. Moreover there had been some difficulty in explaining the fine structure of He+ and, as it happened, 1/2 integer quantum numbers proved to be the missing link. Thus, it would have been difficult to interpret spin as a property of the valence electron given this division between the hydrogen, alkali, and other atoms. Any attempt to do so would have created an unnatural distinction between different kinds of atoms, with hydrogen having only three quantum numbers. Once hydrogen was understood as having essentially the same characteristics as other atoms, however, it became possible to see spin as a universal feature of the valence electron.

Goudsmit and Uhlenbeck's argument, however, was purely formal; no dynamics were given for the fourth quantum number. And, because of Pauli's relativistic correction to the core, Landé's g factor could no longer be given a physically coherent interpretation.[3] In conjunction with the idea of spin, however, Goudsmit and Uhlenbeck suggested that the Landé factor which originally was $g = 2$ for the core should now apply only to the electron itself.

The problem was how to find some way of *physically* reconciling the Landé factor with this new semiclassical argument. At that point Ehrenfest directed them to the earlier article by Max Abraham which stated that an electron considered as a rigid sphere with only surface charge, does have $g = 2$. In other words, such a sphere could be given a classical interpretation. But, far from providing a workable physical interpretation, their semiclassical reasoning when combined with quantum numbers and a classical image of the atom led to a stark conflict with relativity. If they assumed that the electron was an extended object with a classical radius $r_0 = e^2/Mc^2$ and if it rotated with an angular momentum $\hbar/2$, then the surface velocity would be about ten times the velocity of light. Hence for an electron with a magnetic moment of $e\hbar/2Mc$, its magnetic energy would have to be so large that to keep its mass the radius would have to be at least ten times r_0.

This was not the only difficulty with the notion of a spinning electron. The idea of spin as an interpretation of Pauli's formalism had been given earlier that year by Kronig (in an unpublished note written in February 1925). He found that the formula for fine structure splitting in hydrogen, however, calculated from the semiclassical treatment of spin precession, was twice the amount required by observation. Because Goudsmit and Uhlenbeck did not derive this formula, they were unaware of the problem until it was pointed out in a letter from Heisenberg. What is especially interesting about this difficulty is how it goes straight to the core of the problem they were trying to solve, and how it arose from the ongoing tension in attempts to merge quantum and classical ideas. Recall that Pauli had simply assigned the Landé quantum numbers of the atomic core to the electron itself; but because there is no physical model corresponding to this idea, it is difficult to envision how the "core" quantum number could be coupled to the orbital quantum number of the electron. This is what led Pauli to talk about an intrinsic two-valuedness of the motion of the electron and Bohr to speculate about a new force described as an *unmechanischer zwang* or nonmechanical strain. Goudsmit and Uhlenbeck maintained that these ideas could be replaced by a hypothesis about the structure of the electron, yet they never actually *explained* how the basic difficulty could be solved by coupling the rotational and orbital motion of the electron.

Although Heisenberg told them about just such a spin-orbit coupling that gave the right solution but contained the mysterious factor of two, it was Einstein who supplied the hint on which they based their answer. In the coordinate system where the electron is at rest the electric field **E** of the moving atomic core produces a magnetic field $[\mathbf{E} \times \mathbf{v}]/c$ according to the transformations of relativity theory. It is with respect to this magnetic field that the spin of the electron has its two orientations, and the energy difference—the doublet splitting—could then be calculated by first order perturbation theory. The erroneous factor of two still remained in the calculations but there were now several features that recommended the spin hypothesis. In conjunction with Goudsmit and Uhlenbeck's (1925a) earlier work on the hydrogen spectrum, they went on to show (1926) how the combination of the Sommerfeld relativistic effect with the spin-orbit coupling leads to the fine structure of the hydrogen levels, just as they had surmised in the first paper on spin. That, together with the fact that they had successfully synthesized the ideas of Landé and Pauli, was enough to convince Bohr and eventually Heisenberg of the reality of spin. The factor of two was simply explained away—it would undoubtedly disappear once a better calculation was made. Bohr wrote to Erhenfest that he had become a "prophet of the electron magnet gospel."

Before discussing the connections between spin and relativity let us pause briefly and take stock of two aspects of this theoretical history—the influences on Goudsmit's and Uhlenbeck's formulation of spin and the reactions by Bohr and Heisenberg. In his article on the doublet riddle Forman (1968) claims that the sharp distinction between the spectrum of hydrogen and the spectra of many electron atoms was responsible for elevating the problems of the complex structure of spectral lines to a central role in atomic physics. Prior to the introduction of spin there were two competing and contradictory conceptions of the origins of the structure of spectral lines, the magnetic or rump theory and the relativistic account. The problem was that within either of the two conceptions certain features of the relativistic doublets in the structure of x-ray and alkali spectra remained a riddle. Theoretical problems with the core model (mentioned above) indicated that it was no longer viable, yet the attempt to account for complex structure relativistically using quantum numbers called into question the distinction between one-electron atoms and those with many. Heisenberg wanted to show that once the domains of these two conceptions had been properly delineated they could be considered complementary rather than contradictory. Ultimately, however, it was the spin hypothesis that fused the two accounts into one. Forman claims that by proposing the spin hypothesis Goudsmit and Uhlenbeck were obliged to grasp one horn of the dilemma, namely, the riddle of the relativistic doublets had to be explained within the magnetic or core theory; a problem that weighed heavy on them (157). Moreover, the fact that their explanation displaced the doublet riddle from its place as one of the central problems of atomic physics appeared to Goudsmit and Uhlenbeck a principle objection to their story. One can however interpret the situation somewhat differently. Although they were undoubtedly aware of these issues, Uhlenbeck's own recollections suggest that the deeper theoretical issues were not especially vexing, rather, they simply wanted some way of making sense of the asymmetry between hydrogen and other atoms. In the beginning stages of their work Goudsmit, the more senior of the two, had explained to Uhlenbeck that the rump model was used for all atoms except hydrogen which was described by the Sommerfeld theory; a situation that suggested hydrogen was somehow "a horse of a different colour" (1976, 46). Uhlenbeck's skepticism was eventually shared by Goudsmit who got the idea of examining what the fine structure of hydrogen would have to be like if it were an alkali atom. This idea led to their first paper in 1925 which included a modification of Sommerfeld's quantum number assignments for atomic levels, a result that also explained what was considered an anomalous line observed by Paschen in the spectrum of ionized helium.

The implications of this work for their later work on spin was significant. In 1946 Goudsmit wrote that the idea of spin arose when "we were saturated with a thorough knowledge of the structure of atomic spectra, had grasped the meaning of relativistic doublets and just after we had arrived at the correct interpretation of the hydrogen atom." The work on the hydrogen atom clearly laid the foundation for the spin hypothesis; it was now possible to interpret Pauli's fourth quantum number as an intrinsic feature of the electron, something that would have been impossible if the two kinds of atoms continued to be seen as having essentially different natures. In that sense one can see how the simple symmetry argument rather than attempts to deal with the underlying physical problems provided the impetus for solving the complex structure problem, a difficulty that, once resolved, supplied a conceptual framework for the spin hypothesis.[4] Although Goudsmit and Uhlenbeck were surely interested in providing a more complete dynamics for spin, it was not a pressing problem nor was it one that necessarily detracted from their enthusiasm for the successes the spin hypothesis enjoyed.

What then were the specific features that eventually convinced Bohr and Heisenberg of the reality of spin? About three months after the spin paper, in December 1925, Bohr was making a train trip to Leiden and stopped in Hamburg to meet Pauli who asked him about his views on spin. At the time Bohr remarked that although it was interesting he did not see how an electron moving in the electric field of the nucleus could experience the magnetic field necessary for producing fine structure. Upon reaching Leiden he was again asked by Erhenfest and Einstein what he thought about spin. When Erhenfest told him that Einstein had resolved the problem of the magnetic field, thereby producing an effective spin-orbit coupling, Bohr was immediately convinced. The problematic factor of two would presumably disappear once better calculations were made. What exactly was it about this solution that convinced Bohr and caused him to virtually ignore the factor of two? Serwer (1976, 251) claims that the spin hypothesis allowed for the retention of mechanical models as a way of solving problems in atomic physics. Yet, if we look at the note appended to Goudsmit's and Uhlenbeck's 1926 paper on spin and the structure of spectra, there is evidence to suggest that the fundamental issue was Bohr's conviction about the more *general* relationship between classical physics and the quantum theory.

While visiting Leiden Bohr encouraged Goudsmit and Uhlenbeck to look again at the spectrum of hydrogen to see whether the Sommerfeld relativistic effect with the spin-orbit coupling led to the fine structure levels they had predicted in their earlier paper. Goudsmit was able to show this immediately which, according to Uhlenbeck's recollections (1976), appeared decisive for Bohr. They published a paper in *Nature* in 1926 that explained the analogy between the multiplet structure of optical and x-ray spectra while seemingly retaining the principle of the successive build up of atoms. Bohr added a note to their article claiming that, in spite of the "incompleteness of the conclusions that can be derived from models," the spin hypothesis promises to be a welcome addition to ideas about atomic structure. He goes on to point out that it opens up a "very hopeful prospect of our being able to account more extensively for the properties of elements by means of mechanical models, at least in the qualitative way characteristic of the applications of the correspondence principle" (1926, 265). Perhaps the most telling remark is his closing sentence which expresses the timeliness of this possibility which arose when there were great prospects for a quantitative treatment of atomic problems by the new quantum mechanics, a theory that "aims at a precise formulation of the correspondence between classical mechanics and the quantum theory" (ibid.). Spin was a classical idea that was presented in a quantum framework and solved quantum problems. It represented for Bohr the perfect example of how quantum and classical physics could merge, and given his commitment to quantum theory the relativistic problem of the factor of two appeared relatively insignificant.

For Heisenberg the situation was somewhat different. The factor of two represented much more of a problem, but he also believed that a proper solution for the hydrogen spectrum was possible within the new quantum mechanics (which could presumably be treated relativistically). After some persuasion by Bohr, however, together with the failure to obtain complex structure from quantum mechanics, he eventually resigned himself to the spin hypothesis. To that extent spin represented the only possible way of recovering complex structure, and although it still gave incorrect results it was better than any of the alternatives. It incorporated the important half-integral quantum numbers and provided a basis for the application of quantum ideas to the problem of atomic spectra.[5] Thus, for Bohr it was ultimately the connection with quantum theory that figured in changing the status of spin from a calculational device to a real property, while for Heisenberg it was instrumental success in accounting for complex structure. It would be a mistake to say that Heisenberg embraced the reality of spin in the same way that Bohr did. The latter saw it as having deep theoretical connections and explicating what was significant about the correspondence principle. Heisenberg saw no way around its instrumental usefulness, a usefulness that translated into an acceptance of it as real. But even his tacit acceptance is significant in the evolution of spin into a real property of the electron. It is in exactly contexts like these that theoretical histories are especially significant; they make apparent, in the way that experimental histories do, how decisions about what is real are often less than straightforward—the product of different attitudes that eventually converge, for different reasons, to a common point.

Spin and Relativity—The Puzzle Continues

Although Pauli remained unconvinced by the spin hypothesis he was more or less converted a few months later when the problematic factor 2 was explained by Thomas (1926). He showed that the earlier calculations of the precession of the spin were performed in the electron's rest frame without taking account of the precession of the electron's orbit around its normal. Including this relativistic effect reduced the angular velocity of the electron (as seen by the nucleus) by the needed factor of 1/2 and hence gave the correct answers for the magnitude of the doublet splitting.

In his analysis of the differences between Pauli's and Heisenberg's attitudes toward spin, Serwer (1977) claims that Pauli was less concerned with the factor of two than he was with how spin fit the development of quantum theory. He wanted to be able to derive the spectrum of hydrogen, two separate series spectra for Helium and the anomalous Zeeman effect from a single quantum mechanical Hamiltonian. In addition their professional situations with respect to the core model figure importantly; Heisenberg thought it might still bring results while Pauli saw it as an overly complex theory that used arbitrary adjustments from classical mechanics.

Yet it is somewhat difficult to reconcile Serwer's account with the obvious change in Pauli's attitude after Thomas's paper appeared. I claimed above that Pauli's opposition to spin was due to both its classical character and its conflict with relativity. Although Thomas's calculations helped to persuade him that there was something essentially right about Goudsmit and Uhlenbeck's idea, and that spin must be considered real, one could also maintain that he never fully accepted the *dynamical idea* of a rotating electron; and so from his perspective no explanation of the fundamental nature of spin had been given. His initial argument against the "core" theory was compelling for him because it showed that once relativistic refinements were made to the core it proved to be incompatible with experience. Relativistic considerations were also important for the spin hypothesis; once the relativistic corrections could be successfully incorporated in a way that solved the fundamental problem involving spin-orbit coupling Pauli was much more sympathetic. That is not to say that he thought spin to be a relativistic effect; instead he saw it as an essentially quantum mechanical property but one that should not conflict with relativity. And it is that feature that ultimately explains his views on the nature of spin. Although he could accept the formal argument ascribing to the electron an angular momentum in a given direction of $\pm 1/2$ and a magnetic moment of twice its angular momentum, the dynamical picture of a rotating particle that appeared to complete this picture remained unpersuasive.

Pauli was able to bifurcate the formal and physical ideas about spin because quantum theory appeared to require the substitution of abstract mathematical symbols for concrete pictures. Many years later in 1955 Pauli claimed that after a "brief period of spiritual and human confusion caused by a provisional restriction to 'Anschaulichkeit'" a general agreement was reached about replacing physical notions with abstract mathematics as in the case of the psi function. But nowhere was this more striking than the replacement of the concrete picture of rotation with mathematical features of the representations of the group of rotations in three-dimensional space. Although Pauli's initial acceptance of spin is linked to Thomas's relativistic correction, it was because of Pauli's conviction that spin was essentially a quantum mechanical property that he could accept its validity without a satisfactory dynamical image. In that sense it was very much a philosophical attitude about the theoretical nature of quantum mechanics that allowed Pauli to finally accept spin as a real thing.

It was Pauli himself who first introduced the idea of spin into wave mechanics, something that would eventually lead to the relativistic formulation given by Dirac. Darwin had suggested that one could derive the appropriate "spin" properties by applying the usual rules of quantum theory to the classical treatment of a rotating rigid sphere. The key was to use a two-valued wave function. That is, the electron should be understood as a wave of two components, similar to that of light. There was no justification for this procedure within the general quantum theory, however, where it was assumed that transformation and wave functions are always single valued. Although the analogy with light was incorrect, Pauli thought that in order to introduce electron spin into wave mechanics it would be necessary to attribute two components to the ψ wave without supposing that they had the character of rectangular components of a vector, as in the case of light. In the same way that wave mechanics provides only the probability for attributing to a photon a plane polarization in one of two directions, we can also know only the probability of the two possible values of spin in either of two directions. Pauli wrote out the two simultaneous equations that the two components of the ψ wave must satisfy for a given direction D. By looking at the way in which the two components are transformed when the direction is changed it became clear that they would not transform as vector components. Once again Pauli provided a purely formal argument introducing a new mathematical entity now called a spinor. The two component wave function satisfies a Schrödinger equation

$$\sum_\beta H_{\alpha\beta}\psi_\beta = i\hbar \frac{\partial \psi_\alpha}{\partial t} (\alpha = 1, 2) \text{ or } H\psi = i\hbar \frac{\partial \psi}{\partial t}, \quad (1)$$

where it is understood that H is a 2×2 matrix and ψ is a 1×2 matrix. H contains a spin-orbit coupling—treated as a perturbation—with a coefficient inserted by hand that was designed to fit the Thomas factor, but without any theoretical justification.

Essentially what Pauli did was develop a theoretical framework for the extra degree of freedom by linking it with the additional spectral lines of hydrogen observed in the anomalous Zeeman effect. But, there was still no real *explanation* of the spin variables. Despite its nonrelativistic form, Pauli's account represented a significant advance over Goudsmit-Uhlenbeck in that it was truly quantum mechanical in nature, incorporating a description of what Pauli had earlier referred to as the "not classically describable two-valuedness" of the electron. He had effectively abandoned all ties with classical theory. Rather than trying to assimilate the classical treatment of a model to the quantum theory as Bohr wanted to do, he instead forged a direct link between the empirical facts and a quantum representation. For Pauli the reality of spin was intimately linked to its status as a purely quantum phenomenon.

By this point in 1926 there appeared to be no question that spin was real despite the problematic issues surrounding its physical nature. What is philosophically interesting about the history up to this time is not only the different attitudes about the "reality" of spin but also that it was possible to fully accept something as real without the kind of theoretical understanding typical of most classical entities. Even for those who currently interpret quantum mechanics as an instrumentalist theory or a framework for calculation, it is difficult to deny that properties like spin are essential and real features of elementary particles, features that can only be made sense of in quantum mechanical terms. Although the theory provides a framework for attributing reality to these kinds of properties/entities, it fails to tell us what their physical nature is like. Given that the reality of spin was determined largely independently of any "physical" understanding, I now want to look at whether the remaining portion of its history reveals any links between reality and understanding that were absent in 1926.

The formulation of a relativistic wave equation for the electron came with the work of Dirac in 1928, a topic that brings us back to the problem I mentioned at the beginning, namely, how we should understand the connection between spin and relativity. That is, does its connection with relativity provide the missing piece of the puzzle that enables us to more fully comprehend its nature? An answer to the question of whether spin is a "consequence" of relativity, or more precisely, the Dirac equation, depends ultimately on how one interprets the question. Although Dirac goes beyond Pauli by incorporating spin into a relativistic theory of the electron, I want to claim that spin as a "physical property" is no better understood than it was in Pauli's account. That is, the Dirac theory adds nothing new to our knowledge of how or why this property is an intrinsic or internal feature of the electron. Yet, there is an important sense in which the Dirac equation suggests a seemingly intimate connection between quantum mechanics, relativistic invariance, and spin. The evidence for such a view stems from the fact that the Dirac equation automatically yielded exactly the magnetic moment of the electron needed to account for the observations of atomic spectra. And it did so without "building in" the electron magnetic moment in any way and without any parameter adjustment. But does it follow that spin can be considered a "relativistic effect"? To see the connection between these various lines of argument I want to look briefly at the formulation of the Dirac equation itself and how it extends Pauli's ideas into a relativistic theory.

THE RELATIVISTIC SPINNING ELECTRON

Although Dirac had an interest in the problem of incorporating spin and relativity into a coherent theory, his primary motivation was the more general one of trying to develop a relativistic wave equation. The problem of spin was something that could be addressed once a first-order relativistic theory of the electron was in place. The Klein-Gordon equation provided a relativistic treatment without taking account of spin, but Dirac objected that it was in disagreement with the general principles of quantum mechanics. For example, the expression for charge density violated the requirement that the total charge in any region had to be 0 or $-e$ (charge conservation). This problem stemmed from the physical interpretation of ψ on the K-G account; it provided no answer to the following question: What is the probability of any dynamical variable at a specific time having a value lying between specified limits when the system is represented by a given wave function ψ_n? Only in the case of position can the question be resolved, but not with any other dynamical variable. In the non-relativistic theory the wave equation is linear in the operator $W = \partial/\partial t$ (where t is time) while in the Klein-Gordon equation it is quadratic in $\partial/\partial t$. In the latter case if ψ is known at a particular time the derivative $\partial \psi/\partial t$ is arbitrary so that ψ is arbitrary at later times. Because the electric charge density is a function of ψ, it can also be arbitrary a short time later, a clearly unacceptable result. Hence, the relativistic theory should also be linear if it is to accord with a general interpretation of quantum mechanics.

Thus, according to Dirac the difficulties encountered by previous theories resulted either from their disagreement with relativity or, alternatively, with the general transformation theory of quantum mechanics. Supposedly a theory that satisfied both these requirements would also yield, without any arbitrary assumptions, a solution to what Dirac called the "duplexity" phenomena; the fact that the observed number of stationary states for an electron in an atom was twice the number given by theory—a problem that motivated the introduction of spin.

Prior to his work on the electron Dirac developed a formulation of quantum mechanics that was based on a general transformation theory. The impetus was to remove what he saw as the false assumption from classical theory (namely, the commutation of dynamical variables) and replace it with a more general scheme that would allow the whole of atomic theory to follow in a natural way. He began by using equations of motion closely analogous to the equations of classical mechanics but the important difference was that the dynamical variables were the new q numbers that obeyed quantum conditions but not the commutative law of multiplication. Because the principle of relativity demanded that time should be treated on the same footing as the other variables, Dirac also assumed it had to be a q number. (Later work on the electron did not treat time as a q number.) All of this was done without knowing anything about the dynamical variables except the algebraic laws they were subject to. What Dirac needed was a general definition of the function of a q-number variable, a way of relating them to c numbers that would yield comparisons with experimental values.

In Heisenberg's matrix mechanics one finds matrices that satisfy the algebraic relations (quantum conditions, equations of motion, and so forth) that constitute the quantum theory. What Dirac did was formulate a theory of the more general schemes of matrix representations and the laws of transformation from one scheme to another. Without going into the technical details suffice it to say that the outcome was a new way of interpreting the Schrödinger wave function.[6] He also succeeded in providing a general way to obtain c numbers from the theory. This transformation theory formed the basis for his physical interpretation of quantum mechanics. Although he began with classical analogies, later formulations enabled him to apply the theory to cases for which no classical counterpart existed, thereby making the theory truly quantum mechanical in nature.

It was Dirac's commitment to this transformation theory that provided the foundation for his work on the relativistic electron; and although one could argue that he saw spin and relativity as two separate issues (Kragh 1990), one forfeits an important piece of the puzzle by separating them conceptually. Dirac may have wanted to isolate the problem of spin from that of finding a relativistic wave equation but it appears clear that the latter difficulty, for which he so desperately sought a solution, was one that was intimately connected with spin. And, although Dirac didn't explicitly bring spin into the argument, the historical details show that it played a significant role nevertheless.

In typical Dirac style of mathematical manipulation he began with the following equation for a free electron

$$i\hbar \frac{\partial \psi}{\partial t} = c\sqrt{M_0^2 c^2 + p_1^2 + p_2^2 + p_3^2}\, \psi \qquad (2)$$

and wanted to see whether the square root could be arranged in a linear form in the momenta. It was difficult to see how the square root of four quantities could be linearized, but if successful he would be able to avoid the mathematical problems associated with the square root operator—it appeared to yield a differential equation of infinite order. The answer came by "playing around" with the Pauli spin matrices, which were

$$\sqrt{p_1^2 + p_2^2 + p_3^2} = \sigma_1 p_1 + \sigma_2 p_2 + \sigma_3 p_3. \qquad (3)$$

Dirac thought that if equation (2) could be generalized to four squares instead of two he could get the kind of linearization he wanted

$$\sqrt{p_1^2 + p_2^2 + p_3^2 + (M_0 c)^2} = \alpha_1 p_1 + \alpha_2 p_2 + \alpha_3 p_3 + \alpha_4 M_0 c. \qquad (4)$$

The set of conditions for the coefficients for this equation were similar to those fulfilled by the Pauli spin matrices. But because Pauli only had three matrices Dirac needed a fourth, and when he was unable to find one he decided that he would instead extend the argument to quantities that could be represented by four rows and columns. So, the wave function had four components obeying a system of four simultaneous equations in partial derivatives that replaced the single nonrelativistic equation. When Dirac looked at how the equations of propagation and the components of the wave function were transformed under a change of coordinates, he found that the equations were Lorentz invariant. The transformation formulae for the four components were not those of a space-time vector but belonged to "spinorial" transformations.

Kragh (1990) claims that Dirac simply reduced the physical problem to a mathematical one; yet if one looks at the context provided by Dirac's papers during that period it becomes apparent that the argument is, in fact, a curious mix of mathematical and physical reasoning. The demand for linearization resulted in the use of 4×4 matrices as coefficients which required a four component wave function. But the linearity was intimately connected to *physical* problems associated with the Klein-Gordon equation, specifically, the violation of charge conservation and the physical interpretation of ψ. The equation had to be linear if it was to coincide with the general nonrelativistic theory; the wave equation at time t_1 should determine the wave equation at time t_n. But, the more significant point relates to the problem Dirac refers to at the beginning of his first paper in 1928, a problem that he doesn't take up there but influences the evolution of his thought, namely, that a true relativistic theory—a merging of relativity and quantum mechanics—would require a particular form for the wave equation. The Klein-Gordon equation refers equally well to an electron with charge e as one with $-e$. In the case of large quantum numbers some solutions of the wave equation are wave packets moving in the way a particle with charge $-e$ would move classically, while others move in the way a particle of charge e would move classically. For this second class of solutions W (the operator $W = i\hbar(\partial/\partial t)$) has a negative value. In the classical theory one can arbitrarily exclude those solutions but in quantum theory perturbations can cause transitions from states with W positive to those with W negative which would look like the electron suddenly changing its charge from e to $-e$. A true relativistic wave equation should be such that its solutions split up into two noncombining sets referring to charge $-e$ and e.

But how does this connect to spin? We know that in the Pauli theory there are only two components of ψ. This was because spin required the splitting of the ψ function into two components. We also know, however, that Pauli simply added the spin factor on in an ad hoc way; yet Dirac claimed that in his theory spin and the magnetic moment of the electron emerge in a completely natural way. Simply put, spin shows itself as necessary for conservation of angular momentum. If we consider an electron moving in a central field then the calculations show that m, the orbital angular momentum, is not conserved, it is not a constant of motion. If we add to the orbital angular momentum a term whose eigenvalues are $+\hbar/2$ and $-\hbar/2$, then m plus this new term (which is just the spin angular momentum) together define the total angular momentum which is a constant of motion. Pauli had shown that spin required a splitting of the wave function into two components while Dirac's relativistic equations requires a further splitting of each of these two components (the latter not being required in the classical approximation).

I claimed above that although Pauli did not see spin as a relativistic effect, it was the compatibility with relativity displayed by the Thomas factor that convinced him it was real. Although spin was not well understood physically, quantum theory appeared to require that we replace our desire for physical models with abstract mathematics. Hence one could uphold the conviction that spin was something real and measurable. And, although the Dirac theory provided a fuller theoretical treatment of spin than previous accounts, there was still a sense in which it never fully transcended its place in the paper world of mathematics.

Spin remains a mysterious property, yet many have claimed that Dirac has shown us that it is a necessary consequence of the relativistic wave equation. It may well be that characterizing spin as a requirement of relativity legitimates it in a way that eases some of the conceptual confusion. And there is certainly a sense in which the Dirac theory gives us information about spin that is not present in earlier accounts. It tells us that the spinor wavefunctions are endowed with a spin angular momentum of $\hbar/2$. The analysis of the representations of the Lorentz group tells us that the quantum mechanical wave functions must be certain types of spinors characterized by a value of the mass and an integer or half-integer spin. And it supplies the mathematical description of the kinematics of a free particle with spin 1/2 as well as equations for the dynamics of a charged particle in a field which yield the value for the gyromagnetic ratio of the electron. The mathematical formalism of the Dirac equation and group theory require the existence of spin to guarantee conservation of angular momentum and to construct the generators of the rotation group. In that sense spin becomes part of a coherent theoretical framework rather than an ad hoc hypothesis required to account for specific effects. Insofar as the mathematics (the linearity requirement) seems to require the existence of spin to restore conservation and solve the problem of the arbitrariness of the wave equation, it emerges as a property that bridges the gap between the mathematics and the physics.

But ultimately the issue of whether spin is a relativistic effect depends on how one defines the conceptual boundaries of the question. Dirac's theory works well only for spin 1/2 particles; but because a consistent interpretation of the Dirac equation (and Klein-Gordon) requires the framework of quantum field theory, which gives rise to spin 0 fields, it appears that spin could not be a consequence of relativity.[7] In fact, several authors have claimed (Sakurai 1967 and others) that by working out the nonrelativistic limit of the Dirac theory one can incorporate spin and obtain the correct magnetic moment. Moreover, we have certainly seen in Pauli's work how the basic idea of spin appears to be a truly quantum phenomenon. Yet, it is also the case that any fully articulated notion of spin requires kinematical properties like the Thomas precession and dynamical effects such as spin-orbit coupling, both of which are relativistic. So, even though spin itself can be given a nonrelativistic formulation, once one enters the domain of relativity theory spin takes on a relativistic character. And clearly it is the Dirac theory that provides us with an account of how spin fits into the larger picture of a relativistic quantum mechanics. That picture, however, is nothing more than an additional angular momentum that is *somehow* intrinsic to the electron. In that sense the association with relativity has not enhanced the "physical" understanding of spin due to its failure to provide an appropriate etiology.

One might, however, want to argue that spin is given a proper physical explanation from within the framework of group theory. In the 1930s Wigner showed how one could use the Poincaré group (or the inhomogenous Lorentz group) to get a description of the purely kinematical properties like spin for quantum relativistic systems.[8] Each relativistic wave function corresponded to some unitary representation of the Poincaré group and as such one can sometimes say that an elementary particle is associated with a unitary irreducible representation of the group.[9] Given that definition, an elementary particle can be characterized by its mass and spin. Spin turns out to be simply a group invariant characterizing the unitary representation of the relativity group associated with the wave equation. Consequently, one thinks of spin not as the physical rotation associated with a particle but rather as a symmetry, a way of mathematically stating that a system can undergo a certain rotation.

Even before the work of Wigner one could understand spin as a group theoretical property; it emerged out of Dirac's formulation of the wave equation. We saw above that the positive energy solutions for the equation provided a description of a free spin 1/2 particle that was consistent with the requirements of special relativity. Spin seemed to appear as a consequence of the transformation law of the solutions under rotation which in turn were determined by the transformation law under homogenous Lorentz transformations. In other words, spin appeared intimately connected to the relativistic invariance of the Dirac equation. In fact, one can think of the structure and properties of any quantum field as dictated by the representation of the homogenous Lorentz group under which it transforms. A representation is defined as the set of matrices that satisfy the group multiplication law

$$D(\lambda)D(\lambda) = D(\lambda\lambda)$$

One of the most general irreducible representations of the Lorentz group is called a "spinor" which is really the spin 1/2 representation of the group.

To calculate the behavior of the Dirac equation under the Lorentz group, we need to define how spinors transform under some representation $S(\Lambda)$ of that group. Once this is done, one then needs to construct invariants under the group. To show this one defines a new field

$$\overline{\psi} \equiv \psi^\dagger \gamma^0,$$

which, under a Lorentz transformation, obeys the conditions required to form invariants. Hence we find that $\overline{\psi}\gamma^\mu\psi$ is a genuine vector under the Lorentz group, which is tantamount to claiming that the γ^μ matrices transform as vectors under the spinor representation of the group. And, because γ^μ transforms as a vector, the Lagrangian corresponding to the Dirac equation

$$L = \overline{\psi}(i\gamma^\mu\partial_\mu - M)\psi$$

is invariant under the Lorentz group.[10] As far as group theory is concerned then the wave function ψ carries the representation (1/2, 0) (0, 1/2) of the homogenous Lorentz group.

The question of whether this group theoretical account gives us a physical understanding of spin is not unlike the question of whether spin is a relativistic effect. It depends on the context in which the question is raised. If one is satisfied that a particle is simply a unitary irreducible representation of, say, the inhomogenous Lorentz group, then a similar explanation of spin will no doubt be acceptable. Even if one is willing to concede, however, that our contemporary physical understanding of spin is just the group theoretic one, that no more can be said from the point of view of physical theory, one still might want to raise a general philosophical question about the relationship between mathematics and the world; between the mathematical formulation of our theories and the physical world they supposedly represent. Put differently, should we rest content that our understanding of many physical aspects of the world is given simply through the mathematical structures in which these properties are represented? In some sense that is the ultimate metaphysical question for philosophy of science but obviously due to its sheer enormity it cannot be addressed here.

One of my goals in this chapter was to analyze the factors responsible for spin being thought of as a real property of the electron. Those considerations ranged from its connection with relativity which appeared to mark a turning point for Pauli, to its rather curious hybrid nature of quantum and classical features that was influential for Bohr. The affinity between spin and Dirac's relativistic wave equation appeared to further strengthen its position by removing the last traces of any ad hocness in its formulation. Hence the origin of the misnomer that spin is a consequence of relativity theory. We have also seen how the theoretical history of spin revealed its ontological status as something that evolved over time rather than having one decisive moment that marked its birth or acceptance. That history similarly revealed that its "reality" had little to do with a physical understanding of its nature. Those conclusions speak to the methodological importance of theoretical histories as a way of revealing the intricacies involved in providing ontological arguments. Too often philosophical debates about the reality of particular entities focus on specific conditions that are taken as defining what counts as "real." By focusing less on definitional aspects and more on the evolution of properties and ideas within a conceptual/physical framework, our philosophical arguments will gain historical accuracy and hence greater credibility as an explication of scientific practice.

In closing then what can be said about spin from a contemporary perspective. Although it is a measurable feature of our physical world its reality consists of a curious mixture of mathematical and physical properties. The conceptual puzzlement surrounding the physical nature of spin was not in 1926, nor is it now, enough to shake the conviction that it is a fundamental feature of the electron. Just as Pauli was forced to sacrifice physical understanding to the formalism of quantum mechanics, so too with our physical intuitions; we must rest content with a mathematical description of certain kinds of entities and properties. I began by claiming that I wanted to investigate the conditions under which spin was transformed from a property existing on paper to a fundamental feature of the physical world. In some sense we have come full circle, although spin is physically real and measurable, its understanding brings us back to the paper world of mathematics.

Acknowledgments

Support of research by the Social Sciences and Humanities research Council of Canada is gratefully acknowledged. I would like to thank Jed Buchwald and Andy Warwick for suggestions on how to improve an earlier draft. My interest in this topic arose from a question posed by Steven Weinberg about the connection between spin and relativity.

Notes

1. The importance of history for both ontology and epistemology has been highlighted by both Kuhn and Hacking. In an essay on the concepts of cause in the development of physics Kuhn (1971) traces four main stages in the evolution of causal notions. Each theoretical stage represented a different view of what counted as a cause or how to provide a causal explanation. More recently Ian Hacking (1992) has advocated a view that he identifies as "historical epistemology." The idea is that our knowledge of things must take into account aspects of their historical evolution. I want to extend this line of thought to include metaphysics—that the question of whether something can be counted as real will depend on its history. I realize that I am collapsing the distinction between what is real and what can be counted as real. I do so purposely because I think there is no other way to speak about what is real apart from the cognitive constraints we use to determine and isolate such things.

2. The core angular momentum R was equal to $1/2$.

3. This was because it depended on the core having an angular momentum $R = 1/2$ which gave a value 2 for the Landé g factor instead of 1.

4. Forman (1968) also notes (n. 72) that the scheme for hydrogen proposed by Goudsmit and Uhlenbeck was essentially the same as the unpublished version that Landé had developed and communicated in letters to Pauli in January and February 1925. Forman claims that Goudsmit and Uhlenbeck as much as state that they had forgotten their proposal for hydrogen when the spin hypothesis was introduced, implying that it bore no relation to the spin hypothesis whatsoever. Yet, their own remarks suggest that the theoretical context in which they were immersed in 1925 after the publication of the hydrogen paper was significant. Although the hydrogen paper and spin were not explicitly linked they did think it sufficiently relevant to mention it to Einstein who then encouraged them to make the connections more specific, which Goudsmit was able to do on the spot.

5. For an interesting discussion of the differences between Heisenberg's and Pauli's views on the problems and prospects of the quantum theory see Serwer (1977). His article discusses some of these differences in terms of social pressures and personality.

6. The eigenfunctions of the wave equations were just the transformation functions that enabled one to transform to a scheme in which the Hamiltonian is a diagonal matrix.

7. See Weinberg (1987).

8. The Poincaré group is simply an enlargement of the Lorentz group that results from adding translations.

9. An irreducible representation is one that cannot be split up into smaller pieces, each of which would transform under a smaller representation of the same group. All the basic fields of physics transform as irreducible representations of the Lorentz and Poincaré groups. The complete set of finite dimensional representations of the rotation group O(2) or the orthogonal group comes in two classes, the tensors and spinors.

10. This representation of the Lorentz group is not unitary because the generators are not all represented by Hermitian matrices. They do satisfy a pseudounitarity relation. See Weinberg (1995).

References

Abraham, M. 1903. "Prinzipien der Dynamik des Elektrons." *Annalen der Physik* 10: 105–179.

Buchwald, J. Z. 1994. *The Creation of Scientific Effects: Heinrich Hertz and Electric Waves* (Chicago: University of Chicago Press).

Compton, A. H. 1921. "The Magnetic Electron." *Journal of the Franklin Institute* 192: 145–155.

Dirac, P. A. M. 1928a. "The Quantum Theory of the Electron." *Proceedings of the Royal Society of London A* 117: 610–624.

Dirac, P. A. M. 1928b. "The Quantum Theory of the Electron, part II." *Proceedings of the Royal Society of London A* 118: 351–361.

Dirac, P. A. M. 1928c. "Zur Quantentheorie des Elektrons." In H. Falkenhagen (ed.), *Quantentheorie und Chemie* (Leipzig: S. Hirzel), 85–94.

Dirac, P. A. M. 1928d. "Ueber die Quantentheorie des Elektrons." *Physikalische Zeitschrift* 29: 561–563.

Forman, P. 1968. "The Doublet Riddle and Atomic Physics circa 1924." *Isis* 59: 156–174.

Franklin, A. 1986. *The Neglect of Experiment* (Cambridge: Cambridge University Press).

Galison, P. L. 1987. *How Experiments End* (Chicago: University of Chicago Press).

Goudsmit, S. A. 1946. *Physica B1* 21: 445.

Goudsmit, S. A. 1976. "It Might As Well Be Spin." *Physics Today* 29: 40–43.

Goudsmit, S. A., and G. Uhlenbeck. 1925a. *Physica* 5: 266.

Goudsmit, S. A., and G. Uhlenbeck. 1926. "Spinning Electrons and the Structure of Spectra." *Nature* 117: 264–265.

Hacking, I. 1983. *Representing and Intervening* (Cambridge: Cambridge University Press).

Hacking, I. 1992. "Historical Epistemology" (unpublished manuscript).

Heisenberg, W. 1921. *Zeitschrift für Physik* 8: 273.

Jammer, M. 1966. *The Conceptual Development of Quantum Mechanics* (New York: McGraw-Hill).

Kragh H. 1990. *Dirac A Scientific Biography* (Cambridge: Cambridge University Press).

Kronig, R. 1960. "The Turning Point." In M. Fierz and W. Weisskopf (eds.), *Theoretical Physics in the Twentieth Century* (New York: Wiley Interscience).

Kuhn, T. 1971. "Concepts of Cause in the Development of Modern Physics." In *The Essential Tension* (Chicago: University of Chicago Press).

Landé, A. 1921. *Zeitschrift für Physik* 5: 231.

Pais, A. 1986. *Inward Bound* (New York: Oxford University Press).

Parsons, A. L. 1915. "A Magneton Theory of the Structure of the Atom." *Smithsonian Miscellaneous Collections* 65: no.11.

Pauli, W. 1925. *Zeitschrift für Physik* 31: 373.

Sakurai, J. J. 1967. *Advanced Quantum Mechanics* (Reading, MA: Addison-Wesley).

Serwer, D. 1977. "Unmechanischer Zwang: Pauli, Heisenberg and the Rejection of the Mechanical Atom, 1923–1925." *Historical Studies in the Physical Sciences* 8: 189–256.

Thomas, H. 1926. "The Motion of the Spinning Electron." *Nature* 117: 514.

Uhlenbeck, G., and S. A. Goudsmit. 1925b. "Ersetzung der Hypothese vom unmechnaischen Zwang durch eine Forderung bezuglich des inneren Verhaltens jedes einzelnen Elektrons." *Die Naturwissenchaften* 13: 953–954.

Weinberg, S. 1995. *The Quantum Theory of Fields* (New York: Cambridge University Press).

Weinberg, S. 1987. "Towards the Final Laws of Physics." In *Elementary Particles and the Laws of Physics: The 1986 Dirac Memorial Lectures* (New York: Cambridge University Press).

van der Waerden, B. L. 1960. "The Exclusion Principle and Spin." In Fierz and Weisskopf (eds.). *Theoretical Physics in the Twentieth Century* (New York: Wiley Interscience).

15

What Should Philosophers of Science Learn from the History of the Electron?
Jonathan Bain and John D. Norton

We have now celebrated the centenary of J. J. Thomson's famous paper (1897) on the electron and have examined 100 years of the history of the first fundamental particle. What should philosophers of science learn from this history? To some, the fundamental moral is already suggested by the rapid pace of this history. Thomson's concern in 1897 was to demonstrate that cathode rays are electrified particles and not aetherial vibrations, the latter being the "almost unanimous opinion of German physicists" (293). But were these German physicists so easily vanquished? De Broglie proposed in 1923 that electrons are a wave phenomenon after all, and his proposal was soon multiply vindicated, even by the detection of the diffraction of the electron waves. Should we not learn from such a reversal? Should we not dispense with the simple-minded idea that Thomson discovered our first fundamental particle and admit that the very notion of discovery might be ill-suited to science?

The purpose of this paper is to argue at length that this sort of skepticism is hasty and wholly unwarranted. Nevertheless, a more detailed examination of the history of the electron can give further encouragement to these skeptical smolderings. The transition from classical corpuscle to quantum wave was just the most prominent of the many transformations of theories of the electron over the last century. Thomson's electron of 1897 was a charged, massive corpuscle—an electrified particle—obeying Newtonian dynamics. It was briefly replaced by one in the electromagnetic world picture whose mass arose as an artifact of its electromagnetic field. Einstein's electron of 1905 once again sustained an intrinsic mass but now obeyed a relativistic dynamics. The electron of Bohr's old quantum theory of the 1910s and early 1920s displayed a precarious and ever-growing mix of classical and discrete properties. Pauli's electron of 1925 obeyed a bizarre, nonclassical "exclusion principle" under which no two electrons could occupy the same energy state in an atom. The electron of the new quantum theory of the mid- to late-1920s could be portrayed apparently equally well by Heisenberg's matrices, Schrödinger's waves, and Dirac's q-numbers.[1]

At least in this new theory, the electron maintained some measure of identity as an independent physical system. But even this was lost as the electron continued to mutate into forms ever more remote from Thomson's corpuscles. In Jordan and Wigner's (1928) theory, under second quantization of the single-particle electron wave function, the electron became a mere excitation of a fermionic field. Wigner's (1939) analysis of group properties of elementary particles relegated the electron to a spin-1/2 irreducible representation of the Poincaré group. In the 1967–68 Glashow-Salam-Weinberg theory of electroweak interactions, the electron was an even stranger beast: it had massless left-handed and right-handed parts that united to form a massive particle through interactions with a scalar Higgs field. Finally, in the current standard model of fundamental interactions, the electron is a member of the first of three generations of similar leptonic particles that are related in a nontrivial way to three generations of hadronic quarks. With its public persona displaying more aliases than a master confidence trickster, one may well doubt that we have or ever will unmask the identity of the real electron in our theorizing. Is the lesson of history, then, that we should stop taking our theories of the electron as credible reports of physical reality?

Such concerns have long been a subject of analysis in philosophy of science. They have been given precise form in the "pessimistic metainduction":

> Every theory we can name in the history of science is, in retrospect, erroneous in some respect. The Newtonian theory of gravitation is incorrect, as is the classical theory of electromagnetism, Dalton's atomic theory, classical physical optics, the special theory of relativity, the Bohr theory of the atom, and so on. The errors of these theories may not matter for most practical purposes, but from a contemporary point of view they are all, strictly, false theories. Since all theories in history have been false, . . . we should conclude that all the methods of science do not generate true theories; hence our present scientific theories, which were obtained by the same methods, are false as well. (Glymour 1992, 125–126)[2]

The purpose here is to explain why we believe that the history of electron provides no support for the pessimistic metainduction. In brief, we shall argue that the history of the electron shows that there is something right and that there is something wrong about the pessimistic metainduction. What is right is that the history shows how even the best theories are corrigible. If the history of the electron is typical , then we should expect none of our current theories to be the final theory. But what is wrong is the sad portrait of the sequence of theories in the electron's history as nothing more than a sequence of magnificent failures. Although there proved to be something erroneous in each theory of the sequence, there is also a clear sense in which each accumulates results from earlier members of the sequence and provides an ever-improving account of the nature of the electron. Our case for this claim resides in two theses, which are elaborated in the following sections:

- Thomson, Bohr, Dirac, and the other electron theorists all had good evidence for at least some of the novel properties they announced for the electron and these historically stable properties endure through subsequent theory changes.
- This accumulated stock of enduring properties can be collected into what we shall call the structure of electron theory. At any stage in the sequence of theories, one can specify our best candidate for this structure. It gives that theory's representation of the electron and accounts for the successes of earlier theories of the electron.

Thus we shall argue that the gloss of the history of the electron as just a sequence of false theories is seriously misleading. A closer look at the history reveals a sequence of theories in which an evergrowing, historically stable core of properties of the electron is discerned and in which the deficiencies of earlier theories are identified and corrected as our accounts of the electron are brought into ever closer agreement with the minutiae of experiment.

Historically Stable Properties

As we follow the sequence of theories of the electron starting with Thomson, we find each theory contributing stable properties of the electron that are then retained in the later theories. There are many of these. We catalog just a few of the more prominent and easily describable ones.

Whatever we may now think of Thomson's (1897) theory of the electron as a classical, electrified particle, he did succeed in using it to recover from his experiments on cathode-ray deflection values of the mass-to-charge ratio (m/e) of the electron that agree with the modern value on which the electron literature rapidly settled. He recovered values in the range 0.32×10^{-7} to 1.0×10^{-7} (306) from the theoretical analysis of experiments involving deflection by a magnet and values of 1.1×10^{-7} to 1.5×10^{-7} (309) from the theoretical analysis of experiments involving deflection by an electrostatic field (measured as gram/electromagnetic units of charge). This conforms well with the modern value of m/e of 0.57×10^{-7} in the same system of units—the value used with equal comfort and success in classical electrodynamics and quantum field theory.

Correspondingly, Millikan (1917), using essentially the same classical framework, proclaimed the atomicity of the charge of the electron. He found (238) that electrons all carry the same unit of charge of 4.774×10^{-10} esu.

Once again, this compares favorably with the modern value of 4.803×10^{-10} esu. This value proved stable to within a few percent through the development of the theory of the electron. Indeed it had already arisen in Planck's (1900) famous analysis of heat radiation, which is now taken to mark the birth of quantum theory. Planck concluded by showing that his analysis yielded new values for certain fundamental constants of physics, including the charge of the electron, which he reported as 4.69×10^{-10} esu.

Bohr's (1913) celebrated analysis of bound electrons in atoms and their spectra depended on his conclusion that an electron bound into orbit around the positive charge of the nucleus of an atom did admit stationary states, contrary to the classical theory. Moreover, these states were determined by the condition that the angular momentum of the electron due to its orbital motion was a whole multiple of $h/2\pi$, where h is Planck's constant. While the electron has been embedded in ever more sophisticated theories of emission and absorption spectra, the basis of spectrographic analysis retains these two notions as its foundation, with Bohr's angular momentum quantum number now supplemented by further quantum numbers.[3]

In a communication of October 1925, essentially within the aegis of the soon to be superseded "old quantum theory," Uhlenbeck and Goudsmit (1925) introduced electron spin. They inferred from the splitting of spectral lines in the anomalous Zeeman effect that the electron possesses an intrinsic angular momentum of $h/4\pi$ that had been hitherto neglected and was responsible for a fourth quantum number in the theory of line spectra. The equivalent characterization of the electron as a spin-1/2 particle persists in all later, mainstream theories of the electron.

In 1925, Pauli suggested that atomic electrons obey an "exclusion principle" that prohibits more than two electrons from occupying the same atomic energy level. A year later, Pauli's phenomenological rule was formalized by Fermi (1926), and independently by Dirac (1926), as a new type of nonclassical statistics that govern ensembles of particles obeying the rule. Particles, such as the electron, governed by these statistics came to be known as "fermions." Fermi-Dirac statistics entered into quantum field theory in the form of anticommutators in Jordan and Wigner's (1928) extension of second-quantization techniques to fermions. In 1940, the fermionic character of electrons became even more firmly entrenched into electron theory when Pauli proved the spin/statistics theorem. He demonstrated that particles with half-integer spin must obey Fermi-Dirac statistics on pain of violations of causality. Hence, if the electron has spin 1/2, it must obey Fermi-Dirac statistics if it is to be described by a causal theory of quantum fields.

That these investigations into the properties of the electron produce an evergrowing list of stable properties should come as no surprise. In each case, the property discerned results from careful experiment, theoretical analysis, or both, and in each case the investigator had strong evidence for that property. This is not the place to analyze the strategies used to mount evidential cases for microentities such as electrons. In principle, each instance could be different and the investigator could need to mount evidential cases of qualitatively different character for each property. It turns out, however, that this is not the case. As one of us has argued elsewhere,[4] we can discern methods that are used repeatedly to mount the evidential case. One method requires a multiplication of experiments that massively overdetermine some fundamental numerical property of the electron. For example, one evaluates the mass-to-charge ratio revealed by many different manifestations of the electron—such as the deflection of cathode rays in different experimental arrangements or the normal Zeeman effect. That one recovers essentially the same value in all these circumstances is strong evidence that each is a manifestation of the same particle, the electron, and that electrons do carry inertial mass and charge and in the ratio recovered. A second strategy applied to the electron is known in the philosophy of science literature by many names, including eliminative induction or demonstrative induction. In it, one maps out as large a class of candidate theories as possible and then shows that some item of evidence, usually experimental, forces selection of just one theory from that class as the only one that is compatible with this item of evidence. The force of this method is that it not only gives strong evidence for the theory selected, but it also gives direct evidence against the theory's competitors.

Both methods are instances of inductive inference and thus can and did sometimes fail. But should their occasional failure make us complete skeptics about the results of all such investigation and the possibility that we can detect and correct the failures? It should not, just as a few successes should not delude us into the belief that we are infallible.

STRUCTURE

How is it possible for the sequence of theories of the history of the electron to display this growing list of historically stable properties? One of us has argued elsewhere that this can be explained by urging that the theories of the sequence have a common feature.[5] This common feature, the structure, is preserved through the changes of theory and is, in retrospect, that for which the investigators of the electron do have strong evidence. It is by no means assured that a sequence of theories will admit such a common feature. For a sequence of theories with historically stable properties, however, such as the theories of the electron, this view predicts that we will be able to identify a common feature of nontrivial content sufficient to support these properties.

Ideally we would like to be able to set out in simple terms the structure that holds together the sequence of theories of the electron. But that would be impudent and impossible, for it would require us to say what the final, incorrigible theory of the electron must be. But the history of the electron shows us that our theories are always corrigible. Although we cannot display the structure, we can certainly display our best candidate for that structure, recognizing that its form and content are likely to change as understanding grows. At any one time in the development of theories of the electron, we can read our best candidate from the latest theory. It is simply the smallest part of the latest theory that is able to explain the successes of earlier theories. We have followed this prescription and, in the remainder of this section, we will list the best candidates that result for the last 100 years of theories of the electron. We identify these best candidates in the Hamiltonian or Lagrangian for the electron in the corresponding theory.

There is an uncanny stability in this string of best candidates. Except for one brief period in the late 1920s, the structure stays remarkably constant. Changes are not so much changes in the mathematical description of the electron but rather in the framework in which that description sits, or (in the later period) in additions to the vehicles through which the electron interacts with other elements of the physics ontology. Prior to the 1920s, the classical electron Hamiltonian remains unchanged excepting adjustments for relativity theory. After the 1920s, once the Dirac Hamiltonian/Lagrangian is fixed, its form remains unchanged in all subsequent descriptions of the electron. What changes is the list of interactions the electron experiences. And each type of interaction is itself given by a separately definable structural feature.

The basic sequence of developments involves six modifications:

First, virtually all the properties of the electron discovered at the advent of wave/matrix mechanics prior to the incorporation of spin can be recovered from the Hamiltonian of an electron in an electromagnetic field:

$$H = (\mathbf{p} - e\mathbf{A})^2/2m + e\phi, \qquad (1)$$

where $\mathbf{p}$ is the momentum, e is the charge and m is the mass of the electron. $\mathbf{A}$ and ϕ are the vector and scalar electromagnetic potentials.

Embedding this Hamiltonian into a classical (nonquantum, nonrelativistic) dynamics yields the electrostatic interactions Millikan needed for his oil drop experiment and those that Thomson called upon to explain the deflection of cathode rays by electric and magnetic fields. In the old quantum theory, the same Hamiltonian describes the interaction of the electron with the electric field of the atomic nucleus. It does so in sufficient measure to give us the stationary electron states from which the atomic spectra are recovered.

It also accounts for the effects of external electric and magnetic fields on these states, which are in turn associated in the spectra with the Stark and normal Zeeman effects. If, following Schrödinger (1926), this Hamiltonian is inserted into the time-independent Schrödinger equation for a spinless, massive particle using the identification $\mathbf{p} \to -i\hbar\nabla$, we once again recover stationary states capable of returning much of the known atomic spectra. We are, in addition, freed from the old quantum theory's puzzle of how such stationary states are possible.

Second, the classical relativistic Hamiltonian for a particle with mass m and charge e in the presence of an electromagnetic field is

$$H = [(\mathbf{p} - e\mathbf{A}/c)^2 c^2 + m^2 c^4]^{1/2} + e\phi. \tag{2}$$

The change from (1) does not reflect the discovery of some new property peculiar to the electron but does accommodate the relativistic behavior of energy and momentum in all its forms.[6] The adjusted Hamiltonian (2) allowed a more precise accounting of atomic spectra. Most famously, following the approach of Sommerfeld (1915, 1916) in the old quantum theory, the relativistic corrections introduced a precessional motion in the electron's elliptical orbit, eradicated a degeneracy in the energy levels of the Bohr atom, and allowed explanation of the fine structure of the hydrogen emission spectrum. Correspondingly, a relativistic Hamiltonian could be employed in Schrödinger's (1926) wave mechanics. One could recover results in gross agreement with the experimental hydrogen spectrum from a wave equation obtained by substituting the identifications $\mathbf{p} \to -i\hbar\nabla$ and $H \to i\hbar\partial/\partial t$ into (2), for an electron described by a wave equation $\psi(\mathbf{x}, t) = \psi(x)e^{-iEt/\hbar}$ in a coulomb potential, $\mathbf{A} = 0$, $\phi = e/4\pi r$ (i.e., an electron in a hydrogen atom). The Hamiltonian (2) fails, however, to account for the line splitting of the anomalous Zeeman effect. Uhlenbeck and Goudsmit (1925) accounted for this splitting by positing the internal spin of the electron. While the other shifts in electron theory responded to a deeper understanding of the theoretical context in which electrons were set, intrinsic spin was the first new property peculiar to the electron discovered since Thomson.

Third, spin could be accommodated to varying degrees of satisfaction by adding spin coupling terms to (2); but these terms are incomplete as long as they only reflect the two degrees of freedom associated with the angular momentum Hilbert space of a spin-1/2 particle. The simplest and fullest modification of (2) that accommodates spin was accomplished by Dirac (1928) using Dirac spinors with four degrees of freedom.[7] In modern notation (in units where $\hbar$ and c are set equal to 1 and with spacetime signature $(1, -1, -1, -1)$), the Dirac equation is

$$(i\gamma^\mu \partial_\mu - m)\psi(x) = 0 \text{ for } \mu = 0, 1, 2, 3, \tag{3}$$

and the Lagrangian density for which (3) is the Euler-Lagrange equation is

$$\mathcal{L}_{\text{Dirac}} = \bar{\psi}\,(i\gamma^\mu \partial_\mu - m)\psi. \tag{4}$$

Here ψ is a 4-component Dirac spinor, γ^μ are 4×4 anticommuting matrices, and $\bar{\psi} = \gamma^0 \psi^\dagger$. The modification of (3) to account for classical electromagnetic interactions follows the prescription $\partial_\mu \to D_\mu \equiv \partial_\mu + ieA_\mu$ (in analogy with the classical case). In this modified form, the nonrelativistic limit of (3) yields the magnetic moment estimated by Uhlenbeck and Goudsmit due to internal spin as well as the fine-structure spectrum of hydrogen unaccounted for by Schrödinger. The new properties that (3) adds to the electron are spatiotemporal in nature. The electron of (3) is now characterized by a new type of spatiotemporal transformation property that the electron of (2) does not possess. The electron of (2) transforms under Poincaré transformations as a scalar; that of (3) as a 4-component spinor. The electron of (1), in contrast, transforms under Galilean transformations as a scalar.

Fourth, in Dirac's original (1928) theory, $\psi(x)$ is considered a wave function for a single-particle electron. To explain the negative-energy solutions allowed by (3), Dirac (1930) suggested that the vacuum state consists of a negative-energy electron sea. This introduces two conceptual changes into the description of the electron. First, the single-particle Dirac theory must now be considered a many-particle theory. Second, the creation and annihilation of electrons is now possible. The transition of a positive-energy electron to the state occupied by a hole in the sea appears as the annihilation of an electron-hole pair. If a negative-energy electron in the sea absorbs enough energy that its total energy becomes positive, it makes the transition to a positive-energy state, leaving behind a hole. This appears as the creation of an electron-hole pair.[8]

The quantized field interpretation of the electron was proposed by Jordan and Wigner (1928) and employed the Lagrangian (4) that had been introduced in the Dirac theory. Dirac (1927) had previously quantized the electromagnetic field by a process that became known as second quantization. He identified the coefficients of the Fourier expansion of the electromagnetic field as photon creation/annihilation operators obeying commutation relations. Jordan and Wigner interpreted solutions $\psi(x)$ to the Dirac equation as fields and then applied the second-quantization method of Dirac to the electron field. They thus introduced electron creation/annihilation operators, which, owing to Fermi-Dirac statistics, obey anticommutation relations. They did not consider an electron interacting with an electromagnetic field.[9]

The first fully consistent quantum field-theoretic account of the electron that incorporates electromagnetic interactions is quantum electrodynamics (QED). Formally, the move to QED does not require alteration of Dirac's Lagrangian (4) but the addition of new terms to it to accommodate interactions with the electromagnetic field. The electromagnetic field is given by a local abelian U(1) gauge field $A_\mu(x)$, which couples to the electron field $\psi(x)$ with a strength given by the electron charge e. There is a standard recipe for describing such gauge field interactions that amounts to adding two new terms to the Lagrangian density under consideration. To the Dirac Lagrangian density, we add a piece due to the electromagnetic field and an interaction piece:

$$\begin{aligned}\mathscr{L}_{QED} &= \mathscr{L}_{Dirac} + \mathscr{L}_{Maxwell} + \mathscr{L}_{int} \\ &= \overline{\psi}(i\gamma^\mu \partial_\mu - m)\psi - \frac{1}{4}(F_{\mu\nu})^2 - e\overline{\psi}\gamma^\mu A_\mu \psi \\ &= \overline{\psi}(i\gamma^\mu D_\mu - m)\psi - \frac{1}{4}(F_{\mu\nu})^2,\end{aligned} \quad (5)$$

where $F_{\mu\nu} = \partial_\mu A_\nu - \partial_\nu A_\mu$ is the electromagnetic field tensor.[10] QED corrects the Dirac theory in predicting the Lamb shift in the hydrogen spectrum and the anomalous magnetic moment of the electron. The gauge field recipe amounts to a new way, consistent with the properties of a Dirac electron, of embedding the electron in an electromagnetic field and thus maintaining electromagnetism in the list of interactions experienced by it.

Fifth, the electron is embedded into an electroweak field by means of a local symmetry-breaking mechanism. Again, there is a standard recipe for describing such interactions. Formally, the modification has the appearance of adding to the QED Lagrangian density an additional term describing the symmetry-breaking mechanism, although the implementation of the mechanism requires that the modification be a bit more subtle than this. With the addition of the weak force, although the structure of the electron itself remains basically unaltered, given by the Dirac Lagrangian, the gauge fields the electron couples to now have peculiar symmetries. A Lagrangian density is constructed in such a way as to (a) account for parity violations of the weak force, (b) account for the massive vector boson mediators of the weak force, and (c) preserve the massless nature of the photon field and produce the QED interaction term.[11] The Lagrangian density that accomplishes this contains four gauge fields (one abelian U(1) and three nonabelian SU(2)), a massless spin-1/2 fermion field representing the electron, and a scalar Higgs field. After symmetry breaking, the gauge fields combine linearly to form three massive gauge fields identified as the weak gauge fields (the two $W^\pm$ boson fields and the Z^0 boson field) and a massless gauge field identified as the photon field. The electron field acquires a mass and couples to the photon field via the required QED interaction term. Formally, the Electroweak Lagrangian density is given by

$$\mathcal{L}_{\text{Electroweak}} = \mathcal{L}_F + \mathcal{L}_G + \mathcal{L}_{\text{int}} + \mathcal{L}_S, \qquad (6)$$

where $\mathcal{L}_F$ is the Lagrangian density for a massless spin-1/2 fermion field (having the same form as $\mathcal{L}_{\text{Dirac}}$ in (5) without the mass term), $\mathcal{L}_G$ is the Lagrangian density for an abelian U(1) gauge field and a nonabelian SU(2) gauge field (each having the same general form as $\mathcal{L}_{\text{Maxwell}}$ in (5)), $\mathcal{L}_{\text{int}}$ describes the interaction between the gauge fields and the fermion field (having the same general from as $\mathcal{L}_{\text{int}}$ in (5)), and $\mathcal{L}_S$ is the Lagrangian density for a scalar Higgs field that couples to the fermion field via a Yukawa-type interaction.[12]

Sixth, for the standard model, the Lagrangian density is again modified by adding new terms. In this case, the new terms are for a hadron (quark) sector of the Electroweak Lagrangian density and the three terms of (nonabelian SU(3)) quantum chromodynamics (QCD): one for fermion (quark) fields, one for the gluon gauge fields, and one for the quark/gluon interaction term:

$$\mathcal{L}_{\text{Standard Model}} = \mathcal{L}_{\text{Electroweak-lep}} + \mathcal{L}_{\text{Electroweak-had}} + \mathcal{L}_{\text{QCD}}, \qquad (7)$$

where $\mathcal{L}_{\text{Electroweak-lep}}$ and $\mathcal{L}_{\text{Electroweak-had}}$ are of the form (6) and $\mathcal{L}_{\text{QCD}}$ is the Lagrangian density for a nonabelian SU(3) gauge theory (having primarily the same general form as $\mathcal{L}_{\text{QED}}$ in (5)).[13]

To summarize, in terms of properties, the third modification adds a new type of spacetime transformation property to the electron. It consistently describes the electron as a relativistic particle with the property of internal spin (Schrödinger had the relativistic part but not the spin part; Uhlenbeck and Goudsmit had the spin part but not the relativistic part). The fourth modification describes a new way, consistent with the third, of adding electromagnetism to the list of interactions experienced by the electron. In addition, the move from the third to the fourth constitutes a conceptual change in describing the electron, from a purely single-particle description to a field-theoretic/many-particle description.[14] This move adds interactions in which electrons are created and destroyed to the list. The fifth adds the weak force to this list. (It also indicates some of the properties the electron possesses at high energies; namely, at such energies, it decouples from the Higgs field and becomes a massless fermion field.) The sixth adds the property of membership in one of three generations of leptons that have a symmetrical relationship with three generations of quarks.[15]

Again, we emphasize that the development of the first through the sixth involves primarily a preservation and augmentation of structure as given by the Hamiltonian/Lagrangian of the electron. In much of the development the structure is preserved while changes are due to alteration in the theoretical context within which the structure is set: the transition from classical to relativistic space-times and from classical physics through the various forms of quantum theory. The augmentation involves addition: the novel property of spin and an expansion of the list of interactions sustained by the electron.

Conclusion

What, then, should philosophers of science learn from the parade of theories that is a century of the history of the electron? The mere fact that the century has seen a succession of different theories is not, by itself, grounds for pessimism or optimism. What would properly raise our suspicions is the opposite: a vigorous program of investigation into nature in which later researchers find no occasion to correct their predecessors. Our optimism or pessimism should rely on our examination of the details of the changes in electron theories. If these theories were to form a sequence of disconnected portraits, each merely answering to the transient expedients of the moment and each eradicating the content and successes of the earlier theories, then we could be excused for suspecting that we have just replaced one error with another as we pass from one theory to the next. But we do not have such a sequence. We have good reason to see our sequence of theories as correcting errors of former members while preserving their successes and providing richer and improved representations. We have shown that we can discern a growing core of historically stable properties of the electron in the sequence of theories and that this core is supported by a stable evidential base. Whatever we may now think of the details of Millikan's picture of the electron, for example, his experiments on the discreteness and magnitude of electron charge are reliable. Moreover, we have shown that this growing stock of properties can be integrated into a structure that, at each stage of theorizing, captures the essential properties of the electron then known and explains the successes of the earlier theories.

If we are licensed to fit any induction to the history of the electron, then it should not be the pessimistic induction. It should be an "optimistic induction": Physicists are fallible and their evidential base never complete, so that we cannot expect any theory to be error-free or final.[16] We have seen a sequence of theories each of which identifies and corrects errors of its predecessor while preserving a growing core of stable properties. Thus we should expect that errors remaining in our present theories will be identified and corrected by theories to come as we continue to improve our understanding of the electron.

Acknowledgments

We are grateful to Tony Duncan and Laura Ruetsche for helpful discussion.

Notes

1. Such is the received view. Muller (1997) has recently argued that Heisenberg, Jordan, and Dirac's 1925 matrix mechanics and Schrödinger's 1926 wave mechanics were not equivalent until the completion of von Neumann's 1932 work.

2. See Putnam (1978, 24–25) for the original statement. The current proponent of the argument is Laudan (1984, 1981). Some responses to Laudan are given in Kitcher (1993, 136) and Psillos (1996, 1994). For critiques of these positions, see Bain (manuscript).

3. They are, primarily, a principal quantum number n (energy), an orbital magnetic quantum number m (angular momentum in the z direction), a spin quantum number s, and a spin magnetic number m_s (spin in the z direction).

4. Norton (2000).

5. Bain (1998).

6. Hamiltonian (1) proceeds from the classical result that the kinetic energy of a particle of momentum p and mass m is $p^2/2m$, whereas (2) proceeds from the relativistic result that the particle's total energy is $[p^2c^2 + m^2c^4]^{1/2}$, where m is now the rest mass.

7. Dirac's original motivation in part was to find a first-order equation for which a positive definite probability density could be identified.

8. Dirac initially identified the holes as protons but later (in 1931) identified them as a new type of particle: positrons.

9. Nor did they address the problem of the interpretation of the negative energy solutions to the Dirac equation. This had to wait for the papers of Fock (1933) and Furry and Oppenheimer (1934). These authors continue the work of Jordan and Wigner, interpreting solutions to the Dirac equation as fields and quantizing these via the second quantization method. They introduce creation/annihilation operators for positron fields, however, in addition to those for electron fields. The resulting charge-symmetric field theory is then equivalent to Dirac's many-particle Hole theory, accounting for negative energy states without recourse to the negative-energy electron sea.

10. The general gauge field description of interactions (abelian and nonabelian cases) was given first in 1954 by Yang and Mills and became theoretically respectable after it was shown by 'tHooft in the early 1970s to be renormalizable. Nevertheless, the simple abelian case of QED was well established already in the papers of Tomonaga, Schwinger, Feynman, and Dyson in the 1940s.

11. Briefly, to address (a), the two charged weak gauge fields $W^\pm$ should couple only to the left-handed component or the right-handed component of the electron (these are already well-defined in Dirac's (1928) theory). The Electroweak theory assigns the left-

handed component to an SU(2) doublet (the other component of which is a left-handed electron-neutrino) and the right-handed component to an SU(2) singlet. To address (b), this SU(2) symmetry must be spontaneously broken via a Higgs scalar field (this is the only way to obtain massive gauge bosons in a Yang-Mills theory: in standard Yang-Mills theory, mass terms for the gauge fields would ruin the gauge invariance of the Lagrangian). Since SU(2) doublets and singlets cannot be coupled, there can be no mass term for the electron field in the initial Lagrangian. The Higgs field is thus coupled not only to the gauge fields, but also to the left- and right-handed massless components of the electron to produce an electron mass term after symmetry breaking. Finally, to address (c), a U(1) symmetry is introduced that does not get broken by the Higgs.

12. In particular,

$$\mathcal{L}_F = i\bar{\psi}_R \gamma^\mu \partial_\mu \psi_R + i\bar{\psi}_\mathcal{L} \gamma^\mu \partial_\mu \psi_\mathcal{L},$$

where ψ_R and $\psi_\mathcal{L}$ are the right- and left-handed components of the massless spin-1/2 fermion field;

$$\mathcal{L}_G = -\frac{1}{4}(\partial_\mu A_\nu^a - \partial_\nu A_\mu^a + g f^{abc} A_\mu^b A_\nu^c)^2 - \frac{1}{4}(\partial_\mu B_\nu - \partial_\nu B_\mu)^2,$$

where B_μ and A_μ^a ($a = 1, 2, 3$) are the U(1) and SU(2) gauge fields, g is the coupling constant associated with the gauge fields A_μ^a and f^{abc} are SU(2) structure constants;

$$\mathcal{L}_{int} = -g'\bar{\psi}_R \gamma^\mu B_\mu \psi_R - \bar{\psi}_\mathcal{L} \gamma^\mu \left(\frac{1}{2} g' B_\mu + \frac{1}{2} g A_\mu^a \sigma^a\right)\psi_\mathcal{L},$$

where g' is the coupling constant associated with the gauge field B_μ and σ^a are the Pauli matrices; and

$$\mathcal{L}_S = D_\mu \phi^\dagger D^\mu \phi - \mu^2 \phi^\dagger \phi + \lambda(\phi^\dagger \phi)^2 - G_e(\bar{\psi}_\mathcal{L} \phi \psi_R + \bar{\psi}_R \phi^\dagger \psi_\mathcal{L}),$$

where ϕ is the scalar Higgs field with mass μ and self-coupling constant λ. $\mathcal{L}_S$ couples to the fermion field by means of a Yukawa-type interaction with coupling constant G_e. The derivative operator D_μ couples the gauge fields B_μ and A_μ^α to the Higgs field according to

$$D_\mu \phi = \left(\partial_\mu - \frac{i}{2} g' B_\mu - \frac{i}{2} g A_\mu^a \sigma^a\right)\phi.$$

After symmetry breaking, the electron charge is recovered as $e = gg'(g^2 + g'^2)^{-1/2}$ and the electron mass is recovered as $m_e = G_e \mu (2\lambda)^{-1/2}$.

13. In particular,

$$\mathcal{L}_{QCD} = \bar{\psi}_i^a (i\gamma^\mu D_\mu - m)^{ab} \psi_i^b - \frac{1}{4}(\partial_\mu A_\nu^\alpha - \partial_\nu A_\mu^\alpha + g f^{\alpha\beta\gamma} A_\mu^\beta A_\nu^\alpha)^2.$$

Here the fermionic quark fields ψ_i^a are SU(3) triplets with $a, b = 1, 2, 3$ labeling the local SU(3) "color" symmetry. The index $i = 1, \ldots, 6$ labels the global "flavor" symmetry (up, down, strange charm, top, bottom). A_μ^α ($\alpha = 1, \ldots, 8$) are the SU(3) gluon gauge fields.

14. The sense in which the field-theoretic description in interacting QFT is "dual" to the particle description is a topic of some contention. If by duality is meant "to every field there corresponds a particle and vice versa," the duality thesis is simply incorrect. But demoting duality should not tempt us into fundamentalism of either the field or the particle kind.

15. This membership property is nontrivial insofar as the addition of hadron-electroweak couplings serves to cancel potential divergences arising from certain lepton-electroweak couplings (namely what are known as axial vector current anomalies).

16. In speaking of the fallibility of physicists and errors in their theory, we do not have in mind outright blunders. We refer to a more serious problem. While Lorentz developed a most reasonable classical model for the electron as a charged sphere, it was, by later lights, erroneous, since it failed to accommodate quantum properties. The error occurred because physics is an enterprise that makes inductions from experience; physicists must therefore routinely take inductive risks. Lorentz's is one that did not work out. We are arguing, in effect, that the continuing growth of historically stable properties is evidence that such risks are not always in vain.

REFERENCES

Bain, J. (1998). *Representations of Spacetime: Formalism and Ontological Commitment*, Ph.D. thesis, University of Pittsburgh.

Bohr, Niels. 1913. "On the Constitution of Atoms and Molecules." *Philosophical Magazine* 26: 1–26.

Dirac, P. A. M. 1926. "On the Theory of Quantum Mechanics." *Proceedings of the Royal Society (London)* A112: 661–677.

Dirac, P. A. M. 1927. "The Quantum Theory of Emission and Absorption of Radiation." *Proceedings of the Royal Society (London)* A114: 243–265.

Dirac, P. A. M. 1928. "The Quantum Theory of the Electron. I." *Proceedings of the Royal Society (London)* A117: 610–624.

Dirac, P. A. M. 1930. "A Theory of Electrons and Protons." *Proceedings of the Royal Society (London)* A126: 360–365.

Fermi, E. 1926a. "Sulla Quantizzatione del Gas Perfetto Monoatomico." *Rendiconti della Reale Accademia dei Lincei* 3: 145–149.

Fermi, E. 1926b. "Zur Quantelung des Idealen Einatomigen Gases." *Zeitschrift für Physik* 36: 902–912.

Fock, V. 1933. "Zur Theorie der Positronen." *Akademiia Nauk. Doklady*: 267–271.

Furry, W., and J. Oppenheimer. 1934. "On the Theory of the Electron and Positron." *Physical Review* 45: 245–262.

Glymour, C. 1992. "Realism and the Nature of Theories." In M. Salmon et al. *Introduction to the Philosophy of Science* (Englewood Cliffs, NJ: Prentice-Hall), 104–131.

Jordan, P. and E. P. Wigner. 1928. "Über das Paulische Äquivalenzverbot." *Zeitschrift für Physik* 47: 631–651.

Kitcher, P. 1993. *The Advancement of Science* (Oxford: Oxford University Press).

Laudan, L. 1981. "A Confutation of Convergent Realism." *Philosophy of Science* 48: 19–49.

Laudan, L. 1984. "Explaining the Success of Science: Beyond Epistemic Realism and Relativism." In J. Cushing et al. (eds.) *Science and Reality* (Notre Dame: Notre Dame Press), 83–105.

Millikan, R. A. 1917. *The Electron: Its Isolation and Measurement and the Determination of Some of Its Properties*. 2nd ed. 1924 (Chicago: University of Chicago Press), 1917.

Muller, F. A. 1997. "The Equivalence Myth of Quantum Mechanics—Part I." *Studies in History and Philosophy of Modern Physics* 28: 35–61.

Norton, John D. (2000). "How We Know About Electrons." In Robert Nola and Howard Sankey (eds.) *After Popper, Kuhn and Feyerabend: Recent Issues in Theories of Scientific Method* (Dordrecht: Kluwer), 67–97.

Pauli, W. 1925. "Über den Zusammenhang des Abschlusses der Elektronengruppen im Atom mit der Komplexstruktur der Spektren." *Zeitschrift für Physik* 31: 765–783.

Pauli, W. 1940. "The Connection between Spin and Statistics." *Physical Review* 58: 716–722.

Planck, Max. 1900. "Zur Theorie des Gesetzes der Energieverteilung im Normalspectrum." *Verhandl. der Deutschen Physikal. Gesellsch.* 2: 237–245.

Psillos, S. 1994. "A Philosophical Study of the Transition from the Caloric Theory of Heat to Thermodynamics: Resisting the Pessimistic Meta-Induction." *Studies in History and Philosophy of Science* 25: 159–190.

Psillos, S. 1996. "Scientific Realism and the 'Pessimistic Induction.'" *Philosophy of Science* 63: S306–S314.

Putnam, H. 1978. *Meaning and the Moral Sciences* (London: Routledge).

Salam, A. 1968. "Weak and Electromagnetic Interactions." In N. Svaratholm (ed.) *Elementary Particle Theory* (Stockholm: Almquist and Forlag).

Schrödinger, E. 1926. "Quantisierung als Eigenwertproblem." *Annalen der Physik* 79: 361–376, 489–527; 80: 437–490; 81: 109–139.

Sommerfeld, A. 1915. "Die Feinstruktur der wasserstoff- und wasserstoffähnlichen Linien." *Akademie der Wissenschaften, München, Physikalische-mathematische Klasse, Sitzungsberichte*: 459–500.

Sommerfeld, A. 1916. "Zur Quantentheorie der Spetrallinien." *Annalen der Physik* 51: 1–94, 125–167.

Thomson, J. J. 1897. "Cathode Rays." *Philosophical Magazine* 44: 293–316.

Uhlenbeck, G. E., and S. Goudsmit. 1925. "Ersetzung der Hypothese vom unmechanischem Zwang durch eine Forderung bezüglich des inneren Verhaltens jedes einzelnen Elektrons." *Die Naturwissenschaften* 13: 953–954.

Weinberg, S. 1967. "A Model of Leptons." *Physical Review* 19: 1264–1266.

Wigner, E. P. 1939. "On Unitary Representations of the Inhomogeneous Lorentz Group." *Annals of Mathematics* 40: 149–204.

16

The Role of Theory in the Use of Instruments; or, How Much Do We Need To Know about Electrons to Do Science with an Electron Microscope?

Nicolas Rasmussen and Alan Chalmers

Two extreme positions can be identified concerning the role of theory in the use of instruments in science. The first accords with the positivist ideal that observation should provide a theory-neutral basis from which theory can be confirmed or rejected. If the use of instruments is to conform to that ideal then that use should not involve appeal to theory. Opposition to this positivistic view, and to the insistence that observation generally, and the use of instruments in particular, are "theory-independent" is now commonplace. It was made popular in the 1960s by Karl Popper and Thomas Kuhn, although such opposition dates back at least to Pierre Duhem.[1] The opposite extreme, implicit in "theory-dominant" counterarguments, is that all instrumental data are theory laden. Accordingly, it is often supposed that an explicit, and preferably mathematical and precise, theory of an instrument's operation and of specimen-instrument interaction is necessary for interpreting and evaluating the results that it yields. The outputs of instruments do not by themselves yield information about the systems they are used to investigate, it is supposed, but only do so when combined with an explicit theory of the functioning of the instruments.[2]

Classic examples can be invoked that lend plausibility to each of these extremes. On the positivist side it can be noted that one does not need a theory of how litmus paper works to use it to distinguish acids from alkalis. One did not need theory to appreciate that Faraday's primitive electric motors worked and that the direction of rotation was reversed when the poles of the magnet or the battery connections were interchanged. These motors constituted theory-neutral facts that Faraday's theoretical competitors as well as Faraday himself needed to accommodate with their theories. On the other side, for example, J. J. Thomson needed to assume the Lorentz force law to derive the ratio of charge to mass of the particles constituting the cathode rays that were deflected in his discharge tubes. Similarly, the observable tracks in a bubble chamber have little relevance for microparticle physics until they are interpreted with the aid of theory.

Our view is that the relationships between theory and instruments as they are used in science are multifarious and more interesting and complex than the situations captured by the two extremes characterized above and typically lie somewhere in between them. While we see the need to counter the excesses of the theory-dominant viewpoint by closer attention to and appreciation of what can be achieved through experimental practice, we are well aware that theoretical considerations at a variety of levels can be important or essential in various ways. In this chapter we explore and illustrate various kinds of instrument/theory relationships by looking at early uses of the electron microscope in biology and physics. Focusing on the electron microscope is appropriate not only because a wide range of theory/instrument relationships were, and are, involved in its use but also because it has been invoked by a number of philosophers to support what we will show to be inadequate and partial views of the role of theory in the use of instruments.[3] Before proceeding to these detailed, historically embedded cases, we first introduce some of the forms we think theory/instrument relationships can take by means of a few more simple examples.

Experimentalists have a range of techniques for establishing the reality of observed effects by way of purposeful and controlled intervention via instruments—interventions which may require only a minimal appeal to low-level theory. Conclusions arrived at by way of an instrument can be checked by accessing them in alternative ways, by using alternative instruments, or by using no instruments at all. Galileo could check the veracity of sightings of distant terrestrial objects through his telescope against those resulting from direct observation of the same objects close up. He did not need a theory of the telescope to do that. Dense bodies observable with an electron microscope can also be seen through a fluorescent microscope, which greatly reduces the chances that they are artifacts, as Ian Hacking has noted. All that is required for this argument to be convincing is that the two microscopes involve quite different physical processes, making it unlikely for the differing processes to produce identical artifacts. No detailed theory of the workings of either microscope is necessary here.[4] Such reasoning has its correlates in everyday perception. Macbeth could check for the reality of the dagger he thought he saw by trying to touch it as well.[5] It is worth noting that such use of the unaided senses does not depend on a theory of how those senses work; otherwise we would be in the paradoxical position that science could not get started until we had a scientific theory of how our eyes work!

The foregoing examples notwithstanding, it is undoubtedly the case that the use of instruments often does presuppose significant theory. The use of x-ray diffraction to image crystal structure and to measure lattice spacings in crystals makes explicit use of the Bragg theory of diffraction and would not be possible without it. From a positivist perspective, the fact that the use of instruments presupposes theory poses a threat to their trustworthiness. Theory is meant to be substantiated by appeal to the observable facts, while at the same time the observable facts can only be vindicated by appeal to theory. There is no doubt that circularity can sometimes arise as a result of such a circumstance. One of us (AC) recalls, from his schoolteaching days, a group of high school students investigating the dependence of the deflection of a current-carrying coil on the current passing through it by an experimental arrangement that involved measuring the current by an ammeter whose inner workings consisted of a coil suspended between the poles of a magnet. Here the answer to the question investigated was already presupposed in taking the reading of the ammeter to represent the strength of the current. It is not difficult to see how the circularity could have been avoided in this particular case. All that was needed was a current measuring device based on some different effect of a current, such as heating or electrolysis.

This kind of consideration has inspired one would-be general answer to the threat perceived to be generated by the theory-dependence of instruments. The suggestion is that no problems arise due to that theory-dependence, provided the theory being investigated differs from the theory presupposed in the instruments used. Thus, biologists can use electron microscopes and mass spectrographs because the physical theories presupposed are quite different from, and substantiated independently of, the biological phenomena those instruments are used to investigate. Peter Kosso, a philosopher who has developed this point about electron microscopes, refers to a state of affairs in which a theory being investigated in an experiment is presupposed in an account of the workings of the instruments or instrument/specimen interaction in that very experiment as "nepotism." Kosso roundly condemns the practice and goes on to argue that good experimental science must avoid it.[6]

Once again, scientific practice proves to be more versatile than is captured by such general recommendations. There are examples from good science where a striking match between theory, including the theory of an instrument, and the data produced by that instrument serves to vindicate both the theory and the instrumental data at one and the same time without vicious circularity. It has been well documented that Galileo's Aristotelian rivals doubted both the claim that Jupiter has moons and the veracity of the telescope when used to view the heavens. It has also been documented that it was the detailed match between the quantitative and qualitative consequences of the hypothesis that Jupiter has four moons and the telescopic data that convinced Galileo's rivals to accept both. How else could the detailed match be explained?[7] As we shall see in our subsequent study, this counterexample to what Kosso has termed nepotism, involving a significant and productive, rather than problematic circularity is by no means uncommon.

So far we have discussed the epistemological question of how the use of instruments can be justified. We have argued that justification sometimes involves an appeal to significant theory and sometimes does not, and that when it does, the nature of the appeal can take a variety of forms. A related second question concerns the nature of the historical path that culminates in a situation where an instrument can be confidently used. Is it theory or observation by way of instruments that leads the way? Once again, we insist that there is no general answer to that question. Sometimes it is instrumental probing and sometimes it is theory that leads the way. More usually, it is some complex and evolving interaction between the two. The introduction of the telescope into science and the justification of its use preceded, and did not require, an optical theory of the telescope. By contrast, it was theoretical considerations that paved the way for the construction of the first maser. Abbe's theory of diffraction led, eventually, to the construction of microscopes with improved resolving power. But once the improved microscopes were constructed and employed, Abbe's theory was not, and did not need to be invoked to in order to justify the veracity of the images produced.

If we consider instruments such as the mass spectrometer and oscillograph, the historical story is both more complicated and more typical in that theory and experiment appear to have advanced together. Discharge tube phenomena were not well understood in the mid-nineteenth century, nor did experimentalists have a great deal of control over them. An important element of theory necessary for understanding the phenomena was the Lorentz force equation specifying the force on charged bodies moving in electric and magnetic fields. That equation was arrived at theoretically by J. J. Thomson, O. Heaviside and H. A. Lorentz. The most direct way of vindicating it experimentally was by the deflection of cathode rays, *once it was appreciated that those rays were beams of charged particles*. The production of the rays themselves depended on progress in vacuum technology. So Thomson's famous deflection experiments of 1897 served simultaneously to reinforce the view that the Lorentz force equation applied to convection as well as conduction currents, and that cathode rays were indeed beams of charged particles, which also enabled the charge to mass ratio of the particles to be estimated, all in one stroke. Once the laboratory arrangements that constituted the phenomena to be investigated for the likes of Thomson were well understood and controllable, those very setups, in the guise of oscilloscopes and mass spectrographs, became instruments for investigating other phenomena rather than the phenomena themselves.

We have said enough to illustrate our point that the roles of theory in the design and use of instruments are manifold. Let us turn to a detailed study of early uses of the electron microscope. Even though the argument here is that plenty of good science has been done with the electron microscope without taking into account any electron physics, or any other theory pertaining to the instrument, in interpreting results it still might be useful to say a bit about how the electron microscope works. Then we can get on with trying to prove that often this doesn't matter!

A transmission electron microscope is made by directing a high-voltage electron beam from a cathode through an evacuated column, where the beam passes through a specimen and several magnetic fields which act as lenses, to a plate or luminescent screen at the far end. (Electrostatic lenses were also tried on a number of early microscopes.) The ultimate magnification obtained is a function of the focal lengths and positions of all lenses along the beam path, just as in a light microscope. With an electron lens that is an electromagnet solenoid, the focal length varies with current in the coil. In electron microscopes, focal lengths are generally varied electronically and the lens positions are fixed, whereas in a light microscope the opposite is the case, but the consequences are equivalent and as noted can be treated by the same optical theory. And because resolution of a microscope—the minimum separation between points distinguishable in the image—is limited by the wavelength of radiation used for imaging, electron microscopes are capable of much greater resolving power than light microscopes through the shorter wavelengths employable. In practice, by the early 1950s the better electron microscopes with accelerating voltages between 50 and 100 kV were obtaining a resolution on the order of 10 Angstroms (Å), several hundred times better than the best light microscopes. Most significant for the interpretation of results is the mechanism of image formation: the image on the screen of a transmission electron microscope represents the distribution of the scattering (without absorption) of incident electrons by atoms inside the specimen—that is, with amorphous specimens, a map of greater and lesser electron transmission in the specimen indicated by shadow and light on a plate or phosphor screen. The first electron microscopes emerged out of cathode-ray oscillograph technology during the 1930s; their construction was guided by rather finely developed electromagnetic and optical theory, and they started to become available commercially at the beginning of World War II.[8] But construction of an instrument and its use in scientific research are different matters as we already have noted with regard to Abbé and the light microscope.

Biology One: Rockefeller Reticulum

The endoplasmic reticulum (ER) was the first entirely new and unanticipated component of cellular anatomy to be revealed by electron microscopy. Because its novelty aroused skepticism, the case for its existence and properties was carefully worked out as an exemplar of sound electron microscopical epistemology by the main founders of what we now know as "cell biology"—Keith Porter, George Palade, and Albert Claude—at the Rockefeller Institute in New York. This entity first made its appearance in 1944, in one of the very earliest efforts to look at cells of higher organisms with the new microscope in the days before thin sectioning allowed imaging of tissue slices from whole animals. The specimens were intact, individual animal cells grown in tissue culture on supporting film, barely thin enough in their margins that some internal detail could be resolved. Porter and Claude observed a structure quite distinct from the larger bodies expected on the basis of light microscopy (for example, nuclei, mitochondria), a network or "lace-like reticulum" of filaments permeating what had generally been conceived as the structureless, homogeneous protoplasm of the cell body. "At higher magnifications," they could see "vesicle-like bodies, i.e. elements presenting a center of less density, and ranging in size from 100 to 150 mµ, . . . along the strands of the reticulum."[9] The best pictures of the reticulum came from cells preserved with osmium vapor, a traditional fixative for light cytology with the special advantage that this reactive metal heightens contrast in biological specimens. The initial reasons for taking seriously the possibility that the ER is real had to do with the similarity of its appearance in cytoplasm of osmium-fixed cells to that in the condensed, ill-preserved cytoplasm of cells prepared by another method, fixation in chromic acid. Moreover, the membranes implied in all "vesicular" structures are not easily explicable by the spontaneous aggregation of cytoplasmic particles, making its dismissal as an artifact harder.

For a year Porter used an electron microscope to look at whole-mounted osmium-fixed cells, cultured on supporting films—most of them from tumors of various kinds because such cells grow readily in vitro—and he found reticulum in all of them.[10] After another year, in 1947, Porter was convinced that tumor cells had a different arrangement of ER than noncancerous cells,[11] and that the vesicular structure of the small units along the reticulum implied a secretory activity.[12] Thus the ER was ubiquitous, and there appeared to be a correlation between appearances of the reticulum and certain physiological activities, namely secretion and rapid growth. The presence of the ER in all cells was one reason to believe simultaneously in its existence and its importance, since if a part of a living thing is essential to life it must be ubiquitous. Also, the fact that despite variations in preparative procedures something like it can always be found implies, by a basic inferential rule of biological electron microscopy that dates back to ninteenth century cytology (and of which Ian Hacking has made much), that it is probably not an artifact.[13] Only in osmium preparations of one kind or another, however, did Porter find the reticulum with the form he regarded as characteristic, which is only to say that he had more faith in osmium than in his other fixatives. And its appearance changed with increasing exposure to osmium fixation. This reliance on osmium could count against the reasoning from ubiquity, since the ubiquity of ER could be explained as a ubiquitous reaction of protoplasm with the fixative. It should be noted that treatments of organic specimens with reactive metals like osmium, popular because they increase the contrast of organic structures through differential uptake based on the different chemical composition of the structures, so drastically alter the distribution of mass in the specimen that any calculations of densities of different parts of the specimen based on scattering power as manifest in micrographs, even if practicable, would provide no relevant information about the distribution of mass in the specimen before metal staining. Some early biological electron microscopists with physics backgrounds did indeed find the urge to apply densitometric calculations to their micrographs irresistible— uselessly, as it turned out.

By early 1948 fresh grounds for belief in the ER's existence in living cells, besides its ubiquity in electron micrographs of whole cells fixed by various procedures, had been obtained from another instrument that was novel at the time, the phase microscope, which operates at visible light wavelengths but which produces contrast based on even minute differences in refractive index rather than color or opacity.[14] Similar appearances came from living cells viewed by dark-field microscopy in which objects even below the resolution limit reflect oblique light against a dark background. A mass of bright strands was visible by phase microscopy in the cytoplasm of living cultured cells, and these presumptive elements of ER showed no dramatic clumping or other change when observed during the course of osmium fixation. Naturally, the appearance of what was taken to be ER by phase and dark-field microscopy in live cells in culture differed greatly from the appearance of the object by electron microscopy in similar osmium fixed cells, but that was to be expected because the two kinds of microscope make images by such different mechanisms. Not that any particulars of the physics of image formation in either kind of microscope were involved in ascertaining properties of the entity in question. All that mattered was that the microscopes operated according to different mechanisms and that something identifiable as ER could be seen with both. It also mattered a great deal that with the phase microscope the specimens were living, doing away with the possibility that the ER seen by electron microscopy was entirely attributable to the effects of fixation or other postmortem change. This reasoning assumed that the dark strings and vesicles in fixed cells, visualizable in detail by the electron microscope, could be legitimately identified with the vague shadows and bright strands near the phase microscope's limit of resolution. Still, something ought to have been visible in the cytoplasm by phase microscopy if the reticulum is not an artifact of electron microscopic technique—and something was. Comparisons of all these kinds of images allowed something to be identified as ER in all, so observation of the entity by means of different kinds of instrument was added to the initial observations by one instrument (the electron microscope) of the entity in cells prepared by different procedures.[15]

The next change in Porter's interpretation of the reticulum came in 1950 or early 1951 around the time he and Palade were making their first plastic-embedded thin sections of tissue. (Plastic provides an embedment sufficiently strong to support sections cut thin enough to penetrate with an electron microscope's beam, which in turn makes possible the imaging of cells taken from an organism's body tissues where they naturally occur—not just cells growing independently under artificial culture conditions.) The reticulum in whole-mounted cultured cells had looked like a network of fibers dotted with vesicles; later, with different fixative techniques, like a network of tubes with wider vessels along it. In cells within sectioned tissue there were no filaments, and instead of a loose web of tubes and vessels the main membranous structures appeared to be closely packed tubules and sheets in varying arrangements. If these different-looking membranous structures could be identified with the ER as it appeared in whole mounted cells, it now appeared that finding the entity in cells prepared for electron microscopy from tissue (as opposed to cultured cells) confirmed its existence. And as though to answer doubts about its genuine existence (perhaps especially the worry noted above, that ubiquity might simply indicate a universal reaction between protoplasm and osmium), doubts perhaps sharpened by the reconceived character of the entity required for mapping it between whole mount and thin section electron micrographs, Porter now pointed out that in cells cultured and fixed together in the same dish, the reticulum looked different in different cells.[16] This nonuniformity can be much more plausibly explained by presuming differences in physiological state among the cells (for example, stage of cell division) and supposing a role for the ER related to those differences, than by supposing that an artifact induced by fixation in all protoplasm could be so terribly sensitive to even the subtlest physiological differences. So here was an intuitive case for the entity's existence, an argument to the best explanation that traded on an untested notion of the physical chemistry of artifact formation as crude and undiscriminating—the detailed theory of osmium reaction with cell constituents being woefully insufficient for any rigorous assessment of likelihood of artifact formation in this case.[17]

By 1951 the grounds for belief in the endoplasmic reticulum were of three types. There was evidence from dark-field and phase microscopy that changes gross enough to see with light are not induced during osmium fixation of cultured cells, which vindicated fixation procedures to some limited extent. There was the appearance in living cells under dark-field and phase microscopy of cytoplasmic structures at least consistent with the ER seen in electron micrographs of fixed cells. And there was the reticulum's presence in both thin section and whole mount electron micrographs (a consilience weakened by the fact that osmium fixation was necessary for its clear demonstrability in both, by the finding that its appearance differs with varying osmium exposure, and perhaps also by the difference in its appearance in these two sorts of picture, but strengthened by the argument from different appearances in similar cells under identical fixation conditions just mentioned). If these arguments now appear compelling, it is with benefit of hindsight, but taken together they did have some force. And more important, perhaps, the evidence continued to mount.

The progress in the next few years that made the endoplasmic reticulum's existence convincing to the general community of life scientists involved two additional lines of evidence: identification of the viewed entity with a fraction isolated by centrifugation and characterized chemically, and correlations with function according to the traditional logic of anatomy and physiology. By early 1952 Porter and Palade had done substantial exploratory work along the latter lines, searching for ER in the many animal tissue types that had suddenly become accessible to the electron microscope through the new thin sectioning preparation procedures. Among other things, they saw that the cell types with the greatest volume of ER tended to be those that were devoted to producing protein-rich substances, for instance secretion cells in glands.[18] This suggested that the reticulum is involved in protein production and/or export. This sort of morphological logic, which makes sense of an organic structure by finding correlations between variations in its appearance and covariations in physiological function, not only explains what a structural entity is in terms of that correlated function but also (as one of the authors, NR, elsewhere has pointed out) it implicitly reinforces belief that the entity exists. The force of this inference depends on the intuitive implausibility of the converse supposition already noted, viz. that an artifact would coincidentally show variations in structure so closely attuned to physiology.[19]

From 1952 the morphological case for the endoplasmic reticulum was carried further by linking the ER seen by electron microscopy in cultured cells, and more importantly in the new thin sections of tissue, with the masses of information from a century of light microscopy on stained and sectioned tissue carried out by classical histologists and recorded in anatomical atlases.[20] To forge this link the reticulum had to be identified with some part of the cell apparent in classical light microscopy preparations. Once one knew which elements in the preestablished picture of cells corresponded to the electron microscopic reticulum, the old light microscopic observations could be counted as confirming to some extent the existence of the ER roughly as pictured, at much higher resolution, by electron microscopic technique, and one could also use this light microscopical evidence to cast light on the function of the ER. To accomplish the requisite mapping of electron microscopic observations to light microscopic ones, Porter compared similar cultured cells stained for light microscopy with cells fixed in osmium and imaged by electron microscopy. (The convenient experiment encountered by Ian Hacking on his laboratory tour, in which the self-same specimen is imaged by electron and light microscopy, is usually impossible due to necessary differences in preparation procedures and/or destructive effects of the instruments on the specimens.[21]) It was much like the 1944–45 experiments where ER was first observed, only far more systematic. Porter found that the parts of his cultured cells that appeared rich in ER by electron microscopy corresponded roughly to regions that could be seen by light microscopy to take basic dyes like toluidine blue, and that the fine structure of the distribution of basophilic material appeared to correspond, within the resolution limits of light microscopy.[22] It thus appeared that the light microscopic basic-staining material, familiar to classical histologists using such stains, might be identical to the electron microscopic endoplasmic reticulum.

But to seal this identity claim it was necessary to compare *tissues* prepared with basic stains for light microscopy with the same tissues in electron micrographs because the established light microscopic literature dealt with tissues, not cultured cells. Thus one first had to know what the reticulum looked like in tissue sections (and that it actually existed in tissue). We will not elaborate here on the prerequisite evidence Porter and Palade generated to establish that what they called the endoplasmic reticulum in electron micrographs of whole mounted tissue culture cells was in fact the same as the different-looking thing they identified as ER in electron micrographs of sectioned tissue. Suffice it to say that they made sections of cultured cells to compare with sectioned tissue cells.[23] With this evidence finally in hand, Porter and Palade could compare electron microscopical thin sections of a given tissue with traditional histological sections of the same tissue stained with basic dye, with high confidence that what they were describing as ER by electron microscopy was still the same entity. In these comparisons they found that in tissue cells with much basophilic cytoplasm the quantity of endoplasmic reticulum was also great, that in tissue cells with a certain distinctive distribution of basophilic material the pattern of reticulum in thin section was similar, and that when a distinctive distribution of basophilic material was altered by treatment of the animal with drugs the appearance of reticulum in thin sections was altered in similar ways.[24] Light and electron microscopic appearances could now be correlated reliably.

There was another line of investigation into the endoplasmic reticulum pursued by separating the contents of homogenized cells in an ultracentrifuge and analyzing the biochemical properties of the fractions in the test tube. Indeed, this biochemical line of research into cell structure at the Rockefeller Institute predated the electron microscopical one. The cell fractionation work actually gave rise to the electron microscopic cytology already described because it was necessary to counter doubts that fractions were natural kinds of cell components, as opposed to arbitrary and heterogeneous collections of cell fragments. Thus some of the first specimens Claude and Porter looked at with an electron microscope were fractions from the ultracentrifuge to show that different fractions contained distinctly different entities. So, in the first instance, cell fractionation experiments did not validate the existence of the endoplasmic reticulum but the reverse: images of the endoplasmic reticulum supplied evidence that a certain fraction (the "microsome fraction") corresponded to a distinct entity existing in cells before fractionation procedures. Much later (from the late 1950s through the 1960s), once the microsomes and reticulum were solidly identified with one another, fractionation biochemistry supplied lots of information about how the ER functions in protein synthesis and transport.[25] Yet all of this biochemical evidence from fractions cannot be said to supply additional reasons for believing in the ER as pictured by electron microscopy, except by the implicit logic mentioned above: viz., since the behavior of the microsomal cell fraction show such good correlations with physiological changes (for example, nutritional conditions), which often correlate also with changes in the appearance of the ER by electron microscopy, it is implausible to suppose that it might be an artifact. Even entities showing such consistent behavior are sometimes rejected as artifacts.[26] At any rate, the ultracentrifuge work lies largely outside the scope of this chapter because the flow of validation, as to the existence and form of the ER, went primarily in the other direction.

To summarize, then, the existence, form, and possible function in protein traffic of the endoplasmic reticulum was established with the electron microscope using epistemologies of experiment that hinged on comparison between observations, not on theories of the instrument and how information is transmitted by it. Electron images of cells were compared with phase and other light microscope images; comparisons between electron microscope images of cells prepared in various ways were carried out; and comparisons of the ER in electron images were made from cells with various anatomical and physiological differences, some produced by intervention. The experimental logic no doubt derives from prior biology using microscopes, especially pathological and physiological anatomy. At no point was anything gained by calculating or otherwise deducing properties of the ER from its images using the physical theory of electron optics and specimen-beam interaction. The electron microscope and its characteristics remain fully in the background, in reasoning from differences between observations made of various objects or the same object under varying conditions, just where they belong (provided there is no reason to suppose that the microscope behaves differently from observation to observation). Porter's epistemology is much like that Galileo employed in establishing the existence of Jupiter's moons—only Galileo had no second type of telescope by which to plot their position on different nights, and no way of staining his objects. What matters is essential agreement among separate observations and the coherence of subtle differences (in position of moons, in form of ER) against that background of agreement.

Biology Two: Muscle at MIT

It might be argued that the work on endoplasmic reticulum described above was not strictly speaking experimentation in that only the existence and form of an entity was established by a glorified sort of mapping procedure. Although we would resist the notion that experiments always have to address theory, we next offer another story from biological electron microscopy in which theory was the preeminent concern. This second biological case describes the role played in the elucidation of the mechanism of muscle contraction by Jean Hanson, Hugh Huxley, and several other researchers associated with Francis Schmitt's group in the Biology Department of MIT in the mid-1950s. A theory of muscle contraction was elaborated and along with competing theories tested against electron microscopic observations, without significant recourse to theory of electrons or electron microscope function.

Some background on muscle physiology is necessary here. Since the beginning of the century, the dominant theory of muscle held that it contracts through a conformational change in long protein molecules within it, which would knot or wind to assume a more compact shape and thus shorten, probably driven by a change in pH.[27] From the 1920s through the 1940s, the major protein myosin had been extracted from muscle, and a large amount of biochemical work had accumulated around the problems of how these proteins were combined in muscle, and how during contraction changes in chemical energy were related to the conformational change of these hypothesized protein fibers. But even more than the biochemistry of energy transactions in muscle and the associated mass of confusing data, visible changes in the banding pattern of striated muscle tissue during contraction were the greatest puzzle that theories of muscle aimed to explain. The relaxed skeletal muscle of vertebrates shows striations with different staining and polarized light characteristics, the chief ones are called A (anisotropic) and I (isotropic) bands, the I band being marked across the center by a dark Z line and the A by a lighter H zone. On contraction the Z line remains the same while the H zone disappears, and the A band grows relative to the I and apparently at its expense: it looks as if the A splits and each half moves toward its adjacent Z, impinging increasingly on the I zone until at maximum contraction this may altogether disappear. In essence all of this was observed in the nineteenth century. Other bands are also visible under certain conditions, and considerable attention was devoted to these appearances as well as to the higher order architecture of the fibrils and fibril bundles making up the muscle fibers, but this level of detail will suffice for the present story. By the early 1940s, it was generally accepted that there were long molecular chains containing myosin stretching great distances in muscle, and that along these filaments regularly repeating complexes between myosin and other substances were responsible for the appearance of the bands seen by light microscopy; but exactly how these molecules could account for the banding patterns was mysterious.

By the end of the 1940s, new methods and findings were dramatically reinvigorating research into muscle. Significant developments included the wartime demonstration by biochemists that what had been considered the single myosin protein was two separately purifiable proteins, one of which retained the name "myosin" while the other was called "actin." In the test tube, both proteins would reassemble as filaments, either separately or in a complex if mixed. Soon after the war, x-ray diffraction results indicated that the pattern of muscle was nothing but the sum of the patterns of pure actin and myosin lying along the same axis, implying that both proteins exist as a combined filament or parallel filaments in muscle's native state. Wartime electron microscope studies of Schmitt's own group had indicated that muscle contains indefinitely long "myosin filaments" stretching continuously through all the repeating bands.[28] Schmitt's group even found evidence, in favorable electron micrographs, of a "knotted or beaded appearance" along the fibrils as the theory of contraction by folding or knotting would predict. A knotting theory of contraction was gaining credibility from this early electron microscopical work, and from early postwar results in both electron microscopy and biochemistry, though some data were pointing to the possibility that the main, indefinitely long fibrils in muscle were actually actin. Around 1950 Schmitt's version of the standard model was apparently being vindicated by most of the researchers in this suddenly active field, at least in its very general outlines, though contentious questions remained such as the dimensions of the primary filaments, their composition (actin or myosin or both together? was there a real periodicity within the filaments?), and also their length and continuity. But there was still general agreement on a single main type of filament that contracted through conformational shift.[29]

With the introduction of plastic embedding in 1949–50, Schmitt's lab, like Porter's and other groups, turned great efforts to making thin sectioning work and to seeing what it would reveal.[30] This technical change drove an intensification and diversification of work on muscle fine structure, yet no convergence on the controversial issues was forthcoming over the following few years. Indeed, new findings appeared to confuse the picture with novel possibilities such as the spring-like elastic segments seen by some in the primary filaments.[31] So it was in late 1952 when Schmitt's group was joined by Hugh Huxley, followed in early 1953 by Jean Hanson, both postdoctoral visitors from England who wanted to imbibe electron microscopic skills at the source. Using fresh muscle fibrils prepared a certain way so that they can be made to contract slowly at will, Hanson had been using visible light to observe banding changes during contraction with a newly improved phase microscope in London.[32] Huxley's thesis work at Cambridge had compared x-ray diffraction patterns of muscle in relaxed and rigor mortis states, confirming findings that the major outlines of molecular structure in the muscle were not much changed in contraction, and that they could be interpreted as the shift with respect to one another of myosin and actin filaments which remained parallel, rather than as the reconfiguration (folding, winding) of the protein filaments themselves. Huxley then had hypothesized that there were arrays of two kinds of filaments, actin and myosin, which shifted from a less

to a more ordered form of packing on contraction.[33] Soon after arriving at MIT Huxley made transverse (crosswise) thin sections of muscle which showed, end-on, the expected regular hexagonal array of two kinds of filaments in the A band, except in the H zone which had only the thicker filaments.[34] Thus Huxley was exploring the idea of two sets of parallel filaments that might move relative to each other, but the filaments' chemical identification and changes during contraction were uncertain. When Hanson arrived at MIT, she and Huxley began a period of intensive collaboration aimed at correlating light and electron microscopy. Light microscopy on unfixed biological material could explore the dynamic aspect of muscle contraction or changes during biochemical manipulation, while electron microscopy could reveal microstructural detail at each step.[35]

Their joint experiments clinched the argument that the "A substance" was myosin and that the main lattice of filaments, present in both A and I zones, was made of actin. Watching individual isolated muscle fibrils with both phase and polarized light microscopy during the treatments that biochemists used to extract myosin from muscle, they were able to see the dark, anisotropic appearance of the A bands disappear before their very eyes. Electron microscopy of such extracted muscle confirmed that the thick filaments Huxley could find only in the A bands had disappeared. Thus muscle is made of thick and thin filaments, the thick filaments of myosin present only in the A bands, the thin filaments of actin extending from the Z line to the middle region of the A band, where they were presumably linked to the actin filaments of the next Z line by finer connector filaments across the H zone (as yet invisible, but necessary to explain how the fibril stayed together after myosin extraction). Exactly what happens in contraction, however, was still not yet clear from this structural interpretation.[36] To test the idea that the I band shrinks during contraction simply because the myosin filaments in the A band draw the actin filaments in, Hanson and Huxley needed to see whether the myosin filaments of the A band remain unchanged in length, and this could not be measured precisely by electron microscopy because of magnification calibration problems and variable degrees of shrinkage introduced by fixation of the muscle specimens. So the two went on to make phase contrast pictures of individual glycerin-soaked muscle fibrils, which were made to contract slowly, showing that during contraction the A bands remain constant in length, all shortening being at the expense of the I bands. Furthermore, Hanson and Huxley found that when myosin is extracted from fibrils at various stages of contraction, no further contraction could be induced and the entire fibril has the appearance of the I zone in an unextracted specimen.[37] Thus it was concluded that the A bands are made of myosin filaments (the thicker ones in electron micrographs) of constant length, overlapping thin actin filaments which are much longer, and that both filaments must be present for contraction to occur. The new Hanson-Huxley theory of the mechanism of contraction held that the actin filaments move along the myosin filaments, pulling the Z lines anchoring them closer together. This "sliding filament" model of contraction not only had the explanatory power to account for banding changes in skeletal muscle but also brought into agreement the x-ray crystallography and electron microscope results indicating a dual filament array, most biochemical data on muscle and the constancy of A band width observed by light microscopy on stained or live material viewed with special optics.[38] Apart from direct evidence of two filament types, most of these data could also be made to fit with one or another variant of the older, single-filament conformational shift model.

For the present discussion we will leave aside the simultaneous publication of a sliding filament model produced by physiologists employing only light microscopic techniques.[39] It is unlikely that such a model would ever have been widely accepted without the direct visualization of two distinct species of parallel filaments that electron microscopy allowed. Even electron microscopists were not immediately converted, however, to sliding filaments by Huxley's pictures. At the July 1954 International Conference on Electron Microscopy in London, for instance, Hanson and Huxley presented new transverse thin sections showing more sharply the compound array of large (110 Å) and small (40 Å) dots in the A band, that is, end views of the thin and thick filaments.[40] But other electron microscopists present there saw different things in their own muscle pictures due to a combination of different specimen choice and preparative procedure. In particular, many were not convinced that the thin filaments were real, which left open the traditional interpretation of indefinitely long filaments of one type, the actin backbone decorated at intervals with myosin.[41] Over the next two years Huxley and Hanson answered their electron microscopist critics mainly by showing, by careful preparative technique, the distinct thin filaments that others had not been able to find in sections of various types of muscle tissues.[42] They were building the case for the sliding filament model by showing the structure entailed in it to be ubiquitous in muscle. There is no need to follow these debates in any detail since the sort of epistemology involved has already been described in the case of the ER. But what of the conflicting appearances of single, continuous filaments that other electron microscopists were getting? In 1957 Huxley effectively clinched the case with a paper not only showing thin filaments between the thick A band filaments in the same muscle where some others had seen only a single filament type but also one-filament views such as their critics obtained by deliberate degradation of his micrograph resolution.[43] Lingering doubts gradually disappeared as more microscopists were able to distinguish the two filament types. By the early 1960s debate centered on the mechanism of sliding between the two sets of filaments rather than their bare existence.

The sliding-filament theory triumphed over the rival conformational shift theory because increasingly detailed microanatomy revealed structures of muscle more consistent with it than with its rival, and because it had superior explanatory power in accounting for such larger scale phenomena as the changing banding pattern during contraction. In establishing a sliding filament model's greater consistency with observable microstructure, the demonstration of a second set of finer filaments in muscle tissue after many variations on preparation procedure for thin sectioning was especially crucial. The pivotal experimental work establishing said explanatory power involved correlating muscle contraction as observed in real time by light methods, on individual muscle fibers, and fine structure as observed in thin sections of muscle fiber by means of electron microscopy. In both kinds of experiment, the methodology is essentially comparative, just as in the case of the endoplasmic reticulum, only here most of the comparison is of similar specimens (muscle fibers) in various controlled experimental conditions of contraction or preparation, rather than comparisons among the many tissue types offered by organic diversity as natural experiments. That is, here the method is essentially physiological rather than anatomical. In the strategy of correlating stages of muscle contraction and appearances by electron microscopy and light microscopy techniques, as in that of comparing observations of similar specimens made by different preparation methods, the electron microscope and the theory of its operation again remain firmly in the background (or, as phenomenologists might say, bracketed). With no reason to suppose that variations in the function of the instrument affected the observations being compared, microscope function was a constant that could be cancelled from both sides of any comparative equation.

SOLID STATE PHYSICS: CRYSTALS OF CAMBRIDGE

The past two examples were contributions of the electron microscope to biology. It might be suggested that the experimental reasoning of life scientists differs from that of physical scientists (whether due to biology's imprecision, lack of articulated mathematical theory, different and generally more complex nature of subject matter, or what have you) in de-emphasizing the theory of instrumentation. Or perhaps, following Duhem and Kosso after him, it might be argued that biologists don't need to worry about physical instruments because the theory of such instruments does not overlap the biological theories they are testing. We now will counter these lines of argument by offering the example of an important electron microscopic contribution to solid-state physics. In this episode we see at work some of the theory-independent approaches to validating experimental conclusions that were in evidence in the biological examples. We will also see theory of the instrument and of specimen/instrument interaction playing a role, but in ways that are not captured by standard philosophical accounts, and which shows ways in which a "nepotistic" match between theory and otherwise difficult-to-interpret observation can serve to validate both at the same time. Before making our case about the roles of theory, we will begin with a bare description of what was done in his 1956 experiment by James Menter of the Tube Investments Research Laboratories in Cambridge, England.[44]

Menter prepared crystals of various metal derivatives of the organic molecule phthalocyanine, ground and suspended in ethyl alcohol. Drops of the suspensions were evaporated on specimen grids covered with a supporting film riddled (intentionally) with a large number of holes. Small portions (to minimize heating) of the crystals were then examined in an electron microscope operating in transmission mode at a beam energy (80 kV), corresponding to a wavelength of 0.0417 Angstroms. Transmitted electrons were focused in the usual way and photographed at a magnification of 77,000×. The best pictures were those formed by electrons passing through holes in the supporting film covering the crystals so that the image was not overlaid by the structure of the supporting film. Many of the images showed well-defined parallel lines or bands, and these were interpreted as images of planes of molecules within the crystal. Edge dislocations appeared as the boundaries across which the parallel lines showed a small change in orientation. Imperfections interpretable as direct evidence of screw dislocations were also directly visible in the images. These were the first "direct" observations of dislocations. They had previously been predicted on theoretical grounds to explain why crystals were weaker and subject to a greater degree of plastic flow than theory predicted for regular crystalline structures.[45] There was also indirect macroscopic evidence, in the form of observable edge pits, presumed to correspond to edge dislocations and spiral growth patterns presumed to arise from screw dislocations.

Now we will discuss the role theory did or did not play in leading to, interpreting, and justifying these findings. To begin with, what motivated the experiment, theory or unexpected observation? Menter himself describes what led him to realize the possibility of using the transmission electron microscope to observe dislocations in crystals. It was a chance observation he made while using the electron microscope to study crack propagation in glass.

One micrograph showed a fine crack with a highly irregular path, and the width of the crack at its tip was of the same order of magnitude as the resolution limit of the microscope (better than 10 Å). This observation suggested that with this resolution, one might, by choosing a suitable crystal, be able to follow the propagation of a crack through a regular lattice provided the lattice was resolvable in the electron microscope. Indeed, if the lattice of a crystal was resolvable, it should be possible to observe directly various types of imperfection in the lattice and to study directly their relation with all those properties of crystals that have been shown, hitherto only inductively, to depend on the presence of imperfections.[46]

It is certainly the case that Menter's starting assumption, that there are dislocations in crystals to be looked for and that they have an important bearing on basic physical properties of those crystals, was based on theory. At the time, the relevant theory had been only indirectly borne out by macroscopic observations, but well enough for physicists, if not metallurgists, to be firmly convinced of the existence of dislocations and their role in determining properties of solids, especially the ductility and plasticity of metals. (Some of the physicists involved in the early work on dislocations, when interviewed in the 1980s, reported a reluctance on the part of metallurgists to accept the existence and role of dislocations prior to their direct observation.[47]) Speculations about the limit of the resolving power of the microscope were also known to Menter. But the crucial realization that a transmission electron microscope might be used to view dislocations directly was not deduced from theory. Menter simply viewed the image of a narrow crack, and, knowing the magnification of his microscope, was able to recognize that at its narrowest point, the crack imaged must have a width of the same order as the spacing of molecules in crystals, the latter magnitude being measurable by x-ray diffraction experiments. The crack shown on the plate could have been an artifact of the microscope rather than a feature of the glass under investigation. Menter did not think so, and immediately set about the attempt to observe dislocations. Here it was important to establish that the bands on the image did directly represent molecular arrangements rather than artifacts or some more indirect effect. Menter was able to demonstrate convincingly that this was the case using a variety of strategies.

Menter interpreted the observable bands on the plates as images of planes of molecules in the platinum phthalocyanine crystals. More specifically, he interpreted them as images of the (201) planes. He justified that interpretation in a number of independent but mutually reinforcing ways. First, Menter was able to calculate the spacing of the planes presumed to be imaged on the plates simply from a knowledge of the magnification of the microscope. Magnification was measured directly in the standard way by including synthetic latex particles of known size (3650 ± 80 Å) with the specimen, yielding a micrograph magnification of 77,000 × with an accuracy of about 2 percent. Using this value for the magnification, the spacing of the (201) planes was found to be 12.01 ± 0.2 Å. This compared favorably with the value calculated from x-ray diffraction data and Bragg's theory of the 1930s, namely, 11.94 Å. The fact that independent measurements based on different processes and instruments, x-ray diffraction and electron imaging, yielded the same spacing counts as strong evidence that what was measured related to something in the crystals and was not an artifact of each of the quite different measuring techniques, in line with the methodology we have seen at work in the previous two examples. But this was by no means the end of Menter's case.

The next point requires an appreciation of the difference between a diffraction pattern and an image. It will help to begin clarifying this distinction with reference to an optical diffraction grating. If a grating is illuminated perpendicularly with a parallel beam of monochromatic light, then the light emerging from the grating on the other side will be diffracted so that a region of a screen placed in the path of the transmitted beam will be illuminated over an area that is much greater than the area that would be illuminated by the original parallel beam passing undeflected. The parallel lines on the grating will give rise to bright and dark bands on the screen. A bright band is formed at a location on the screen when light from each of the gaps in the grating constructively interferes when reaching that location, while dark bands occur when the contributions from each gap destructively interfere. It is important to recognize that the diffraction pattern on the screen is not an image of the diffraction grating. Light from each of the gaps contributes to each of the bright bands on the screen. Should one of the lines on the grating be imperfect, this will not show up as an imperfect band on the screen. Rather, it will cause a blurring of the pattern as a whole. Though a diffraction pattern is not an image of the grating, an image of the grating can be formed on the screen simply by placing a suitable lens between grating and screen. The relationship between object and image distances will be related to the focal length of the lens in the way dictated by geometrical optics.

Turning now to x-ray diffraction, diffraction patterns can be formed by diffraction through crystals with the regularly spaced planes of molecules playing the role of the grating; however, images cannot be formed simply because there is no effective way of focusing x-rays (or at least, there wasn't in

Menter's day). Thus it is possible to calculate with some accuracy the spacing of planes in crystals from the spacing of the bands in an x-ray diffraction pattern, but it is not possible to form images of crystals using x-rays. The beam of an electron microscope can be used in the same way, making a diffraction pattern closely analogous to that made in x-ray diffraction. Electrons, in contrast to x-rays, however, can readily be focused by lenses such as those in a transmission microscope so images as well as diffraction patterns can in principle be formed. It was this possibility that Menter was able to capitalize on.

Let us return to Menter's line of argument correlating diffraction patterns and images of the platinum phthalocyanine crystals. With the microscope arranged to observe diffraction patterns rather than images, Menter illuminated in turn various regions of the specimen support on which crystals had been deposited. Some of these illuminations resulted in diffraction patterns characteristic of diffraction from (201) planes, indicating that the crystals illuminated in those cases were suitably oriented. (This conclusion was itself reinforced by macroscopic observation of these crystals, deducing their orientation from the crystal structure and its relationship to observable crystal habit that had been worked out in detail by Robertson).[48] Menter then operated the microscope in image-forming mode and found that the regions of the grid which, when illuminated, led to an image of bands corresponding to a 11.9 Å spacing corresponded precisely to those regions that had yielded diffraction patterns caused by diffraction from (201) planes. For Menter this was sufficient to relate the structures he imaged to structures in the crystals: "the fact that a crystal showing the (201) reflection in the diffraction pattern also shows the lines [in the image] 11.9 Å apart, together with the crystal habit data of Robertson, was taken as conclusive evidence that the line structure is associated with the (201) planes."[49]

A third cross-check involved physically bending a crystal and noting that the lines in the image themselves bent accordingly. (This is reminiscent of Hacking's slogan "don't just peer, interfere," invoked to stress the importance of controlled practical intervention in the case of optical microscopy.)[50]

Menter had powerful arguments, then, for treating the bands in his diffraction images "in first approximation as a projection of the sheets of molecules in the (201) planes." Having established this, it was a straightforward step to interpret interruptions in the regular pattern of the bands as images of dislocations. None of the arguments cited involve a detailed account of the interaction between the electron microscope and the crystal specimens, although they do presuppose an account of the atomic structure of crystals and how regular arrays of them can give rise to Bragg reflection. This was necessary, for example, to make possible the calculation of the separation of crystal planes from the spacing between the observable interference fringes of x-rays diffracted by them, and to a lesser degree, to appreciate the extent to which x-ray diffraction, electron diffraction, and electron microscopy are different processes, for it is that extent which gives force to the arguments based on the agreement between results arising from the different processes. In the second half of his paper, however, Menter did propose and substantiate a theoretical account of some aspects of image formation. In effect, he simply followed a hint implicit in some work of Hillier and Scherzer, transferring the Abbé theory of the optical microscope to the electron microscope.[51]

When a light wave passes through a small aperture it is diffracted into what would be the shadow region were the wave to be undiffracted. The amplitude of the diffracted wave varies with the angle of diffraction. There is a peak in amplitude in the straight-through direction, and a number of subsidiary peaks, the first order, second order peaks, and so on, whose amplitude decreases as the angle with the straight-through direction increases. The angle at which these peaks occur is a function of the width of the aperture and the wavelength of the light. Abbé pointed out that if the diffracted waves are to be brought together to form an image, then ideally all of the diffracted light, including that corresponding to the peaks of the various orders, should be integrated into the image. A loss of any one of the peaks results in a loss of information and a consequent blurring of the image. An absolute prerequisite for the formation of an image is that the first order peak be included. If the straight-through wave only is focused, no image results.

Menter applied these considerations of Abbé to the electron microscope, treating the electron beam as a wave and the thin crystals as analogous to a diffraction grating. The appropriateness of this could be tested in a straightforward way simply by varying the size of an aperture intercepting the beam emerging from the specimen. An aperture of diameter 10 μ was sufficient to cut off the first order diffraction peak for diffraction from the (201) planes, and with this setting no image was formed and no bands appeared. With the aperture increased to 30 μ the first diffraction peak was collected and the image complete with bands duly appeared. So here a theory of the working of the instrument is tested by experiment and duly supported. Note, however, that to argue in this way, as Menter certainly did, is to presuppose that the bands do indeed represent images of the crystal planes, which is precisely what Menter was at pains to justify in the first half of his article. We appear to have a clear case of a match between theory of the instrument and an interpretation of observations serving to validate both, that is, a clear case of nepotism of the kind forbidden by Kosso.

That this is indeed the case is brought out when we look at the details of one of Menter's arguments, one that he himself considered to be particularly telling. The argument involved electron images obtained using bent crystals. When bent platinum phthalocyanine crystals were imaged with the objective aperture set at 10μ a series of broad, diffuse fringes (called Bragg fringes) were obtained. They arise as follows. Because the specimen crystal is bent, only some regions of it will be oriented at the appropriate angle for Bragg reflection. Much of the intensity of the beam emerging from those regions will be concentrated in the first order, second order, and so forth diffracted beams, and these will not be collected by an objective aperture of 10 μ for reasons we discussed above. Consequently, the intensity for those regions is depleted and the image on the screen correspondingly dark. Hence the fringes. Suppose now that the aperture is widened to 30μ. The diffraction beams of the second and higher orders will still not be accepted, so the dark fringes will still appear (although not as dark as with the smaller aperture which cut out the first order beam as well). But now, with the first order diffraction beam accepted, an image of the crystal planes giving rise to the diffraction can be formed. These duly appeared in Menter's images, superimposed on the diffuse Bragg fringes. The opening sentence of the paragraph in which Menter reports this result presents it as "evidence confirming the general theory of the formation of the image [which] has been obtained from a study of the broad diffuse fringes frequently observed in bent crystals."[52] So the formation of these images superimposed on fringes in just the way the theory of the instrument predicts is a striking confirmation of that theory. But since the whole discussion presupposes Bragg diffraction by crystal planes, the result can equally well be interpreted as evidence for the interpretation of the images as representing those planes. And this, in effect, is precisely what Menter concludes in the final sentence of the very same paragraph. "This experiment proves conclusively that it is only when the crystal is in a suitable orientation for Bragg reflection to occur that the closely spaced lines are seen," with the implication that those lines are images of the planes that are "suitably orientated."[53] We could hardly ask for a clearer example of what Kosso calls "nepotism." And yet, there is nothing wrong with Menter's argument in this highly regarded paper. It is a particularly compelling one, both for the interpretation of Menter's images and for his theory of image formation. Certainly his scientific peers accepted his case on both counts. For instance, in the summarized proceedings of a conference on electron microscopy in 1956, at which Menter presented "a most stimulating description . . . of his work on the direct observation of crystal lattices and their imperfections," we read simply that "[i]mperfections in the lattices were demonstrated in the micrographs." But we also read that, as far as Menter's theory of the formation of images is concerned, "[e]xperiment has justified in general terms the validity of his approach."[54] Kosso's strictures are quite out of keeping with common scientific practice and with some of the rather straightforward epistemological strategies we have described here.

Menter's paper is instructive in part because it involved some theoretical considerations about the functioning of the electron microscope. A second important paper reporting the direct observation of dislocations, by P. B. Hirsch, R. W. Horne, and M. J. Whelan working at the Cavendish Laboratory in Cambridge, submitted for publication just three months after Menter's, contains no theory of specimen-instrument interaction beyond assuming the received account of Bragg reflection from lattice planes, but their results were impressive and persuasive nevertheless.[55] Parts of the case made by the authors that their images did indeed represent dislocations (in this case in aluminum foil) resemble strategies used by Menter. They refer to congruencies between the electron microscope images and x-ray diffraction results, and they make practical interventions to test that their images behaved in accordance with their being formed by Bragg reflection from and diffraction by crystal planes. For instance, it is a consequence of the Bragg theory that the position of the images will remain unchanged when either incident beam or specimen are tilted by angles less than a degree or so. Hirsch and the others did not give the theoretical detail, but Menter did in his article.[56] This prediction was tested and borne out. Hirsch and his coauthors conclude that such experiments "show that the visibility of the lines is due to Bragg contrast, and that they represent a definite property of the specimen."[57] Once again, a congruence between theoretical treatment and observation helps to bolster both the former and the interpretation of the latter.

A telling feature of the results that went beyond Menter's findings was that the dislocations were observed to move, and moreover, move in precisely the way that prevailing understanding of dislocations predicted they should move. The presentation of the results at the 1956 conference referred to above included a film showing the movements. As Braun remarks "even metallurgists were forced to believe in dislocations now."[58] (The movement of dislocations was crucial. Such things as the ductility and plasticity of metals were to be explained in terms of the transmission of dislocations through a solid, since the forces between adjacent layers of atoms or molecules are much too strong to permit easy relative movement of those layers.) Witnessing the dislocations move exactly as expected was clearly persuasive. As Hirsch and his coauthors put it, the "behaviour of these moving lines is identical with that expected of dislocation lines," so that these observations "leave little doubt that the lines represent single dislocation lines." At the same time this feature of the experiment "represents direct proof of the Mott-Frank screw dislocation mechanism in cross slip."[59] A coincidence here between a detailed prediction of the theory of dislocations and the observed features of electron microscope images serves to vindicate, at one and the same time, both the theory and the relevant interpretation of those images, analogous to the way Galileo's observations of Jupiter served to vindicate both the claim that Jupiter has moons and the veracity of the telescope. No theory of the instrument is required in either case.

The Hirsch group at the Cavendish did eventually develop a detailed account of the interaction of instrument and specimen to good effect, as we shall see. But before we discuss that aspect of their work, it is instructive to clarify the relationship between their experiments on aluminum and Menter's earlier ones. The platinum phthalocyanine crystals studied by Menter were not of great interest in themselves. They were selected because they were well suited to Menter's objective of resolving lattice planes using the electron microscope. In crystals of platinum phthalocyanine, the platinum atoms, with relatively high atomic number, are at the center surrounded by the atoms of the organic component with much lower atomic number. Scattering from the platinum atoms is much more intense than by the others, as x-ray diffraction had already revealed, so that, as far as scattering experiments are concerned, the crystals could be considered as regular arrays of platinum atoms. The surrounding atoms, however, served to increase the spacing of the platinum atoms, which is why Menter considered himself to have a good chance of resolving them. As we have seen, the 12 Å spacing of the (201) planes lay just within the resolving power of Menter's electron microscope. Hirsch's interest lay squarely with metals, rather than with their derivatives, having studied them initially by way of x-rays ever since he had first come to the Cavendish as a research student in 1946. Apart from their obvious technological importance, metals were at the center of theoretical interest. The spacing of atoms in pure metals, however, is considerably less than the 12 Å separating the platinum atoms in Menter's crystals, so it was not at all clear that dislocations in pure metals would be observable with the electron microscope. The theoretical conjecture that led Hirsch to believe they might be observable, and which gave him the encouragement to turn to the experiments we have described, turned out to be mistaken.[60] The fact that the experiments were successful nevertheless adds support to the case we are making here.

As an aside, we note that it may be due to the special interest and importance of metals that Ernest Braun, in his article on the history of dislocations, overlooks Menter's experiment completely and attributes the first observations of dislocations to the Cavendish group. Hirsch himself acknowledged Menter in 1980, in some recollections of the period, referring to his "beautiful pictures of dislocations in platinum pthalocyanine."[61] From those recollections it appears that Hirsch first examined their aluminum foil looking at electron diffraction patterns. It was Menter's suggestion that they try micrographs! (Menter had been a research student at Cambridge, in the Research Laboratory for the Physics and Chemistry of Rubbing Solids, which was only later moved to the Cavendish, at which time the "Rubbing" was dropped from its name. Menter moved the Tube Investments Research Laboratories in Hinxton Hall, also in Cambridge, in 1954.)

As we have noted, having achieved their initial experimental success, Hirsch and his collaborators turned to a theory that would show how their achievement had been possible. In his theoretical treatment, Menter had given a general account of how images of lattice planes are formed by Bragg diffraction and had applied Abbé's theory to the resolution of those images. He then simply assumed that a change in orientation of the observed lines corresponded to dislocations. This was not adequate to deal with image formation in metal foils, where lattice planes could not be resolved as directly. Within a year of the report of their initial observations of dislocations in alluminum, Whelan and Hirsch had published the necessary theory, which analyzed the details of how electron beams on either side of a stacking fault interfere to yield an image of dislocations.[62] Details of their theory were borne out by further experiments with the electron microscope. Once again, however, we find our authors taking this to bear out both their theory of the instrument and the claim that the images represent dislocations. "The general agreement obtained between theory and experiment confirms that the fringes [observable on the micrographs] are due to stacking faults and that the theory developed . . . is basically correct."[63] We do not wish to deny that a theory of specimen-instrument interaction can be of crucial importance. As Howie and Whelan point out in a subsequent article, "much of the available information must inevitably be lost . . . unless a sufficiently good theory exists by which the observed and often complex image contrast effects can be interpreted," and they proceed to press their theoretical analysis and its fit with experiment further.[64]

While a theory of instruments can be important our two main points have been that it is often unnecessary, and when it is necessary may be developed to support a theory of the specimen even while a theory of the specimen supports it, in a way that Kosso refers to and rashly condemns as "nepotism."

Physics Versus Biology?

We have shown that it is often unnecessary to consider the theory of an instrument in interpreting experimental findings with an instrument, whether or not a theory is being tested in the experiment and whether or not the theory being tested in the experiment is in principle independent of the theory of the instrument. We have further shown, in the last example, that the theory of the specimen can throw light on the theory of the instrument.[65] Even though we have seen that comparative experimental methodologies that bracket all theory of instrumentation are common both in life and physical sciences, might we not still argue for the relevance of the theory of instruments as a special hallmark of the physical sciences?

Menter and Hirsch were, indeed, in their own way concerned about discussing the theory of the electron microscope's function. There is a grain of truth in this thought, having to do with the complexity of the objects under investigation: it is more likely in physics that some sort of detailed theoretical treatment of instrument-specimen interactions will be possible. (Let us leave aside for the moment the different question about whether, and if so in what circumstances, theoretical treatments may be necessary or desirable). Only when an object is simple or homogeneous does one have at least a fighting chance of modeling its interactions with experimental apparatus with some degree of precision. This condition was met by Menter with his choice of crystal, but it is often unsatisfiable by biologists obtaining their specimens from living organisms, the unavoidably complex objects of their inquiry. The point is that understanding how the instrument works is only a start; one then has to apply the theory of the instrument to explain in detail how it interacts with the specimen to use such theory to interpret data. Precisely working out specimen-beam interactions (for instance) with mathematically formulated physical theory can quickly become so intractable and/or doubtful, due to unknowns about the specimen and simplifying assumptions, as to discourage any experimental physicist on grounds both of reliability and efficiency. Hence with complex specimens, such as amorphous or heterogeneous materials, physical scientists using electron microscopy will resort to much the same tactics as biologists, varying observation conditions as widely as possible and comparing results.[66]

Conversely, in situations where they believe the specimen to be sufficiently simple and well-characterized, biologists may attempt to extract quantitative information by engaging in careful analysis of specimen-beam interaction, with the help of electron-optical theory, in much the same manner as physicists. But they can easily overstep, building quantitative castles on the sandy foundations of measurements taken from micrographs of specimens whose chemical and structural alterations during preparation procedures cannot be judged. Wariness of such quantitative overreach is one of the main factors that made Porter's Rockefeller school of cell biology, with its emphasis on comparative methodology as described above, prevalent over Sjöstrand's rival school, which used quantitative information derived from electron micrographs to decide on such matters as the biochemical makeup of membranous structures.[67] Even where considerable exact knowledge of the specimen is available, a mixture of methodologies is generally most fruitful; for example in the muscle investigation described above, quantitative information from x-ray diffraction was used to constrain interpretation of data from a primarily qualitative, comparative microscopical approach. It should be noted that muscle is an unusual biological specimen in that, while a tissue, it is also a regular (indeed, almost crystalline) subcellular structure produced naturally in macroscopic quantities by the highly specialized muscle cells. Thus muscle is a particularly suitable material for methodologies like those used by physical scientists studying simple materials.

Another nice example of biophysical hybrid methodology, a combination of the specific interpretive methods discussed so far in this by no means exhaustive story, comes from the 1960s when electron diffraction began to find serious application in biology. Here, New York anatomist H. C. Anderson investigated the role of certain inclusion-bearing vesicles in the extracellular matrix of growing limbs, common in the region where cellular cartilage calcifies to mineral, cell-free bone. He compared thin section electron micrographs of limb tissue prepared by standard fixation and metal staining methods, with tissue prepared the same way after treatment with a chelating agent that strips away minerals, and found that the vesicles lost their crystalline inclusions with demineralization. So far, this evidence that the vesicles play a role in calcification follows from the typical comparative methodology for biology such as we have seen above, with similar specimens prepared differently but imaged similarly, indicating that any difference in appearance must be due to differences in preparation. Admittedly, here there is something of a theory concerning the mechanism of the chelation treatment, but from the very need to carry the experiment further it appears that this was not sufficient in itself to secure the inference that the vesicle inclusions were calcium. To clinch the case, Anderson prepared sections untreated with metal fixatives or stains—thus unaltered by heavy minerals—and viewed them in an electron microscope fitted with an attachment that can make electron-diffraction images of a tiny, selected area of the imaged field. The vesicles, in contrast to other parts of the tissue section, gave diffraction patterns that matched those of hydroxyapatite, a form of calcium. Here, only chemical identification could be obtained by crystallography rather than detailed structural information because the crystals were small and randomly oriented (effectively, powder), and also probably impure chemically, but identification was all the experimental design required.[68] Even within this experiment we see the limitations on physical methodology imposed by complexity of experimental object.

Conclusions: Theory and the Various Styles of Experiment

In this casual browsing through the history of electron microscopy, we have seen that experimenters use data interpretation methods to suit the object under investigation and their purpose of inquiry, and that these do not map neatly to discplines or subfields of investigation but depend in more complex ways on the experimenter's epistemic situation and resources. Explicit theory of microscope function and specimen-beam interaction sometimes comes into play in image interpretation but by no means always. Particularly, but not exclusively in biology, a comparative method predominates. Here, the specimens themselves, or the conditions of their observation, are altered, and the different appearances manifest from comparing the different observations are explained by reference to whatever has been changed. All that matters for this comparative methodology is there be no reason to suspect the instrument to behave differently when one switches between specimens, or when one manipulates a given specimen. Otherwise the theory of the instrument does not enter into the reasoning at all. We have suggested that this comparative type of methodology can be the only viable choice, even where theory of an instrument is well developed, when the specimen is too complex or too incompletely characterized for actual calculation of specimen-beam interactions from the theory. But it may be reasonable to prefer the comparative method, in case the purpose of the inquiry requires no quantification, even where the nature of the specimen might permit use of a quantitative, theory-intensive interpretive methodology. After all, the wise inquirer seeks no more exactness than needed to answer a question, or than the subject matter admits, as Aristotle reminded his logician friends when introducing his lectures on ethics.[69] (And in a related display of uncommon wisdom, Aristotle insisted that the proper method of reasoning, not just the degree of rigor, varies between sciences because of differing subject matter.[70]) Again, the comparative method is no definitive mark of life science, though it is true enough that biology has objects and purposes that recommend qualitative, comparative approaches more often. Admittedly, traditions within fields may exaggerate preferences for certain experimental methodologies, for instance, qualitative ones in life sciences and theory-intensive, quantitative ones in branches of physical sciences; indeed we find interesting variations among research traditions within a given branch of science, for instance the more image-oriented bubble chamber school of particle physicists versus the statistical, spark-chamber school as described by Galison.[71] But among electron microscopists, as among experimentalists in general, biologists use quantitative theory of instrument/specimen interaction when the situation permits and calls for it, just as physicists often use qualitative, comparative methods.

As the story about dislocations indicates, even when there is extensive overlap among theories of the instrument and the object under investigation, this is not necessarily a cause for worry. If there is a match between the precise predictions of some speculative theory, which may be or include a theory of instrument/specimen interaction, and the interpretation of some reproducible but otherwise mysterious observations, then why should this match not be taken as confirming both the theory and the interpretation of the observations? We have already seen Galileo, Menter, and Hirsch arguing in this way. There is nothing wrong with this from an epistemological point of view. The lack of independence that Kosso fears under the name of nepotism need not be a problem. This is not to deny that serious circularities can arise—only that they do not necessarily arise when theories being tested and theories of instruments are related and simultaneously open to revision or test. Nor is it to deny that there are circumstances in which a detailed specification of instrument/specimen interaction is vital, as we illustrated with reference to the development of the Cavendish program for examining dislocations in metals. But even here we found reason to suspect that nepotism is endemic, even in the best of science.

Another, basic misconception that our studies aim to counter is the lingering, still-common supposition that the sole purpose of experiment is to test theories. Experiment can take on a life of its own when experimentalists attempt to exploit the opportunities offered by the available resources, even where theoretical payoff is unlikely. Both Menter and the Cavendish group endeavored to take advantage of the possibilities opened up to them by the electron microscope. They endeavored to visualize dislocations, the existence of which they never doubted. They were not initially testing any theory, and a failure to observe dislocations would probably only have sent Hirsch back to his x-ray technology and Menter back to his study of crack propagation. Hirsch himself attributes the main cause of his successful observation of dislocations in aluminum to the fact that the group had "access to a new generation of electron microscope at the right time."[72]

All of this is to suggest that there is much more to the experimenter's craft than dreamt about in many philosophies of science, particularly those that suppose that the link between experimental data and objects or systems in the world is forged by way of deductive connections between complete, mathematically formulated theories, making possible the tracing of the causal path from object via experimental apparatus to detecting or measuring device. Some experiments might be captured at least approximately by this characterization. But much more work needs to be done on how experimenters reason even in such experiments, especially in the various kinds of experiments beyond the classic type designed to test a theory. Making lists of the sorts of things experimenters have in their bags of tricks, as Franklin and Hacking especially have done to good effect,[73] is only a start. The conceptual tools experimenters bring to bear should be related systematically to the subjects and purposes of their inquiry, as well as to their technical and cultural resources. And we must also not assume that an adequate taxonomy of experimental logic will be a static structure. Keeping the historical dimension of the experimenter's craft in mind is essential, to help avoid overhasty universal claims about the essence of experimentation, and to help find patterns of change that may characterize different sciences in different contexts. The new enthusiasm among philosophers of science to study epistemology as scientists do it is unquestionably laudable. For this enthusiasm to bring philosophy of science substantially closer to its own subject matter, however, it must be tempered with the realization that science might be much more heterogeneous and complex than philosophers have long been imagining.

NOTES

1. Pierre Duhem, *The Aim and Structure of Physical Theory,* trans. Phillip Wiener (New York: Atheneum, 1962), part II, chap. 4.

2. Dudley Shapere, "The Concept of Observation in Science and Philosophy," *Philosophy of Science* 49 (1982): 485–525. Peter Kosso, "Dimensions of Observability," *British Journal for the Philosophy of Science* 39 (1988): 449–467; idem., "Science and Objectivity," *Journal of Philosophy* 86 (1989): 245–257 and idem., *Observability and Observation in Science,* (Dordrecht: Kluwer Academic Publishers, 1989), esp. chap. 4.

3. Ibid. For other examples of appeal by philosophers to examples involving the microscope in general and the electron microscope in particular see Ian Hacking, *Representing and Intervening* (Cambridge: Cambridge University press, 1983), chap. 11; Allan Franklin, *The Neglect of Experiment* (Cambridge: Cambridge University Press, 1986), chap. 6.

4. Hacking, *Representing*, 200–201.

5. On intuitive, everyday roots of interpretive methods in electron microscopy, see Nicolas Rasmussen, *Picture Control: The Electron Microscope and the Transformation of Biology in America, 1940–1960* (Stanford: Stanford University Press, 1997), chap. 6.

6. Kosso, "Dimensions," 460–463.

7. For more details of the history and epistemology of Galileo's use of the telescope, see Alan F. Chalmers, *Science and its Fabrication* (Minneapolis: University of Minnesota Press, 1990), 50–56, Neil Thomason, "The Power of ARCHED Hypotheses: Feyerabend's Galileo as a Closet Rationalist," *British Journal for the Philosophy of Science* 45 (1994): 255–264, and idem., "Elk Theories—A Galilean Strategy for Validating a New Scientific Discovery," in Peter J. Riggs, ed., *Natural Kinds, Laws of Nature and Scientific Methodology* (Dordrecht: Kluwer, 1996), 123–144. Thomason's papers contain discussions of the relations between theory and evidence that are germaine to our position in this paper.

8. John Reisner, "An Early History of the Electron Microscope in the United States," *Advances in Electronics and Electron Physics* 73 (1989): 134–233; Gregory Kunkle, "Technology in the Seamless Web: 'Success' and 'Failure' in the History of the Electron Microscope," *Technology and Culture* 36 (1994): 80–103. Lin Qing, *Zur Frühgeschichte des Elektronenmikroskops* (Stuttgart: Verlag für Geschichte der Naturwissenschaften und Technik, 1995).

9. Keith Porter, Albert Claude, and Ernest Fullam, "A Study of Tissue Culture Cells by Electron Microscopy," *Journal of Experimental Medicine* 81 (1945): 233–246.

10. Porter to Pickels, 3 June 1946, carton 1, second box, in the Keith Porter Papers in the archives of the University of Colorado (hereafter, KPA). See also Hans-Joerg Rheinberger, "From Microsomes to Ribosomes: 'Strategies' of 'Representation' 1930–1955," *Journal of the History of Biology* 28 (1995): 49–89.

11. Murphy, "Reports to the Board . . . April 19, 1947," 81, unsorted KPA.

12. Ibid., 80.

13. Hacking, *Representing*, chap. 11. See Nicolas Rasmussen, "Making a Machine Instrumental: RCA and the Wartime Beginnings of Biological Electron Microscopy," *Studies in the History and Philosophy of Science* 27 (1996): 311–349.

14. Porter, [April ?] 1948, "Annual Report—1948," folder "Report to Dr. Murphy 1948," 4, unsorted KPA.

15. Cf. Franklin, *Neglect* , chap. 6.

16. Porter, [April ?] 1951, "Report of Dr. Porter ," unsorted KPA.

17. As Porter himself acknowledged; see K. Porter and F. Kallman, "The Properties and Effects of Osmium Tetroxide as a Tissue Fixative with Special Reference to its Use for Electron Microscopy," *Experimental Cell Research* 4 (1953): 127–141. There is no reason to regard this method-dependent situation as a scandal of epistemological inconsistency, once it is grasped that experimentalists can behave as pragmatists not striving to represent nature as it "really" is independent of methods of inquiry.

18. Porter and Gasser, [April ?] 1952, "Report of Dr. Porter . . . 1951–1952," unsorted KPA.

19. N. Rasmussen, "Facts, Artifacts, and Mesosomes: Practicing Epistemology with the Electron Microscope," *Studies in History and Philosophy of Science* 24 (1993): 227–265.

20. K. Porter, "The Submicroscopic Structure of Protoplasm," *Harvey Lectures* 51 (1956): 175–228.

21. Hacking, *Representing,* 200–201.

22. K. Porter, "Observations on a Submicroscopic Basophilic Component of Cytoplasm," *Journal of Experimental Medicine* 97 (1953): 727–750; idem., "Electron Microscopy of Basophilic Components of Cytoplasm," *Journal of Histochemistry and Cytochemistry* 2 (1954): 346–375.

23. Porter, "Electron Microscopy of Basophilic Components;" G. E. Palade and K. R. Porter, "Studies on the Endoplasmic Reticulum. I. Its Identification in Cells in Situ," *Journal of Experimental Medicine* 100 (1954): 641–656.

24. Ibid.

25. Rasmussen, *Picture Control* , chap. 4; also see Rheinberger; Albert Claude "Studies on Cells: Morphology, Chemical Constitution, and Distribution of Biochemical Function," *Harvey Lectures* 43 (1950): 121–164.

26. See Rasmussen, "Facts, Artifacts."

27. See Otto Meyerhoff, *Chemical Dynamics of Living Phenomena* (Philadephia: J. B. Lippincott, 1924), chap. 4, for a summary of early twentieth century muscle work.

28. For instance see Albert Szent-Györgyi, *The Chemistry of Muscular Contraction* (New York: Academic Press, 1947); William Astbury, "X-Ray and Electron Microscope Studies, and Their Cytological Significance, of the Recently-Discovered Muscle Proteins, Tropomyosin and Actin," *Experimental Cell Research, Supplement* 1 (1949): 234–246. Also see C. E. Hall, Marie Jakus, and F. O. Schmitt, "An Investigation of Cross Striation and Myosin Filaments in Muscle," *Biological Bulletin* 90 (1946): 32–50.

29. See Rasmussen, *Picture Control,* chap. 4.

30. Ibid.

31. Delbert Philpott and Albert Szent-Györgyi, "The Series Elastic Component in Muscle," *Biochimica Biophysica Acta* 12 (1953): 128–133. See also John Farrant and E. Mercer, "Studies on the Structure of Muscle. II. Arthropod Muscle," *Experimental Cell Research* 3 (1952): 553–563; H. S. Bennett and K. R. Porter, "An Electron Microscope Study of Sectioned Breast Muscle of the Domestic Fowl," *American Journal of Anatomy* 93 (1953): 61–106.

32. F. O. Schmitt, *The Never-Ceasing Search* (Philadelphia: American Philosophical Society, 1991), 154; J. T. Randall, "Emmeline Jean Hanson," *Biographical Memoirs of the Fellows of the Royal Society* 21 (1975): 313–344. Randall recalls that Huxley went in September

1952 and Hanson in February 1953, while Schmitt's records list them on the lab roster from November 1952 and July 1953 respectively, which may simply date their official MIT registration. Both left in spring 1954.

33. Randall. H. E. Huxley, "X-Ray Analysis and the Problem of Muscle," *Proceedings of the Royal Society B* 141 (1953): 59–62.

34. Cf. Rasmussen, *Picture Control,* chap. 4.

35. Randall.

36. E. J. Hanson and H. E. Huxley, "The Structural Basis of Contraction in Striated Muscle," *Symposia of the Society for Experimental Biology and Medicine* 9 (1955): 228–263.

37. E. J. Hanson and H. E. Huxley, "Changes in the Cross-Striations of Muscle During Contraction and Stretch and their Structural Interpretation," *Nature* 173 (1954): 973–976.

38. Muscle as prepared for electron microscopy was of course no longer in a state capable of contraction, so a given specimen could only be observed at one stage of contraction; moreover, different preparations even of initially identical biological material could easily undergo different degrees of shrinkage during preparation. X-ray crystallography of muscle only gave information on cross-sectional structure.

39. A. Huxley and R. Niedergerke, "Structural Changes in Muscle During Contraction," *Nature* 173 (1954): 971–973. See Rasmussen, *Picture Control,* chap. 4.

40. Hanson and Huxley, "Structural Changes," and idem., "Structural Basis of Contraction."

41. See "Discussion" following Hanson and Huxley, "Structural Basis of Contraction" (580–582). Also see Alan Hodge, "Electron Microscopic Studies of Insect Flight Muscle," in *Proceedings of the Third International Conference on Electron Microscopy, 1954* (London: Royal Microscopical Society, 1956), 572–576; idem., "Studies on the Structure of Muscle. III. Phase Contrast and Electron Microscopy of Dipteran Flight Muscle," *Journal of Cell Biology* 1 (1955): 361–338.

42. E. J. Hanson and H. E. Huxley, "Preliminary Observations on the Structure of Insect Flight Muscle," in *Stockholm Conference on Electron Microscopy, Proceedings* (New York: Academic Press, 1957), 202–204.

43. H. E. Huxley, "The Double Array of Filaments in Cross-Striated Muscle," *Journal of Cell Biology* 3 (1957): 631–646.

44. J. W. Menter, "The Direct Study by Electron Microscopy of Crystal Lattices and Their Imperfections," *Proceedings of the Royal Society of London, Series A* 236 (1956): 119–135.

45. G. I. Taylor, "The Mechanism of Plastic Deformation of Crystals. Part I. Theoretical," *Proceedings of the Royal Society of London, Series A* (1934) 145: 362–387, and "The Mechanism of Plastic Deformation of Crystals. Part II. Comparison with Observations," *Proceedings of the Royal Society of London, Series A* 145 (1934): 388–405.

46. Menter, 119–120.

47. See Ernest Braun, "Mechanical Properties of Solids," in Lillian Hoddeson, Ernest Braun, Jurgen Teichmann and Spencer Weart (eds.), *Out of the Crystal Maze: Chapters from the History of Solid State Physics* (Oxford: Oxford Univerity Press, 1992), 317–358, especially 333 and 336.

48. J. M. Robertson, "An X-ray Study of the Structure of the Pthalocyanines. Part I. The Metal-Free, Copper, and Platinum Compounds," *Journal of the Chemical Society, London* (1935): 615–621; idem., "Part II. Quantitative Structure Determination of the Metal-Free Compound," *Journal of the Chemical Society, London* (1936): 1195–1209.

49. Menter, 123.

50. Hacking, *Representing*, 189.

51. James Hillier, "A Discussion of the Fundamental Limit of Performance of an Electron Microscope," *Physical Review* 60 (1941): 743–745 and O. Scherzer, "The Theoretical Resolution Limit of the Electron Microscope," *Journal of Applied Physics* 20 (1949): 20–29.

52. Menter, 129.

53. Menter, 130.

54. C. E. Challice, "Summarized Proceedings of a Conference on Electron Microscopy-Reading, July 1956," *British Journal for Applied Physics* 8 (1957): 259–269, quotations on 268 and 269.

55. P. B. Hirsch, R. W. Horne, and M. J. Whelan, "Direct Observation of the Arrangement and Motion of Dislocations in Aluminium," *Philosophical Magazine* 1 (1956): 677–684.

56. Menter, 127–128.

57. Hirsch et al., 679.

58. Braun, 348.

59. Ibid., 679 and 682.

60. See P. B. Hirsch, "Direct Observations of Dislocations by Transmission Electron Microscopy: Recollections of the Period 1946–56," *Proceedings of the Royal Society, Series A* 371 (1980): 160–164, 163; also Braun, 348–349.

61. Hirsch.

62. M. J. Whelan and P. B. Hirsch, "Electron Diffraction from Crystals Containing Stacking Faults," *Philosophical Magazine* 1 (1956): 1121–1142 and 1303–1204.

63. Ibid., 1303.

64. A. Howie and M. J. Whelan, "Diffraction Contrast of Electron Microscope Images of Crystal Lattice Defects: III. Results and Experimental Confirmation of The Dynamical Theory of Dislocation Image Contrast," *Proceedings of the Royal Society*, 267(1962): 206–230, quotation on 206.

65. Rasmussen, "Facts, Artifacts."

66. Peter. W. Hawkes, *Beginnings of Electron Microscopy. Advances in Electronics and Electron Physics, Supplement* 16 (1985).

67. For an example, see N. Rasmussen, "Mitochondrial Structure and the Practice of Cell Biology in the 1950s," *Journal of the History of Biology* 28 (1995): 381–429.

68. H. C. Anderson, "Vesicles Associated with Calification in the Matrix of Epiphyseal Cartilage," Journal of Cell Biology 41 (1969): 59–72.

69. Ethics 1094b.

70. Met 1025 b ff; PA 640a.

71. Peter Galison, *Image and Logic: A Material Culture of Microphysics* (Chicago: University of Chicago Press, 1997).

72. Hirsch, 163.

73. Franklin, *Neglect,* chap. 6. See also idem., *Experiment Right or Wrong* (Cambridge: Cambridge University Press, 1990), 104; also see Ian Hacking, "The Self-Vindication of the Laboratory Sciences," in Andrew Pickering, ed., *Science as Culture and Practice* (Chicago: University of Chicago Press, 1992), 29–64.

Index

Abegg, Richard, 342
Abraham, Max, 89, 153, 155
 and development of the theory of spin, 428–429
Aether and Matter, 115
Anderson, Carl, 317
Annalen der Physik, 314
Armstrong, H. E., 103, 114
Arndt, Fritz, 349
Arnold, Harold, 328
Aston, F. W., 62, 215
AT&T, 328
Atombau und Spektrallinien, 237
Atoms, 2. *See also* Particles
 appropriation by chemists, 367–369
 and chemical bonds, 385–386, 454
 and chemical valence, 342–345, 352–353
 and composition of matter, 201–202
 and Coulomb's law, 371
 discovery of, 38–39, 202–203
 and electron bonds, 345–348, 371–372
 and group theory, 376–379
 ionic charge of, 57–58
 and J. J. Thomson, 141, 204–206
 and Joseph Larmor, 183–184
 and mechanical resonance, 350–352
 in metals, 491
 and molecular orbitals, 386–390
 nuclear, 308–310
 Owen Richardson on, 245–247
 and Pauli exclusion principle, 454
 and Paul Villard, 137–138, 141
 plum-pudding model of, 22–23
 and polyelectronic molecules, 376–379
 research on structure of, 22–23, 38–40, 245
 and Schrodinger equation, 457
 and the theory of resonance, 381–386
 vibration of, 176
 and the vortex atomic theory, 199–200
Audions, 328
Avogadro's number, 49
Ayrton, W. E., 119

Baedeker, Karl, 263, 274–275
Bardeen, John, 327, 331, 332–336
Beketov, Nikolai, 210
Bell Laboratories, 327, 330, 331
Benzene, 349–352, 382
Berichte, 339
Berzelius, Jons, 196
Bethe, Hans, 310
Biological electron microscopy, 471–483
 and physics, 493–495
Bjerrum, J., 370
Blackbody radiation, 270–273, 278
Blackett, P. M. S., 317
Bloch, Felix, 279, 330
Bodlander, Guido, 342
Bohr model, 62, 63–64, 215, 227, 245, 247–248
Bohr, Niels, 236, 237
 and chemical bonds, 454
 and development of the theory of spin, 434–435
 and electron gas theory of metals, 265–266
 on free electrons, 269
 and Hall effect, 276
 and nuclear atoms, 308
 and the periodic table, 343–344
Boltzmann, Ludwig, 256, 258, 266
Bond, W. G., 108
Born, Max, 352, 378, 393–394

Bose electron, 311–312
Bragg fringes, 489–490
Bragg, W. H., 244
Branch, Gerald, 341, 351
Brattain, Walter, 327, 332–335
Braun, Ernest, 491–492
Braun, Ferdinand, 65
Brillouin, Marcel, 268
Broca, M., 113, 147–148
Brown, F. C., 233
Bucherer, Alfred, 208
Burton, H., 347, 349–350
Bush, Vannevar, 332
Butlerov, A. M., 385

California Institute of Technology (Caltech), 341, 374
Calvin, Melvin, 341
Canalstrahlen, 152
Carnelley, Thomas, 206
Cathode Rays, 21
Cathode rays, 21
  and discovery of the electron, 404, 415–416
  early research on, 27–28, 110–111
  and electrons, 43, 92, 204–206
  French scientists' research on, 147–150, 158–159
  and Goldstein rays, 143
  J. J. Thomson's work on, 36–45, 65, 112–115, 419
  Lord Kelvin on, 28–29
  and m/e value, 36–45
  Oliver Lodge's work on, 79–80
  oscilloscope, 119–120
  Paul Villard's work on, 138–139, 140–145
  research on structure of, 22, 40–45, 135, 139
  response to research by Paul Villard on, 148
  response to research on, 109–112, 157–160
  structure of, 147–151
  wave-like nature of, 42
  and wireless telegraphy, 118–119
  and x-rays, 105–106

Cavendish Laboratory, 21, 30, 35, 102, 110, 124–125, 144, 158, 235–236, 490–491
Chadwick, James, 307
Chemical News, 110
Chemical Reviews, 348, 375
Chemistry, organic, 339–354
  and chemical affinity, 339–340
  and chemical bonds, 339–340
  and chemical valence, 342–345, 352–353
  and discovery of the electron, 340–342
  and electron bonds, 345–348
  and mesomerism, 348, 350–351
  and the periodic table, 343–344
  and physics, 363–365
  reductionist, 352, 364–365
  structural formulas, 340(figure)
Chemistry, quantum, 363–365
  and chemical valence, 367–369, 376–379
  and Coulomb's law, 371
  development of, 373–376, 394
  and development of the theory of spin, 429–436
  and electrons, 367–369
  and group theory, 376–379
  and molecular orbitals, 386–390
  and the Pauli exclusion principle, 372–373
  and the periodic table, 376
  and physics, 365–367, 379–381
  and polyelectronic molecules, 376–379
  and quantum mechanics, 368–369, 369–371
  and relativity, 439–446
  and scientific realism, 366
Clark, G. L., 379
Claude, Albert, 472–478
Clausius, Rudolf, 202
Cohn, Emil, 212
Comparative methodology, 495–496
Comptes Rendus, 138, 144, 157
Compton cross-section, 319
Compton, Karl T., 234
  and development of the theory of spin, 428
Conduction of Electricity through Gases, 23–24, 34, 37, 61, 63, 147

INDEX   505

Confinement of nuclei, 309
Corpuscles, 3–6, 22, 37
  acceptance of early theories on, 101–102
  and discovery of the electron, 404
  versus electrons, 83, 85, 86, 115–118, 145, 159–160
  J. J. Thomson's work on, 57–59, 77, 89
  John Ambrose Fleming on, 116–118
  and Paul Villard, 140–141
  and the proutean theory, 203–210
  and wireless telegraphy, 118–119
Corpuscular Theory of Matter, The, 58, 62, 123
Cosmic rays, 314–317
Coulomb's law, 343–344, 371
Coulson, Charles, 353, 384
Crookes, William, 27, 28–29, 103, 105
  and discovery of the electron, 403–404
  and electrical discharge of gases, 117
  and proutean hypothesis, 197–198
Cross-section formulae, 319–320
Crystal sets, 329–330
Crystal structure, 484–491

D'Albe, Edmund E. F., 105, 120–121
Dampier-Whetham, William C. D., 125
Darwin, Charles G., 267, 272
Davisson, Clinton, 329
Davydov, Boris, 330–331
Davy, Humphry, 196
de Broglie, Louis, 329, 374
Debye, Peter, 266
de Forest, Lee, 328
Dickinson, Roscoe, 374
Dictionary of Scientific Biographies, 373
Different contributions view, 411
Dirac, Paul, 195, 216, 352, 365, 454, 456
  and development of the theory of spin, 437, 439
  and relativity, 439–442
  and structure of the electron, 457–459
Discovery. *See also* Electrons, discovery of
  contrasting views of, 408–414
  definitions of, 405–408
  different contributions view of, 411
  important classification view of, 409–410
  manipulation-and-measurement view of, 409

  social constructivist view of, 410–411
  strong, 418–420
  true belief view of, 411–412
Drude, Paul, 115, 255, 257–260, 268–269
  and blackbody radiation, 270
Duddell, W. Dubois, 119
Duhem, Pierre, 467
Dumas, Jean-Baptiste, 196
Dynamical Theory of The Electric and Luminiferous Medium, A, 201

Ebert, H., 81
Ecole Normale Superieure, 144
Ehrenfeld, Richard, 211
Ehrenfest, Paul, 258, 271, 272, 432
Ehrenfest, Tatiana, 258, 272
Einstein, Albert, 211–212, 243
  and degenerate gas, 278
  and electron gas theory of metals, 269
  and quantum theory, 277
  and radiation law, 271
Electrical discharges. *See also* Electricity
  in gases, 51–52
  in ionization, 47–49
  J. J. Thomson's work on, 62
  of matter in general, 58
  positive, 62
  research on, 27–28
Electrical engineering, 119–121
Electrical Review, The, 111
Electrician, The, 36, 90
  and acceptance of corpuscles, 115–117
  commentaries on J. J. Thomson's work, 110–111
  and discovery of electricity, 105–109
Electricity. *See also* Electrical discharges
  and discovery of the electron, 415
  early research on, 103–105
  and electromagnetic theory, 207–208
  and electrons, 211–214
  and invention of the transistor, 332–336
  John Ambrose Fleming on, 123–124
  and Joseph Larmor, 181–182
  and Owen Richardson, 229–236
  and positive electrons, 206–209
  and the vortex atomic theory, 206
Electricity and Matter, 62, 206

Electromers, 343
Electronic Interpretations of Organic Chemistry, 341
Electronic Theory of Valency, The, 345, 368–369
Electrons, 78, 121
Electrons. *See also* Nuclear electrons; Particles
  acceptance of early theories on, 86–91, 102–103, 115–118, 120–124, 216, 453–461
  appropriation by chemists, 363–365, 367–369
  and Arnold Sommerfeld, 215
  asymmetry of, 24
  audiences for various theories of, 144–145
  beams, 307
  and benzene, 349–352
  and blackbody radiation, 278
  bonds, 345–348, 371–372
  Bose, 311–312
  and cathode rays, 90
  charged particles within, 261–262, 315, 319–320
  and chemical bonds, 385–386
  and chemical valence, 342–345, 352–353
  and composition of matter, 239–242
  and contrasting views of discovery, 408–414
  versus corpuscles, 83, 115–118, 145, 159–160
  and Coulomb's law, 343–344, 371
  credit for discovery of, 124–126, 187–188, 412, 414–418, 420–422, 451–453
  and development of the theory of spin, 427–436
  discovery of, 2–9, 21–27, 37–38, 64–66, 110–111, 124–126, 187–188, 403–404
  displacement, 346–347
  early motivations for research on, 6–9, 86–91
  early research on, 9–12, 82–85, 112–115, 195–196
  and electricity, 211, 229–236
  and electromagnetic theory, 207–208, 215–216, 256–260
  and electron gas theory of metals, 280–281
  and electron microscopy, 471–478
  free, in gases, 51–52, 241, 266–267, 269, 273
  free nuclear, 313–314
  and group theory, 376–379
  Hamiltonian of, 456–457
  and Heisenberg's n-p nuclear model, 310–312
  J. J. Thomson's work on, 3–6, 24–27, 77, 110–111, 403–404, 414–418, 451
  and Joseph Larmor, 181–188
  and Langrangian density, 459–460
  and light, 176–177
  and mechanical resonance, 350–352
  mass/charge (m/e) values, 106–107, 110–112, 145–146, 179–180, 185, 213
  in metals, 260–263, 266–267
  and molecular orbitals, 386–390
  negative, 313
  nuclear, 307–320
  and the nucleus, 319–320
  Oliver Lodge's work on, 78–80
  operational reality of, 327–329
  Owen Richardson on, 238
  Paul Villard's work on, 155
  as philosophical objects, 12–15
  and polyelectronic molecules, 376–379
  positive, 206–208, 260
  and the proutean theory, 201, 203–210
  and quantum chemistry, 363–365
  and quantum mechanics, 350–352
  and relativity, 439–446
  and Schrodinger equation, 457
  and spin, 427–446, 457–458
  stable properties of, 453–455
  structure of, 183–184, 455–461
  and the theory of resonance, 381–386
  thermal emissions of, 229–236
  and Tolman-Stewart experiment, 266–267
  unshared, 387
  usefulness of, 327
  vapor theory, 263–265

velocity of ions in, 271–272
and the vortex atomic theory, 198–200
Electron, The, 124
Electron Theory of Matter, The, 236–249
Electron Theory, The, 120
Elements of Chemical Philosophy, 196
Emerson, Benjamin, 210
Emission of Electricity from Hot Bodies, The, 236
Encyclopaedia Britannica, 123
Endoplasmic reticulum (ER), 472–478
Equipartition theorem, 268–270
Ether, 1–2, 211, 213, 237–238
Ewing, J. Alfred, 108
Experiments
  and cathode ray oscilloscope, 119–120
  and comparative methodology, 495–496
  on existence of electrons, 112–115
  and extended working hypotheses, 55–60
  on ionic charge, 47(figure), 47–49
  by J. J. Thomson, 109–112, 142–143
  and metaphysics, 425–427
  not supported by theory, 24–25
  and observable versus unobservable entities, 173
  by Owen Richardson, 229–233
  by Paul Villard, 143–144, 154
  by Pieter Zeeman, 174–176
  and realism versus unrealism, 171–172, 425–427
  and scientific discovery, 171–173
  and use of instruments, 467–497

Fairbank, W. M., 373
Faraday's laws of electrolysis, 79
Faraday tube, 30–32, 62, 84–85, 342
Fermi, Enrico, 278, 307–308, 318, 454
FitzGerald, George, 86, 181–183
  on corpuscles, 110–111, 116
  on electrons, 116
Fleming, John Ambrose, 102, 104, 328
  on corpuscles, 114–115, 116–118
  on electricity, 123–124
  and wireless telegraphy, 114, 118–119
  work on cathode rays, 105–106
Fowler, Ralph Howard, 278

Frankland, Edward, 339, 353
Free radicals, 349
Frenkel, Yakov, 216
Fry, Harry S., 343, 380

Gans, Richard, 263
Gases
  and blackbody radiation, 270–273
  degenerate, 277–278
  electric discharge through, 27–28, 31–32, 51, 81, 107–109, 260
  electrification of, 23–24, 31–32, 261–263
  and electron gas theory of metals, 280–281
  and electron vapor theory, 263–265
  and equipartition theorem, 268–270
  and free electrons, 273–274
  ionization and electrical conduction in, 21, 33(figure), 45–47, 55–60, 88–89, 147, 229–236, 256–260
  and *m/e* value, 36–45
  and osmotic pressure, 258
  specific heats of, 349
  structure of, 106, 152
  thermal emissions of electrons in, 229–236
  and vacuum tubes, 328–329
  velocity of ions in, 46, 51
Geissler, Heinrich, 27
  and the different contributions view of discovery, 411
General Chemistry, 341
Germer, Lester, 329
Glazebrook, R. T., 121, 125
Goldstein, Eugen, 27
Goldstein rays, 138–139
Goudsmit, Samuel, 427–436, 454
Graham, Thomas, 201
Grassmann, Robert, 203
Group theory, 376–379
Guthrie, Frederick, 118

Haga, Hermann, 263
Hall, Edwin Herbert, 256, 273
Hall effect, 256, 260–261, 273–277
Hamiltonian, 456–457

Hanson, Jean, 478–483
Harkins, William, 214
Heaviside, Oliver, 108, 255
Heilbron, J. L., 247
Heisenberg, Werner, 279–280, 307–308
  and cosmic rays, 314–317
  and development of the theory of spin, 429, 432–436
  and free electrons, 312–314
  n-p nuclear model of, 310–312
Heitler, Walter, 319–320, 349, 370–372
  and Fritz London, 390–395
  and group theory, 376–379
  and the Heitler-London paper, 373–376, 380–381, 387–389
  and Robert Mullikan, 387–389
Hertz, Heinrich, 29, 80–82, 255
Hertz, Paul, 263
Herzberg, Karl, 370
History of Science in its Relations with Philosophy and Religion, 125
Hittorf, J. W., 27
  and the different contributions view of discovery, 411
Holes, 327
  early research on, 329–332
  and postmodern technologies, 336
Holman, Silas, 199–200
Houston, William, 279
Houtermans, Fritz, 310
Huckel, Erich, 351
Hughes, Edward D., 348
Hund, Friedrich, 351, 386, 389
Hunt, Sterry, 206
Huxley, Hugh, 478–483
Hydrogen, 213, 370, 371–372
  and chemical valence, 376–377
  and group theory, 376–379
  and Humphry Davy, 196
  and the Pauli exclusion principle, 372–373
  and Paul Villard, 140–141, 148, 155
  and Pieter Zeeman, 180

Important classification view of discovery, 409–410
Incandescent filament electrical discharges, 52–55, 118, 265. *See also* Light
Ingold, Christopher, 341, 345, 347–348, 349–352
  and the Ingold notation, 348(figure)
Ingold, Hilda Usherwood, 345, 347, 349
Instruments, scientific
  and circularity problem, 469–470
  and endoplasmic reticulum (ER) study, 472–478
  justification for use of, 470
  and muscle physiology, 478–483
  positivistic view of, 467, 468–469
  and realism versus unrealism, 468
  and solid-state physics, 483–491
  theory-dependent view of, 467–468
Introduction to Quantum Mechanics with Applications to Chemistry, 341
Ions
  and electrons, 177–180
  and H. A. Lorentz, 176–180
Ions, Electrons, Corpuscles, 153
Isomers, 343
Iwanenko, Dmitri, 313

Jaumann, G., 105
Jeans, James, 266
  and positive electrons, 206–207
Jessup, A. C., 206
Jessup, A. E., 206
Jewett, Frank, 328
Jones, Harry, 331
Jordan, P., 458
Journal de Physique, 150, 154, 157

Kaufmann, Walter, 22, 44, 77, 78, 187, 203, 259–260
  and acceptance of the electron, 86–91
  and the proutean theory, 208, 209–210
  research on electron activity, 80–82
Keesom, Willem Hendrik, 264, 277
Kelly, Mervin J., 331, 332
Kelvin, Lord, 28–29
  and the true belief view of discovery, 411–412
  and the vortex atomic theory, 198–200
Kermack, W. O., 346

Kerr effect, 174
Kirrmann, Albert, 346
Kleeman, R. D., 244
Klein-Nishina formula (K-N), 314–315
Koenigsberger, Johann, 276
Kronig, R., 432
Kruger, Friedrich, 263
Kuhn, Thomas, 467

Langevin, Paul, 146–147, 153–154
Langmuir, Irving, 234, 343
Langrangian density, 459–460
Lapworth, Arthur, 346–347
Larmor, Joseph, 38, 77, 86–91, 201, 408
  on corpuscles, 204
  on electrons, 83–84, 181–188, 255
  ether theory of, 110–111, 115
  and Pieter Zeeman, 181–187
Le Bon, Gustave, 145
Le Journal de Physique et Le Radium, 154
Lenard, Phillip, 29, 45, 88, 415
  and cathode rays, 201
  and discovery of the electron, 404, 418
  and ionic charge, 107
Lennard-Jones, John E., 341, 393–394
Le Radium, 154
Lewis, Gilbert N., 344–345, 368
Light, 176–177, 244. *See also* Incandescent filament electrical discharges
  and solid-state physics, 487–488
Linstead, R. P., 352–353
Lockyer, Norman, 113–114, 196, 204
Lodge, Oliver, 1–2, 78–80, 102, 113
  and discovery of the electron, 106–107, 121
  and Pieter Zeeman, 179–180
  and positive electrons, 206–207
London, Fritz, 349, 351, 369–373
  and group theory, 376–379
  and the Heitler-London paper, 373–376, 380–381, 387–389
  and Robert Mullikan, 387–389
  and Walter Heitler, 390–395
Lorentz, H. A., 77, 81–82, 408
  and acceptance of the electron, 86–87
  and development of the theory of spin, 445
  and electromagnetic theory, 238–239
  and electron gas theory of metals, 260–263, 274
  on electrons, 83, 255
  and electron velocity, 271–272
  and free electrons, 270
  on ionic charge, 106
  and Pieter Zeeman, 177
  and *Theory of Electrons,* 237
  theory of ions, 176–180
  and Tolman-Stewart experiment, 267
Losanitsch, Sima, 210
Lowry, Thomas, 345–346
Lucas, Howard, 341

Magnetic spectrum, 42, 332–334
  and electromagnetic theory, 207–208, 210–211
  experiments by Pieter Zeeman, 174–176
  and free electrons, 273–274
  and Joseph Larmor, 181–182, 183–185
  and oscillation of the ion, 178–179
Manipulation-and-measurement view, 409
Marconi, Guilelmo, 114, 117, 118–119
Marignac, Jean, 196
Massachusetts Institute of Technology (MIT), 332, 478
Matter, composition of, 196–198
  and corpuscles, 203–210
  and electromagnetic theory, 210–211, 237–239, 315
  and molecules, 201–202
  and positive electrons, 206–208
Maurer, Robert J., 276
Maxwell equations, 238
Maxwell, James Clerk, 104, 105, 181
  on gases, 265–266
McClelland, J. A., 35, 51, 229
McLaren, Samuel Bruce, 270
Meitner-Hupfeld anomaly, 318
m/e (mass/charge) values
  and discovery of the electron, 416–417
  in incandescent filaments, 52–55
  J. J. Thomson's work on, 36–45, 60–61, 109–112, 145–146
  and Pieter Zeeman, 179–180, 185
  research on, 44–45, 213

Index

Mendeleev, Dmitri, 196
Menter, James, 484–491, 493
Mesomerism, 348, 350–351
Metallernes Elektrontheori, 237
Metals
  and blackbody radiation, 270–273, 278
  charged particles within, 261–262
  conduction in, 278–279
  electric behavior of, 258–259
  and electron gas theory of metals, 265–266, 274, 280–281
  and equipartition theorem, 268–270
  and free electrons, 266–267
  and Tolman-Stewart experiment, 266–267
Metaphysics, 425–427
Meyer, Victor, 197
Microscopes, transmission electron, 471
  and endoplasmic reticulum (ER) study, 472–478
  and muscle physiology, 478–483
  and solid-state physics, 483–491
Mie, Gustav, 212–214
Millikan, R. A., 60–61, 124, 328
  and discovery of the electron, 453–454
Molecular orbitals, 386–390
Mott, Nevill, 330, 331
Mulliken, Robert, 351, 370
  early work, 386–387
  and the Heitler-London paper, 387–389
  and molecular orbitals, 386–390
Muscle physiology, 478–483

Nature, 36, 113, 236, 237, 390
Nature of the Chemical Bond, The, 381, 383
Nernst, Walther, 267, 278
Neutrons, 310, 312. *See also* Nuclear electrons
Nichols, Ernest Fox, 267
Nobel Prize, 121, 263
  Clinton Davisson and, 329
  J. J. Thomson and, 24, 62
  Philip Lenard and, 88
Notes on Recent Researches in Electricity and Magnetism, 27–28
Nuclear electrons. *See also* Electrons; Neutrons
  early research on, 308–310
  free, 316

Nuclei, 307
  confinement, 309
  and electrons, 319–320
  energy non-conservation, 309
  magnetic moments of, 309
  spin and statistics of, 309, 311–312
  structure of, 317

Observable versus unobservable entities, 173, 406–407
Occhialini, G. P. S., 317
Ohl, Russell, 330
Onnes, Kamerlingh, 174, 268–270, 277
On the Charge of Electricity carried by the Ions produced by Rontgen Rays, 21
On the Masses of the Ions in Gases at Low Pressures, 21
Orbitals, molecular. *See* Molecular orbitals
Organic Chemistry, 341
Organic chemistry. *See* Chemistry, organic
Osmotic pressure, 258

Pais, Abraham, 216
Palade, George, 472–478
Parson, Alfred, 215
Particles. *See also* Atoms; Electrons
  and cathode rays, 413
  and discovery of the electron, 416–417
  electrons as, 328–329
  positively charged, 330
Pauli exclusion principle, 372–373, 376, 454. *See also* Pauli, Wolfgang
Pauling, Linus, 341, 350, 351, 368
  and group theory, 376–379
  and the Heitler-London paper, 392–393
  on quantum chemistry, 374–376
  and the theory of resonance, 381–386
Pauli, Wolfgang, 216, 278–279, 307–308, 317, 330. *See also* Pauli exclusion principle
  and development of the theory of spin, 427–432, 436–438
Pearson, Gerald, 333
Peierls, Rudolf, 279, 327, 330
Pellat, H., 152–154

Periodic table, 342, 376
  and Niels Bohr, 343–344
Perrin, Jean, 35, 36, 148–148, 263
Perrot, Adolphe, 33–34
Perspectives in Organic Chemistry, 383
Pessimistic metainduction, 452, 461
Pfander, Alexander, 369
Philosophical Magazinen, 42, 144, 186
Philosophical Transactions, 181
Photoelectric effect, 234–235, 244, 314–317
Physics, 363–365
  and biology, 493–495
  and chemistry, 365–367, 379–381
  and development of the theory of spin, 426–427
  solid-state, 483–491
Physiology, muscle, 478–483
Planck, Max, 211–212, 234, 242–243, 259
  and radiation law, 272
Plucker, Julius, 27
Plum-pudding model, 22–23, 26, 62
Poincaré, Henri, 263
Popper, Karl, 467
Porter, Keith, 472–478
Poynting, John Henry, 107
Practical Electrical Engineering, 124–125
Preece, William, 122
Prevost, Charles, 346
Protons, 195, 307, 310–314
Protyle, 195–196
Proutean theory, 197–203
Prout, William, 196–197

Quantum chemistry. *See* Chemistry, quantum
Quantum mechanics, 329, 330, 341, 353
  and benzene, 349–352
  development of, 368–369, 451
  and development of the theory of spin, 434
  early research on, 369–371
  and molecular orbitals, 386–390
  and reductionism, 365
  and the theory of resonance, 381–386
Quantum theory, 242–245
  and development of the theory of spin, 439–442
  of solids, 331

Quantum Theory of Radiation, The, 319–320

Radiation, 244–245
  electromagnetic, 315
Radioactivity, 62
Ramsay, William, 210
Rayleigh-Jeans law, 270, 271
Rayleigh, Lord, 30, 112
Rays of Positive Electricity, 62
Realism versus unrealism, 171–172, 425–427
  and use of instruments, 468
Recent Researches, 30–31, 33(figure), 34
Reductionism
  and chemistry, 364–367
  and physics, 142
Reinganum, Max, 268–269
Relativity, 439–446
  and Paul Dirac, 439–442
Remick, Edward, 341
Resonance, mechanical, 350–352, 381–386
Rice, Francis, 352
Richardson, Owen, 208
  and Bohr's theory, 245, 247–248, 273
  and composition of matter, 239–242
  early work, 227–228
  and *The Electron Theory of Matter*, 236–249
  and electron vapor theory, 263–264
  and J. J. Thomson, 241–242
  and quantum theory, 242–245
  research on electricity, 229–236
  research on radiation, 228–229
  and Richardson's law, 229
  and the structure of the atom, 245–247
Richarz, F., 81
Riecke, Eduard, 256–257, 276
Riemann, Bernhard, 203
Roberts, John D., 341
Robinson, Robert, 346–347
Rodebush, Worth, 379–380
Rontgen rays, 45, 200. *See also* X-rays
Rontgen, Wilhelm, 34–35
Rowland, Henry, 274
Rucker, Arthur, 113

INDEX

Rutherford, Ernest, 35, 46, 51, 54, 62
  on corpuscles, 146
  and hydrogen nucleus, 307
  and nuclear atoms, 308–309
Rydberg, Janne, 210

Schmitt, Francis, 478–483
Schottky, Walter, 330, 333
Schrodinger equation, 457
Schrodinger, Erwin, 278, 370–371, 380
Schuster, Arthur, 103, 113, 202, 260
  and composition of matter, 197
  and discovery of the electron, 404
Science: The Endless Frontier, 332
Scientific discovery method, 171–173
Semiconductors, 276, 329, 330, 333
Sidgwick, Nevil, 341, 344–345, 368–369
Silicon, 330
Slater, John, 331
Social constructivist view of discovery, 410–411
Sodium, 174–176, 232
Sommerfeld, Arnold, 215, 237, 279, 330, 331, 374
Spin and statistics of nuclei, 309, 311–312, 373, 454, 457–458
  development of the theory of, 426–427
  and relativity, 433–434, 439–446
Stas, Jean-Servais, 196
Stewart, Balfour, 198
Stewart, Thomas Dale, 267
Stoney, G. Johnstone, 37, 79, 81
  and electrons, 180–181
  and structure of gases, 106
Strache, H., 211
Structure and Mechanism in Organic Chemistry, 348
Structure of Matter, The, 352
Strutt, R. J., 42
Stuhlmann, O., 244
Sutcliffe, Brian, 339
Sutherland, William, 51, 204
Swinton, Campbell, 110, 145

Tait, Peter G., 198
Telephones, 328–329
Teller, Edward, 352

Temperature and energy, 256–257
Tetrode, Hermann, 277
Theoretical history, 426
Theory of Electrons, 89, 237
Theory of Organic Chemistry, The, 341
Theory of Resonance, The, 341, 382
Thermal emission of electrons, 229–236
Thompson, Silvanus Phillips, 102, 110
  on corpuscles, 121–124
  and J. J. Thomson, 121–124
Thomsen, Julius, 197
Thomson cross-section, 319
Thomson, Elihu, 102, 111, 112
Thomson, G. P., 404
Thomson, J. J., 3–6
  and acceptance of corpuscles, 101–102, 115–118, 121–124
  and atoms, 36–37, 204–206
  and cathode rays, 21–22, 30–32, 36–45, 65, 77, 80, 112–115, 419, 453
  and corpuscles, 22, 37, 57–59, 77, 101–102, 141, 145–146
  and discovery of the electron, 21–27, 64–66, 84, 89–92, 403–404, 414–418
  and electrical discharge of gases, 107–109
  on electrification of gases, 31, 33(figure)
  and electromagnetic theory, 207–208
  and electron vapor theory, 263
  experiments on cathode rays, 30–32, 40(figure), 40–45, 109–115
  extended working hypotheses of, 55–60
  on Faraday tubes, 30–32
  on free electrons, 269
  and H. Pellat, 152–154
  and the important classification view of discovery, 409–410
  and incandescent filaments, 52–55
  on ionic charge, 45–50, 56–57
  later years, 124
  and the manipulation-and-measurement view of discovery, 409
  and m/e value, 27–28, 36–45, 40(figure), 60–61, 145–146
  and Nobel Prize, 24, 62, 121
  and Owen Richardson, 241–242
  and Paul Villard, 139–145, 148–152, 156–160

INDEX

and Pieter Zeeman, 179–180
plum-pudding model of the atom by, 22–23, 26, 62, 342
and positive electrons, 206–208
and the proutean theory, 203–204
and reductionist physics, 142
research style, 25–26, 109–115
response to research by, 109–112, 145
significance of early work by, 23, 26–27, 420–422
and Silvanus Phillips Thompson, 121–124
and the social constructivist view of discovery, 410–411
and structure of atoms, 245
and the true belief view of discovery, 412
and use of instruments, 470
and the vortex atomic theory, 198–200
on x-rays, 35–576
Thomson, William. *See* Kelvin, Lord
Thorpe, Jocelyn, 345, 352
Tolman, Richard Chase, 267
Tolman-Stewart experiment, 266–267
Townsend, John, 35, 54, 112
on ionization and electrical conduction in gases, 147
Transistor, 327
invention of the, 332–336
Treatis on Electricity and Magnetism, 104, 105
True belief view of discovery, 411–412
Turner, Edward, 196

Uhlenbeck, G., 427–436, 454
Unseen Universe, The, 198, 212
Usherwood, Edith Hilda. *See* Ingold, Hilda Usherwood

Vacuum tube amplifier, 327, 329–330
Valence and the Structure of Atoms and Molecules, 345
Valence, chemical, 342–345, 352–353, 367–369, 391–392
and polyelectronic molecules, 376–379
and the theory of resonance, 381–386
van der Waals forces, 370, 373

van't Hoff, Henricus, 258
van Vleck, John, 373–374, 375–376, 394–395
Vapor theory, electron, 263–265
Varley, W. Mansergh, 119–120
Velocity of ions, 46, 51
Villard, Paul
and atoms, 137–138
on cathode rays, 138–139, 140–145
on corpuscles, 154–155
early work, 135–138
on Goldstein rays, 138–139
and H. Pellat, 152–154
and J. J. Thomson, 139–145, 148–152, 156–160
response to research by, 148, 152–156
Vine, Benjamin H., 276
von Laue, Max, 374
von Neumann, Carl, 203
Vortex atomic theory, 198–200

Weber's electromagnetic theory, 80, 255
Weber, Wilhelm, 202
Weyl, Hermann, 213
Wheaton, Bruce R., 243–244
Wheland, George, 341, 382–386
Wiechert, Emil, 21–22, 44, 77, 203
and acceptance of the electron, 86–87
and discovery of the electron, 418
and the important classification view of discovery, 409–410
Wiedemann-Franz law, 271
Wien, Wilhelm, 45, 152, 270
and electron velocity, 271
Wigner, Eugene, 331, 374, 452, 458
and development of the theory of spin, 444
and group theory, 376–379
Wilde award, 153, 154
Wilson, C. T. R., 35, 46, 54, 112
Wilson, E. Bright, Jr., 214, 341
on quantum chemistry, 374
Wilson, Harold, 51, 60, 233–234
and electron vapor theory, 263–264
and positive electrons, 208
Wireless telegraphy, 114, 118–119
Wooldridge, Dean, 331

X-rays, 35, 45–46
  and cathode rays, 105–106
  Elihu Thomson's work on, 112
  and the proutean theory, 200
  and solid-state physics, 486–488

Y-rays, 314–317, 318

Zeeman effect, 7–8, 202, 455. *See also* Zeeman, Pieter
  and development of the theory of spin, 429–430
Zeeman, Pieter, 37, 38, 82, 103. *See also* Zeeman effect
  early work, 174
  on electrons, 83
  experiments on magnetism, 174–176
  and H. A. Lorentz, 176–180
  and J. J. Thomson, 179–180
  and Joseph Larmor, 181–187
  and the Kerr effect, 174
  and m/e value, 179–180, 185
  and structure of gases, 106
Zeleny, John, 51
Zermelo, Ernst, 258
Zollner, Friedrich, 203

RETURN TO ➡ **PHYSICS LIBRARY**
351 LeConte Hall   642-3122

LOAN PERIOD 1 1-MONTH	2	3
4	5	6

ALL BOOKS MAY BE RECALLED AFTER 7 DAYS
Overdue books are subject to replacement bills

DUE AS STAMPED BELOW

JAN 0 6 1996

JAN 1 1 1996

Rec'd UCB PHYS

This book will be held
in PHYSICS LIBRARY
until JUL 0 2 2001

JUL 3 0 2001
AUG 0 6 2001
AUG 1 9 2002
FEB 2 0 2006
MAR 3 0 2009

JAN 0 4 2012

FORM NO. DD 25

UNIVERSITY OF CALIFORNIA, BERKELEY
BERKELEY, CA 94720